Plant Proteomics

There have been several advancements made in high-throughput protein technologies, creating immense possibilities for studying proteomics on a large scale. Researchers are exploring various proteomic techniques to unravel the mystery of plant stress tolerance mechanisms. *Plant Proteomics: Implications in Growth, Quality Improvement, and Stress Resilience* introduces readers to techniques and methodologies of proteomics and explains different physiological phenomena in plants and their responses to various environmental cues and defence mechanisms against pathogens. The main emphasis is on research involving applications of proteomics to understand different aspects of the life cycle of plant species, including dormancy, flowering, photosynthetic efficiency, accumulation of nutritional parameters, secondary metabolite production, reproduction and grain yield as well as signalling responses during abiotic and biotic stresses. The book takes a unique approach, encompassing high-throughput and sophisticated proteomic techniques while integrating proteomics with other "omics."

Features:

- Integrates the branch of proteomics with other "omics" approaches, including genomics and metabolomics, giving a holistic view of the overall "omics" approaches.
- Covers various proteomic approaches for the identification of biological processes, future perspectives, and upcoming applications to identify diverse genes in plants.
- Presents readers with various proteomic tools for the improvement of plant growth, quality, and resilience against climate change, and pathogen infection.
- Enables researchers in identifying novel proteins that could be used as targets to generate plants with improved traits.

Prof. Aryadeep Roychoudhury is a Professor in the Discipline of Life Sciences, Indira Gandhi National Open University, New Delhi, India.

Plant Proteomics
Implications in Growth, Quality Improvement, and Stress Resilience

Edited by
Aryadeep Roychoudhury

CRC Press
Taylor & Francis Group
Boca Raton London New York

CRC Press is an imprint of the
Taylor & Francis Group, an **informa** business

First edition published 2024
by CRC Press
2385 NW Executive Center Drive, Suite 320, Boca Raton FL 33431

and by CRC Press
4 Park Square, Milton Park, Abingdon, Oxon, OX14 4RN

ISBN: 9781032395852 (hbk)
ISBN: 9781032395920 (pbk)
ISBN: 9781003350453 (ebk)

DOI: 10.1201/b23255

Typeset in Times
by KnowledgeWorks Global Ltd.

Contents

About the Editor

Prof. Aryadeep Roychoudhury is currently working as a Professor in the Discipline of Life Sciences, Indira Gandhi National Open University, New Delhi, India. Earlier, he served as an Assistant Professor at the Post Graduate Department of Biotechnology, St. Xavier's College (Autonomous), Kolkata, West Bengal, India. He earned his B.Sc. (Hons.) in Botany from Presidency College, Kolkata, and M.Sc. in Biophysics and Molecular Biology, University of Calcutta, West Bengal, India. He earned his Ph.D. from Jadavpur University, Kolkata, in the area of stress biology in plants. Following his Ph.D. work, he joined the University of Calcutta as a Research Associate (post-doctorate), pursuing translational research on transgenic rice. He is presently involved in active research in the field of abiotic stress responses in plants with perspectives to the physiology, molecular biology, and cell signalling under diverse stress conditions. He has 23 years of research experience in the concerned discipline. He has handled several government-funded projects as principal investigator and supervised five Ph.D. students as principal investigator. Prof. Roychoudhury has published over 250 articles in peer-reviewed journals and chapters in books of international and national repute. He has already edited 14 books with Wiley, Elsevier, Springer, Taylor & Francis Group, and Nova, and has also handled special issues as Guest Editor for several renowned international journals. He is a regular reviewer of articles in high-impact, international journals, Life Member of different scientific associations and societies, and the recipient of the Young Scientist Award 2019, conferred upon him by International Foundation for Environment and Ecology, at the University of Allahabad, Prayagraj, Uttar Pradesh. His name is included in the Stanford University's List of World's Top 2% Influential Scientists.

Contributors

Shadma Afzal
Department of Biotechnology
Motilal Nehru National Institute of Technology
Prayagraj, India

Harika Amooru
ICAR – Indian Agriculture Research Institute
New Delhi, India

Mirza Jaynul Baig
Crop Physiology and Biochemistry Division
ICAR – National Rice Research Institute
Cuttack, India

Aditya Banerjee
Department of Biotechnology
St. Xavier's College (Autonomous)
Kolkata, India

Lambodar Behera
Crop Improvement Division
ICAR – National Rice Research Institute
Cuttack, India

Prafulla K. Behera
Department of Biodiversity and Conservation
  of Natural Resources
Central University of Odisha
Koraput, India

Jinal Paresh Bhavsar
Institute of Science, Nirma University
Ahmedabad, India

Nilanjan Chakraborty
Department of Botany
Scottish Church College
Kolkata, India

Nidhi Choudhary
Department of Biotechnology
Birla Institute of Scientific Research
Jaipur, India

B. Balaji
ICAR – Indian Agriculture Research Institute
New Delhi, India

E. Dharani
ICAR – Indian Agriculture Research Institute
New Delhi, India

Sivapragasam E.
ICAR – Indian Agriculture Research
  Institute
New Delhi, India

Uday G.
University of Agricultural Sciences
Dharwad, India

Purnendu Ghosh
Birla Institute of Scientific Research
Jaipur, India

Raj Kumar Gothwal
Department of Bioenginecring
Birla Institute of Technology, Birla Institute of
  Scientific Research
Jaipur, India

Shamrao Jagirdhar
Department of Plant pathology
University of Agricultural Sciences
Dharwad, India

Riya Jain
Department of Botany
Dayalbagh Educational Institute (Deemed to be
  University)
Agra, India

Shailendra K. Jhaand
ICAR – Indian Agriculture Research Institute
New Delhi, India

Priyanka K.
Department of Genetics and Plant Breeding
University of Agricultural Sciences
Dharwad, India

Rita Kundu
Department of Botany
University of Calcutta
Kolkata, India

Shuvobrata Majumder
Department of Plant & Microbial
    Biotechnology
DBT – Institute of Life Sciences (ILS)
Bhubaneswar, India

Shivangi Mathur
Department of Botany
Dayalbagh Educational Institute (Deemed
    to be University)
Agra, India

Monalisha Mishra
Department of Biotechnology
Motilal Nehru National Institute of
    Technology
Prayagraj, India

Soumita Mitra
Department of Botany
Scottish Church College
Kolkata, India

Soumya Mohanty
Crop Improvement Division
ICAR – National Rice Research Institute
Cuttack, India

Nimmy M. S.
ICAR – Indian Agriculture Research
    Institute,
New Delhi, India

Jayanta K. Nayak
Department of Anthropology
Central University of Odisha
Koraput, India

Erica Zinnia Nehra
DY Patil University
Navi Mumbai, India

Sampat Nehra
Birla Institute of Scientific Research
Jaipur, India

Suraj K. Padhi
Department of Biodiversity and Conservation
    of Natural Resources
Central University of Odisha
Koraput, India

Asmita Pal
Department of Botany
University of Calcutta
Kolkata, India

Darshan Panda
Crop Physiology and Biochemistry Division
ICAR – National Rice Research Institute
Cuttack, India

Debabrata Panda
Department of Biodiversity and Conservation
    of Natural Resources
Central University of Odisha
Koraput, India

Ashish Kumar Pathak
University Institute of Biotechnology
Chandigarh University
Mohali, India

Seema Pradhan
Department of Plant & Microbial
    Biotechnology
DBT – Institute of Life Sciences (ILS)
Bhubaneswar, India

Gaurav Prajapati
Department of Biotechnology
Motilal Nehru National Institute of
    Technology
Prayagraj, India

Suriya Prakash
ICAR – Indian Agriculture Research Institute
New Delhi, India

Rajiv Ranjan
Department of Botany
Dayalbagh Educational Institute (Deemed to be
    University)
Agra, India

Shruti Rohatgi
Department of Botany
Dayalbagh Educational Institute (Deemed to be
    University)
Agra, India

Aryadeep Roychoudhury
Discipline of Life Sciences
School of Sciences
Indira Gandhi National Open University
New Delhi, India

Pradeep Krishnan S.
ICAR – Indian Agriculture Research Institute
New Delhi, India

Shardha H. B.
University of Agricultural Sciences
Dharwad, India

Shrizcharan S.
ICAR – Indian Agriculture Research Institute
New Delhi, India

Aarushi Sachdeva
Nirma University
Ahmedabad, India

Shubhadarshini Sahoo
Department of Biotechnology
Motilal Nehru National Institute of Technology
Prayagraj, India

Lekshmy Sathee
ICAR – Indian Agriculture Research Institute
New Delhi, India

Namisha Sharma
Department of Plant & Microbial Biotechnology
DBT – Institute of Life Sciences (ILS)
Bhubaneswar, India

Aruna Shekhar N. C.
TATA Memorial Centre
Navi Mumbai, India

Shivaleela
Department of Agricultural Entomology
University of Agricultural Sciences
Raichur, India

Deeksha Singh
Department of Botany
Dayalbagh Educational Institute (Deemed to be
    University)
Agra, India

Nand K. Singh
Department of Biotechnology
Motilal Nehru National Institute of Technology
Prayagraj, India

Parul Sinha
TATA Memorial Centre
Navi Mumbai, India

Pooran Singh Solanki
Department of Biotechnology and
    Bioengineering
Birla Institute of Scientific Research, Birla
    Institute of Technology
Jaipur, India

Suvarna
Department of Genetics and Plant Breeding
University of Agricultural Sciences
Raichur, India

Baishnab C. Tripathy
Jawaharlal Nehru University
New Delhi, India

Santosh Kumar Upadhyay
Department of Botany
Punjab University
Chandigarh, India

Alok Kumar Varshney
Department of Biotechnology and
    Bioengineering
Birla Institute of Scientific Research, Birla
    Institute of Technology
Jaipur, India

1 Plant Proteome Analyses in Response to Seed Development

Ashish Kumar Pathak and Santosh Kumar Upadhyay

1.1 INTRODUCTION

Seed evolution is a landmark for photosynthetic organisms. The embryo that will one day become the new plant that results from sexual reproduction is contained within the growing seed. As a result, the seed holds an extremely important position during the entire life cycle of the higher plant. Since most crop species are grown from seed, seeds from plants like cereals and legumes are significant sources of nourishment. To feed a burgeoning population, we must exploit the maximum genetic potential of crop species; this requires knowing seed growth, dormancy, and germination (Wang et al., 2015; Nabuuma et al., 2022).

Seed development starts with the fertilisation of egg cells with the gamete released from the pollen sac. The ensuing stages of embryonic development are triggered by this event, which also reprogrammes the cellular state. Fertilisation causes significant changes in nuclear architecture and chromatin structure, which leads to the transition from the non-embryonic to the embryonic state. Numerous positive regulators of embryo initiation, most notably transcription factors, have been discovered in addition to repressive chromatin control (de Vries and Weijers, 2017; Verma et al., 2022). The development of a seed can be classified into two phases that are conceptually distinct from one another. The first stage is called morphogenesis during which the body plan of the embryo is established. The other phase is seed maturation during which the seed accumulates storage reserves, acquires desiccation tolerance, and undergoes dormancy (Harada, 1997).

There are essentially two different types of seeds. The first type is known as "orthodox" seeds that can withstand a significant level of desiccation by the time they have reached the end of their maturation and still keep their germination potential. The other type of seeds are known as recalcitrant seeds, and they have a high water content when they are mature and are unable to tolerate being dried out. This latter category of seeds skips over the stages of drying out and metabolic quiescence because they initiate the germination process as soon as the maturation phase is complete (Catusse et al., 2008). Dormancy is one of the most important characteristics of seeds, as it prevents viable seeds from germinating during the challenging growing season. It is essential to plant ecology and agriculture because it enables seeds to survive periods that are unfavourable for the establishment of seedlings. The major endogenous factors that act in the control of seed dormancy are plant hormones, primarily abscisic acid (ABA) and gibberellin (GA). ABA positively regulates the induction and maintenance of dormancy (Tuan et al., 2018). For instance, the regulation of seed maturation and the transition from embryo to seedling phase require the presence of all four transcription factors that are part of the ABA signalling pathway. These transcription factors include Fusca 3 (FUS3), abscisic acid insensitive 3 (ABI3), leafy cotyledon 1 (LEC1), and LEC2. The heterochronic phenotypes that are caused by mutations in any of these transcriptional factors as well as changes in the ways that these factors interact with one another include abnormal seed maturation and decreased dormancy (Holdsworth et al., 2008).

Seed dormancy is followed by seed germination. Germination encompasses all of the events that begin with the dormant, dry seed taking in water. The resumption of essential processes, such

DOI: 10.1201/b23255-1

as transcription, translation, and DNA repair, is one of the first signs that germination has begun. This is then followed by cell elongation, and finally, at the time that radicle protrusion occurs, cell division is resumed (Bentsink and Koornneef, 2008). Germination of seeds is essential to the cultivation of virtually all plant species used in agriculture. During their development, the seeds of the vast majority of species acquire the ability to germinate. This ensures that seed germinates quickly after it is sown, which is important for crop production because it reduces the risk of contamination. However, this can lead to precocious germination in a few species, such as maize, wheat, and rice. A decrease in grain quality and significant economic losses can result from premature germination of the grain (Kermode, 2005).

The overall protein content of a cell at a given time can be referred to as the "proteome," and it can be defined as the collection of proteins that are characterised in terms of their localisation, interactions, post-translational modifications, and turnover (Roychoudhury et al., 2011; Aslam et al., 2017). Proteomics has been widely used in evaluating the mechanisms of many biological processes and has proven to be very effective as more and more genome sequence information becomes available. The functional characterisation of plants is becoming increasingly reliant on the use of proteome analysis as a powerful tool. This analysis is accomplished by isolating and identifying proteins, determining their function and the functional network they belong to. Rapid progress has been made in a variety of methods for separating proteins and determining their identities, including two-dimensional electrophoresis (2DE), nano-liquid chromatography, and mass spectrometry. There have recently been developments in some new methods, such as top-down mass spectrometry and tandem affinity purification (Aslam et al., 2017). These methods have made it possible to conduct high-throughput analyses of the functions of proteins in plants as well as the functional networks formed by those proteins (Hirano et al., 2004). It has been widely used for studying the seed biology (Chen et al., 2021; Islam et al., 2022). In the following section of this chapter, we are going to talk about how the field of proteomics has been used to better understand pollen development and growth, dormancy, germination, and seed development.

1.2 PROTEOMICS IN RESPONSE TO POLLEN DEVELOPMENT AND GROWTH

The stages of flower growth, pollination, and fertilisation are crucial for controlling the creation and development of seeds. Microsporogenesis and microgametogenesis are the two processes that make up pollen formation. A tetrad of four haploid microspores is produced by the meiotic division of the diploid pollen mother cell, also referred to as the microsporocyte. Each post-meiotic microspore undergoes an asymmetric mitotic division after being released from the tetrad to become mature bicellular pollen (Owen and Makaroff, 1995). The interaction between the pollen and the pistil at the cellular and molecular levels is the first crucial factor in the successful formation of seed. If the environment is favourable, the "compatible" pollen will grow. The sperm cells are released when the pollen tube (PT) expands and penetrates the style into the ovule.

A reference map of the Arabidopsis mature pollen proteome was created. The identified proteins were involved in metabolism, energy generation, or cell structure. These were similar to those found for the pollen transcriptome, and this similarity is consistent with the idea that in addition to the messenger RNAs (mRNAs), the mature pollen grain contains proteins required for germination and rapid PT growth (Holmes-Davis et al., 2005). Multiple isoforms of the same proteins exist, which shows that these proteins undergo some sort of post-translational modification during pollen development (Kerim et al., 2003). At the proteome level, tomato pollen formation is a tightly regulated sequential process. In extremely early embryonic stages, heat stress-related proteins are highly prevalent, indicating a significant function in stress prevention (Chaturvedi et al., 2013). The extracellular matrix of pollen contains elements necessary for effective pollination. The majority of the relevant genes were found to be located in two clusters after the

TABLE 1.1

Selected List of Proteomic Studies for Understanding Pollen Growth and Development

Title of the Article	Plant Studied	Reference
Proteome mapping of mature pollen of *Arabidopsis thaliana* proteomics	*Arabidopsis thaliana*	Holmes-Davis et al. (2005)
Proteome analysis of male gametophyte development in rice anthers	*Oryza sativa*	Kerim et al. (2003)
Cell-specific analysis of the tomato pollen Proteome from pollen mother cell to mature pollen provides evidence for developmental priming	*Solanum lycopersicum*	Chaturvedi et al. (2013)
Gene families from the *Arabidopsis thaliana* pollen coat proteome	*Arabidopsis thaliana*	Mayfield et al. (2001)
Comparative proteomic analysis reveals a dynamic pollen plasma membrane protein map and the membrane landscape of receptor-like kinases and transporters important for PT growth and interaction with pistils in rice	*Oryza sativa*	Yang and Wang (2017)
Exploration of rice pistil responses during early postpollination through a combined proteomic and transcriptomic analysis	*Oryza sativa*	Li et al. (2016)

identification of all proteins in the Arabidopsis pollen covering. The other cluster comprises six lipid-binding oleosin genes, including GRP17, a gene that aids in effective pollination, while the first cluster encodes six lipases (Mayfield et al., 2001). Ability of PTs to transfer sperm cells into ovules for fertilisation depends on the synchronisation of PT growth, guidance, prompt growth, and rupture mediated by PT-pistil contact. Plasma membrane (PM) proteins of PTs are pioneers in mediating PT integrity and contact with pistils. The proteomics of rice revealed functional tilt towards signal transduction, transporters, wall remodelling/metabolism, and membrane trafficking of both the PM-related proteins (PMrP) and differentially expressed PMrP sets (Yang and Wang, 2017). Results show significant protein rearrangements at the PM and endomembranes both before and after the pollen grains begin to form tubes (Pertl et al., 2009). After pollination, reactive oxygen species (ROS) were greatly accumulated in the stigma, and numerous changes were made to the number of redox homeostasis system genes. Changes in the genes encoding certain E3 ligases may be a hint that protein ubiquitination is crucial for pollination-related cell signal transmission (Li et al., 2016). In brief (Table 1.1), the translational and post-translational machinery of the cell coordinates with the dynamic pollen development and tube growth processes. Once the pollen reaches the ovule, where it is intended to land, the embryogenesis process begins, followed by fertilisation.

1.3 PROTEOME ANALYSES IN RESPONSE TO SEED DEVELOPMENT

The embryo proper, a tissue storing nutrients for the embryo, and the seedcoat protective covering make up the three essential components of embryonic plants that are found in seeds. Depending on the variety of plant species, plants acquire different amounts of proteins, carbohydrates, and oil during seed formation (Weber et al., 2005). Hence, changing the expression of vast protein sets in a highly coordinated manner controls the intricate process of embryogenesis. The process of histodifferentiation in the zygote and during seed development is largely regulated by protein expression (Han et al., 1997). In a few plant species, researchers have looked at seed proteomics to learn more about how seeds grow and develop (Shi et al., 2010; Wang et al., 2015). In this section, we give an overview of what they found (Table 1.2).

TABLE 1.2
Overview of Results of Proteomic Studies

Title of the Article	Plant Studied	Reference
A Proteomic analysis of seed development in *Brassica campestris* L	*Brassica campestris*	Li et al. (2012)
Proteome profiling of early seed development in *Cunninghamia lanceolata* (Lamb.) Hook	*Cunninghamia lanceolata*	Shi et al. (2010)
Proteomics of *Medicago truncatula* seed development establishes the time frame of diverse metabolic processes related to reserve accumulation.	*Medicago truncatula*	Gallardo et al. (2003)
Integrated seed proteome and phosphoproteome analyses reveal interplay of nutrient dynamics, carbon-nitrogen partitioning, and oxidative signalling in chickpea.	*Cicer arietinum*	Sinha et al. (2020)

Use of proteomics in understanding seed development in rice has been detailed by Deng et al. (2013). In brief, large-scale analyses of proteome changes related to rice seed development reveal that the coordination of numerous metabolic, molecular, and cellular processes, including the transition from central carbon metabolism to alcoholic fermentation, is required for endosperm growth, embryogenesis, and starch accumulation. When it comes to certain metabolic, molecular, or cellular processes, the expression patterns of proteins during embryo and endosperm development are closely synchronised with the time and order of seed developmental events. It has also resulted in the discovery of several novel proteins, including those involved in RNA metabolism and stress tolerance that are connected to controlling seed growth.

A proteome analysis of growing *Brassica campestris* L. seeds at various stages of embryogenesis was conducted to acquire insight into the dynamics of proteins during seed development. The metabolism-related proteins were the most prevalent, indicating a high demand for resources for the quick development of the embryo. Also, there were a lot of proteins involved in protein processing, which shows importance of protein renewal while the seed was growing. The rest were involved in oxidation/detoxification, energy, defence, transcription, protein synthesis, transporter, cell structure, signal transduction, secondary metabolism, transposition, DNA repair, storage, etc. (Li et al., 2012). Early embryogenesis was used to conduct a proteomic analysis of Chinese fir seeds in six developmental stages. Proteins with different kinetics of appearance revealed their involvement in chromatin modification and programmed cell death, suggesting that the proteins may be crucial in determining the number of zygotic embryos produced as well as in regulating patterning and shape remodelling. Other proteins involved in energy production, protein storage, disease and defence, the cytoskeleton, and embryo development were discovered through the study (Shi et al., 2010). *Medicago truncatula* seed growth during germination and protein deposition was studied. The study found a buildup of carbon metabolism enzymes (sucrose synthase, starch synthase) and embryonic photosynthesis proteins (chlorophyll a/b binding), which may provide cofactors for protein/lipid synthesis or CO_2 refixation during seed filling. The methionine metabolism-related enzymes *S*-adenosylmethionine synthetase and *S*-adenosylhomocysteine hydrolase were discovered to accumulate differently. These enzymes may play a part in the change from a highly active to a quiescent state during seed formation (Gallardo et al., 2003). An integrated proteomics and phosphoproteomics investigation in six sequential chickpea seed developmental stages was done to investigate nutrient dynamics throughout embryonic and cotyledonary stages. During seed filling, metabolic and photosynthetic proteoforms accumulated. Increased abundance of oxidative and serine/threonine signalling proteins showed redox sensing and signalling during seed development (Sinha et al., 2020).

1.4 PROTEOMICS IN RESPONSE TO DORMANCY

It is generally agreed that dormancy is an adaptive trait because it enables seeds to remain dormant until such time as the conditions necessary for germination are met. Depending on the species, the lipids, carbohydrates, and proteins needed for germination accumulate in the embryo, endosperm, or aleurone layer during seed maturation. The embryo is now quiescent and tolerant of desiccation. It is possible for the seeds of many plant species, including Arabidopsis (*Arabidopsis thaliana*), to enter a state of deep dormancy after they have reached their physiological maturity. During this time, the dormant (D) seeds, when sown, either will not germinate at all or will germinate very slowly in comparison to the non-dormant (ND) seeds of the same species. Using the dormant (D) accession, proteomics was used to characterise the mechanisms that control seed dormancy in Arabidopsis. According to the findings, proteins connected to metabolic functions that might play a role in germination can potentially accumulate during after-ripening in the dry state. The processes preventing ND seeds from germinating by applying ABA are distinct from those preventing germination of D seeds ingested in basal media (Chibani et al., 2006). Using the deep dormant rice variety FH7185 as a resource, researchers were able to create a comprehensive proteome map of the rice seed by employing two-dimensional polyacrylamide gel electrophoresis (2D-PAGE) and isobaric tags for relative and absolute quantification (iTRAQ) of proteins in dormant and after-ripening seeds. A wide variety of proteins involved in metabolic pathways and plant hormone signal transduction were found to accumulate during seed dormancy. According to the results, the level of dormancy exhibited by FH7185 was not directly influenced by the presence of either ABA or GA. Therefore, the hormone is not the only factor that is responsible for regulating and coordinating the maintenance and release of seed dormancy. Other regulatory factors also play a role in this process (Xu et al., 2019). The seed proteome response of *Fraxinus mandshurica* Rupr. to dehydration was investigated under four different conditions. These conditions established the possible involvement of detoxifying enzymes, transport proteins, and nucleotide metabolism enzymes in the reaction of seeds to dehydration (Zhang et al., 2015). Seed dormancy breaking in Norway maple involves proteins that are involved in a variety of processes; however, the proteasome proteins, *S*-adenosylmethionine synthetase, glycine-rich RNA-binding protein, ABI3-interacting protein 1, EF-2, and adenosylhomocysteine are of particular importance (Pawłowski, 2009). Adenosine kinase, methionine synthase, and glycine-rich RNA-binding proteins are important in dormancy breaking of seed of *Acer pseudoplatanus* (Pawłowski and Staszak, 2016). When comparing ND seeds to dormant seeds in *Leymus chinensis*, the proteins involved in chromatin structure and dynamics, intracellular trafficking, sectioning, and vesicular transport, as well as cytoskeleton, were all found to be elevated. On the other hand, the levels of proteins that are involved in the processing and modification of RNA, the formation of cell walls, membranes, and envelopes, as well as signal transduction pathways, all dropped (Hou et al., 2019). *Zanthoxylum nitidum* seeds have intermediate physiological dormancy and require three months of cold stratification (4°C) or GA (300 mg L^{-1} soaking for two days) treatment to stimulate germination. Mature *Z. nitidum* seeds were treated with cold stratification and GA solution. iTRAQ-coupled LC-MS/MS characterised the differently expressed proteins; most were involved in respiration-related metabolism and amino acid biosynthesis (Lu et al., 2018). Significant differences in the proteins found in *Prunus campanulata* seeds before and after dormancy were found. By using mass spectrometry and sequence comparison, these proteins were identified as dehydrin, prunin 1 precursor, prunin 2 precursor, and prunin 2 (Lee et al., 2006). According to these findings (Table 1.3), the process of developing dormancy and eliminating it involves multiple stages and is controlled by a complex network. On the other hand, the molecular mechanisms underlying the regulation of seed dormancy remain obscure, and each species possesses a unique molecular dynamism.

TABLE 1.3

Seed Dormancy Studies

Title of the Article	Plant Studied	Reference
Proteomic analysis of seed dormancy in Arabidopsis.	*Arabidopsis thaliana*	Chibani et al. (2006)
Proteomic analysis reveals different involvement of proteins during the maintenance and release of rice seed dormancy.	*Oryza sativa*	Xu et al. (2019)
Proteome analysis of dormancy-released seeds of *Fraxinus mandshurica* Rupr. in response to re-dehydration under different conditions.	*Fraxinus mandshurica*	Zhang et al. (2015)
Proteome analysis of Norway maple (*Acer platanoides* L.) seeds dormancy breaking and germination: influence of abscisic and gibberellic acids.	*Acer platanoides*	Pawłowski (2009)
Analysis of the embryo proteome of sycamore (*Acer pseudoplatanus* L.) seeds reveals a distinct class of proteins regulating dormancy release	*Acer pseudoplatanus*	Pawłowski and Staszak (2016)
Physiological and proteomic analyses for seed dormancy and release in the perennial grass of *Leymus chinensis*	*Leymus chinensis*	Hou et al. (2019)
Proteomic analysis of *Zanthoxylum nitidum* seeds dormancy release: Influence of stratification and gibberellins	*Zanthoxylum nitidum*	Lu et al. (2018)
Protein changes between dormant and dormancy-broken seeds of *Prunus campanulata* Maxim.	*Prunus campanulata*	Lee et al. (2006)

1.5   PROTEOME ANALYSES IN RESPONSE TO GERMINATION

A crucial stage in the growth of life cycle of plants is seed germination. Embryonic cell transition occurs from quiescence to a metabolic state of high activity. Seeds must be hydrated in environment that supports metabolism, such as appropriate temperature and the presence of oxygen, in order for germination to take place. This water uptake occurs in three phases: an initial rapid period (phase I, imbibition), a plateau phase with little change in water content (phase II, increased metabolic activity), and a subsequent increase in water content that coincides with radicle emergence and the resumption of growth (Bewley, 1997). Germination involves numerous processes, including proteolysis, macromolecule synthesis, respiration, alterations to subcellular structures, and cell elongation.

It was shown that the ability for germination in Arabidopsis seeds is substantially programmed during the process of seed dormancy, as inhibition of transcription by alpha-amanitin or actinomycin D was unable to prevent germination (Rajjou et al., 2004) (Table 1.4). During the early stages of imbibition, the Arabidopsis translatome continues to reflect an upregulation in embryonic maturation programme until a particular checkpoint in the development process. One of the most important aspects of seed germination is the process of selective mRNA translation (Galland et al., 2014). RNA-binding proteins, or RBPs, serve a variety of roles in controlling translation (Sajeev et al., 2022). Apart from that, GAs do not take part in a significant number of the activities that are involved in germination prior to the radicle protrusion (Gallardo et al., 2002). Proteins involved in seed imbibition (actin isoforms, WD-40 repeat proteins) and seed dehydration (e.g. cytosolic glyceraldehyde-3-phosphate dehydrogenase) were discovered to be playing a role in Arabidopsis seed germination (Gallardo et al., 2001).

Comparative 2-DE maps revealed that 148 proteins were exhibited in distinct ways during the germination of rice seeds. Storage proteins like globulin and glutelin, proteins involved in seed maturation like "early embryogenesis protein" and "late embryogenesis abundant protein," and proteins involved in desiccation like "abscisic acid-induced protein" and "cold-regulated protein" made up the majority of the down-regulated proteins. The majority of the glycolysis-related proteins,

TABLE 1.4

Seed Germination Studies

Title of the Article	Plant Studied	Reference
The effect of α-amanitin on the Arabidopsis seed proteome highlights the distinct roles of stored and neosynthesized mRNAs during germination	*Arabidopsis thaliana*	Rajjou et al. (2004)
Dynamic proteomics emphasizes the importance of selective mRNA translation and protein turnover during Arabidopsis seed germination.	*Arabidopsis thaliana*	Galland et al. (2014)
The mRNA binding proteome of a critical phase transition during Arabidopsis seed germination	*Arabidopsis thaliana*	Sajeev et al. (2022)
Proteomics of Arabidopsis seed germination. A comparative study of wild-type and gibberellin-deficient seeds.	*Arabidopsis thaliana*	Gallardo et al. (2002)
Proteomic analysis of Arabidopsis seed germination and priming	*Arabidopsis thaliana*	Gallardo et al. (2001)
Proteomic analysis of rice (*Oryza sativa*) seeds during germination.	*Oryza sativa*	Yang et al. (2007)
Constructing the metabolic and regulatory pathways in germinating rice seeds through proteomic approach	*Oryza sativa*	He et al. (2011b)
Quantitative proteomics reveals the role of protein phosphorylation in rice embryos during early stages of germination.	*Oryza sativa*	Han et al. (2014)
Quantitative proteomics reveals the role of protein phosphorylation in rice embryos during early stages of germination	*Oryza sativa*	Han et al. (2014)
Proteomic dissection of seed germination and seedling establishment in *Brassica napus*	*Brassica napus*	Gu et al. (2016)
Comparative proteome analysis of embryo and endosperm reveals central differential expression proteins involved in wheat seed germination.	*Triticum aestivum*	He et al. (2015)
Proteomic analysis of oil mobilization in seed germination and postgermination development of *Jatropha curcas*	*Jatropha curcas*	Yang et al. (2009)

including UDP-glucose dehydrogenase, fructokinase, phosphoglucomutase, and pyruvate decarboxylase, were upregulated during germination (Yang et al., 2007). For the first 48 hours after imbibition of rice, the expression of storage proteins and some proteins linked to seed development and desiccation was down-regulated. Proteins linked to catabolism were upregulated after ingestion (He et al., 2011a). Regulation of redox homeostasis and regulation of gene expression are both critical for the germination of rice seeds (He et al., 2011b). The role of protein phosphorylation events in the initial phases of rice seed germination suggests that seed germination is probably triggered by brassinosteroid signal transduction (Han et al., 2014). The details of proteomics application in understanding rice germination have been summarised in the review by He and Yang (2013). In summary, when rice seeds germinate, reserves are mobilised via both the fermentation route and the core carbon metabolism pathways (glycolysis and Krebs cycle). Several controls are present in the embryo during seed germination. In addition to sugar and energy, amino acid production may help cells tolerate anoxia.

In *Brassica napus*, protein expression patterns demonstrated that heterotrophic metabolism may be initiated during seed germination and that defensive mechanisms may begin (Gu et al., 2016). In barley seed proteome, after imbibition, the quantity of many proteins involved in germination inhibition and desiccation tolerance rapidly dropped. Following this, there was a reduction in the proteins responsible for storing fat, protein, and nutrients, which was consistent with the induction and activation of mechanisms for nutrient mobilisation to feed the developing embryo. The biochemical activity of dormant seeds differed significantly from that of germination-stage seeds in terms of sulphur metabolic enzymes, endogenous alpha-amylase/trypsin inhibitors, and histone proteins (Tan et al., 2013). Bread wheat embryo and endosperm proteomics were studied at four germination phases. Differentially expressed proteins from the embryo were involved in glucose metabolism, proteometabolism, amino acid metabolism, nucleic acid metabolism, and stress-related proteins, whereas those from the endosperm were involved in protein storage, carbohydrate metabolism, inhibitors, stress response, and protein synthesis (He et al., 2015). During germination, *Jatropha curcas* seeds mobilised oil for early seedling development. Oil mobilisation involves oxidation, glyoxylate cycle, glycolysis, citric acid cycle, gluconeogenesis, and pentose phosphate pathway (Yang et al., 2009). In brief (Table 1.3), during the shift of seeds from dormant to germination phase, the stored nutrients are mobilised. The stored RNA in the dormant seeds is sufficient for the initiation of germination. However, it involves a number of physiological, biochemical, and morphological changes as it transits from heterotrophic to autotrophic growth.

In conclusion, insights into the processes of metabolic and molecular regulation during seed development have been greatly expanded and clarified by proteomic research. New facets of the control mechanisms driving seed development are revealed by comparing the proteome changes globally across several seed developmental stages. However, the seed proteome represents a highly complex system, only a part of which has been sufficiently described to date. The systems biology of pollen and pistil development, dormancy, germination, and seed development is still lacking. The overlapping study of omics technique might provide more clear knowledge.

REFERENCES

Aslam B, Basit M, Nisar MA, Khurshid M, Rasool MH (2017) Proteomics: Technologies and Their Applications. J Chromatogr Sci **55**: 182–196

Bentsink L, Koornneef M (2008) Seed Dormancy and Germination. Arab B **6**: e0119–e0119

Bewley JD (1997) Seed Germination and Dormancy. Plant Cell **9**: 1055–1066

Catusse J, Job C, Job D (2008) Transcriptome- and Proteome-Wide Analyses of Seed Germination. C R Biol **331**: 815–822

Chaturvedi P, Ischebeck T, Egelhofer V, Lichtscheidl I, Weckwerth W (2013) Cell-Specific Analysis of the Tomato Pollen Proteome from Pollen Mother Cell to Mature Pollen Provides Evidence for Developmental Priming. J Proteome Res **12**: 4892–4903

Chen X, Börner A, Xin X, Nagel M, He J, Li J, Li N, Lu X, Yin G (2021) Comparative Proteomics at the Critical Node of Vigor Loss in Wheat Seeds Differing in Storability. Front. Plant Sci. **12**: 707184

Chibani K, Ali-Rachedi S, Job C, Job D, Jullien M, Grappin P (2006) Proteomic Analysis of Seed Dormancy in Arabidopsis. Plant Physiol **142**: 1493–1510

de Vries SC, Weijers D (2017) Plant Embryogenesis. Curr Biol **27**: R870–R873

Deng ZY, Gong CY, Wang T (2013) Use of Proteomics to Understand Seed Development in Rice. Proteomics **13**: 1784–1800

Galland M, Huguet R, Arc E, Cueff G, Job D, Rajjou L (2014) Dynamic Proteomics Emphasizes the Importance of Selective mRNA Translation and Protein Turnover during Arabidopsis Seed Germination. Mol Cell Proteomics **13**: 252–268

Gallardo K, Job C, Groot SPC, Puype M, Demol H, Vandekerckhove J, Job D (2001) Proteomic Analysis of Arabidopsis Seed Germination and Priming. Plant Physiol **126**: 835–848

Gallardo K, Job C, Groot SPC, Puype M, Demol H, Vandekerckhove J, Job D (2002) Proteomics of Arabidopsis Seed Germination. A Comparative Study of Wild-Type and Gibberellin-Deficient Seeds. Plant Physiol **129**: 823–837

Gallardo K, Le Signor C, Vandekerckhove J, Thompson RD, Burstin J (2003) Proteomics of *Medicago truncatula* Seed Development Establishes the Time Frame of Diverse Metabolic Processes Related to Reserve Accumulation. Plant Physiol **133**: 664–682

Gu J, Chao H, Gan L, Guo L, Zhang K, Li Y (2016) Proteomic Dissection of Seed Germination and Seedling Establishment in *Brassica napus*. Front. Plant Sci. **7**: 1–19

Han B, Hughes DW, Galau GA, Bewley JD, Kermode AR (1997) Changes in Late-Embryogenesis-Abundant (LEA) Messenger RNAs and Dehydrins during Maturation and Premature Drying of *Ricinus communis* L. seeds. Planta **201**: 27–35

Han C, Yang P, Sakata K, Komatsu S (2014) Quantitative Proteomics Reveals the Role of Protein Phosphorylation in Rice Embryos during Early Stages of Germination. J Proteome Res **13**: 1766–1782

Harada JJ (1997) Seed Maturation and Control of Germination BT– Cellular and Molecular Biology of Plant Seed Development. *In* BA Larkins, IK Vasil, eds, Springer Netherlands, Dordrecht, pp 545–592

He D, Han C, Yang P (2011a) Gene Expression Profile Changes in Germinating Rice. J Integr Plant Biol **53**: 835–844

He D, Han C, Yao J, Shen S, Yang P (2011b) Constructing the Metabolic and Regulatory Pathways in Germinating Rice Seeds Through Proteomic Approach. Proteomics **11**: 2693–2713

He D, Yang P (2013) Proteomics of Rice Seed Germination. Front. Plant Sci. **4**: 246

He M, Zhu C, Dong K, Zhang T, Cheng Z, Li J, Yan Y (2015) Comparative Proteome Analysis of Embryo and Endosperm Reveals Central Differential Expression Proteins Involved in Wheat Seed Germination. BMC Plant Biol **15**: 97

Hirano H, Islam N, Kawasaki H (2004) Technical Aspects of Functional Proteomics in Plants. Phytochemistry **65**: 1487–1498

Holdsworth MJ, Bentsink L, Soppe WJJ (2008) Molecular Networks Regulating Arabidopsis Seed Maturation, After-Ripening, Dormancy and Germination. New Phytol **179**: 33–54

Holmes-Davis R, Tanaka CK, Vensel WH, Hurkman WJ, McCormick S (2005) Proteome Mapping of Mature Pollen of *Arabidopsis thaliana*. Proteomics **5**: 4864–4884

Hou L, Wang M, Wang H, Zhang W-H, Mao P (2019) Physiological and Proteomic Analyses for Seed Dormancy and Release in the Perennial Grass of *Leymus chinensis*. Environ Exp Bot **162**: 95–102

Islam N, Krishnan HB, Natarajan S (2022) Quantitative Proteomic Analyses Reveal the Dynamics of Protein and Amino Acid Accumulation during Soybean Seed Development. Proteomics **22**: 2100143

Kerim T, Imin N, Weinman JJ, Rolfe BG (2003) Proteome Analysis of Male Gametophyte Development in Rice Anthers. Proteomics **3**: 738–751

Kermode AR (2005) Role of Abscisic Acid in Seed Dormancy. J Plant Growth Regul **24**: 319–344

Lee C-S, Chien C-T, Lin C-H, Chiu Y-Y, Yang Y-S (2006) Protein Changes between Dormant and Dormancy-Broken Seeds of *Prunus campanulata* Maxim. Proteomics **6**: 4147–4154

Li W, Gao Y, Xu H, Zhang Y, Wang J (2012) A Proteomic Analysis of Seed Development in *Brassica campestris* L. PLoS One **7**: e50290

Li M, Wang K, Li S, Yang P (2016) Exploration of Rice Pistil Responses during Early Post-Pollination Through a Combined Proteomic and Transcriptomic Analysis. J Proteomics **131**: 214–226

Lu Q, Zhang ZS, Zhan RT, He R (2018) Proteomic Analysis of *Zanthoxylum nitidum* Seeds Dormancy Release: Influence of Stratification and Gibberellin. Ind Crops Prod **122**: 7–15

Mayfield JA, Fiebig A, Johnstone SE, Preuss D (2001) Gene Families from the *Arabidopsis thaliana* Pollen Coat Proteome. Science **292**: 2482–2485

Nabuuma D, Reimers C, Hoang KT, Stomph T, Swaans K, Raneri JE (2022) Impact of Seed System Interventions on Food and Nutrition Security in Low- and Middle-Income Countries: A Scoping Review. Glob Food Sec **33**: 100638

Owen HA, Makaroff CA (1995) Ultrastructure of Microsporogenesis and Microgametogenesis *Inarabidopsis thaliana* (L.) Heynh. Ecotype Wassilewskija (Brassicaceae). Protoplasma **185**: 7–21

Pawłowski TA (2009) Proteome Analysis of Norway Maple (*Acer platanoides* L.) Seeds Dormancy Breaking and Germination: Influence of Abscisic and Gibberellic Acids. BMC Plant Biol **9**: 48

Pawłowski TA, Staszak AM (2016) Analysis of the Embryo Proteome of Sycamore (*Acer pseudoplatanus* L.) Seeds Reveals a Distinct Class of Proteins Regulating Dormancy Release. J Plant Physiol **195**: 9–22

Pertl H, Schulze WX, Obermeyer G (2009) The Pollen Organelle Membrane Proteome Reveals Highly Spatial–Temporal Dynamics during Germination and Tube Growth of Lily Pollen. J Proteome Res **8**: 5142–5152

Rajjou L, Gallardo K, Debeaujon I, Vandekerckhove J, Job C, Job D (2004) The Effect of α-Amanitin on the Arabidopsis Seed Proteome Highlights the Distinct Roles of Stored and Neosynthesized mRNAs during Germination. Plant Physiol **134**: 1598–1613

Roychoudhury A, Datta K, Datta SK (2011) Abiotic Stress in Plants: From Genomics to Metabolomics. *In* N Tuteja, SS Gill, R Tuteja, eds, Omics and Plant Abiotic Stress Tolerance, Bentham Science Publishers Sharjah, UAE, pp 91–120

Sajeev N, Baral A, America AHP, Willems LAJ, Merret R, Bentsink L (2022) The mRNA-Binding Proteome of a Critical Phase Transition during Arabidopsis Seed Germination. New Phytol **233**: 251–264

Shi J, Zhen Y, Zheng R-H (2010) Proteome Profiling of Early Seed Development in *Cunninghamia lanceolata* (Lamb.) Hook. J Exp Bot **61**: 2367–2381

Sinha A, Haider T, Narula K, Ghosh S, Chakraborty N, Chakraborty S (2020) Integrated Seed Proteome and Phosphoproteome Analyses Reveal Interplay of Nutrient Dynamics, Carbon–Nitrogen Partitioning, and Oxidative Signaling in Chickpea. Proteomics **20**: 1900267

Tan L, Chen S, Wang T, Dai S (2013) Proteomic Insights into Seed Germination in Response to Environmental Factors. Proteomics **13**: 1850–1870

Tuan PA, Kumar R, Rehal PK, Toora PK, Ayele BT (2018) Molecular Mechanisms Underlying Abscisic Acid/Gibberellin Balance in the Control of Seed Dormancy and Germination in Cereals. Front. Plant Sci. **9**: 668

Verma S, Attuluri VPS, Robert HS (2022) Transcriptional Control of Arabidopsis Seed Development. Planta **255**: 90

Wang W-Q, Liu S-J, Song S-Q, Møller IM (2015) Proteomics of Seed Development, Desiccation Tolerance, Germination and Vigor. Plant Physiol Biochem **86**: 1–15

Weber H, Borisjuk L, Wobus U (2005) Molecular Physiology of Legume Seed Development. Annu Rev Plant Biol **56**: 253–279

Xu H, Lian L, Jiang M, Zhu Y, Wu F, Jiang J, Zheng Y, Tong J, Lin Y, Wang F, et al. (2019) Proteomic Analysis Reveals Different Involvement of Proteins during the Maintenance and Release of Rice Seed Dormancy. Mol Breed **39**: 60

Yang M-F, Liu Y-J, Liu Y, Chen H, Chen F, Shen S-H (2009) Proteomic Analysis of Oil Mobilization in Seed Germination and Postgermination Development of *Jatropha curcas*. J Proteome Res **8**: 1441–1451

Yang P, Li X, Wang X, Chen H, Chen F, Shen S (2007) Proteomic Analysis of Rice (*Oryza sativa*) Seeds during Germination. Proteomics **7**: 3358–3368

Yang N, Wang T (2017) Comparative Proteomic Analysis Reveals a Dynamic Pollen Plasma Membrane Protein Map and the Membrane Landscape of Receptor-Like Kinases and Transporters Important for Pollen Tube Growth and Interaction with Pistils in Rice. BMC Plant Biol **17**: 2

Zhang P, Liu D, Shen H, Li Y, Nie Y (2015) Proteome Analysis of Dormancy-Released Seeds of *Fraxinus mandshurica* Rupr. in Response to Re-Dehydration Under Different Conditions. Int J Mol Sci **16**: 4713–4730

2 Advances in Plant Proteomics toward Improvement of Crop Productivity and Stress Resistance

Govinda Reddy Uday, Halehundi Bellegowda Shardha,
Krishna Reddy Priyanka, and Shamrao Jagirdhar

2.1 INTRODUCTION

Abiotic/biotic stress factors severely impede agricultural yield as well as plant growth and development. Acclimation to these abiotic stresses is mediated through profound changes in gene expression which result in changes in composition of plant transcriptome, proteome, and metabolome. Drought, salinity, temperature (freezing/heat), light intensity, and heavy metal contamination are the most common abiotic stressors that significantly impede not only plant growth but also crop quality (Ahmad et al. 2008; Roychoudhury et al. 2008). Plant stress response is a dynamic process that is dependent on stress intensity and stress duration.

A rough estimate places the number of individuals who experience hunger amounting to around 925 million (Karimizadeh et al. 2011). In order to eliminate that unsightly patch of hunger from the lovely face of humanity, we must considerably boost food production and supply by combining many factors and adopting plant breeding methods for crop advances (Beddington et al. 2012). A key challenge for plant breeders working on crop improvement projects is the small gene pool of domesticated crop species. A critical step in improving significant crop features is the identification of possibly beneficial genes from the animal and plant kingdoms. These genes are typically discovered through molecular biology research, including genomics and proteomics. Numerous finished and ongoing plant sequencing projects, such as those on *Arabidopsis thaliana*, rice (Goff et al. 2002; Yu et al. 2002), and soybean, are nearing the point where they can give modern breeders the blueprints to access a large number of genes. Plant genetic information (DNA) is converted to protein through the intermediate process of transcription and translation. Studying the proteins produced by those genes is one of the many cutting-edge methods for figuring out how genes are expressed globally and how they function; this field of science is called proteomics.

The word "proteome" is derived from PROTEins expressed by a genOME. Proteomics is the study and characterization of the entire set of proteins present in the cell at any one time (Wilkins et al. 1995). Proteomic investigations identify the functional players for mediating particular biological processes, whereas genetic researches help scientists understand what is theoretically feasible. In contrast to the genome, which is static in nature, the proteome is dynamic. Post-translational modifications (PTMs) are first introduced in the study of proteins, and this knowledge is crucial for understanding the biological processes (Nat et al. 2007). The genome sequence projects and/or transcript abundance alone cannot explain the PTMs that are crucial for plant growth and development and/or in response to various stress conditions (Roychoudhury et al. 2011). Proteomics knowledge provides functional genomics with completeness toward understanding the process (Gygi et al. 2000). Thus, proteomics research may help identify prospective proteins whose abundance variations are quantitatively correlated with alterations in specific physiological parameters that describe the stress tolerance of a genotype.

Proteomics comprises protein profiling, protein quantification, post-translation modifications as well as protein/protein interactions (Baginsky 2009). Proteomics deals with protein expression patterns, identification, PTMs, and interactions between proteins under both stressful and non-stressful circumstances (Hashiguchi et al. 2010). Plant stress proteomics offers the capacity to locate potential candidate genes that could be utilized to genetically modify plants to withstand stress (Rodziewicz et al. 2014).

High-throughput technologies like proteome analysis make it possible to take a very different approach. A novel method for identifying the genes and pathways essential for stress responsiveness and tolerance is provided by proteomic analysis. Proteins are packed with data that have proven to be particularly useful for the description of biological processes; proteomics serves as the primary link between genomics, transcriptomics, and metabolomics (Roychoudhury and Banerjee 2015). Protein interaction networks, PTM profiles, and quantitative expression profiles are examples of proteomic data that are highly instructive.

2.2 CROP PROTEOMICS FOR ABIOTIC STRESS TOLERANCE AND ITS CONTRIBUTION

Abiotic stress factors severely impede agricultural yield as well as plant growth and development. Acclimation to these abiotic stresses is mediated through profound changes in gene expression which result in changes in composition of plant transcriptome, proteome, and metabolome. Drought, salinity, temperature (freezing/heat), light intensity, and heavy metal contamination are the most common abiotic stressors that significantly impede not only plant growth but also crop quality (Ahmad et al. 2008). Plant stress response is a dynamic process which is dependent on stress intensity and stress duration (Roychoudhury et al. 2013). Proteins have direct stress-adaptation roles that result in changes to the cytoskeleton, plasma membrane, cell cytoplasm, and intracellular compartment composition, including conversion of cell cytoplasm to water. Therefore, study on how plants respond to stress at the protein level can greatly advance our understanding of the physiological processes behind plant stress tolerance. Thus, proteomics research may help identify prospective proteins whose abundance variations are quantitatively correlated with alterations in specific physiological parameters that describe the stress tolerance of a genotype.

2.2.1 DIFFERENTIAL-EXPRESSION PROTEOMICS

Differential-expression proteomics is based on the comparison of compositions of different proteomes. It is an approach used for the description of sets of proteomes differing both in protein quality and quantity, and it is aimed at protein identification and relative quantitation. However, protein function cannot be determined by the differential-expression proteomics approach (protein identification and quantitation) alone because a single protein can have a wide range of functions depending on its subcellular localization, PTMs, or interacting partners (protein–protein interactions).

2.2.2 PROTEOMICS APPROACH FOR ABIOTIC STRESS TOLERANCE

2.2.2.1 Salinity Tolerance

Increased salt concentrations in soil (mostly Na^+ and Cl^- and also Ca^{2+}, K^+, $(CO_3)_2$, (NO_3), and $(SO_4)_2$) led to a drop in soil water potential, which in turn affects the ability of roots to absorb water. Na^+ concentrations are actively kept low in the cytoplasm of plant cells by ATP-dependent ion pumps, which transport them into the vacuolar compartment (Basu and Roychoudhury 2014). Soil water with higher NaCl concentrations also experiences osmotic stress and an imbalance in intracellular ion homeostasis. Glycophytes used in proteomics research include tobacco and *A. thaliana* (Ndimba et al. 2005) as model plants (Table 2.1).

TABLE 2.1
Identification and Specific Roles of Different Proteins in Salt Tolerance

Crop Plants	Identified Proteins	Role in Salt Tolerance	References
Rice (*Oryza sativa* L.)	APX (ascorbate peroxidase), DHAR (dehydroascorbate reductase), SOD (superoxide dismutase)	Improved leaf sheath and leaf blade	Abbaasi et al. (2004)
Pea (*Pisum sativum* L.)	Cu-ZnSOD-II	Superoxide- and H_2O_2-mediated oxidative damage	Hernandez et al. (1995)
Sorghum (*Sorghum bicolor* L.)	Malate dehydrogenase, APX	ROS scavenging	Ngara et al. (2012)
Soybean (*Glycine max* L.)	LEA (late embryogenesis-abundant proteins)	Seed and hypocotyl development	Aghaei et al. (2009)
Potato (*Solanum tuberosum* L.)	Osmotin-like protein	Osmotic stress tolerance	Aghaei et al. (2008)
Wheat (*Triticum aestivum* L.)	Glutamine synthase, glycine dehydrogenase	Improved protein biosynthesis	Caruso et al. (2008)

It is common to find an increased production of the glycolysis and carbohydrate metabolism enzyme, fructose bisphosphate aldolase (ENO) in glycophytes (crops), which denotes a greater requirement for energy (Abbasi and Komatsu 2004). ROS scavenging enzymes (APX, DHAR, Trxh, peroxiredoxin, and SOD) are another significant category of elevated proteins that point to oxidative stress (Kim et al. 2005; Singh and Roychoudhury 2021).

Pang et al. (2010) compared the proteome alterations brought by salt stress in the halophyte *Thellungiella halophila* and the glycophyte *A. thaliana*. Salt stress caused significantly fewer disturbances in the proteome of *T. halophila* than in *A. thaliana* (37 and 88 proteins, respectively, displaying differential abundance under stress).

2.2.2.2 Cold Stress

A decrease in the rate of enzyme-catalyzed processes causes metabolic imbalances linked to oxidative stress, and cold as a stress factor is connected with major changes in energy metabolism. Cold significantly affects proteosynthesis. Increased levels of RNA-binding protein cp29 have been repeatedly reported (Amme et al. 2006). This protein is localized in chloroplast stroma, its activity could be regulated by phosphorylation, and it is involved in plastid mRNA processing. Kosmala et al. have studied proteomics response to cold (2°C) in two genotypes of meadow fescue (*Festuca pratensis*) differing in their frost tolerance. When protein abundance in these two genotypes exposed to cold was compared, significant differences (at least 1.5-fold differences) were observed for several components of thylakoid-membrane-associated photosynthetic apparatus including light-harvesting complexes, oxygen-evolving complex OEC (oxygen-evolving enhancer protein 1 OEE1), cytochrome b6/f complex iron–sulfur subunit or Rieske Fe–S protein.

Additionally, exposure to cold promotes the accumulation of COR/LEA proteins, which are very hydrophilic and can act as high-molecular osmoprotectants. Enhanced accumulation of specific dehydration-inducible LEA-II proteins named dehydrins has been repeatedly reported (Kawamura and Uemura 2003). Under cold, enhanced accumulation of several reactive oxygen species (ROS) scavenging enzymes, especially enzymes involved in the metabolism of ascorbate and glutathione (GSH) (enzymes of the so-called ascorbate–glutathione cycle; several isoforms of glutathione-S-transferase GST), have regularly been detected (Banerjee et al. 2017).

In addition, an increased abundance of other enzymes involved directly in ROS scavenging has been reported. An increased abundance of Cu/Zn superoxide dismutase (Cu/Zn-SOD) has been reported by Degand et al. 2009.

2.2.2.3 Heat Stress

Heat stress is associated with an enhanced risk of improper protein folding and denaturation of several intracellular protein and membrane complexes. It has long been well known that heat leads to increased expression of several proteins with chaperone functions, especially several members of large family of heat-shock proteins (HSPs) which are classified into five distinct sub-families according to their molecular weight (HSP110, HSP90, HSP70, HSP60, so-called small HSPs or sHSPs) (Baniwal et al. 2004; Banerjee and Roychoudhury 2018).

Heat-stress response at proteome level has been studied predominantly in rice and wheat during grain filling period, a heat- and drought-tolerant poplar (*Populus euphratica*), and also in wild plant *Carissa spinatum* inhabiting hot and dry valleys in central China (Lee et al. 2007). Another characteristic feature of heat stress is oxidative damage. Up-regulation of several enzymes involved in redox homeostasis such as GST, dehydroascorbate reductase (DHAR), thioredoxin h-type (Trx h), and chloroplast precursors of SOD was reported (Lee et al. 2007).

2.2.2.4 Drought

Drought stress is associated with a reduced water availability and cellular dehydration. As a result, it would be reasonable to anticipate that an osmotic adjustment would cause alterations in cellular metabolism. Bogeat-Triboulot et al. (2007) have investigated parallel changes at transcript and protein levels in a relatively drought-tolerant *P. euphratica* using cDNA microarray and 2D-DIGE analysis, respectively. Changes in gene expression upon drought observed at the transcript level differed from those observed at the protein level and alcohol dehydrogenase (ADH), cold-regulated LTCOR12, asparagine synthetase, cysteine protease, trypsin inhibitors, xylose isomerase, and sucrose synthase were all up-regulated at the transcript level. In contrast, the down-regulation of genes encoding proline-rich cell wall protein, aquaporin, and several photosynthesis-related genes was observed (Rabara et al. 2021).

It is well recognized that decreased root water intake negatively affects the growth of plant cells based on cell wall elongation. Zhu et al. (2007) discovered elevated amounts of many apoplastic ROS-scavenging enzymes, specifically peroxidases, in maize roots. These proteins have been proposed to contribute to an enhanced cell wall loosening upon dehydration stress.

2.2.2.5 Waterlogging

Waterlogging occurs when the amount of water in the soil exceeds its capacity to absorb it, leaving some water on the soil surface. Waterlogging is also characterized by a deficiency of oxygen accessible to plant roots. The development of anaerobic metabolism (pyruvate fermentation instead of glycolysis followed by oxidative subsequent breakdown by oxidative decarboxylation Krebs cycle) in addition to the PCD of a number of cells in the root cortex to create lysigenous aerenchyma that can supply oxygen to roots from the unflooded shoot regions of shoot regions.

Ahsan et al. (2007) and Alam et al. (2010) have analyzed proteome changes as well as changes in in vivo hydrogen peroxide content and lipid peroxidation in tomato leaves and soybean roots, respectively, affected by waterlogging stress. Interestingly, waterlogging has resulted in increased levels of H_2O_2 and lipid peroxidation indicating that this stress factor has an oxidative component. At the proteome level, waterlogging causes changes in the abundance of proteins involved in several processes, like photosynthesis, energy metabolism, redox homeostasis, signal transduction, PCD, RNA processing, protein biosynthesis, disease resistance, stress, and defense mechanisms. A large degradation of RuBisCO's large subunit (LSU) was observed.

Cui et al. (2009) observed differences in the proteome composition of swamp and desert dune ecotypes of common reed (*Phragmites communis*) when sampled in their natural habitats. Desert dune ecotype revealed a higher abundance of several stress-related proteins such as 60-kDa chaperonin α and β subunits and several HSPs, or enzymes of ascorbate and GSH metabolism (APX, GST) indicating that the desert ecotype has to cope with more adverse environmental conditions than swamp ecotype.

2.2.2.6 Translational Plant Proteomics

Translational plant proteomics builds on knowledge generated in model and crop plants, for example, crop disease proteomics can specifically address responses to avirulent and virulent pathogen strains and, vice versa, the responses of susceptible and resistant crop cultivars to pathogen infections. Comparative proteomic studies of compatible and incompatible crop–pathogen interactions have already been used to identify resistance mechanisms in crop pathosystems. Proteomics analysis of wild relatives of crops and also their ancient cultivars that are no longer used in breeding programs could provide potentially useful knowledge on protein expression associated with useful traits or even serve as a resource of proteins and genes for restoring traits lost during domestication and selective breeding.

2.3 CROP PROTEOMICS FOR BIOTIC STRESS TOLERANCE

The proteomics study offers a new approach to discover proteins and pathways associated with crop physiological and stress responses. Thus, studying the plants at proteomic levels could help understand the pathways involved in stress tolerance. Furthermore, improving the understanding of the identified key metabolic proteins involved in tolerance can be implemented into biotechnological applications, regarding recombinant/transgenic formation. Additionally, the investigation of identified metabolic processes ultimately supports the development of anti-stress strategies (Ahmad et al. 2016).

2.3.1 ROLE OF PROTEOMICS IN BIOTIC STRESS TOLERANCE MECHANISMS

2.3.1.1 Resistance to Fungi

Proteomic approaches are useful to study the molecular mechanism involved in the interaction between a plant and its pathogens. *Fusarium graminearum* spores were used to inoculate the wheat spikelet, and the total proteins from the infected spikelet were then exposed to 2-DE for proteome profiling under both healthy and diseased conditions. They observed that the *Fusarium* infection caused a distinct regulation of 41 proteins. According to geneontology (GO) annotation, proteins from the pathways for antioxidant and JA signaling, pathogenesis-related response, amino acid synthesis, and nitrogen metabolism were up-regulated, whereas proteins from the system for photosynthesis were down-regulated (Zhou et al. 2006).

After 6, 12, and 24 hours of spike inoculation with the fungus *Fusarium*, Wang et al. (2005) discovered that the expression of the genes involved in carbon metabolism and photosynthesis significantly decreased. They studied the proteome response of the resistant wheat cultivar Wangshuibai.

To investigate the molecular processes behind the defensive responses of the *Setosphaeria turcica* race-13-resistant maize line, A619 Ht2, a comparative proteome research utilizing 2-DE and MS, was carried out (Zhang et al. 2014). Numerous proteins, including those involved in protein transport and storage, disease defense, and energy metabolism, were discovered to have changed amounts in response to northern leaf blight. Some defense-related proteins, including β-glucosidase, SOD, polyamines oxidase, and PPIases, were increased upon *S. turcica* inoculation, but proteins involved in photosynthesis and metabolism were down-regulated. These findings suggest that direct release of defense proteins from maize, modulation of primary metabolism, and alteration of photosynthesis and carbohydrate metabolism may be the basis of their resistance to *S. turcica*.

Using a plant miRNA microarray technology, the miRNA expression patterns in maize in response to *E. turcicum* stress have been examined (Wu et al. 2014). A total of 118 miRNAs were found, including the previously unidentified miRNAs miR530, miR811, miR829, and miR845. Additionally, in response to *E. turcicum* infection, miR811, miR829, miR845, and miR408 showed differential regulation. At the post-transcriptional level, stress-responsive miRNAs controlled the metabolic, morphological, and physiological changes in maize seedlings. Additionally, miR811 and miR829 were highly resistant to *E. turcicum* and can be exploited in maize breeding programs. Utilizing 2-DE and MS/MS, a comparative proteome study of leaves from virus-infected and healthy plants was carried out (Li et al. 2011). The majority of the proteins that were discovered to have differential expression between control maize and maize that had been infected with RBSDV belonged to metabolic pathways.

When rice was infected with sheath blight brought on by *Rhizoctonia solani*, Lee et al. (2006) employed tandem mass spectral analysis and two-dimensional gel electrophoresis to find the proteins that were produced differently. For the first time, the trigger of 3-hydroxysteroid dehydrogenase/isomerase was found in the resistant plants, demonstrating the enzyme function in the plant defensive system against *R. solani*. They also discovered that a genetic marker for resistance to sheath blight was physically near four induced proteins.

2.3.1.2 Resistance to Bacteria

The most significant rice disease brought by bacterial pathogens is called bacterial blight, which is caused by *Xanthomonas oryzae pv. oryzae (Xoo)*. Research has been done to examine how the *Xoo* virus affects rice proteins globally. Reduced central carbon catabolism is accompanied by increased signal transmission, including multiple suspected resistance (R) genes, putative receptor-like kinases, and PR, which are linked to disease resistance, pathogenesis, and the control of cell metabolism (Mahmood et al. 2006).

Plant growth-promoting rhizobacteria (PGPR), whose expansion is induced by root exudates, aid plants by absorbing nutrients and producing phytohormones. In a proteomics research, it was discovered that *Pseudomonas fluorescens* and *Sinorhizobium meliloti* accumulate proteins associated with photosynthesis and defense (Chi et al. 2010).

2.3.1.3 Response to Virus

Viruses that are spread to rice by either plants or insects result in physiological modifications such as chlorosis and a suppression of photosynthesis. The main changes brought about by viral infection via insects as the vector include carbon metabolism. A glycolytic enzyme called glyceraldehyde-3-phosphate dehydrogenase (GAPDH) has become a multifunctional protein in a number of non-metabolic activities and is up-regulated in a number of plant species after viral infection (Alexander and Cilia 2016). A virus infection can also cause changes in amino acid levels, which offer the building blocks for viral replication, plant defense, ROS buildup for chlorotic damage, and respiration as a source of energy. Proteome and metabolome investigations have verified and explained these results.

During rice stripe virus (RSV, *Tenuivirus* genus) infection, chloroplast proteins are destroyed, chlorophyll a and chlorophyll b production is blocked, and 26S proteasome is up-regulated (Wang et al. 2015). Additionally, the rice black-streaked dwarf virus (RBSDV, Reoviridae family, *Fijivirus* genus) may produce more H_2O_2, which might restrict light absorption and impair photosynthesis.

Changes in the protein profiles of sugarcane mosaic virus (SCMV)-resistant and susceptible maize during SCMV infection have been analyzed using a DIGE-based proteomics approach (Wu et al. 2013). A total of 93 proteins were found to be differentially expressed after virus inoculation, and these were primarily involved in energy, metabolism, stress and defense responses, photosynthesis, and carbon fixation. SCMV-responsive proteins were also identified and analyzed in the maize lines Siyi and Mo17 in response to SCMV infection by using 2-DE and MALDI-TOF-MS/

MS. Most of the identified proteins were present in chloroplasts, chloroplast' membranes, and the cytoplasm. Further study of the roles of these proteins in maize-virus interactions will be valuable.

2.3.2 USE OF PROTEOMICS TO STUDY PLANT-PATHOGEN INTERACTIONS

Plant-pathogen interactions first used proteomics approximately two decades ago, when electrophoretic and protein identification methods were less advanced than they are now. Ekramoddoullah and Hunt used 2D PAGE in 1993 to differentiate between the proteins in sugar-pine seedlings that were susceptible to or resistant to the fungus that causes white pine blister rust. After inoculating pine seedlings with *C. ribicola*, the foliage proteins were isolated and sorted using a 2D gel electrophoresis. Two acidic proteins were found in relatively high concentrations; the larger of the two, measuring 36.7 kDa, was dramatically reduced in susceptible seedlings at day three and significantly increased in resistant seedlings at day nine. The nature of the infection process results in the susceptible and resistant seedlings as explained by the differential expression of these two proteins. Additionally, in resistant seedlings, only susceptible seedlings showed a decrease in the proteins involved in this crucial cellular process, while resistant seedlings showed an increase in a small number of proteins involved in protein biosynthesis and a decrease in others involved in the same process (Ekramoddoullah and Hunt 1993).

Due to its tiny genome size, *Medicago truncatula* is regarded as the model plant for proteomic investigations (454–526 Mbp). Globally, several articles on the proteome analysis of *M. truncatula* have been published. In order to identify the *M. truncatula* response to the pathogenic *Pseudomonas aeruginosa* N-acyl homoserine lactone (AHL) signals, a proteomic technique was developed. It was discovered that one-third of the responsive components differ from the AHL produced by the nitrogen-fixing bacterium *S. meliloti*. Proteomics allowed it to be proven that plants are able to differentiate between AHL exposure times and concentrations from structurally identical sources. Plants create "signal mimic" chemicals in response to AHL that affect the bacterial activity governed by quorum sensing.

The etiological agent of soft rot disease in ornamental plants and vegetables is *Erwinia chrysanthemi*. There are 55 spots on the 2D-PAGE gel from which 25 were uniquely present, according to the results of the proteome study of the secretome (extracellular proteins) of uninduced *E. chrysanthemi*. The main proteins detected in the unreduced state included cellulase (Cel5), protease (PrtA, B, C), flagelin (Fli C), and other intracellular proteins. On the other hand, 14 distinct proteins from the secretome of culture induced by a heat-soluble extract made from *Chrysanthemum* leaves were effectively identified. Pectin lyases, pectin acetylesterase, pectin methylesterase, and polygalacturonase make up the majority of these proteins. The type-I system of the secretome plays a significant role in the secretion of proteases, while the type-II system is responsible for the secretion of esterase and the Avr-like protein AvrL, according to the study of mutants. Similar to *Xanthomonas campestris*, the cause of the disease known as "black rot" in conifers, AvrL is similar to the antiviral protein (Kazemi and Hugouvieux-Cotte-Pattat 2004).

Using 2D PAGE and MALDI-TOF-MS, Watt et al. (2005) performed a proteome study on the *X. campestris pv campestris*. They observed 97 separate protein spots. These spots were excised, the tryptic-digested fragments were analyzed by MALDI-TOF-MS, and 68 different proteins were identified. The findings indicated that a large number of these extracellular proteins are engaged in deteriorative processes and crucial elements enabling the infection of sensitive plant hosts.

A facultative disease of soybeans called *Phytophthora sojae* causes root and stem rot, which results in a global loss of 1–2 billion US dollars in soybean production. In order to ascertain the impact of *P. sojae* infection on soybean hypocotyls, the proteomic investigation was conducted showing 46 differentially expressed proteins using 2D-PAGE and MALDI-TOF/TOF. In the tolerant soybean line "Yudou25," the expression levels of 26 proteins were significantly impacted at different times (12 up-regulated and 14 down-regulated). In comparison, just 20 proteins substantially differed in the sensitive soybean line "NG6255" (11 up-regulated and 9 down-regulated). These proteins had a 26% energy control role, 15% to protein destination and storage function, 11% disease defense function, 11% metabolism function, 9% protein synthesis function, 4% secondary metabolism function, and 24% function that was unclear. The findings of this study provide a window into the array of strategies *P. sojae* uses to infect and colonize a host.

We can now consider proteomics to be a mature platform for proteome analysis during plant–pathogen interaction. Unknown protein functions are being revealed by proteome-wide functional categorization utilizing bioinformatics techniques. Scientists can better understand the activities of proteins and the intricate regulatory networks that govern the basic biological processes due to proteomics.

2.3.2.1 Plant–Virus Interaction

For the success of plant infection, viruses must first be transmitted either mechanically or by a vector (transmission), replicate in plant cells (replication), subsequently move through plasmodesmata to neighboring cells (cell-to-cell movement), and, finally, attain the vascular tissue to circulate systemically through the phloem to the sink tissues of the host (vascular movement). After being unloaded from the phloem, viruses establish systemic infection through new cycles of replication and cell-to-cell/vascular movement. In both compatible (susceptible host) and incompatible (resistant host) interactions, viruses use plant host proteins to complete the steps of the infection process and suffer the influences of plant host proteins as a counteraction against the infection. The genes that encode these proteins have been studied extensively in numerous host–virus systems, mainly using transcriptional analysis (Whitham et al. 2006).

Recently, 2DE and subsequent MALDI-TOF MS have been performed to analyze the induced expression of nuclear proteins in *Capsicum annuum* cv. Bugang (hot pepper) infected by tobacco mosaic tobamovirus (TMV) (Lee et al. 2006). *C. annuum* cv. Bugang is hypersensitive response resistant against TMV-P0 and susceptible to TMV-P1.2 strains. A hypothetical protein and five annotated nuclear proteins were identified in hot pepper infected by TMV-P0, including four defense-related proteins [14-3-3 protein (regulator of proteins involved in response to biotic stresses), 26S proteasome subunit (RPN7) (postulated to be involved in programmed cell death), mRNA-binding protein (may interact with viral RNA or interfere with plant RNA metabolism), and Rab11 GTPase (responsible for membrane trafficking/recycling and endocytosis/exocytosis)] and a ubiquitin extension protein.

Diaz-Vivancos et al. (2006) used proteomic approaches to study the changes in enzymatic activity and protein expression in the antioxidative system within the leaf apoplast of *Prunus persica* cv. GS305 (peach) on plum pox potyvirus (PPV) infection. PPV infection provoked oxidative stress in peach leaf apoplast by increasing the antioxidant enzymatic activities and H_2O_2 contents. The 2DE of apoplastic fluids from peach leaves infected with PPV, and subsequent MALDI-TOF MS analyses, revealed the identification of four proteins of the 22 analyzed: one thaumatin-like and three mandelonitrile lyases (MDLs). Thaumatins are proteins involved in the plant response against fungal infection and may equally be expressed in peaches as a response to PPV infection. MDLs are flavoproteins involved in the catabolism of (R)-amygdaline; however, to define their role in the peach plant–PPV interaction, further investigations must be performed.

Another study on plant–virus interaction was performed by Rahoutei et al. (1999). These authors demonstrated that the pepper mild mottle tobamovirus Spanish strain S (PMMoV-S) inhibits photosystem II electron transport, disturbing the oxygen-evolving complex, composed of the three proteins PsbP, PsbO, and PsbQ, present within plant thylakoid membranes. PMMoV-S infection results in a lower expression of PsbP and PsbQ in the susceptible host *Nicotiana benthamiana* Domin (tobacco) relative to that in healthy control plants. Proteomic analysis was also performed to study the compatible interaction between *Oryza sativa* (rice) and rice yellow mottle sobemovirus (RYMV) (Delalande et al. 2005). This analysis led to the identification of a phenylalanine ammonia-lyase, a mitochondrial chaperonin-60, and an aldolase C, but the role of these proteins during RYMV infection of rice remains to be determined.

2.3.2.2 Plant–Bacterium Interactions

To circumvent plant defenses and establish a successful colonization of the host plant, bacteria rely on a variety of secretion mechanisms. Bacteria have been shown to have five secretion systems (types I–V), which may be identified by the proteins that make up each system (Lee and Schneewind 2001). The type III secretion system (TTSS), which is implicated in some of the deadliest diseases in animals and plants, is the primary secretion system employed by pathogenic bacteria during infection. By directly injecting proteins into the host cell, known as effectors or virulence factors, bacteria are able to disrupt biological functions. The proteins exported by this system are more varied, despite the fact that TTSS is conserved and necessary for pathogenicity in Gram-negative bacteria (Galan and Collmer 1999). Avirulence (Avr) proteins, which have been discovered in a number of plant diseases, are the best studied TTSS effectors. Various phytopathogenic bacterial species have also been shown to produce additional effectors, such as the Xanthomonas outer protein (Xop) in Xanthomonas, the Hrp outer protein (Hop) in Pseudomonas, and the Pseudomonas outer protein (Pop) in Ralstonia (based on a former genus classification) (Arlat et al. 1994). The type II secretion system, which is involved in the production of extracellular enzymes, poisons, and virulence factors, is another crucial component of bacterial pathogenicity. It is anticipated that various infections would exhibit striking variations in the quantity and configurations of these enzymes.

The investigation of outer membrane proteins of the soft rot pathogen *Dickeya dadantii* (syn. *E. chrysanthemi*) by 2DE and MALDI-TOF MS studies also utilized plant extracts as a stress condition (Babujee et al. 2007). A number of proteins were discovered, including the porin OmpA, which is involved in binding to particular host cell receptor molecules, the porin HrcC, a member of the PulD pIV superfamily, which is involved in the outer membrane translocation of type II and type III secretion pathways, and the porins KdgM and KdgN, which are specific for oligogalacturonate-0. By comparing *E. chrysanthemi* wild-type and osmoregulated periplasmic glucan (OPG)-defective mutant cells, which exhibit a lack of virulence, by 2DE, the proteome of this organism was further examined. Differentially expressed proteins that are vital for cellular functions including protein folding and degradation and glucose metabolism were found in the mutant cells (Bouchart et al. 2007). The scientists came to the conclusion that *E. chrysanthemi* responds to OPG deficiency by energizing cellular mechanisms that shield the cell from external stimuli, indicating that the opgG strain has trouble perceiving its surroundings. When compared to *X. axonopodis* pv. citri, *Xylella fastidiosa*, the cause of citrus variegated chlorosis, did not significantly alter the expression of the HSP, according to a 2DE-mediated proteomic research (Martins et al. 2007). However, it was discovered that *X. fastidiosa* produced a number of stress-inducible proteins that were activated in *X. citri* under stressful circumstances, including HspA and GroeS. The constitutive production of these proteins, according to the scientists, may aid *X. fastidiosa* in adjusting to rapid environmental changes and stresses.

Mahmood et al. (2006) used a proteomic method to study the involvement of defense-responsive proteins in the interaction between rice and *X. oryzae* pv. oryzae. Three days after the inoculation with races of *X. oryzae* pv. oryzae, cytosolic and membrane proteins of rice leaf blades were separated. Twenty proteins out of the 366 showed differential expression in response to bacterial inoculation. Four defense-related proteins, PR-5, probenazole-inducible protein (PBZ1), SOD, and Prx, were all induced for both compatible and incompatible *X. oryzae* pv. Oryzae races, with PR-5 and PBZ1 being more rapid and exhibiting higher induction in incompatible interactions as well as in the presence of jasmonate. This was clearly demonstrated by analyses.

2.3.2.3 Plant–Fungus Interactions

Considerable advances have been achieved in the last 10 years in the identification of the determinants of plant–fungus interactions. Currently, more than 25 fungal genomes have been elucidated, including human and plant pathogens, such as *Aspergillus fumigatus* and *Magnaporthe grisea*, respectively. A key challenge in modern fungal biology is to analyze the expression, function, and regulation of the entire set of proteins encoded by the revealed fungal genomes.

Proteomic analyses have also been used to study wheat leaf rust, caused by the fungus *Puccinia triticina*. Rust diseases cause a significant annual decrease in the yield of cereal crops worldwide. In order to better understand this problem at the molecular level, the proteomes of both host and pathogen were evaluated during disease development. A susceptible line of wheat infected with a virulent race of leaf rust was compared with mock-inoculated wheat using 2DE (with isoelectric focusing, pH 4–8) and MS analysis (Rampitsch et al. 2006). The fungus differentially expressed 22 different proteins during pathogen infection, including proteins with known and hypothetical functions.

Another approach, which has been frequently employed for the study of fungal proteins, involves the analysis of the exoproteome, also known as the secretome (Phalip et al. 2005). In this context, *F. graminearum*, a devastating pathogen of wheat, maize, and other cereals, was grown on hop (*Humulus lupulus*) cell walls. Using 1DE and 2DE, followed by MS analyses, 84 fungal-secreted proteins were identified. Among the identified proteins were cellulases, glucanosyltransferases, endoglucanases, phospholipases, proteinases, and chitinases. It was observed that 45% of the proteins observed in *F. graminearum* grown in the presence of hop cells were strictly involved in cell wall degradation and indirectly related to carbon and nitrogen absorption. When this same fungus was grown in a medium containing glucose, however, the enzyme patterns were totally different, showing that fungi are capable of regulating their secretion according to the presence of substrate.

2.3.2.4 Plant–Nematode Interactions

Phytonematodes, which constantly attack plants and severely harm agricultural crops that are vulnerable to them, generate significant economic losses globally. *Meloidogyne* spp., *Heterodera* spp., and *Globodera* spp. are obligate sedentary endoparasites that are among the most dangerous plant–parasitic nematodes. The juvenile larvae (J2) of these organisms enter plant roots, and after three molts, they mature into adult forms that continuously reproduce. As a result, the root system is severely altered, which significantly reduces nutrient and water intake and results in plant death (Curtis 2007).

Several nematode-expressed sequence tag (EST) libraries have been created recently, mostly to uncover parasitic nematode-specific genes. Roughly 100,000 ESTs from *Meloidogyne*, *Globodera*, and *Heterodera* species have been sequenced. Even though there are numerous ESTs, only a small number of these genes are known to be involved in parasitism, even though many of the transcripts exhibit variable expression throughout the parasitic phases. Although to a lesser extent, proteomic methods have also helped to identify potential phytonematode parasitome possibilities.

2.3.3 Development of Proteomics-based Fungicides

2.3.3.1 New Protein-based Strategies to Classical Chemical Fungicide Design

Drugs have historically been derived from plant and animal products, from human endogenous ligand derivatives, or from synthetic or semi-synthetic compounds. Chemical substances are the foundation of traditional approaches to treating fungal plant diseases. Despite the successes, new standards for the reckless use of hazardous substances found in nature forego employing this technique. Control methods based on conventional fungicides have detrimental side effects, primarily linked to environmental damage and the development of drug resistance.

The scientific community is addressing several developments in the design of chemical fungicides by compiling the available genomic and proteomic data. The development of biosynthetic fungicides has become a new area of emphasis (Collado et al. 2007). Using of foreign or altered natural substances may offer a species-specific means of managing plant diseases by specifically inhibiting those proteins involved in the infection cycle, according to a thorough examination of fungal biology (Pinedo et al. 2008). Due to their biodegradability, high selectivity, and poor integration into the food chain, the usage of these chemicals has a minor negative impact on the environment.

New terminologies associated with chemical "-omics" have emerged in the post-genomic age. The use of tiny chemicals to specifically disrupt gene function is referred to as "genetic chemical" in scientific literature. Chemogenomics refers to the application of this idea throughout the entire genome. Chemoproteomics is the term for the application of chemogenomics to protein targets; however, TRAP (targeted-related affinity profiling), which is defined as the use of biology to guide chemistry, provides a more precise description (Beroza et al. 2002). The gathering of proteomic data on fungi that cause plant diseases may encourage the creation of novel, ecologically friendly fungicides.

2.3.4 ROLE OF TRANSCRIPTION FACTORS (TFs) IN BIOTIC STRESS TOLERANCE

As the primary controllers of gene expression, transcription factors (TFs) influence key elements of plant function, such as hormone and environmental responses, cell differentiation, and organ development (Nath et al. 2019). By binding to the local and distal cis-elements of a particular gene in various biological situations, TFs modify gene expression. Recent studies have shown how TF interactions, local DNA structure, and genomic characteristics might affect TF binding to DNA, according to Inukai et al. (2017). The capacity of TFs to bind particular DNA sequences and interact with other proteins in transcriptional complexes that control the expression of a large number of genes is essential to their function. Therefore, understanding the molecular effects of TFs is crucial for future research. In plants, ~10% of genes encode TFs (Gonzalez 2015), which take part at different stages for a specific function. Several TF databases are now available and provide comprehensive information about TF families in different species. In the Plant Transcription Factor Database (plant TFDB v5.0, Center for Bioinformatics, Peking University), 134 WRKY, 180 NAC, 145 MYB, 172 ERF, and 166 bZIP genes have been identified in sorghum (Figure 2.1). TFs are promising genetic

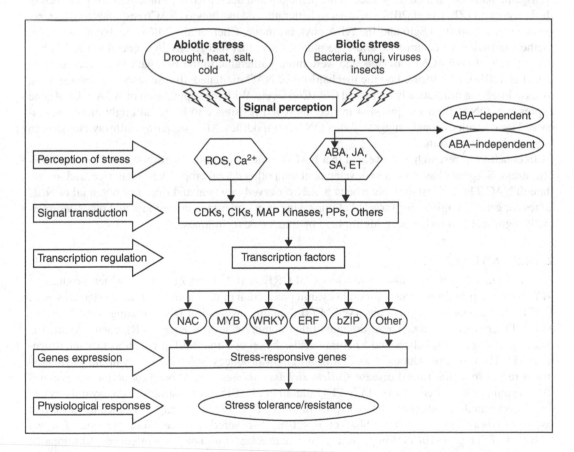

FIGURE 2.1 A schematic model of the signaling pathways involved in abiotic and biotic stress responses.

engineering targets because of their function as master regulators of several stress-related genes. Numerous TF families, such as WRKY, MYB, NAC, and bZIP, have been linked to stress reactions, and numerous TF genes are linked to increased tolerance in model and crop plants (Wang et al. 2016). Approximately 30 TF families with 1922 TFs involved in various activities are estimated to exist in *Arabidopsis* (Aglawe et al. 2012). Sorghum has a total of 2448 TFs documented, along with 1611 in rice and 3337 in maize. Numerous research conducted over the past two decades have discovered TF genes and described how they react to biotic and abiotic stressors. Due to the fact that many of these genes are stress-responsive and regulate a large number of downstream genes, increasing plant stress tolerance by modifying the expression of TF genes has become a popular study topic. Therefore, it may be possible to develop crops with greater stress tolerance (Hoang et al. 2017). The overexpression of multiple TF genes has allowed for significant advances in this arena.

2.3.4.1 NAC TFs

It is well known that NAC TFs offer diverse crop resilience to abiotic stressors. NAC TFs are now known to have a role in biotic stress responses as well, according to recent investigations. For instance, Zhang and Huang (2013) showed that NAC TFs were induced in response to a sorghum infestation with greenbugs. Numerous *NAC* genes were found to be expressed in response to green-bug infestation, according to RNA-sequencing data. The authors came to the conclusion that *SbNAC* genes serve a variety of purposes in response to green insect infestation and may thus have a role in the genetic resistance to green bugs. Wheat NACs have also been linked with the response to powdery mildew disease in addition to these reactions. For instance, TaNAC6 overexpression in transgenic plants increased resistance to the pathogen and decreased the amount of the pathogen in the haustorium (Zhou et al. 2018). In order to better understand how HvNAC6 responds to *Blumeria graminis f. sp. hordei* (Bgh) and the ABA phytohormone, Chen et al. (2013) adopted a transgenic method in barley. In contrast to wild-type plants, those that had HvNAC6 silenced via RNAi had lower levels of HvNAC6 transcripts and were more vulnerable to the Bgh pathogen. According to Liu et al. (2018), who focused on the mechanisms of NAC-mediated disease responses in rice, blast disease infection dramatically increased ONAC66 levels. With the suppression of ABA-related gene expression, ONAC66 overexpression in rice improved resistance to bacterial blight and blast disease. These results strongly suggested that ONAC66 modifies ABA signaling pathway elements to support disease resistance.

In conclusion, research has revealed that NAC TFs are critical in the response to biotic and abiotic stress. Sorghum has made less development than other cereal crops like wheat, rice, and maize, though NAC TFs, for instance, have been widely observed in wheat and rice. The potential of NAC genes for genetic engineering targeted at improving stress tolerance in sorghum and other commercially significant crops has been highlighted by transgenic techniques.

2.3.4.2 MYB TFs

The first plant MYB gene discovered was COLORED1 (C1) from *Zea mays*, which produces a MYB domain protein necessary for anthocyanin production in the aleurone of maize (Pazares et al. 1987) The existence of highly conserved MYB domains involved in DNA binding is the basis for MYB TF characterization. These domains often comprise several repeats (R), each of which is made up of three alpha helices and 52 amino acids, the second and third of which form a helix-turn-helix (HTH) structure (Dubos et al. 2013). Sorghum produces 3-deoxyanthocyanidin phytoalexins in response to the fungal disease *Colletotrichum sublineolum*, whose production necessitates the sorghum yellow seed 1 (Y1) MYB TF (Baldoni et al. 2015). Interestingly, 3-deoxyanthocyanidin was elevated in transgenic maize expressing this MYB TF, increasing resistance to leaf blight pathogen (Ibraheem et al. 2015). Shan et al. (2016) discovered that the overexpression of wheat R2R3-MYB TF gene *TaRIM1* improved resistance to *Rhizoctonia cerealis* infection. Additionally, the authors demonstrated how *TaRIM1* attachment to the MYB-binding site altered the expression of defense genes in transgenic wheat plants. Chrysanthemum has been used as an illustration of MYB TF involvement in the response to insect invasion. Aphid infection was discovered to trigger the *CmMYB19* gene, and transgenic plants that overexpressed *CmMYB19* inhibited aphid growth by accumulating lignin (Wang et al. 2017).

In conclusion, MYB TFs are critical in the response to a variety of environmental stresses, and research from several species has illuminated the mechanisms by which MYB genes are engaged in biotic stress responses.

2.3.4.3 WRKY TFs

One of the biggest families of transcriptional regulators, WRKYs control a variety of plant functions. The WRKY TF family has undergone significant research since the first member was discovered in sweet potato (*Ipomoea batatas*; SPF1) 25 years ago (Ishiguro and Nakamura 1994). Recent studies have demonstrated that WRKY TFs are crucial in controlling the defensive responses to pathogen attacks. Another study found that overexpressing the rice gene WRKY67 made transgenic rice plants more resistant to the two major rice diseases, blast and bacterial blight, making WRKY67 a prime candidate for rice improvement (Yang et al. 2018). According to Wang et al. (2017), TaWRKY49 silencing improved resistance to the disease compared to non-silenced plants, but TaWRKY62 silencing increased susceptibility to stripe rust disease in wheat. WRKY TFs are also implicated in the responses to plant viruses. For instance, WRKY8 controls the ABA and ET signaling pathways in *Arabidopsis* in order to impart resistance against the infection of the tobacco mosaic virus (TMV-cg), which infects crucifers (Chen et al. 2013). In addition, WRKY TFs could act as a mediator of communication between biotic and abiotic stress response pathways. For instance, Lee et al. (2018) discovered that OsWRKY11 can improve biotic and abiotic stress-related genes that are involved in pathogen defense and drought tolerance in rice.

2.4 LIMITATIONS OF CROP PROTEOMICS

2.4.1 LARGE-SCALE IDENTIFICATION OF CROP PROTEINS

Even with the highly advanced technologies and state-of-the-art proteome databases, full proteome coverage has not been achieved even in model plants. Approximately 300,000 non-redundant peptides matching about 25,000 unique proteins are currently available via MASCP Gator, representing 70% of the expected *Arabidopsis* proteome.

Most plants have experienced polyploidization of their genome during evolution and important crops such as wheat have allopolyploid genomes. Genomes of polyploid species, especially those in which polyploidization has occurred recently, are more difficult to sequence and assemble because of gene redundancies.

2.4.2 TARGETING OF LOW-ABUNDANCE PROTEINS FOR USEFUL CROP TRAITS

Regulatory proteins that have important roles in plant responses to abiotic and biotic stress are often difficult to detect even with the highly advanced MS instruments because of their low abundance. Pre-fractionation of protein extracts combined with RuBisCO depletion can facilitate the detection of LAPs. Protein fractionation and enrichment generate additional variability in biological samples, which is not compatible with high-throughput quantitative proteomics methods. Recently, combinatorial peptide ligand libraries (CPLL) have proven to be effective in removing abundant proteins from plant extracts to increase proteome coverage.

2.4.3   POST-TRANSLATIONAL MODIFICATIONS OF REGULATORY PROTEINS AS TARGETS FOR CROP BREEDING

High-throughput phosphoproteomics has revealed the degree of protein phosphorylation in plants such as *Arabidopsis* and rice, as well as the modulation of their phosphoproteome under various conditions. Although the broader relevance of PTMs in crop breeding strategies remains to be established, a recent study demonstrated that the phosphorylation status of a pathogen resistance protein can discriminate between stem rust susceptible and resistant barley varieties. Characterization of the phosphoproteome remains challenging because protein phosphorylation is generally rapid and transient.

2.4.4   DEVELOPMENT OF BIOMARKERS USING CROP PROTEOMICS

Isozymes have long been useful markers in plant breeding programs, but with the development of advanced breeding methods, they were rapidly displaced by more economic molecular and functional markers. The potential of crop proteomics for the development of novel markers for plant breeding has not been fully realized yet, although such markers may be diagnostic for quantitative trait loci or link to useful cis-regulatory elements. The concept of using plant proteomics for the identification of proteins as biomarkers was first introduced with the large-scale proteome analysis of *Arabidopsis* organs and later for the reconstruction of structural and metabolic transitions of maize leaf development and differentiation. The information currently available from model plants has not been exploited yet for crops, and biomarkers have only been developed for food composition and quality of certain crops.

2.5   CONCLUSIONS AND FUTURE PERSPECTIVES

To help with the biological advancement of agricultural yield, proteins that regulate crop design and/ or stress tolerance in a variety of conditions must be found in order to address the present concerns of food insecurity. Modern proteomic methods are the greatest option for scientists to identify these proteins. However, certain drawbacks are restricting the application of proteomics. Proteomic snapshots can be challenging due to the dynamic and interacting nature of proteins. Furthermore, their tight association may not always imply a functional link between two proteins. In order to increase our understanding of protein expression during plant-pathogen interactions, more sensitive analytical tools and efficient methodologies for large-scale data comparison are required. However, new techniques and apparatus are being developed, and proteomic tools are advancing quickly. Future proteomic research, along with functional validation and bioinformatical analysis, may, from our perspective, offer fresh perspectives on plant disease resistance and pathogenicity.

Proteomics will aid in the development of biomarkers closely linked to the trait of interest and will assist plant breeders in creating new stress-resistant varieties of plants in addition to assisting in the understanding of the proteins, pathways, and metabolites underlying stress-responsive mechanisms. Additionally, it will aid in the discovery of novel genes involved in stress response mechanisms that plant biotechnologists may focus upon in order to create transgenic plants that are resistant to stress.

REFERENCES

Abbasi F M and Komatsu S, 2004, A proteomic approach to analyze salt-responsive proteins in rice leaf sheath. *Proteomics*, 4: 2072–2081. doi: 10.1002/pmic.200300741

Aghaei K, Ehsanpour A and Komatsu S, 2008, Proteome analysis of potato under salt stress. *Journal of Proteome Resources*, 7: 4858–4868.

Aghaei K, Ehsanpour A A, Shah A H and Komatsu S, 2009, Proteome analysis of soybean hypocotyls and root under salt stress. *Amino Acids*, 36: 91–98. doi: 10.1007/s00726-008-0036-7

Aglawe S B, Fakrudin B, Patole C B, Bhairappanavar S B, Koti R V and Krishnaraj P U, 2012, Quantitative RTPCR analysis of 20 transcription factor genes of MADS, ARF, HAP2, MBF and HB families in moisture stressed shoot and root tissues of sorghum. *Physiology of Molecular Biology in Plants*, 18: 287–300.

Ahmad P, Abdel Latef A A, Rasool S, Akram N A, Ashraf M and Gucel S, 2016, Role of proteomics in crop stress tolerance. *Frontiers in Plant Science*, 7: 1336.

Ahmad P, Sarwat M and Sharma S, 2008, Reactive oxygen species, antioxidants and signaling in plants. *Journal of Plant Biology*, 51: 167–173.

Ahmad P, Sarwat M and Sharma S, 2008, Reactive oxygen species, antioxidants and signaling in plants. *Journal of Plant Biology*, *51*, 167–173. doi: 10.1007/BF03030694

Ahsan N, Lee D G, Lee S H, Kang K Y, Bahk J D and Choi M S, 2007, A comparative proteomic analysis of tomato leaves in response to waterlogging stress. *Physiology of Plants*, 131: 555–570.

Alam I, Lee D G, Kim K H, Park C H, Sharmin S A and Lee H, 2010, Proteome analysis of soybean roots under waterlogging stress at an early vegetative stage. *Journal of Biosciences*, 35: 49–62.

Alexander M M and Cilia M, 2016, A molecular tug-of-war: Global plant proteome changes during viral infection. *Current Plant Biology*, 5: 13–24.

Alfano J R and Collmer A, 1997, The type III (Hrp) secretion pathway of plant pathogenic bacteria: Trafficking harpins, Avr proteins, and death. *Journal of Bacteriology*, 179: 5655–5662.

Amme S, Matros A, Schlesier B and Mock H P, 2006, Proteome analysis of cold stress response in *Arabidopsis thaliana* using DIGE-technology. *Journal of Experimental Botany*, 57: 1537–46.

Arlat M, Van Gijsegem F, Huet J C, Pernollet J C and Boucher C A, 1994, PopA1, a protein which induces a hypersensitivity-like response on specific Petunia genotypes, is secreted via the Hrp pathway of *Pseudomonas solanacearum*. *The EMBO Journal*, 13: 543–553.

Babujee L, Venkatesh B, Yamazaki A and Tsuyumu S, 2007, Proteomic analysis of the carbonate insoluble outer membrane fraction of the soft-rot pathogen *Dickeya dadantii* (syn. *Erwinia chrysanthemi*) strain. *Journal of Proteome Research*, 6: 62–69.

Baginsky S, 2009, Plant proteomics: Concepts, applications and novel strategies for data interpretation. *Mass Spectrometry Reviews*, 28: 93–120.

Baillo E H, Kimotho R N, Zhang Z and Xu P, 2019, Transcription factors associated with abiotic and biotic stress tolerance and their potential for crops improvement. *Genes*, *10*(10): 771.

Baldoni E, Genga A and Cominelli E, 2015, Plant MYB transcription factors: Their role in drought response mechanisms. *International Journal of Molecular Sciences*, *16*: 15811–15851.

Banerjee A and Roychoudhury A, 2018, Small heat shock proteins: Structural assembly and functional responses against heat stress in plants. In: Ahmad P, Ahanger M A, Singh V P, Tripathi D K, Alam P and Alyemeni M N (Eds.). *Plant Metabolites and Regulation Under Environmental Stress*, Elsevier (Academic Press), Pp. 367–376.

Banerjee A, Wani S H and Roychoudhury A, 2017, Epigenetic control of plant cold responses. *Frontiers in Plant Science*, 8: 1643.

Baniwal S K, Bharti K, Chan K Y, Fauth M, Ganguli A and Kotak S, 2004, Heat stress response in plants: A complex game with chaperones and more than twenty heat stress transcription factors. *Journal of Biosciences*, 29: 471–87.

Basu S and Roychoudhury A, 2014, Expression profiling of abiotic stress-inducible genes in response to multiple stresses in rice (*Oryza sativa* l.) Varieties with contrasting level of stress tolerance. *BioMed Research International*, 2014(706890): 12.

Beddington J, Asaduzzaman M, Clark M, Bremauntz A, Guillou M and Jahn M, 2012, The role for scientists in tackling food insecurity and climate change. *Agriculture & Food Security*, 1: 10.

Beroza P, Villar H O, Wick M M and Martin G R, 2002, Chemo proteomics as a basis for post-genomic drug discovery. *Drug Discovery Today*, 7: 807–814.

Bogeat-Triboulot M B, Brosché M, Renaut J, Jouve L, Le Thiec D and Fayyaz P, 2007, Gradual soil water depletion results in reversible changes of gene expression, protein profiles, ecophysiology, and growth performance in *Populus euphratica*, a poplar growing in arid regions. *Plant Physiology*, 143: 876–892.

Bouchart F, Delangle A, Lemoine J, Bohin J P and Lacroix J M, 2007, Proteomic analysis of a non-virulent mutant of the phytopathogenic bacterium *Erwinia chrysanthemi* deficient in osmoregulated periplasmic glucans: Change in protein expression is not restricted to the envelope, but affects general metabolism. *Microbiology*, 153: 760–767.

Caruso G, Cavaliere C, Guarino C, Gubbiotti R, Foglia P and Lagana A, 2008, Identification of changes in *Triticum durum* L. leaf proteome in response to salt stress by two-dimensional electrophoresis and MALDI-TOF mass spectrometry. *Analytical and Bioanalytical Chemistry*, 391: 381–390.

Chen Y J, Perera V, Christiansen M W, Holme I B, Gregersen P L, Grant M R, Collige G D and Lyngkjær M F, 2013, The barley HvNAC6 transcription factor affects ABA accumulation and promotes basal resistance against powdery mildew. *Plant Molecular Biology*, 83: 577–590.

Chen L, Zhang L, Li D, Wang F and Yu D, 2013, WRKY8 transcription factor functions in the TMV-cg defense response by mediating both abscisic acid and ethylene signaling in *Arabidopsis*. *Proceedings of the National Academy of Sciences of the United States of America*, 110.

Chi F, Yang P and Han F, 2010, Proteomic analysis of rice seedlings infected by *Sinorhizobium meliloti* 1021. *Proteomics*, 10: 1861–1874.

Collado I G, Sanchez A J and Hanson J R, 2007, Fungal terpene metabolites: Biosynthetic relationships and the control of the phytopathogenic fungus *Botrytis cinerea*. *Natural Product Reports*, 24: 674–686.

Cui S, Hu J, Yang B, Shi L, Huang F and Tsai S N, 2009, Proteomic characterization of *Phragmites communis* in ecotypes of swamp and desert dune. *Proteomics*, 9: 3950–67.

Curtis R H, 2007, Plant parasitic nematode proteins and the host–parasite interaction. *Briefings in Functional Genomics and Proteomics*, 6: 50–58.

Dani V, Simon W J, Duranti M and Croy R, 2005, Changes in the tobacco leaf apoplast proteome in response to salt stress. *Proteomics*, 5: 737–45.

Degand H, Faber A M, Dauchot N, Mingeot D, Watillon B and Van Cutsem P, 2009, Proteomic analysis of chicory root identifies proteins typically involved in cold acclimation. *Proteomics*, 9: 2903–7.

Delalande F, Carapito C, Brizard J P, Brugidou C and Van Dorsselaer A, 2005, Multigenic families and proteomics: Extended protein characterization as a tool for paralog gene identification. *Proteomicsm*, 5: 450–460.

Diaz-Vivancos P, Rubio M, Mesonero V, Periago P M, Barcelo A R, Martinez-Gomez P and Hernandez J A, 2006, The apoplastic antioxidant system in Prunus: Response to long-term plum pox virus infection. *Journal of Experimental Botany*, 57: 3813–3824.

Dubos C, Stracke R, Grotewold E, Weisshaar B and Martin C and Lepiniec L, 2013, MYB transcription factors in *Arabidopsis*. *Trends in Plant Science*, 15: 573–581.

Ekramoddoullah A K M and Hunt R S, 1993, Changes in protein profile of susceptible and resistant sugar-pine foliage infected with the whitepine blister rust fungus *Cronartium ribicola*. *Canadian Journal in Plant Pathology*, 15(4): 259–264.

Galan J E and Collmer A, 1999, Type III secretion machines: Bacterial devices for protein delivery into host cells. *Science*, 284: 1322–1328.

Goff S A, Ricke D, Lan T H, Presting G, Wang R and Dunn M, 2002, A draft sequence of the rice genome (*Oryza sativa L. ssp. japonica*). *Science*, 296: 92–100.

Gonzalez D H, 2015, *Plant Transcription Factors: Evolutionary, Structural and Functional Aspects*. Elsevier: London, UK,

Gygi S P, Rist B and Aebersold R, 2000, Measuring gene expression by quantitative proteome analysis. *Current Opinion in Biotechnology*, 11: 396–401.

Hashiguchi A, Ahsan N and Komatsu S, 2010, Proteomics application of crops in the context of climatic changes. *Food Research International*, 43: 1803–1813.

Hernandez J A, Olmos E, Corpas F, Sevilla F and Rio L A, 1995, Salt-induced oxidative stress in chloroplast.

Hoang X L T, Nhi D N H, Thu N B A, Thao N P and Tran L S P, 2017, Transcription factors and their roles in signal transduction in plants under abiotic stresses. *Current Genomics*, 18: 483–497.

Ibraheem F, Gaffoor I, Tan Q, Shyu C R and Chopra S A, 2015, Sorghum MYB transcription factor induces 3-deoxyanthocyanidins and enhances resistance against leaf blights in maize. *Molecules*, 20: 2388–2404.

Inukai S, Kock K H and Bulyk M L, 2017, Transcription factor-DNA binding: Beyond binding site motifs. *Current Opinion in Genetics & Development*, 34: 110–119.

Ishiguro S and Nakamura K, 1994, Characterization of a cDNA encoding a novel DNA-binding protein, SPF1, that recognizes SP8 sequences in the 5′ upstream regions of genes coding for sporamin and β-amylase from sweet potato. *Molecular Genetics and Genomic*, 244: 563–571.

Karimizadeh R, Mohammadi M, Ghaffaripour S, Karimpour F and Shefazadeh M K, 2011, Evaluation of physiological screening techniques for drought-resistant breeding of durum wheat genotypes in Iran. *African Journal of Biotechnology*, 10: 12107–12117.

Kawamura Y and Uemura M, 2003, Mass spectrometric approach for identifying putative plasma membrane proteins of *Arabidopsis* leaves associated with cold acclimation. *The Plant Journal*, 36: 141–54.

Kazemi P N G and Hugouvieux-Cotte-Pattat N, 2004, The secretome of the plant pathogenic bacterium *Erwinia chrysanthemi*. *Proteomics*, 4(10): 3177–3186.

Kim D W, Rakwal R, Agrawal G K, Jung Y H, Shibato J and Jwa N S, 2005, A hydroponic rice seedling culture model system for investigating proteome of salt stress in rice leaf. *Electrophoresis*, 26: 4521–4539.

Kosmala A, Bocian A, Rapacz M, Jurczyk B and Zwierzykowski Z, 2009, Identification of leaf proteins differentially accumulated during cold acclimation between *Festuca pratensis* plants with distinct levels of frost tolerance. *Journal of Experimental Botany*, 60: 3595–609.

Kushalappa A C and Gunnaiah R, 2013, Metabolo-proteomics to discover plant biotic stress resistance genes. *Trends in Plant Science*, 18(9): 522–531.

Lee D G, Ahsan N, Lee S H, Kang K Y, Bahk J D and Lee I J, 2007, A proteomic approach in analyzing heat-responsive proteins in rice leaves. *Proteomics*, 7: 3369–3383.

Lee J, Bricker T M, Lefevre M, Pinson S R M and Oard J H, 2006, Proteomic and genetic approaches to identifying defence-related proteins in rice challenged with the fungal pathogen *Rhizoctonia solani*. *Molecular Plant Pathology*, 7(5): 405–416.

Lee H, Cha J, Choi C, Choi N, Ji J, Park S, Lee S and Hwang D, 2018, Rice WRKY11 plays a role in pathogen defense and drought tolerance. *Rice*, *11*.

Lee B J, Kwon S J, Kim S K, Kim K J, Park C J, Kim Y J, Park O K and Paek K H, 2006, Functional study of hot pepper 26S proteasome subunit RPN7 induced by tobacco mosaic virus from nuclear proteome analysis. *Biochemical and Biophysical Research Communications*, 351: 405–411.

Lee V T and Schneewind O, 2001, Protein secretion and the pathogenesis of bacterial infections. *Genes & Development*, 15: 1725–1752.

Liu Q, Yan S, Huang W, Yang J, Dong J, Zhang S, Zhao J, Yang T, Mao X and Zhu X, 2018, NAC transcription factor ONAC066 positively regulates disease resistance by suppressing the ABA signaling pathway in rice. *Plant Molecular Biology*, *98*: 289–302.

Li K, Xu C and Zhang J, 2011, Proteome profile of maize (*Zea may* L.) leaf tissue at the flowering stage after long-term adjustment to rice black-streaked dwarf virus infection. *Gene*, 485: 106–113.

Lodha T D, Hembram P and Nitile T J B, 2013, Proteomics: a successful approach to understand the molecular mechanism of plant-pathogen interaction.

Mahmood T, Jan A, Kakishima M and Komatsu S, 2006, Proteomic analysis of bacterial-blight defense-responsive proteins in rice leaf blades. *Proteomics*, 6: 6053–6065.

Martins D, Astua M G, Coletta H D, Winck F V, Baldasso P A, de Oliveira B M, Marangoni S, Machado M A, Novello J C and Smolka M B, 2007, Absence of classical heat shock response in the citrus pathogen *Xylella fastidiosa*. *Current Microbiology*, 54: 119–123.

Mathesius U S, Mulders M S, Gao M, Teplitski G, Caetano-Anollés G, Rolfe B G and Bauer W D, 2003, Extensive and specific responses of a eukaryote to bacterial quorum-sensing and signals. *Proceedings of National Academy of Science of the United States of America*, 100(3): 1444–1449.

Nath V S, Mishra A K, Kumar A, Matousek J and Jakse J, 2019, Revisiting the role of transcription factor in coordinating the defense response against Citrus Bark Cracking Viroid infection in commercial hop (*Humulus lupulus* L.). *Viruses Journal*, 11: 419.

Nat N V K, Sanjeeva S, William Y and Nidhi S, 2007, Application of proteomics to investigate plant microbe interactions. *Current Proteomics*, 4: 28–43.

Ndimba B K, Chivasa S, Simon W J and Slabas A R, 2005, Identification of *Arabidopsis* salt and osmotic stress responsive proteins using two-dimensional difference gel electrophoresis and mass spectrometry. *Proteomics*, 5: 4185–4196.

Ngara R, Ndimba R, Borch J J, Jensen O N and Ndimba B, 2012, Identification and profiling of salinity stress-responsive proteins in *Sorghum bicolor* seedlings. *Journal of Proteomics*, 75: 4139–4150.

Pang Q, Chen S, Dai S, Chen Y, Wang Y and Yan X, 2010, Comparative proteomics of salt tolerance in *Arabidopsis thaliana* and *Thellungiella halophila*. *Journal of Proteome Research*, 9: 2584–99.

Pazares J, Ghosal D, Wienand U, Petersont A and Saedler H, 1987, Products and with structural similarities. *The EMBO Journal*, 6: 3553–3558.

Phalip V, Delalande F, Carapito C, Goubet F, Hatsch D, Leize E, Dupree P, Dorsselaer A V and Jeltsch J M, 2005, Diversity of the exoproteome of *Fusarium graminearum* grown on plant cell wall. *Current Genetics*, 48: 366–379.

Pinedo C, Wang C M, Pradier J M, Dalmais B R R, Choquer M, Le Pelcheur P, Morgant G, Collado I G, Cane D E and Viaud M, 2008, Sesquiterpene synthase from the botrydial biosynthetic gene cluster of the phytopathogen botrytis cinerea. *ACS Chemical Biology*, 3: 791–801.

Rabara R, Msanne J, Basu S, Ferrer M and Roychoudhury A, 2021, Coping with inclement weather conditions due to high temperature and water deficit in rice: An insight from genetic and biochemical perspectives. *Physiologia Plantarum*, 172: 487–504.

Rahoutei J, Baro´ N M, Garcı´L L, Droppa M, Neme´nyi A and Horvath G, 1999, Effect of tobamovirus infection on the thermoluminescence characteristics of chloroplast from infected plants. *Z Naturforsch Teil C*, 54: 634–639.

Rampitsch C, Bykova N V, McCallum B, Beimcik E and Ens W, 2006, Analysis of the wheat and *Puccinia triticina* (leaf rust) proteomes during a susceptible host–pathogen interaction. *Proteomics*, 6: 1897–1907.

Rodziewicz P, Swarcewicz B, Chmielewska K, Wojakowska A and Stobiecki M, 2014, Influence of abiotic stresses on plant proteome and metabolome changes. *Acta Physiologiae Plantarum*, 36: 1–19.

Roychoudhury A and Banerjee A, 2015, Transcriptome analysis of abiotic stress response in plants. *Transcriptomics*, 3(2): e115.

Roychoudhury A, Basu S, Sarkar S N and Sengupta D N, 2008, Comparative physiological and molecular responses of a common aromatic indica rice cultivar to high salinity with non-aromatic indica rice cultivars. *Plant Cell Reports*, 27(8): 1395–1410.

Roychoudhury A, Datta K and Datta S K, 2011, Abiotic stress in plants: From genomics to metabolomics. In: Tuteja N, Gill S S and Tuteja R (Eds.). *Omics and Plant Abiotic Stress Tolerance*, Bentham Science Publishers, Pp. 91–120.

Roychoudhury A, Paul S and Basu S, 2013, Cross-talk between abscisic acid-dependent and abscisic acid-independent pathways during abiotic stress. *Plant Cell Reports*, 32(7): 985–1006.

Russel M, 1994, Mutants at conserved positions in gene IV, a gene required for assembly and secretion of filamentous phages. *Molecular Microbiology*, 14: 357–369.

Shan T, Rong W, Xu H, Du L, Liu X and Zhang Z, 2016, The wheat R2R3-MYB transcription factor TaRIM1 participates in resistance response against the pathogen *Rhizoctonia cerealis* infection through regulating defense genes. *Scientific Reports*, 6: 1–14.

Singh A and Roychoudhury A, 2021, Gene Regulation at transcriptional and post transcriptional levels to combat salt stress in plants. *Physiologia Plantarum*, 173: 1556–1572.

Wang B, Hajano J U D and Ren Y, 2015, iTRAQ-based quantitative proteomics analysis of rice leaves infected by rice stripe virus reveals several proteins involved in symptom formation. *Virology Journal*, 12: 1–21.

Wang Y, Sheng L, Zhang H, Du X, An C, Xia X, Chen F, Jiang J and Chen S, 2017, CmMYB19 overexpression improves aphid tolerance in chrysanthemum by promoting lignin synthesis. *International Journal of Molecular Sciences*, 18.

Wang J, Tao F, Tian W, Guo Z, Chen X, Xu X, Shang H and Hu X, 2017, The wheat WRKY transcription factors TaWRKY49 and TaWRKY62 confer differential high-temperature seedling-plant resistance to *Puccinia striiformis* f. sp. tritici. *PLoS ONE*, 12: 1–23.

Wang H, Wang H, Shao H and Tang X, 2016, Recent advances in utilizing transcription factors to improve plant abiotic stress tolerance by transgenic technology. *Frontiers in Plant Science*, 7: 1–13.

Wang Y, Yang L, Xu H, Li Q, Ma Z and Chu C, 2005, Differential proteomic analysis of proteins in wheat spikes induced by *Fusarium graminearum*. *Proteomics*, 5: 4496–4503.

Wang J, Zhou J, Zhang B, Vanitha J, Ramachandran S and Jiang S Y, 2011, Genome-wide expansion and expression divergence of the basic leucine zipper transcription factors in higher plants with an emphasis on sorghum. *Journal of Integrative Plant Biology*, 53: 212–231.

Watt S A, Wilke A, Patschkowski T and Niehaus K, 2005, Comprehensive analysis of the extracellular proteins from *Xanthomonas campestris* pv. *campestris* B100. *Proteomics*, 5(1): 153–167.

Whitham S A, Yang C and Goodin M M, 2006, Global impact: Elucidating plant responses to viral infection. *Mol Plant–Microbe Interact*, 19: 1207–1215.

Wilkins M R, Sanchez J C, Gooley A A, Appel R D, Humphery S I and Hochstrasser D F, 1995, Progress with proteome projects: Why all proteins expressed by a genome should be identified and how to do it. *Biotechnology & Genetic Engineering Reviews*, 13: 19–50.

Wu L, Hana Z, Wanga S, Wanga X, Sund A, Zua X and Chen Y, 2013, Comparative proteomic analysis of the plant–virus interaction in resistant and susceptible ecotypes of maize infected with sugarcane mosaic virus. *Journal of Proteomics*, 89: 124–140.

Wu L, Han Z and Wang S, 2013, Comparative proteomic analysis of the plant-virus interaction in resistant and susceptible ecotypes of maize infected with sugarcane mosaic virus. *Journal of Proteomics*, 89: 124–140.

Wu F, Shu J and Jin W, 2014, Identification and validation of miRNAs associated with the resistance of maize (*Zea mays* L.) to *Exserohilum turcicum*. *PLoS ONE*, 9.

Yang T, Zhu X, Yang J, Li X, Dong J, Zhang S, Zhao J, Yang T, Mao X and Zhu X, 2018, OsWRKY67 positively regulates blast and bacteria blight resistance by direct activation of PR genes in rice. *BMC Plant Biology*, 18: 1–13.

Yu J, Hu S, Wang J, Wong G K, Li S and Liu B, 2002, A draft sequence of the rice genome (*Oryza sativa* L. ssp. *indica*). *Science*, 296: 79–92.

Zhang H and Huang Y, 2013, Genome-wide survey and characterization of greenbug induced NAC transcription factors in sorghum (*Sorghum bicolor* (L.) Moench). *In Proceedings of the XXI Annual International Plant & Animal Genome Conference, San Diego*, CA, USA,. 191.

Zhang X L, Si B W, Fan C M, Li H J and Wang X M, 2014, Proteomics identification of differentially expressed leaf proteins in response to *Setosphaeria turcica* infection in resistant maize. *Journal of Integrative Agriculture*, 13: 789–803.

Zhang J H, Sun L, Liu L L, An S L, Wang X, Zhang J, Jin J L, Li S Y and Xi J H, 2010, Proteomic analysis of interactions between the generalist herbivore *Spodoptera exigua* (lepidoptera noctuidae) and *Arabidopsis thaliana*," *Plant Molecular Biology Reporter*, 324–333.

Zhou W, Eudes F and Laroche A, 2006, Identification of differentially regulated proteins in response to a compatible interaction between the pathogen *Fusarium graminearum* and its host, *Triticum aestivum*. *Proteomics*, 6: 4599–4609.

Zhou W, Qian C, Li R, Zhou S, Zhang R, Xiao J and Cao A, 2018, TaNAC6s are involved in the basal and broad-spectrum resistance to powdery mildew in wheat. *Plant Science*, 277: 218–228.

Zhu J M, Alvarez S, Marsh E L, LeNoble M E, Cho I J and Sivaguru M, 2007, Cell wall proteome in the maize primary root elongation zone. II. Region-specific changes in water soluble and lightly ionically bound proteins under water deficit. *Plant Physiology*, 145: 1533–1548.

3 Proteomics Research in Understanding Photosynthesis-related Proteins in Plants

Darshan Panda, Soumya Mohanty, Mirza Jaynul Baig,
Lambodar Behera, and Baishnab C. Tripathy

3.1 INTRODUCTION

Identification and characterization of the photosynthetic proteins are critical in deciphering the intricacies of carbon assimilation and its dynamic responses to altering plant developmental status and abiotic conditions. The primary objectives of this process include protein identification and their specific location in the chloroplast, determination of protein accumulation levels, protein-protein interactions, and various post-translational modifications (PTMs) (Banerjee and Roychoudhury, 2018). The outcome of these investigations will assist us in comprehending the role of proteins in the entire carbon reduction process and will also develop a possibility to investigate their precise biological roles via reverse genetics. Proteomics of photosynthesis includes an array of interdisciplinary studies such as RNA expression, bioinformatics, protein expression and quantification, protein structure determination, PTMs, protein-protein interaction studies, protein localization, biochemical and mutational knock-out studies, understanding protein function in time, space, and physiological state (Figure 3.1).

Facilitation in plant proteomics has been expedited by high throughput techniques such as mass spectrometry (MS) employing electrospray ionization (ESI) and matrix-assisted laser desorption ionization (MALDI) along with data-dependent acquisition (DDA) techniques from different species. In particular, the MS methodology utilizing the linear trap quadrupole (LTQ)-Orbitrap (Hu et al., 2005; Yates et al., 2006; Elmore et al., 2021) and its advanced modification such as LTQ-Fourier-transform ion cyclotron-resonance (FTICR) instrumentations (Maia et al., 2021) have escalated the speed, sensitivity, accuracy, and resolution of MS that enable the identification of several proteins in a biological sample (Adachi et al., 2006; Hu et al., 2021). These advancements in MS have now allowed for extensive characterization of the CO_2 assimilation machinery, identifying several structural and regulatory proteins involved in the adjustment of photosynthesis under various environmental regiments. We first discuss various studies enumerating the identification of the proteins associated with the photosynthetic machinery and its modulators. Proteomics is also employed to study various protein complexes and intriguing protein-protein interactions. This takes into consideration several fundamental biochemical techniques, such as native gel-based isolation of proteins and affinity chromatography. When protein-protein interaction has been detailed in yeast (Gavin et al., 2002; Richards et al., 2021) and *Escherichia coli* (Butland et al., 2005), large-scale protein interaction has not yet been successfully coursed in photosynthetic organisms. Next, we discuss several instances of a recently identified array of protein-protein interactions and several protein complexes associated with photosynthesis. In current times, wide varieties of protocols are available for pursuing comparative and quantitative proteome analysis, e.g., to understand the adaptation of the chloroplasts to several abiotic stresses and to differentiate the concentration of

DOI: 10.1201/b23255-3

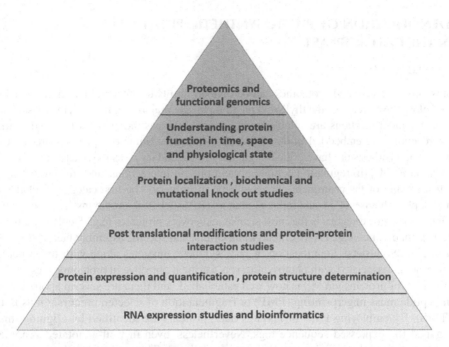

FIGURE 3.1 Proteomics research could be visualized as a pyramid having the lower bases as the broad research fields where fundamental disciplines meet.

the proteins in various photosynthetic cells, such as in the bundle sheath and mesophyll cells of C4 plants. Advancements in these studies have been facilitated in the last decade. Here, we will deliberate on recent investigations related to proteomics of chloroplasts. Furthermore, PTMs such as glycosylation, lipid modifications, and phosphorylation play a significant role in protein stability, subcellular localization as well as in efficacy of protein function. High-resolution MS (HRMS), in particular, is widely and successfully employed to understand protein-specific PTMs (Jorrin-Novo et al., 2019). Although still in its inception, noticeable advancements have been made in the analysis of PTMs of the chloroplast proteins, particularly the phosphorylation of the thylakoid proteome. We will discuss several critical PTMs associated with proteomics of photosynthesis. Eventually, bioinformatics has become an essential tool in biology research. The colossal amounts of data generated from proteomics, metabolomics, transcript analysis, protein structure, and proteomics analysis have provided extensive possibilities to understand chloroplast behavior and photosynthetic regulations under various environmental conditions. We have discussed some of the significant browser-based data resources commonly used for the proteomics study of plastids and delineated several issues and opportunities in the closing section.

Along with the several reviews on major plants (e.g., *Arabidopsis*) and *Chlamydomonas* proteomics (Stauber and Hippler, 2004; Roustan and Weckwerth, 2018), an array of reviews have been published in the last decade that has particularly concentrated on the proteomics analysis of photosynthetic apparatus and non-green plastids. Furthermore, various reviews have discussed technical details of photosynthetic proteomics, giving particular attention to thylakoid membrane proteomics (Whitelegge, 2003; Flannery et al., 2021). Photosynthesis is regulated by the redox potential, which is modulated by *ferredoxin-thioredoxin* reductase. Recently, affinity chromatography (AC) along with MS has helped us understand this regulatory mechanism and has been extensively reviewed (Buchanan and Luan, 2005). Taking into consideration reviews and research in eminent journals and books, this chapter will enumerate the proteomics of photosynthesis reported from 2006 and onwards, including breakthrough concepts, various challenges, and possible opportunities relevant to photosynthesis.

3.2 IDENTIFICATION OF PHOTOSYNTHETIC PROTEINS IN THE CHLOROPLAST

3.2.1 BACKGROUND

The photosynthesis-associated proteomes can be divided into two categories based on their localization in the chloroplast stroma or the thylakoid membrane. The majority of the enzymes associated with the light-independent reactions are soluble in the stroma, whereas the proteins involved in the light-dependent reactions are embedded in the thylakoid membrane. Numerous proteins are not centrally involved in photosynthesis but have critical regulatory or secondary roles in (i) activation/deactivation of enzymes via PTMs, (ii) regulating definite protein-protein interactions, and (iii) promoting assembly and disassembly of the prominent photosynthetic machinery. The first category includes protein kinases (PK), phosphatases (PT), and redox regulators (RR) such as thioredoxins. The second category covers CP12. The third category is the most sizable and covers proteases (e.g., DegP, Clp), isomerases (e.g., TPL40), chaperones (e.g., PAA2), metal transporters, and specific assembly factors (e.g., HCF101 involved in 2Fe-2S clusters). Identification of photosynthetic enzymes along with the structural and regulatory proteins associated with photosynthesis is a critical objective in proteomics research.

Identification of proteins via MS is now well established, and there are several approaches, either based on peptide mass fingerprinting (PMF) or fragmentation of selected precursor ions in tandem MS (FSTMS). By employing PMF or FSTMS data, proteins are identified by aligning this information against the expressed sequence tags. Nevertheless, even in well-annotated genomes such as *Arabidopsis*, all of the proteins are accurately predicted. This can be attributed to erroneous intron-exon splice site and N-terminus predictions, or simply due to the escaping of the genes from the annotation programs due to their small size, resulting in an MS data mismatch and finally the predicted protein information (Van Wijk, 2001). The pioneering paper on chloroplast proteomics was published in 2000 and concentrated on the thylakoid proteome (Peltier et al., 2000), followed by other studies concerning the identification of thylakoid-associated and free chloroplast proteins (Ytterberg et al., 2006), inner and outer chloroplast membranes (Froehlich et al., 2003), or covering the entire etioplast or chloroplast (Kleffmann et al., 2007). Bouchnak and his team in the year 2019 quantitatively juxtaposed the proteomes of the total leaf crude cell extract of Arabidopsis (Figure 3.2) and then purified the chloroplast envelope to standardize a novel parameter, calculated enrichment factor (EF), for individual putative envelope protein (Bouchnak et al., 2019).

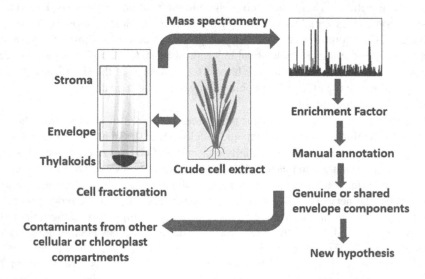

FIGURE 3.2   The protocol developed by Bouchnak et al. (2019) to identify the chloroplast envelope protein of Arabidopsis using mass spectrometry.

3.2.2   PROTEOMICS OF THE CHLOROPLAST

An advisable approach to understanding the proteomics of the photosynthetic mechanism would be a thorough analysis of the leaves and wholesome chloroplasts. However, the dynamic protein content of the chloroplast and, specifically, its higher composition of membrane proteins hinder the success rate of such approaches. For example, a study of the chloroplast-associated proteins of *Arabidopsis* via two-dimensional (2-D) electrophoresis would have to obtain the isolation and visualization of about 5000 proteins followed by their PTMs, which is beyond the scope of single 2-D gels. The greater contents of RuBisCo and LHCII can assuredly suppress the identification of low-concentration polypeptides and other non-polar membrane proteins that are mostly insoluble. To expedite the study of chloroplast polypeptides by proteomics, several approaches utilized purification and fractionation of the chloroplast polypeptides from a particular target site such as the inner envelope (Ferro et al., 2002). To further understand the protein expression pattern during the senescence period, Wilson et al. (2002) used 2-D electrophoresis to measure the changes in the number of chloroplast polypeptides in white clover (Table 3.1).

The image was analyzed by the 2-D gels from the matured green leaves, which covered about 580 protein spots during the senescence. Alterations in expression during the period of senescence were also observed for 295 protein spots that were then classified into four distinct groups. In a recent study, Fish et al. (2022) for the first time developed a novel bioinformatic approach (Figure 3.3) to sectionize the outer membrane proteome of the chloroplast (OMCP) based on their biochemical properties. The objective of this approach was to identify the candidates for the experimental studies, which would be beneficial in confirming the additional targeting signals and related pathways. In their review, the authors have summarized recent developments in the comprehension of well-analyzed OMCP, expanding its recorded number from 118 to 138.

3.2.3   PROTEOME OF THE CHLOROPLAST ENVELOPE

As a conjoin between the cytosol and the chloroplast inner core, the envelope is a complex network of interconnected transport systems that modulates the transportation of essential proteins across the chloroplast. To better understand the plastidial envelope-associated transport systems and to discover novel proteins for future functional studies, Ferro et al. (2002) investigated the proteome of the envelope in spinach by employing an assortment of organic-solvent extraction of highly purified chloroplast envelope membranes along with protein identification via LC-MS. This protocol successfully identified 54 proteins, out of which 41 were putative envelope proteins. Additionally, the technique discovered novel putative envelope proteins with unknown functions, for example, the H^+/Pi transporter (IEP60) and taurocholate transporters. However, the restricted genomic data of spinach further limited the rigorous identification of the envelope proteins.

TABLE 3.1
Analytical Strategies for Investigating Chloroplast Proteomics in Crops Exposed to Abiotic Stress

Sl No	Tissues	Species	Methodology	Critical Findings	References
1	Leaves	Glycine max	2-DE, MALDI-TOF MS	32 differentially expressed proteins from O3-treated chloroplast	Ahsan et al. (2010)
2	Leaves	*Triticum aestivum*	1-DE, LTQ-FTICR, green plants UniprotKB/Swiss-Prot database	Thylakoid proteome, identified 767 unique proteins	Kamal et al. (2013)
3	Leaves	*T. aestivum*	2-DE, LTQ-FTICR, green plants UniprotKB/Swiss-Prot database	65 proteins responded to salt treatment	Kamal et al. (2012)
4	Leaves	*Zea mays*	2-DE, MALDI-TOF MS	20 proteins responded to short-term salt exposure	Zörb et al. (2009)

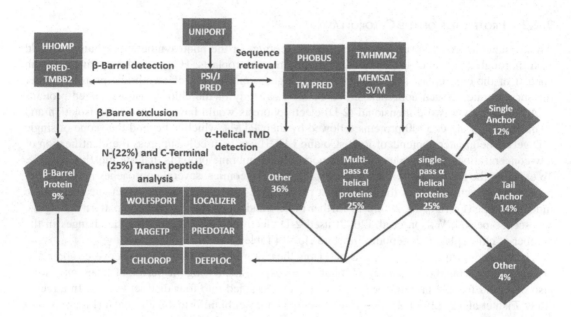

FIGURE 3.3 Multiple bioinformatic approaches categorize the proteome of the chloroplast's outer membrane by targeting critical photosynthesis-associated pathways. Bioinformatic tools were employed for sequence retrieval, β-barrel exclusion, β-barrel detection, N- and C-terminal transit peptide identification, and α-helical transmembrane domain detection (TMD), and the resultant targeting pathways along with the proteome percent composition provided to each category are represented in the flow chart (Drozdetskiy et al., 2015).

Furthermore, Ferro et al. (2003) translated the studies on spinach in *Arabidopsis* by employing an assortment of sophisticated extraction protocols, identifying 113 proteins out of which 90 were putative proteins. However, this approach failed to detect 15 proteins in *Arabidopsis* whose homologs were previously identified in spinach. Thus, this study is considered a near-complete identification of the chloroplast envelope proteome, deciphering not only the functional components of a sophisticated protein network but also essential proteins associated with the pigment and lipid metabolism pathways of the chloroplast. This study has provided a bedrock for formulating ingenious approaches that could integrate studies into other plant systems. Because the annotation of the *Arabidopsis* genomic sequence is complete, workers around the world are following *in silico* analysis to predict the chloroplast envelope proteins in different major crops. Nevertheless, as the outer membrane proteins are devoid of the target peptides, it is difficult to identify them by TargetP or ChloroP (Emanuelsson et al., 1999, 2000). Additionally, so far there is no protocol to differentiate between the unidentified proteins of the inner envelope and thylakoid membranes. Still, a great amount of success has been achieved by Schleiff et al. (2003) who scanned through the *Arabidopsis* genome to identify the candidate genes for the outer chloroplast envelope proteome. They further utilized some characteristic features of these proteins, for instance, beta-barrel structure, in terms of the selection criteria. Similarly, by a methodology called BITS, the *Arabidopsis* proteins were scrutinized via employing a β-barrel prediction, which also includes manual selection, TargetP analysis, and isoelectric points. This analysis revealed exclusive 891 proteins belonging to the outer envelope of the chloroplast. In another complementary study, Koo and Ohlrogge (2002) studied a group of candidates belonging to the inner envelope of the chloroplast of *Arabidopsis*. Considering that the chloroplast inner envelope primarily comprises integral transmembrane proteins, they further screened the TargetP-predicted proteins. Following this, all previously known thylakoid-related proteins were separated from the group of chloroplast membrane proteins, resulting in 541 envelope proteins, which are six times greater compared to the identified proteins in both envelopes of the chloroplast (Ferro et al., 2002, 2003).

3.2.4 Identification of the Thylakoid Proteome

The thylakoid proteomics has been studied in significant depth in *Arabidopsis* over the last few decades, identifying the major proteins of photosynthesis; PS-I, PS-II, cytochrome b_6f, and ATP synthase utilizing MS. However, no data has been generated related to the smaller transmembrane helix proteins and few transiently expressed light stress-related proteins (e.g., Elips). Additionally, members of smaller groups of homologs such as psaH1,2 and several LHCII proteins could not be individually identified, but only in the group. These studies could lay the bedrock of quantitative alterations in the photosynthetic protein composition in plants exposed to varied environmental conditions. The NDH complex is a sizable thylakoid protein complex associated with cyclic electron flow (CEF) and chlororespiration (CR). However, due to its low abundance in the thylakoid membrane, a marginal success rate identification has been achieved by MS. To comprehend the structure of all its subunits, NDH-H was 6× histidine (His) tagged at its N-terminus employing the plastid transformation method (Yu et al., 2020). Furthermore, the functional NDH subcomplex was purified by using Ni2+ affinity chromatography, and its subunit composition was analyzed by MS. This resulted in the identification of six subunits of NAD-H in two different thylakoid proteomics studies (Flannery et al., 2021). Similarly, in several other studies employing the aforesaid method, 60 proteins have been identified in the luminal zone of the thylakoid. However, studies based on luminal transit peptides suggest the existence of more than 180 luminal proteins (Ichikawa and Bui, 2018). This provides us a reason to introspect the sensitivity of MS instrumentation that should provide data clarity on the luminal proteomics and possibly assist us to understand the functions of luminal proteins. In addition, thylakoid proteomic studies have also identified several fibrillin proteins (Sakai and Keene, 2019) located in association with plastid lipid particles called plastoglobules (PGs) (Kessler et al., 1999). Several studies have identified more than 25 proteins associated with PGs with distinct metabolic functions. Further works have analyzed the metabolic functions of PG-associated proteins and delineated their precise topology and location using tomography (Ni et al., 2022). To add more, these data have revealed the transient storage function of PGs, which acts as a repository of quinones, phylloquinone, tocopherols, and other lipid molecules.

In *Arabidopsis*, in addition to the PGs and other proteins of the photosynthetic electron transport chain, more than 150 auxiliary thylakoid proteins have been identified that primarily interact with the thylakoid membrane surface (Van Wijk, 2006). Besides, K. Lilley and associates developed an ingenious procedure (i.e., LOPIT) to assign the position of various proteins to several cellular membrane systems (Huang et al., 2022). They utilized the density gradient centrifugation of assorted cellular membranes, followed by iTRAQ labeling of individual fractions and MS for their precise quantification and identification. This was followed by the principal component analysis to assign proteins to several cellular locations.

3.2.5 Proteome of the Thylakoid Lumen

Shortly after the initial progression in luminal chloroplast proteins (Kieselbach et al., 1998), several researchers concentrated further on this particular part of the chloroplast (Kieselbach et al., 2000). Furthermore, Schubert et al. (2002) studied the luminal proteomics of *Arabidopsis* and identified 80 distinguished proteins. In separate proteomics research on the *Arabidopsis* peripheral thylakoid proteins, Peltier et al. (2002) reported 200 lumenal proteins, of which two families of proteins such as PsbP and FKBP (peptidyl-prolyl *cis-trans* isomerases) are notable.

3.2.6 Identification of the Stromal Proteome

Chloroplast as a semi-autonomous organelle has gene transcriptional and translation machinery, having several mRNA binding and processing proteins that maintain RNA stability and transport them to the translational zone (Monde et al., 2000; Scaltsoyiannes et al., 2022). In recent years, proteomics studies have extensively analyzed the stromal proteomics of *Arabidopsis* (Peltier et al., 2006), maize (Majeran et al., 2005), and rice etioplasts (von Zychlinski et al., 2005; Kleffmann et al., 2007). In C4 plants, the bundle sheath and mesophyll chloroplasts each assemble a distinct category of photosynthetic enzymes and accessory proteins involved in the carbon reduction cycle. Besides, the bundle sheath and mesophyll enzymes of the C4-shuttle and the thylakoid electron transport chain were also identified. Precisely, the objectives of the stromal proteomic analysis of the *Arabidopsis* chloroplast study were to (i) identify the stromal proteome and to determine the expression of the chloroplast paralogues within the protein families, (ii) obtain a quantitative survey of the chloroplast proteome, and (iii) begin building resources to resolve the protein inter-action network in the chloroplasts of *Arabidopsis*. The future challenge lies in understanding the dynamism of protein interactions and the vast protein expression range. Furthermore, the stromal proteome was isolated by 2-D electrophoresis with non-denaturing PAGE (CN-PAGE), followed by SDS-PAGE as the second dimension.

3.2.7 Proteomics of the Cyanobacterial Photosynthetic System

The cyanobacterial species *Synechocystis* sp., strain PCC 6803, is a significant prokaryotic model organism widely studied to understand the proteomics of photosynthesis. Many features qualify *Synechocystis* as a potential organism for studying photosynthetic proteomics. Firstly, the genome is completely sequenced and assessable in the public databases, and secondly, it is com-paratively easier in this organism to generate site-directed mutants to pursue functional studies. *Synechocystis* is equally lucrative from the phylogenetic point for sharing a common ancestry with the higher plants by having a similar three-membranous structure analogous to the chloroplast. On the contrary, *Synechocystis* is remarkably resistant to any kind of mechanical rupture, and thus, the required chemical-mediated fractionation of its photosynthetic compartments is difficult. This might construe why the proteomics studies have so far advanced only in identifying only 3168 genes out of 6803 genes in the genome of *Synechocystis*. In this context, Sazuka and Ohara (1997) enumerated the proteome map of *Synechocystis* for the first time. By employing 2-D elec-trophoresis, 140 protein spots were isolated from the cell extracts and further identified by the tech-nique of micro-sequencing. As this method excluded a larger part of the *Synechocystis* undetectable gene products, Sazuka et al. (1999) reconstructed their protocol in a second study and integrated the isolation of *Synechocystis* cells in the form of soluble fractions, insoluble thylakoid, and other secretory proteins. Further in this analysis, Sazuka et al. (1999) specifically identified 227 spots on 2-D electrophoresis maps developed from the *Synechocystis* proteins and linked them to other 144 independent genes. The results of this study are also accessible at http://www.kazusa.or.jp/cyano/Synechocystis/cyano2D/index.html. In a mirror study by Wang et al. (2019), the peripheral proteomes of the thylakoid membrane of *Synechocystis* were analyzed and 51 gene products were identified. However, in two independent comparative studies by Sazuka et al. (1999) and Wang et al. (2019), a comparatively larger overlapping group was identified. Nonetheless, the list of gene products identified by these groups included 168 proteins. In order to further understand the single compartment proteomes of *Synechocystis*, Norling et al. (1998) employed an assortment of aqueous polymer two-phase (APTP) techniques along with sucrose density centrifugation (SDC) to sepa-rate pristine fractions of the plasmalemma and thylakoid membranes. The protein population of these separated membrane preparations was mapped by employing 2-D electrophoresis and MS. Furthermore, Fulda et al. (2000) studied the photosynthetic proteomics of prokaryotic plasma mem-brane along with the periplasmic space and identified 116 proteins, out of which 40 peripheral and 17 integral plasma membrane photosynthetic proteins have been identified. Additionally, Fulda et al. (2000) analyzed the differential expression patterns of periplasmic photosynthetic genes in response to elevated and depleted salt concentrations in *Synechocystis*. Another study by Norling et al. (1998) further purified and studied the proteins associated with the outer membrane. In a nutshell, the studies of Fulda et al. (2000) identified 125 different genes in *Synechocystis*. Besides, an extensive understanding of the proteome of the PS-II complex is a requirement for comprehending its three-dimensional (3-D) structure and function. Further, a study by Kashino et al. (2002) enumerated the isolation of ultra-pure PS-II complex via AC from the His-tag labeled (CP47) strain of *Synechocystis*. By employing SDS-electrophoresis, 37 protein bands of PS-II complex were successfully resolved, out of which 34 were identified via micro-sequencing and MS. It is also important to note that this PS-II preparation complex itself contains five novel proteins. Further, Kashino et al. (2002) identified 280 photosynthesis-associated proteins that are even less than 10% of the total predicted genes in *Synechocystis* 6803.

3.3 PROTEIN COMPLEXES AND DYNAMIC PROTEIN–PROTEIN INTERACTIONS IN PHOTOSYNTHESIS

3.3.1 TOOLS AND TECHNIQUES

The biological function of proteins is dependent on transient protein-protein interactions. By protocol, MS is employed to identify such interactions along with affinity purification of targeted proteins tagged to specific epitopes (e.g., hemagglutinin) and native gel-based protein separation. The above-discussed protein-protein interaction studies have been carried out in yeast (Krogan et al., 2006) and *E. coli* (Butland et al., 2005) but have not been carried out successfully for photosynthetic organisms.

3.3.2 PROTEIN COMPLEXES AND PROTEIN–PROTEIN INTERACTIONS IDENTIFIED BY NATIVE GELS AND AFFINITY CHROMATOGRAPHY

As discussed previously, the NDH complex in the thylakoid lumen is of lower abundance, and its identification in tobacco has been achieved by tagging His at the N-terminus via plastid transformation technique followed by purification by Ni^{2+} AC, and then its subunit compositions were analyzed by MS (Ayutthaya et al., 2020). Furthermore, the methodology for the 'Blue Native'-gel electrophoresis (also called BN-PAGE) was developed to isolate the membrane complexes of thylakoid (Schagger et al., 1994). With the success rate of this method, BN-PAGE gels have been increasingly used to understand the dynamics and steady state of thylakoid proteomics. To facilitate the migration of both alkaline and acidic proteins and complexes in various proportions to their native mass, the target thylakoid proteins were 'coated', followed by staining with negatively charged Coomassie Blue. In many plant systems, BN-PAGE has been followed to understand the composition of the thylakoid protein complex of *Arabidopsis* (Heinemeyer et al., 2004), maize (Darie et al., 2005), barley (Vincis Pereira Sanglard and Colas des Francs-Small, 2022), and spinach (Danielsson et al., 2006). Besides, BN-PAGE has been employed for various comparative studies, which include the identification of more than 50 proteins of major protein complexes of the thylakoid membrane, including PS-II, PS-I, cytochrome b_6f complex, ATP synthase, and NDH complex. In this chapter, we enumerate one of the pioneering studies employing MS and BN-PAGE to analyze the supramolecular organization of photosystems in *Arabidopsis* utilizing the detergent digitonin. Eventually, nine photosystem supercomplexes have been identified as having molecular masses spanning from 600 to 3200 kDa in BN gels. This study also revealed the protein arrangement patterns. For example, the supercomplexes of molecular weight 1060 and 1600 kDa represent the dimeric and trimeric forms of PS-I that have been found to tightly bound LHCI proteins. However, PS-II is primarily composed of supercomplexes of 850, 1000, 1050, and 1300 kDa. An elucidatory review of the assembly states of PS-I and PS-II photosynthetic complexes has been discussed by Aro et al. (2005). Here, it is essential to note that BN-PAGE is insensitive to the identification of low-abundant proteins and their fine association with other complexes. On the contrary, analysis of thylakoid membranes of bundle sheath and mesophyll chloroplasts of maize employing one-dimensional separation along with BN-PAGE gels in combination with MS/MS analysis, i.e., LTQ-Orbitrap, has successfully identified approximately 500 proteins associated with the thylakoid membranes (Majeran et al., 2008). The colorless native PAGE (CN-PAGE) developed by H. Schagger and G. von Jagow is a milder protocol than BN-PAGE, which maintains a superior protein complex stability and is particularly effective in understanding thylakoid soluble complexes (Ziehe et al., 2018). This has been successfully used to generate a 'snap-shot' of the oligomeric chloroplast proteome of *Arabidopsis* (Peltier et al., 2006). Approximately 10% of the identified 241 proteins in this research were found in a monomeric state, and more than 85% of the chloroplast proteins were particularly reported at a single native mass, within a mass range of ~10 kDa. Furthermore, the rest 15% have been reported at multiple native masses.

3.4   DIFFERENTIAL PROTEIN ACCUMULATION AND DYNAMIC CHANGES OF THE CHLOROPLAST PROTEOME

3.4.1   BACKGROUND

The likelihood of comparing the quantitative accumulation between the different protein samples has largely improved over the last few decades. In the primary stage, the most reliable techniques were based on the analysis of the image of the stained proteins separated on 2-D gels. In this section, we will discuss how the 2-D gel approach has been employed to study the proteomics of photosynthetic apparatus. Recently, an array of new techniques based on comparative quantification in the MS have provided lucrative opportunities to enhance throughput, obtain superior accuracies, and at the same time quantify membrane proteins. To assist such MS-based quantification, proteins are differentially labeled with stable isotopes from the samples used for the comparative studies. This is accomplished through labeling via N-14/N-15 or C-12/C-13 incorporated during cell growth cultures (Gorka et al., 2019). Otherwise, the labeled isotopes can be incorporated after the purification step of the plant protein extraction, employing the cross-linking of smaller isotope-labeled peptides (e.g., cICAT) (Majeran et al., 2005), trypsin-catalyzed ^{18}O labeling (Nelson et al., 2006), D,H-formaldehyde (Ytterberg et al., 2006), or via iTRAQ (Rudella et al., 2006).

Several of these techniques have been used to study the differential accumulation of chloroplast proteins, as will be reviewed below. In recent times, the novel technique of label-free comparative proteomics has emerged as a feasible alternative during the involvement of fast and exceptionally accurate mass spectrometers (Wiener et al., 2004; Menacherry et al., 2022). We can anticipate that this novel technique will be employed to understand the alterations in the photosynthetic apparatus soon. An extensive discussion on quantitative proteomics has been done in the reviews by Goshe and Smith (2003).

3.4.2   BIOGENESIS AND DIFFERENTIATION OF THE PHOTOSYNTHETIC APPARATUS STUDIED BY PROTEOMICS

The D1-reaction center of PS-II has a critical role in the modulation of photosynthesis and has been reviewed thoroughly by Aro et al. (2005). The PS-II reassembly and disassembly have been extensively studied by combining MS with BN-PAGE, western blots, and pulse-chase labeling techniques (Rokka et al., 2005). These researchers have identified the most copious proteins distributed in the thylakoid membrane but have partially quantified them. Additionally, these studies have not identified the photosynthetic auxiliary proteins due to their undetectability via protein staining as a consequence of their lower accumulation levels (Figure 3.4). During chloroplast maturation, the etioplasts matured in one to two days when the dark-grown plants are exposed to light. In this process of transformation, the maturation of the etioplast pro-lamellar thylakoid to the chloroplast lamellar membranes has been addressed in rice (Kleffmann et al., 2007) and maize (Delfosse et al., 2018) by employing 2-D gels, revealing 526 high-quality unique spots (Lonosky et al., 2004). However, in the rice system, the etioplast has been found to allocate a

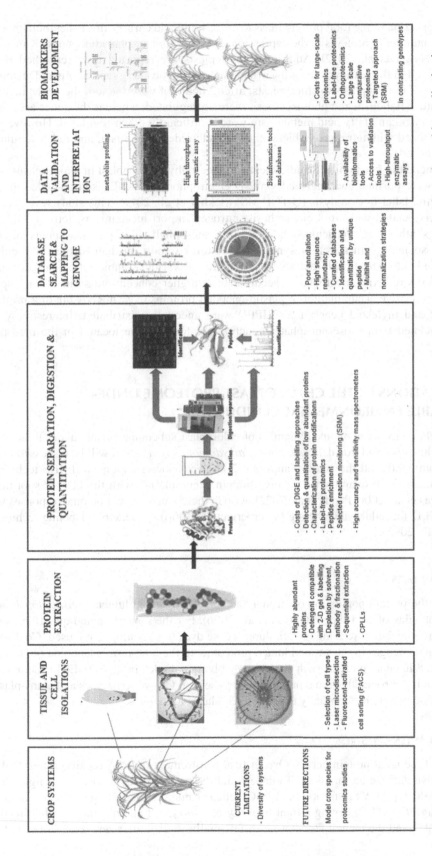

FIGURE 3.4 Thematic diagram showing the workflow of isolation and identification proteomics of photosynthesis-associated proteins and its application in crop improvement.

significant proportion of the total protein mass to amino acid and carbohydrate metabolism along with a greater number of proteins in the expression and regulation of plastidial genes (Kleffmann et al., 2007; Nayak et al., 2022). Additionally, etioplasts are reported to be enriched in the enzymes involved in the tetrapyrrole pathway and photosynthetic energy metabolism. During the transformation of the etioplasts to chloroplasts, after 2 hours of illumination, the genes involved in carbohydrate metabolism and photosynthesis were up-regulated, whereas the genes associated with the amino acid and fatty acid metabolism were significantly down-regulated. However, the enzymes associated with tetrapyrrole biosynthesis, nucleotide metabolism, and redox regulation remained unaltered.

To comprehend the functions of bundle sheath and mesophyll chloroplast in maize leaves, stromal proteins were purified and analyzed via 2-D electrophoresis PAGE, isotope-coded affinity tags (ICAT), and label-free analysis (Q-TOF MS) (Sekhar et al., 2019). This study revealed that the enzymes associated with lipid biosynthesis, nitrogen import mechanisms, tetrapyrrole, and isoprenoid biosynthesis are particularly housed in the mesophyll chloroplasts, whereas an array of enzymes associated with starch biosynthesis and sulfur import mechanisms are particularly accumulated in bundle sheath chloroplasts. Furthermore, the various soluble antioxidative systems particularly peroxiredoxins are found to accumulate at higher concentrations in the mesophyll chloroplasts. A relative accumulation of plastid-encoded proteins (e.g., mRNA binding proteins, EF-Tu, EF-G) and thylakoid generation (VIPP1) were unequally distributed. Interestingly, the enzymes associated with triose-phosphate reduction are found to be located in the mesophyll chloroplasts.

3.5   ALTERATIONS IN THE CHLOROPLAST PROTEOME UNDER VARIABLE ENVIRONMENTAL CONDITIONS

The studies related to the proteome dynamics of chloroplast subcompartments under light stress, cold stress, light stress, and iron deficiency in *Chlamydomonas reinhardtii* will be discussed in this section. Predominantly, all the studies employed 2-D electrophoresis along with IEF techniques in the first dimension to quantify the relative protein accumulation with the help of spot-image analysis (Goulas et al., 2006; Panda et al., 2022). Also in these studies, the proteome responses were analyzed via H,D-formaldehyde labeling (Ytterberg et al., 2006) or metabolic labeling techniques (Naumann et al., 2005).

3.5.1   COLD STRESS

Alterations in the protein population of stroma and soluble thylakoid lumen in *Arabidopsis* were studied after one day of cold shock (Goulas et al., 2006). For the separation and quantification of proteins, differential 2-D-PAGE along with fluorescence dyes (for staining) were used. Cold shock showed marginal changes in the stromal lumen proteome. In the long-term adaptation (40 days) to the cold shock, 8 lumenal proteins such as oxygen-evolving complex proteins (OECP), isomerases, and HCF136 and 35 stromal thylakoid proteins such as several C3 cycle enzymes and many plastid-metabolic functions were differentially expressed (Goulas et al., 2006).

3.5.2   LIGHT INTENSITY STRESS

The thylakoid functional protein network has critical involvement in DNA repair, protein folding, and antioxidative defense mechanisms. To understand the proteomic dynamics under light stress, *Arabidopsis* wild type (WT) and its vitamin C deficient mutant (*vtc2-2*) were studied after the transition to high light (HL) having a light intensity of 1000 μmol photons $m^{-2}s^{-1}$. The thylakoid proteomes of WT and the mutant *vtc2-2* were differentially studied after zero, one, three, and five days of consecutive HL treatment by employing 2-D-PAGE in three independent experiments and then followed by a compound statistical analysis and MS. After five days of HL treatment, both WT and *vtc2-2* mutants amassed anthocyanins, elevated their total vitamin C content, and reduced 10% of PS-II activity, but displayed no bleaching. Furthermore, 45 protein spots were remarkably altered due to genotype, light treatments, or both, and further confirmation was procured from the western blot studies. In this study, the thylakoid gene YCF37, which regulates the PS-I assembly, aldolase, flavin reductase-like protein (FRL), and specific fibrillin genes, was up-regulated in thylakoid. Besides, the *vtc2-2* mutant plants equally showed an up-regulation of steroid dehydrogenase-like proteins. Several other stress-related proteins, thylakoid proteases, and lumenal isomerases remained unaltered, while PsbS concentration escalated in the WT under light stress (Giacomelli et al., 2006; Panda et al., 2020).

3.5.3 Iron Deficiency

Alterations induced in the thylakoid protein population by Fe deficiency in *Beta vulgaris* grown in the hydroponics were studied by employing IEF-SDS PAGE along with BN-SDS PAGE by Andaluz et al. (2006). The data revealed remarkable Fe deficiency-associated changes in the protein population of thylakoid, particularly in the reduced concentration of electron transfer protein complexes and increment in the proteins associated with the C3 cycle (Naumann et al., 2005). In summary, these studies indicate the role of Fe in the re-modulation of the thylakoid proteome, thus regulating the photosynthetic efficiency of plants.

3.5.4 Drought Stress

Drought is a critical constraint for crop growth and productivity that is getting prime focus due to radical global climate changes. Tamburino et al. (2017) studied the response of the chloroplast to extreme drought conditions followed by a recovery cycle in tomato. Proteomics analysis revealed that drought affects the entire chloroplast protein pool, specifically involved in the catabolic reactions. Under drought conditions, around 54 chloroplast-associated proteins are primarily involved in photosynthesis and associated with retrograde signaling between chloroplast and nucleus.

3.6 POST-TRANSLATIONAL PROTEIN MODIFICATIONS OF THE PHOTOSYNTHETIC PROTEOME

3.6.1 Background

PTM of proteins includes reversible and irreversible changes such as glycosylation, lipidation, and phosphorylation, which regulate protein subcellular localization, stability, function, and protein-protein interactions of the newly translated polypeptides (Huber and Hardin, 2004). In current methodologies to understand PTMs, HRMS, along with other techniques such as metal-affinity columns for phosphopeptides, is particularly used. Several comprehensive reviews enumerate these approaches (Jensen, 2004, 2006). Although still in its inception, remarkable advancements have been made in understanding the PTMs of the photosynthesis-associated proteins, particularly phosphorylation of the thylakoid proteins (Table 3.2).

3.6.2 Regulation of Thylakoid State Transitions and PS-II Repair

In current times, MS and proteomics are providing better opportunities to ascertain amino acid-specific phosphorylation in a large population of protein samples (Turkina and Vener, 2007). Determination of the phosphorylation status of the photosynthetic proteins and recording the quantitative changes in the protein phosphorylation as well as thylakoid kinase activities have been

TABLE 3.2
Novel Protocols and Methodologies for the Identification of PTMs of Photosynthesis-associated Proteins

PTMs	Methodology	References
Photophosphorylation	Polymer-supported metal ion affinity capture	Lliuk et al. (2015)
	Functional ligand-binding identification by Tat-based recognition of associating proteins	Meksiriporn et al. (2019)
Glycosylation	Dendrimer-conjugated benzoboroxole	Xiao et al. (2018)
	Activated ion electron transfer dissociation	Riley et al. (2019)
Ubiquitination	Combined fractional diagonal chromatography	Walton et al. (2016)
Sumoylation	Replaced SUMO1 and SUMO2 isoforms with a variant and purification	Rytz et al. (2018)
Acetylation	Peptide prefractionation, immunoaffinity enrichment	Jiang et al. (2018)
Methylation	Combining SCX, IMAC, and H-pH-RPLC	Want et al. (2019)

achieved via MS studies in *Arabidopsis*. Within the context of photosynthesis, progress in the area of thylakoid phosphoproteomics and state transitions has been impressive in the last several years. Comprehensive reviews on thylakoid phosphor-proteomics, state transitions, acclimation, and other PTMs were recently reported (Tikkanen et al., 2006; Rochaix, 2007).

After many studies, a forward genetic approach employing chlorophyll a fluorescence primary screening in *Chlamydomonas reinhardtii* has identified a novel chloroplast-associated serine-threonine protein kinase, Stt7, which is involved in the phosphorylation of the primary light-harvesting protein complex (LHCII) and at the same time for modulating the process state transitions (Depege et al., 2003). Furthermore, the Stt7 orthologue in *Arabidopsis* is identified as STN7. The *STN7* mutant lines completely lost the ability of state transitions, and their thylakoid got locked in state 1, resulting in an impaired LHCII phosphorylation and growth under fluctuating light conditions. This suggests that STN7 plays a significant role in regulating the adaptation of plants under altering environmental conditions (Bellafiore et al., 2005).

3.6.3 N-TERMINAL MODIFICATIONS AND THEIR SIGNIFICANCE

As enumerated by Van Wijk (2004a, b), several N-terminal alterations of photosynthetic proteins are essential for the proper functioning of chloroplasts. Since not many advancements have been made in understanding this, we briefly outline some of the important observations. The nuclear-encoded chloroplast proteome passes through an N-terminal modification system in the stroma after plastidial translocation. For instance, the N-terminal acetylation of the chloroplast proteins is carried out in the nucleoplasm (Pesaresi et al., 2003). Interestingly, the proteins encoded by the chloroplasts are synthesized along with a formylated methionine, which is then removed by the chloroplast deformylase (Giglione and Meinnel, 2001) followed by the removal of the methionine by a methionine-endopeptidase (Ross et al., 2005). With the availability of high throughput and resolving MS such as LTQ-Orbitrap and FTICR, it is now possible to analyze the N-terminal modifications of photosynthetic proteins in greater detail with accuracy.

3.6.4 REVERSIBLE/IRREVERSIBLE MODIFICATIONS OF THE PHOTOSYNTHETIC PROTEINS UNDER OXIDATIVE STRESS

Remarkable advancements have been done in the identification of irreversible protein carbonylation and the reversible methionine sulfoxide reaction, resulting as a consequence of chloroplast under oxidative stress. It is important to note that protein carbonylation is an irreparable oxidative process that leads to a loss of protein function in an array of model systems such as mammals, flies, and worms (Nystrom, 2005). Nonetheless, in *Arabidopsis*, an inceptive increment in the amino acid oxidation during the initial 20 days of its life cycle is followed by a remarkable depletion in the protein-carbonyls just before bolting and flowering (Johansson et al., 2004). Incidentally, protein carbonylation at this specific stage was found to target several photosynthetic polypeptides such as proteins associated with the C2 cycle, RuBisCo, ATP synthase, and Hsp70 (Cheng et al., 2006). It seems reasonable to deduce this data and hypothesize that an extensive oxidative-stress defense operates in the protection of photosynthetic proteins against protein carbonylation. In recent times, the development of the methionine sulfoxide (MetSo) protocol has resulted in the desired modification of the activity and 3-D conformation of proteins, together with those involved in the process of photosynthesis. Conveniently, the MetSo modifications can be successfully reversed with the assistance of methionine sulfoxide reductase (MSR), particularly via MSRA4, MSRB1, and MSRB2, which are relatively highly expressed in photosynthetic organs (Vieira Dos Santos et al., 2005). In this process, the thioredoxins (TRXs) acted as electron donors to the MSR proteins for the operation of MetSO-mediated reduction; however, the specificity between the several TRXs and their corresponding target MSRs is yet to be investigated. The critical role of MSRs in protecting against oxidative damage was enumerated in transgenic *Arabidopsis* engineered with cytosolic/plastidic MSRA (Rouhier et al., 2006). Transgenic *Arabidopsis* plants were produced with MSRA4 expression altered from 95% to 40% via antisense and greater than 600% in wild-type lines (Romero et al., 2004). Under favorable growth parameters, the phenotypes of the transgenics remained unaffected. However, upon exposure to various oxidative-stress conditions such as HL, ozone, and methyl viologen, the rate of photosynthetic efficiency and optimum quantum yield of PS-II (Fv/Fm) along with the chloroplast methionine sulfoxide content remarkably altered. The PMSR4 overexpressors were relatively more resistant to chloroplast-associated oxidative damage, whereas underexpressors were significantly susceptible (Romero et al., 2004).

3.7   CHLOROPLAST PROTEOME DATABASES

In recent years, an array of proteomic studies, in several laboratories, focused on understanding the subcellular components of chloroplast and at the same time quantifying thousands of proteins, providing humongous proteomics data accessible to the public. For instance, the AT_CHLORO database provides information about more than 2000 chloroplast protein subfractions of *Arabidopsis*, covering the proteomics of stroma/grana lamellae, stromal enzymes, and envelope proteins. Additionally, it provides information on protein function, localization, and MS data (Orre et al., 2019). Similarly, the Chloroplast Function Database II is an extensive collection of more than 2000 nuclear genome-encoded chloroplast proteins associated with homozygous, single-gene knockout of T-DNA mutant lines (Milner and Wallington, 2022). The plastid protein database (plprot) is equally a large-scale assemblage of more than 2500 proteins of the etioplasts, matured chloroplasts, chromoplasts, and undifferentiated proplastid organelles of tobacco BY2 cells, which provides BLAST search and the data of the plastid proteomes (Figure 3.5). Also, we can access the Plant Proteomics Database (PPDB), which provides data related to the cellular proteins and their PTMs identified in *Arabidopsis* and maize (Joo et al., 2021). Curiously, the ChloroKB is a web application applicable to the study of chloroplast metabolic networks and associated cellular pathways. Additional databases such as PRIDE and NCBI receive the deposition of unprocessed files generated from MS analysis.

3.8   ENLISTMENT OF MAJOR CHLOROPLAST PROTEOMES

A comprehensive list of the chloroplast proteome is cataloged by Lande et al. (2020) as per separate plant species and then provided in separate non-repetitive sets. The above-mentioned review documented the proteomics of individual plants via non-identical tissue types and protocols. About

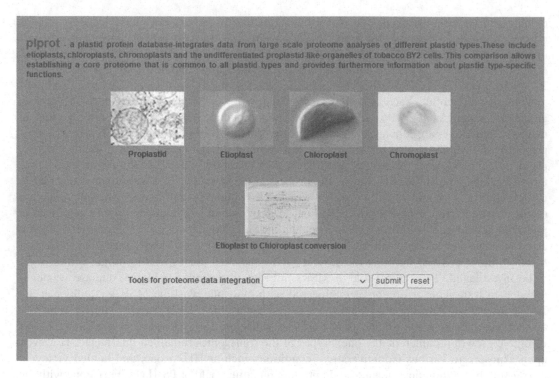

FIGURE 3.5   The screenshot of the plprot-database interface (https://plprot.ethz.ch/).

9626 chloroplast-associated proteins were recognized from plants and *Chlorophycean algae*. Of the enlisted proteins, from different (18) chloroplast proteomes, a sum of 3628 non-redundant proteins were identified. Outlining this study, the extensive non-redundant proteome set was reported in *Arabidopsis* (3629), succeeded by chickpea (2452), maize (1439) and pea (893). Furthermore, to estimate the possibility of overlaps between the chloroplast proteomes identified from non-identical plants, BLAST analyses were followed. At most 378 proteins were reported to be identical across pea, maize, chickpea, and Arabidopsis datasets with greater than 60% homology. Amid these data, some photosynthetic proteins with evolutionarily conserved regions have been recognized. Berglund et al. (2009) identified 40 proteins encoded by the nuclear genome, then translated into the cytosol, and subsequently translocated to chloroplast and mitochondria. Furthermore, NCBI and TAIR databases theorized 76 and 87 proteins to be possibly encoded by the chloroplast genome of chickpea and *Arabidopsis*, respectively.

3.9   PERSPECTIVES

PMF and MS-based identification of proteomes surfaced about a few decades ago, and its primary level application in the study of photosynthesis started in the year 2000. Since then, the influence of proteomics and protein-MS has been critical, with fundamental efforts laid on the identification of photosynthesis-associated proteins in the chloroplast along with understanding the PTMs, protein-protein interactions, and comparative proteomics studies. A majority of these studies are carried out in *Arabidopsis*, the very first plant whose genome was completely sequenced. The protocols followed in all these courses, except for some smaller, very non-polar thylakoid proteins, use MS, but an array of PTMs remain to be brought to light. One of the significant challenges in the study of photosynthesis proteomics is to identify and then characterize the proteins with modulatory functions, and the proteins associated with the chloroplast biogenesis development and adaptation. The escalating amount of reported chloroplast and 'non-chloroplast' proteins via the analysis of the proteome will dispense new opportunities to reteach or redevelop these chloroplast predictors. Surely, although *Arabidopsis* provides a lucrative experimental component, the utilization, of other, sequenced C3 plants such as rice and C4 plants such as maize, should assist us to accelerate the further exploration of the biological dynamism and the analysis of cell and tissue-specific differentiation, relevant to the mechanism of photosynthesis.

REFERENCES

Adachi J, Kumar C, Zhang Y, Olsen JV and Mann M (2006) The human urinary proteome contains more than 1500 proteins, including a large proportion of membrane proteins. Genome Biol 7: R80

Ahsan N, Nanjo Y, Sawada H, Kohno Y and Komatsu S (2010) Ozone stress-induced proteomic changes in leaf total soluble and chloroplast proteins of soybean reveal that carbon allocation is involved in adaptation in the early developmental stage. Proteomics 10: 2605–2619

Andaluz S, Lopez-Millan AF, De las Rivas J, Aro EM, Abadia J and Abadia A (2006) Proteomic profiles of thylakoid membranes and changes in response to iron deficiency. Photosynth Res 89: 141–155

Aro EM, Suorsa M, Rokka A, Allahverdiyeva Y, Paakkarinen V, Saleem A, Battchikova N and Rintamaki E (2005) Dynamics of photosystem II: A proteomic approach to thylakoid protein complexes. J Exp Bot 56: 347–356

Banerjee A and Roychoudhury A (2018) Regulation of photosynthesis under salinity and drought stress. In: Singh VP, Singh S, Singh R, Prasad SM (eds) Environment and Photosynthesis: A Future Prospect, Studium Press, India, pp 134–144

Bellafiore S, Barneche F, Peltier G and Rochaix JD (2005) State transitions and light adaptation require chloroplast thylakoid protein kinase STN7. Nature 433: 892–895

Berglund AK, Spånning E, Biverståhl H, Maddalo G, Tellgren-Roth C, Mäler L and Glaser E (2009) Dual targeting to mitochondria and chloroplasts: Characterization of Thr–tRNA synthetase targeting peptide. Mol Plant Nov 2(6): 1298–1309

Bouchnak I, Brugière S, Moyet L, Le Gall S, Salvi D, Kuntz M, Tardif M and Rolland N (2019) Unraveling hidden components of the chloroplast envelope proteome: Opportunities and limits of better MS Sensitivity*[S.]. Mol Cellr Proteom 18(7): 1285–1306

Buchanan BB and Luan S (2005) Redox regulation in the chloroplast thylakoid lumen: A new frontier in photosynthesis research. J Exp Bot 56: 1439–1447

Butland G, Peregrin-Alvarez JM, Li J, Yang W, Yang X, Canadien V, Starostine A, Richards D, Beattie B, Krogan N, Davey M, Parkinson J, Greenblatt J and Emili A (2005) Interaction network containing conserved and essential protein complexes in *Escherichia coli*. Nature 433: 531–537

Cheng NH, Liu JZ, Brock A, Nelson RS and Hirschi KD (2006) AtGRXcp, an *Arabidopsis* chloroplastic glutaredoxin, is critical for protection against protein oxidative damage. J Biol Chem 281: 26280–26288

Danielsson R, Suorsa M, Paakkarinen V, Albertsson PA, Styring S, Aro EM and Mamedov F (2006) Dimeric and monomeric organization of photosystem II. Distribution of five distinct complexes in the different domains of the thylakoid membrane. J Biol Chem 281: 14241–14249

Darie CC, Biniossek ML, Winter V, Mutschler B and Haehnel W (2005) Isolation and structural characterization of the Ndh complex from mesophyll and bundle sheath chloroplasts of *Zea mays*. FEBS J 272: 2705–2716

Delfosse K, Wozny MR, Anderson C, Barton KA and Mathur J (2018) Evolving views on plastid pleomorphy. In: Concepts in Cell Biology-History and Evolution, Springer, Cham, pp 185–204

Depege N, Bellafiore S and Rochaix JD (2003) Role of chloroplast protein kinase Stt7 in LHCII phosphorylation and state transition in *Chlamydomonas*. Science 299: 1572–1575

Drozdetskiy A, Cole C, Procter J and Barton GJ (2015) JPred4: A protein secondary structure prediction server. Nucleic Acids Res 43(W1): W389–W394

Elmore JM, Brianna DG and Justin WW (2021) Advances in functional proteomics to study plant-pathogen interactions. Curr Opin Plant Biol 63: 102061

Emanuelsson O, Nielsen H and von Heijne G (1999) ChloroP, a neural network-based method for predicting chloroplast transit peptides and their cleavage sits. Protein Sci 8: 978–984

Emanuelsson O, Nielsen H, Brunak S and von Heijne G (2000) Predicting subcellular localization of proteins based on their N-terminal amino acid sequence. J Mol Biol 300: 1005–1016

Ferro M, Salvi D, Brugiere S, Miras S, Kowalski S, Louwagie M, Garin J, Joyard J and Rolland N (2003) Proteomics of the chloroplast envelope membranes from *Arabidopsis thaliana*. Mol Cell Proteom: Epub ahead of print 310: 1035–1040

Ferro M, Salvi D, Riviere-Rolland H, Vermat T, Seigneurin-Berny D, Grunwald D, Garin J, Joyard J and Rolland N (2002) Integral membrane proteins of the chloroplast envelope: Identification and subcellular localization of new transporters. Proc Natl Acad Sci USA 99: 11487–11492

Fish M, Nash D, German A, Overton A, Jelokhani-Niaraki M, Chuong SD and Smith MD (2022) New insights into the chloroplast outer membrane proteome and associated targeting pathways. Int J Mol Sci 23(3): 1571

Flannery SE, Hepworth C, Wood WH, Pastorelli F, Hunter CN, Dickman MJ, Jackson PJ and Johnson MP (2021) Developmental acclimation of the thylakoid proteome to light intensity in Arabidopsis. Plant J 105(1): 223–244

Froehlich JE, Wilkerson CG, Ray WK, McAndrew RS, Osteryoung KW, Gage DA and Phinney BS (2003) Proteomic study of the *Arabidopsis thaliana* chloroplastic envelope membrane utilizing alternatives to traditional two-dimensional electrophoresis. J Proteome Res 2: 413–425

Fulda S, Huang F, Nilsson F, Hagemann M and Norling B (2000) Proteomics of *Synechocystis* sp. strain PCC 6803: Identification of periplasmic proteins in cells grown at low and high salt concentrations. Eur J Biochem 267: 5900–5907.

Gavin AC, Bosche M, Krause R, Grandi P, Marzioch M, Bauer A, Schultz J, Rick JM, Michon AM, Cruciat CM, et al. (2002) Functional organization of the yeast proteome by systematic analysis of protein complexes. Nature 415: 141–147

Giacomelli L, Rudella A and Van Wijk KJ (2006) High light response of the thylakoid proteome in Arabidopsis wild type and the ascorbate-deficient mutant *vtc2-2*. A comparative proteomics study. Plant Physiol 141: 685–701

Giglione C and Meinnel T (2001) Organellar peptide deformylases: Universality of the N-terminal methionine cleavage mechanism. Trends Plant Sci 6: 566–572

Gorka S, Dietrich M, Mayerhofer W, Gabriel R, Wiesenbauer J, Martin V, Zheng Q, Imai B, Prommer J, Weidinger M and Schweiger P (2019) Rapid transfer of plant photosynthates to soil bacteria via ectomycorrhizal hyphae and its interaction with nitrogen availability. Front Microbiol 10: 168

Goshe MB and Smith RD (2003) Stable isotope-coded proteomic mass spectrometry. Curr Opin Biotechnol 14: 101–109

Goulas E, Schubert M, Kieselbach T, Kleczkowski LA, Gardestrom P, Schroder W and Hurry V (2006) The chloroplast lumen and stromal proteomes of *Arabidopsis thaliana* show differential sensitivity to short- and long-term exposure to low temperature. Plant J 47: 720–734

Heinemeyer J, Eubel H, Wehmhoner D, Jansch L and Braun HP (2004) Proteomic approach to characterize the supramolecular organization of photosystems in higher plants. Phytochem 65: 1683–1692

Hu Q, Noll RJ, Li H, Makarov A, Hardman M and Graham Cooks R (2005) The Orbitrap: A new mass spectrometer. J Mass Spectrom 40: 430–443

Hu W, Han Y, Sheng Y, Wang Y, Pan Q and Nie H (2021) Mass spectrometry imaging for direct visualization of components in plants tissues. J Separat Sci 44(18): 3462–3476

Huang DX, Yu X, Yu WJ, Zhang XM, Liu C, Liu HP, Sun Y and Jiang ZP (2022) Calcium signaling regulated by cellular membrane systems and calcium homeostasis perturbed in Alzheimer's disease. Front Cell Develop Biol 10: 3472–3478

Huber SC and Hardin SC (2004) Numerous posttranslational modifications provide opportunities for the intricate regulation of metabolic enzymes at multiple levels. Curr Opin Plant Biol 7: 318–322

Ichikawa M and Bui KH (2018) Microtubule inner proteins: A meshwork of luminal proteins stabilizing the doublet microtubule. BioEssays 40(3): 1700209

Iliuk A, Jayasundera K, Wang WH, Schluttenhofer R, Geahlen RL and Tao WA (2015) In-depth analyses of B cell signaling through tandem mass spectrometry of phosphopeptides enriched by PolyMAC. Int J. Mass Spectro 377: 744–753

Jensen ON (2004) Modification-specific proteomics: Characterization of post-translational modifications by mass spectrometry. Curr Opin Chem Biol 8: 33–41

Jensen ON (2006) Automated phosphorylation site mapping. Nat Biotechnol 24: 1226–1227

Jiang J, Gai Z, Wang Y, Fan K, Sun L, Wang H and Ding Z (2018) Comprehensive proteome analyses of lysine acetylation in tea leaves by sensing nitrogen nutrition. BMC Genom 19(1): 1–3

Johansson E, Olsson O and Nystrom T (2004) Progression and specificity of protein oxidation in the life cycle of *Arabidopsis thaliana*. J Biol Chem 279: 22204–22208

Joo H, Baek W, Lim CW and Lee SC (2021) Post-translational modifications of bZIP transcription factors in abscisic acid signaling and drought responses. Curr Genom 22(1): 4–15

Jorrin-Novo JV, Komatsu S, Sanchez-Lucas R and de Francisco LE (2019) Gel electrophoresis-based plant proteomics: Past, present, and future. Happy 10th anniversary Journal of Proteomics!. J Proteom 198: 1–0

Kamal AHM, Cho K, Choi JS, Bae KH, Komatsu S, Uozumi N, et al. (2013) The wheat chloroplastic proteome. J Proteom 93: 326–342

Kamal AH, Cho K, Kim DE, Uozumi N, Chung KY, Lee SY, et al. (2012) Changes in physiology and protein abundance in salt-stressed wheat chloroplasts. Mol Biol Rep 39: 9059–9074

Kashino Y, Koike H, Yoshio M, Egashira H, Ikeuchi M, Pakrasi HB and Satoh K (2002) Low-molecular-mass polypeptide components of a photosystem II preparation from the thermophilic cyanobacterium *Thermosynechococcus vulcanus*. Plant Cell Physiol 43: 1366–1373.

Kessler F, Schnell D and Blobel G (1999) Identification of proteins associated with plastoglobules isolated from pea (*Pisum sativum* L.) chloroplasts. Planta 208: 107–113

Kieselbach T, Bystedt M, Hynds P, Robinson C and Schröder WP (2000) A peroxidase homologue and novel plastocyanin located by proteomics to the *Arabidopsis* chloroplast thylakoid lumen. FEBS Lett 480: 271–276

Kieselbach T, Hagman Å, Andersson B and Schröder WP (1998) The lumen of the chloroplast's thylakoid membrane: Isolation and characterization. J Biol Chem 273: 6710–6716

Kleffmann T, von Zychlinski A, Russenberger D, Hirsch-Hoffmann M, Gehrig P, Gruissem W and Baginsky S (2007) Proteome dynamics during plastid differentiation in rice. Plant Physiol 143: 912–923

Koo AJK and Ohlrogge JB (2002) The predicted candidates of Arabidopsis plastid inner envelope membrane proteins and their expression profiles. Plant Physiol 130(2): 823–836

Krogan NJ, Cagney G, Yu H, Zhong G, Guo X, Ignatchenko A, Li J, Pu S, Datta N, Tikuisis AP, et al. (2006) Global landscape of protein complexes in the yeast *Saccharomyces cerevisiae*. Nature 440: 637–643

Lande NV, Barua P, Gayen D, Kumar S, Chakraborty S and Chakraborty N (2020) Proteomic dissection of the chloroplast: Moving beyond photosynthesis. J Proteom 212: 103542

Lonosky PM, Zhang X, Honavar VG, Dobbs DL, Fu A and Rodermel SR (2004) A proteomic analysis of maize chloroplast biogenesis. Plant Physiol 134: 560–574

Maia M, Figueiredo A, Cordeiro C and Sousa Silva M (2021) FT-ICR-MS-based metabolomics: A deep dive into plant metabolism. Mass Spectrom Rev137: 563–570

Majeran W, Cai Y, Sun Q and Van Wijk KJ (2005) Functional differentiation of bundle sheath and mesophyll maize chloroplasts determined by comparative proteomics. Plant Cell 17: 3111–3140

Majeran W, Zybailov B, Ytterberg AJ, Dunsmore J, Sun Q and Van Wijk KJ (2008) Consequences of C4 differentiation for chloroplast membrane proteomes in maize mesophyll and bundle sheath cells. Mol Cell Proteom 7: 1609–1638

Meksiriporn B, Ludwicki MB, Stephens EA, Jiang A, Lee HC, Waraho-Zhmayev D, Kummer L, Brandl F, Plückthun A and DeLisa MP (2019) A survival selection strategy for engineering synthetic binding proteins that specifically recognize post-translationally phosphorylated proteins. Nature Commun 10(1): 1–0 https://doi.org/10.1038/s41467-019-09854-y

Menacherry SP, Aravind UK and Aravindakumar CT (2022) Critical review on the role of mass spectrometry in the AOP based degradation of contaminants of emerging concern (CECs) in water. J Environl Chem Eng 25: 108155

Milner MJ and Wallington EJ (2022) Genome editing and identification of targeted heritable mutations in wheat. In: Accelerated Breeding of Cereal Crops Humana, Springer Protocols Handbooks (SPH), New York, NY, pp 225–238

Monde RA, Schuster G and Stern DB (2000) Processing and degradation of chloroplast mRNA. Biochimie 82: 573–582

Na Ayutthaya PP, Lundberg D, Weigel D and Li L (2020) Blue native polyacrylamide gel electrophoresis (BN-PAGE) for the analysis of protein oligomers in plants. Curr Protocol Plant Biol 5(2): e20107

Naumann B, Stauber EJ, Busch A, Sommer F and Hippler M (2005) N-terminal processing of Lhca3 is a key step in remodeling of the photosystem I-light-harvesting complex under iron deficiency in *Chlamydomonas reinhardtii*. J Biol Chem 280: 20431–20441

Nayak L, Panda D, Dash GK, Lal MK, Swain P, Baig MJ and Kumar A (2022) A chloroplast Glycolate catabolic pathway bypassing the endogenous photorespiratory cycle enhances photosynthesis, biomass and yield in rice (*Oryza sativa* L.). Plant Sci 314: 111103

Nelson CJ, Hegeman AD, Harms AC and Sussman MR (2006) A quantitative analysis of *Arabidopsis* plasma membrane using trypsin-catalyzed 18O labeling. Mol Cell Proteom 5: 1382–1395

Ni T, Frosio T, Mendonça L, Sheng Y, Clare D, Himes BA and Zhang P (2022) High-resolution in situ structure determination by cryo-electron tomography and subtomogram averaging using emClarity. Nature Protocols 17(2): 421–444. 17, no. 2 (2022): 421–444

Norling B, Zak E, Andersson B and Pakrasi H (1998) 2D-isolation of pure plasma and thylakoid membranes from the cyanobacterium *Synechocystis* sp. PCC 6803. FEBS Lett 436: 189–192.

Nystrom T (2005) Role of oxidative carbonylation in protein quality control and senescence. Embo J 24: 1311–1317

Orre LM, Vesterlund M, Pan Y, Arslan T, Zhu Y, Woodbridge AF, Frings O, Fredlund E and Lehtiö J (2019) SubCellBarCode: Proteome-wide mapping of protein localization and relocalization. Mol Cell 73(1): 166–182

Panda D, Biswal M, Mohanty S, Dey P, Swain A, Behera D, Baig MJ, Kumar A, Sah RP, Tripathy BC and Behera L (2020) Contribution of phytochrome a in the regulation of sink capacity, starch biosynthesis, grain quality, grain yield and related traits in rice. Plant Arch 20(1): 1179–1194

Panda D, Mohanty S, Das S, Sah RP, Kumar A, Behera L, Baig MJ and Tripathy BC (2022) The role of phytochrome-mediated gibberellic acid signaling in the modulation of seed germination under low light stress in rice (*O. sativa* L.). Physiol Mol Biol Plant 28(3): 585–605

Peltier JB, Cai Y, Sun Q, Zabrouskov V, Giacomelli L, Rudella A, Ytterberg AJ, Rutschow H and Van Wijk KJ (2006) The oligomeric stromal proteome of *Arabidopsis thaliana* chloroplasts. Mol Cell Proteom 5: 114–133

Peltier JB, Emanuelsson O, Kalume DE, Ytterberg J, Friso G, Rudella A, Liberles DA, Soderberg L, Roepstorff P, von Heijne G and Van Wijk KJ (2002) Central functions of the lumenal and peripheral thylakoid proteome of *Arabidopsis* determined by experimentation and genome-wide prediction. Plant Cell 14: 211–236

Peltier JB, Friso G, Kalume DE, Roepstorff P, Nilsson F, Adamska I and Van Wijk KJ (2000) Proteomics of the chloroplast. Systematic identification and targeting analysis of lumenal and peripheral thylakoid proteins. Plant Cell 12: 319–342

Pesaresi P, Gardner NA, Masiero S, Dietzmann A, Eichacker L, Wickner R, Salamini F and Leister D (2003) Cytoplasmic N-terminal protein acetylation is required for efficient photosynthesis in *Arabidopsis*. Plant Cell 15: 1817–1832

Richards AL, Eckhardt M and Krogan NJ (2021) Mass spectrometry-based protein–protein interaction networks for the study of human diseases. Mol Syst Bio l17(1): e8792

Riley NM, Hebert AS, Westphall MS and Coon JJ (2019) Capturing site-specific heterogeneity with large-scale N-glycoproteome analysis. Nature Commun 10(1): 1–3

Rochaix JD (2007) Role of thylakoid protein kinases in photosynthetic acclimation. FEBS Lett 581: 2768–2775

Rokka A, Suorsa M, Saleem A, Battchikova N and Aro EM (2005) Synthesis and assembly of thylakoid protein complexes: Multiple assembly steps of photosystem II. Biochem J 388: 159–168

Romero HM, Berlett BS, Jensen PJ, Pell EJ and Tien M (2004) Investigations into the role of the plastidial peptide methionine sulfoxide reductase in response to oxidative stress in Arabidopsis. Plant Physiol 136: 3784–3794

Ross S, Giglione C, Pierre M, Espagne C and Meinnel T (2005) Functional and developmental impact of cytosolic protein N-terminal methionine excision in Arabidopsis. Plant Physiol 137: 623–637

Rouhier N, Vieira Dos Santos C, Tarrago L and Rey P (2006) Plant methionine sulfoxide reductase A and B multigenic families. Photosynth Res 89: 247–262

Roustan V and Weckwerth W (2018) Quantitative phosphoproteomic and system-level analysis of TOR inhibition unravel distinct organellar acclimation in *Chlamydomonas reinhardtii*. Front Plant Sci 9: 1590

Rudella A, Friso G, Alonso JM, Ecker JR and Van Wijk KJ (2006) Downregulation of ClpR2 leads to reduced accumulation of the ClpPRS protease complex and defects in chloroplast biogenesis in *Arabidopsis*. Plant Cell 18: 1704–1721

Rytz TC, Miller MJ, McLoughlin F, Augustine RC, Marshall RS, Juan YT, Charng YY, Scalf M, Smith LM and Vierstra RD (2018) SUMOylome profiling reveals a diverse array of nuclear targets modified by the SUMO ligase SIZ1 during heat stress. Plant Cell 30(5): 1077–1099

Sazuka T, Yamaguchi M and Ohara O (1999) Cyano2Dbase updated: Linkage of 234 protein spots to corresponding genes through N-terminal microsequencing. Electrophor 20: 2160–2171.

Sakai LY and Keene DR (2019) Fibrillin protein pleiotropy: Acromelic dysplasias. Matrix Biol 80: 6–13

Scaltsoyiannes V, Corre N, Waltz F and Giegé P (2022) Types and functions of mitoribosome-specific ribosomal proteins across eukaryotes. Int J Mol Sci 23(7): 3474

Schagger H, Cramer WA and von Jagow G (1994) Analysis of molecular masses and oligomeric states of protein complexes by blue native electrophoresis and isolation of membrane protein complexes by two-dimensional native electrophoresis. Anal Biochem 217: 220–230

Schleiff E, Eichacker LA, Eckart K, Becker T, Mirus O, Stahl T and Soll J (2003) Prediction of the plant beta-barrel proteome: A case study of the chloroplast outer envelope. Protein Sci 12: 748–759

Schubert M, Petersson UA, Haas BJ, Funk C, Schröder WP and Kieselbach T (2002) Proteome map of the chloroplast lumen of *Arabidopsis thaliana*. J Biol Chem 277: 8354–8365

Sekhar S, Panda D, Kumar J, Mohanty N, Biswal M, Baig MJ, Kumar A, Umakanta N, Samantaray S, Pradhan SK, Shaw BP and Behera L (2019) Comparative transcriptome profiling of low light tolerant and sensitive rice varieties induced by low light stress at active tillering stage. Sci Rep 9(1): 1–4

Sheen J (1999) C-4 gene expression. Annu Rev Plant Physiol Plant Mol Biol 50: 187–217

Stauber EJ and Hippler M (2004) *Chlamydomonas reinhardtii* proteomics. Plant Physiol Biochem 42: 989–1001

Tamburino R, Vitale M, Ruggiero A, Sassi M, Sannino L, Arena S, Costa A, et al. (2017) Chloroplast proteome response to drought stress and recovery in tomato (*Solanum lycopersicum* L.). BMC Plant Biol 17 (1): 1–14

Tikkanen M, Piippo M, Suorsa M, Sirpio S, Mulo P, Vainonen J, Vener AV, Allahverdiyeva Y and Aro EM (2006) State transitions revisited-a buffering system for dynamic low light acclimation of *Arabidopsis*. Plant Mol Biol 62: 779–793

Turkina MV and Vener AV (2007) Identification of phosphorylated proteins. Methods Mol Biol 355: 305–316

Van Wijk KJ (2000) Proteomics of the chloroplast: Experimentation and prediction. Trends Plant Sci 5: 420–425

Van Wijk KJ (2001) Challenges and prospects of plant proteomics. Plant Physiol 126: 501–508

Van Wijk KJ (2004a) Chloroplast proteomics. In: Leister D (ed) Plant Functional Genomics, The Haworth Press, Inc., Book Division, Binghamton, NY, pp 329–358

Van Wijk KJ (2004b) Plastid proteomics. Plant Physiol Biochem 42: 963–977

Van Wijk KJ (2006) Expression, prediction and function of the thylakoid proteome in higher plants and green algae. In: Wise RR and Hoober JK (eds) The Structure and Function of Plastids, Advances in Photosynthesis and Respiration, Vol 23, pp 125–143

Vieira Dos Santos C, Cuine S, Rouhier N and Rey P (2005) The Arabidopsis plastidic methionine sulfoxide reductase B proteins. Sequence and activity characteristics, comparison of the expression with plastidic methionine sulfoxide reductase A, and induction by photooxidative stress. Plant Physiol 138: 909–922

Vincis Pereira Sanglard L and Colas des Francs-Small C (2022) High-throughput BN-PAGE for mitochondrial respiratory complexes. In: Plant Mitochondria: Methods and Protocols, Humana, New York, NY, pp 111–119

von Zychlinski A, Kleffmann T, Krishnamurthy N, Sjolander K, Baginsky S and Gruissem W (2005) Proteome analysis of the rice etioplast: Metabolic and regulatory networks and novel protein functions. Mol Cell Proteom 4: 1072–1084

Walton A, Stes E, Cybulski N, Van Bel M, Iñigo S, Durand AN, Timmerman E, Heyman J, Pauwels L, De Veylder L and Goossens A (2016) It's time for some "site"-seeing: Novel tools to monitor the ubiquitin landscape in Arabidopsis thaliana. The Plant Cell 28(1): 6–16

Wang Q, Liu Z, Wang K, Wang Y and Ye M (2019) A new chromatographic approach to analyze methylproteome with enhanced lysine methylation identification performance. Anal Chim Acta 1068: 111–119

Whitelegge JP (2003) Thylakoid membrane proteomics. Photosynth Res 78: 265–277

Wiener MC, Sachs JR, Deyanova EG and Yates NA (2004) Differential mass spectrometry: A label-free LC-MS method for finding significant differences in complex peptide and protein mixtures. Anal Chem 76: 6085–6096

Wilson R, Goyal L, Ditzel M, Zachariou A, Baker DA, Agapite J, Steller H and Meier P (2002) The DIAP1 RING finger mediates ubiquitination of Dronc and is indispensable for regulating apoptosis. Nat. Cell Biol 4(6): 445–450

Xiao H, Chen W, Smeekens JM and Wu R (2018) An enrichment method based on synergistic and reversible covalent interactions for large-scale analysis of glycoproteins. Nature Commun 9(1): 1–2

Yates JR, Cociorva D, Liao L and Zabrouskov V (2006) Performance of a linear ion trap-Orbitrap hybrid for peptide analysis. Anal Chem 78: 493–500

Ytterberg AJ, Peltier JB and Van Wijk KJ (2006) Protein profiling of plastoglobules in chloroplasts and chromoplasts; a surprising site for differential accumulation of metabolic enzymes. Plant Physiol 140: 984–997

Yu Y, Yu PC, Chang WJ, Yu K and Lin CS (2020) Plastid transformation: How does it work? Can it be applied to crops? What can it offer? Int J Mol Sci 21(14): 4854

Ziehe D, Dünschede B and Schünemann D (2018) Molecular mechanism of SRP-dependent light-harvesting protein transport to the thylakoid membrane in plants. Photosynth Res 138(3): 303–313

Zörb C, Herbst R, Forreiter C and Schubert S (2009) Short-term effects of salt exposure on the maize chloroplast protein pattern. Proteom 9: 4209–4220

4 Proteomic Analyses to Understand Differentially Expressed Proteins during Floral Development

Monalisha Mishra, Shubhadarshini Sahoo, Gaurav Prajapati, Shadma Afzal, Nidhi Chaudhary and Nand K. Singh

4.1 INTRODUCTION

Flowering is a unique and intricately orchestrated process. Flowering is a significant turning point in the life cycle of a plant since it signifies the passage from vegetative development to the reproductive phase. In contrast to many annual plants, only some meristems convert to floral apices, while the majority of them continue to be in an indeterminate vegetative stage, which is necessary for the plant to continue growing [1]. The length of juvenility, which varies depending on the species, is essential in determining when the plant begins to produce fruit because reproductive development only takes place once the plant has attained maturity. The majority of the commercial fruit crops, including grapes, apples, and citrus, are governed by this programme for age-related development. For instance, citrus trees often go through a juvenile stage before beginning to flower, which can take anywhere between two and more than ten years [2]. The stages of flower growth, pollination, and fertilization are crucial for controlling the formation and development of seeds as well as the sexual reproduction of plants. The complex process of flower growth and development can be categorized into the following three stages: (1) floral induction, (2) floral evocation, and (3) flower organ development [1]. These are groups of undifferentiated cells that develop into axillary shoots, internode tissue, lateral organs (such as leaves), and flowers [3]. Therefore, whether flowering takes place is determined by the type of cells that are created, and their eventual developmental fate, as a component of vegetative or reproductive system. Fusion of pollen and egg cells produces seeds, ensuring progeny for the next generation, and hence floral reproductive system plays an essential role in plant sexual reproduction. The time when a flower is fully open and active is known as anthesis. Anthesis describes the time when a flower bud first opens. The style of a flower in anthesis extends significantly beyond the upper perianth. It makes pollination easier. Various floral organs, in notably the petals, move in synchronization when the flower opens. Pollination, described as the interaction of pollen and the stigma, is a crucial stage in the reproductive cycle that influences seed formation [4]. Many genes, proteins, and environmental elements interact and have an impact on pollination. Furthermore, only a small number of flowers and ovules turn into fruits or seeds due to the complexity of the pollen-stigma interaction [5]. With an emphasis on the study of several proteins involved in the pollen-stigma interaction, proteomics has made significant progress in understanding this relationship [6]. Prior research has mainly focused on the protein-expression profile of pollen grains or tubes at various stages of pistil development in a variety of species, including *Glycine max* Linn., *Zea mays* L., and *Oryza sativa* L. [4]. Protein alterations that occur during pollination and post-pollination senescence have, however, received very little investigative focus.

In higher plants, pollination starts a sequence of developmental processes that lead to flower senescence, such as changes in flower coloration, crinkling and fading of the petal edges, petal

DOI: 10.1201/b23255-4

senescence, etc. [7]. Programmed cell death (PCD) occurs in the conducting tissues around the pollen tubes as the pollen grains germinate and the pollen tube develops down the pistil [8, 9]. Additionally, immediately after pollination, petal senescence, a visible sign, develops [10]. Petal senescence results into petal withering and cell death, as well as drying of the flower, an increase in film permeability, and extravasation of micromolecular compounds [11]. Signal transmission starts pollination, which is then controlled by ethylene synthesis to start the physiological process of petal senescence [12]. Numerous physiological events, including the death and destruction of specific pistil cell types, can occur as a result of pollination.

Floral induction involves spatiotemporal interactions between external and internal elements. Temperature and day length are two external stimuli important to flowering. The variation in day length, or photoperiod, is one of the fundamental aspects of the change of the seasons [13]. Garner and Allard were the first to identify the regulating role of day length on the onset of flowering [14]. Depending on their species, cultivars, and geographic location, plants can adjust their flowering time in response to the varying light and dark cycle to be in sync with their environment. While some plants only flower when exposed to short days, others require sustained exposure to lengthy days [15]. For instance, rice is in the former category, while the widely researched model plant *Arabidopsis thaliana* is in the latter. A group of proteins made up primarily of photoreceptors and transcription factors will detect and respond to the numerous environmental signals when plants have reached maturity and are ready to flower.

Gibberellin (GA), vernalization, and photoperiod are the genetic pathways that have been demonstrated in Arabidopsis to integrate environmental and endogenous stimuli in determining the period of flowering onset [16]. At least, three genes together known as the floral pathway integrators constitute a checkpoint for flowering time control which are LEAFY (LFY), FLOWERING LOCUS T (FT), and SUPPRESSOR OF OVEREXPRESSION OF CONSTANS 1 (SOC1). This is where the various pathways converge. The floral meristem identity genes such as FRUITFULL (FUL), APETALA1 (AP1), SEPALLATA4 (SEP4), CAULIFLOWER (CAL), and LFY are activated by these floral pathway integrators, and their activation results in the conversion of vegetative apices to inflorescence meristems [17–21]. The ABCDE model, which specifies floral organ identity by floral whorl-specific combinations of class A, B, C, D, or E genes, is a fundamental hypothesis that has been developed to explain the formation of floral organs in higher plants. Except for the class A gene APETALA2, the class A, B, C, D, and E genes in Arabidopsis all encode MADS-box transcription factors [22]. In plants, the MADS-box gene family is a putative transcription factor gene family which is highly conserved. Their role in the development of floral organs in *A. thaliana* and *Antirrhinum majus* led to their initial discovery [23, 24]. These genes produce proteins that have a 56 amino acid highly conserved domain that is important in recognizing and binding to DNA sequences with CArG boxes as consensus sequence. These proteins have been relevant in several stages of floral development [25–27]. The name of the MADS-box gene family is derived from several of the early homeotic genes that have been identified, including MCM1 from yeast [28], AGAMOUS from Arabidopsis [24], DEFICIENS from *Antirrhinum majus* [23, 29], and SRF from *Homo sapiens* [30].

We now have a better understanding of the molecular processes that surround the change from the vegetative to reproductive phase in angiosperms owing to advances in the study of flowering in model plant systems. Floral organs originate from highly modified leaves [31]. So, during flowering, a portion or sometimes the entire proportion of what would otherwise be vegetative shoots is transformed into flowers. The majority of these new developments in flowering dynamics are due to the use of molecular genetic technology in annual plants, specially *A. thaliana*, which are well suited for this kind of research [32]. It has become increasingly evident that the development of the typical flower parts, i.e., petals, sepals, carpels, and stamens, is majorly conserved across all flowering plants and that the observed variation in flower morphology is accounted for by relatively minor variations in the expression of key developmental genes [31, 33].

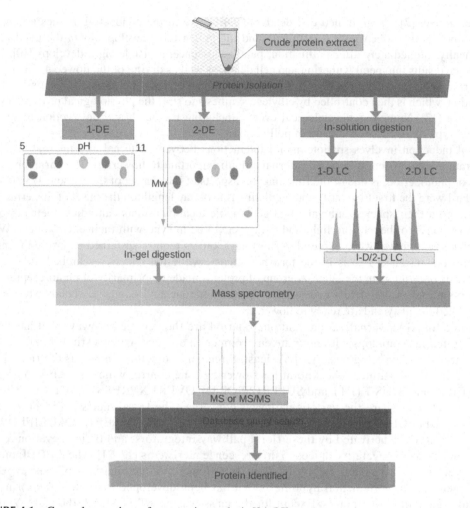

FIGURE 4.1   General procedure of proteomics analysis [34, 35].

Proteomics, a promising coming-of-age technology, can address some of the problems with genomic technologies. Proteomic analysis, which combines the high resolution of two-dimensional electrophoresis (2-DE) and sensitive mass spectrometry, is a potent method for detecting proteins in particular tissues under particular circumstances [6]. A general strategy to perform proteomic analysis is depicted in Figure 4.1. In order to investigate changes in cellular protein expression in response to several biotic and abiotic stresses in plants, the proteomic technique has been used [34, 35]. Asa key component of the functional genomic technique, proteomics is a potent tool for analysing protein complexes. Analysis of protein alterations in response to environmental changes has been done using proteomics. A proteomics approach helps in analysing the physiological changes and function of stress-induced proteins, as evident from the numerous proteomic studies that have been conducted to investigate the effects of pathogen, salt, and drought stresses on plants. These studies have identified a number of stress-associated proteins [36]. Comparing the expression of proteins in the floral bud and mature flower using quantitative proteomics is a unique method for learning more about the protein networks involved in this complex floral development system.

In this chapter, an effort has been put to develop understanding of the stages of floral development, the methods utilized in proteomic investigations, and the procedures involved in carrying out a proteome study. The metabolic pathways and related genes that are active during flower development are explained in this chapter. The majority of secondary metabolites, including flavonoids, carotenoids, and other fragrance compounds, are produced by these pathways and genes. Additionally, this chapter includes details on many proteins that are expressed during the differentiating, enlarging, and anthesis stages of flowers. In this chapter, differentially expressed proteins are also discussed in relation to pre-pollination, pollination, and post-pollination senescence in flowers.

4.2   ABCDE MODEL OF FLOWER DEVELOPMENT

The ABCDE model for flower development represents significant improvements in our comprehension of how the floral organs emerged from the leaves. According to the floral quartet models of floral organ specification, the five classes of homeotic genes—A, B, C, D, and E—define the identity of floral organs [16, 27]. All ABCDE homeotic genes encode MADS-box transcription factors with the exception of the class A gene, APETALA2 (AP2) [37–40]. The class A MADS-box gene in Arabidopsis is AP1 [17, 18], as are the class B genes AP3 and PISTILLATA (PI) [41–43], the class C gene AGAMOUS (AG) [44], and the class D genes SEEDSTICK (STK), SHATTERPROOF1 (SHP1), and SHP2 [20]. To determine ovule identity, the E-class proteins and D-class proteins work together in a bigger complex. SEPALLATA1 (SEP1), SEP2, SEP3, and SEP4 are class E genes that have been identified in the Arabidopsis genome and which demonstrate largely redundant roles in the identification of sepals, petals, stamens, and carpels [45–47]. For the first floral whorl, the A- and E-class protein complex develops sepals as the ground-state floral organs. For the second whorl, the A-, B-, and E-class protein complex specifies petals. For the third whorl, the B-, C-, and E-class protein complex specifies stamens. The floral development model is shown in Figure 4.2. The enormous variety of flower morphologies in angiosperms has been attributed to the diversity of MADS-box genes during evolution. The AGL6 and AGL13 genes from the AGAMOUS-LIKE 6 (AGL6)-clade, which are not included in the classical ABCDE model, may be needed in the development of flower parts, presumably ovules [48]. The E-class function of SEP genes may be shared by AGL6-clade genes, which make up the sister clade of SEP genes [49]. The ABCDE concept may also apply to monocots, according to research on ABCDE genes in monocot species like rice. OsMADS14/RAP1B, OsMADS15/RAP1A, and OsMADS18, three AP1-like MADS-box genes that are all descended from the FRUITFULL (FUL) branch rather than AP1 as in Arabidopsis, have been discovered in the rice genome. OsMADS14 may play a role in stimulating flowering and identifying the floral meristem, according to studies on transgenic plants [50]. SUPERNUMERARY BRACT (SNB) and Os INDETERMINATE SPIKELET1 (OsIDS1), two AP2-like genes, were found to be essential for lodicule development in rice. MULTI-FLORET SPIKELET1 (MFS1), another AP2-like gene, positively regulates SNB and OsIDS1. Furthermore, via controlling modifications in spikelet meristem fate, these rice AP2-like genes control inflorescence architecture [51, 52]. According to studies, OsMADS2 and OsMADS4 are the rice orthologs of PI and were produced by gene duplication event. These findings suggest that OsMADS2 and OsMADS4 both contribute equally to the development of stamens and that OsMADS2 is more crucial for lodicule specification than OsMADS4. In the rice genome, there is only one AP3 ortholog, OsMADS16/SUPERWOMAN1 (SPW1). Overall, it could be concluded that the PI-like OsMADS2 and OsMADS4 genes, as well as the AP3-like OsMADS16 gene, are class B genes in *O. sativa*. According to research, the rice class C genes OsMADS3 and OsMADS58 exhibit some functional conservation with the Arabidopsis class C gene AG. Along with OsMADS13 (a class D gene), the two genes (OsMADS3 and OsMADS58) redundantly mediate the C-function and are crucial for inflorescence meristem determinacy [50]. Additionally, it has recently been discovered that OsMADS16, a class B gene, interacts with the two class C genes to inhibit indefinite growth in the inflorescence meristem. It is interesting to note that the YABBY gene DROOPING LEAF controls the identification of the carpels in rice (DL) [53]. Specifying ovule identity involves the rice class D gene OsMADS13. OsMADS6/MOSAIC FLORAL ORGANS 1 (MFO1), a rice AGL6-clade gene

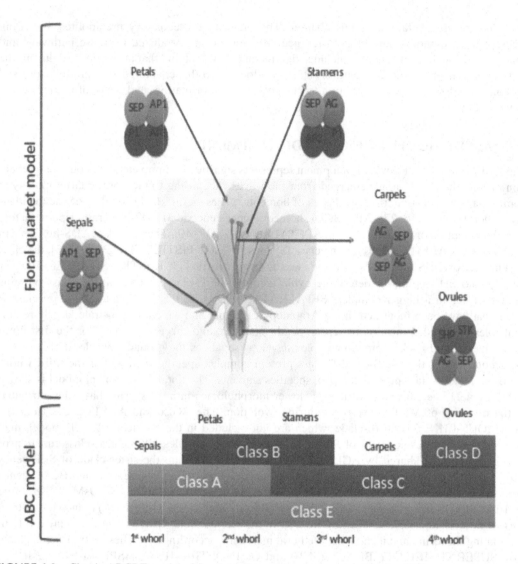

FIGURE 4.2   Classic ABCDE model of floral development and organ specification [27].

that controls floral organ identity, may also have an E-class role [14, 54]. OsMADS17, a gene belonging to the AGL6 lineage, performs a limited but redundant role with MFO1.

4.3   PROTEINS EXPRESSED DURING FLORAL DEVELOPMENT

Through the combination of physiological (hormonal control) and environmental (photoperiodism and vernalization) signals, the dynamic flowering process of *A. thaliana* is driven from many pathways [17]. A complex network of interconnected genes and proteins, with repressor and activator functions that influence one another, is activated by these signals. In Arabidopsis, APETALA1 (AP1) governs the production of the sepal and petal, APETALA3 (AP3) and PISTILLATA (PI), the stamen and petal, and AGAMOUS (AG), the stamen and carpel [55]. All floral organ types are specified by the SEPALLATA1–4 (SEP1–4) proteins, which are also thought to be crucial for the production of quaternary MADS protein complexes [24, 26, 56]. A floral inducer named FLOWERING LOCUS T (FT), also known as a mobile "floral integrator," was found to be crucial to the several pathways that control the formation of flowers. FT is a conserved promoter of flowering that is expressed in the cotyledon and leaf vasculature [57]. It functions downstream of a number of regulatory pathways, including one that facilitates photoperiodic induction through CONSTANS (CO). The photoperiod route, one of the main flowering mechanisms in Arabidopsis, positively controls the nuclear protein CO, which functions upstream of a graft-transmissible signal generated in leaves. CO directly activates genes that perform a variety of metabolic processes [58]. Two of these genes, SUPPRESSOR OF OVEREXPRESSION OF CONSTANS 1 (SOC1), a transcription factor, and FT, which encodes a tiny globular protein linked to the floral repressor TERMINAL FLOWER 1 (TFL1), are known to induce flowering [59–61]. Additionally, these investigations identified two structural genes for ethylene and proline biosynthesis enzymes, ACS10 and P5CS2, as possible CO targets [58, 62]. For FT to induce flowering, a bZIP transcription factor called FD that is selectively expressed in the shoot apex is necessary. A floral meristem identity gene called APETALA1 is transcriptionally activated at the shoot apex by FD and FT, who work as interdependent partners to induce floral transition and to start floral development (AP1) [63]. The circadian clock regulates flower initiation in angiosperms in response to seasonal variations. The length of the day, which is the primary factor of photoperiodism, can regulate plant flowering in the majority of cases. In contrast, A. thaliana, Avena sp., Lolium sp., and Trifolium sp. respond to long day signals. As an illustration, Cannabis sp., Gossypium sp., and Xanthium sp. respond to short-day signals. The MADS DNA binding domains found in floral homeotic proteins are involved in multimeric complexes that partially overlap to regulate flower development [64]. Transcription factors with the MADS domain are crucial to several developmental stages in flowering plants [65]. The change from vegetative to floral state and the identification of floral organs are significantly influenced by members of this family. According to researches AGAMOUS (AG) regulates the expression of a number of known regulators of stamen and carpel formation, including CRABS CLAW (CRC), SPOROCYTELESS (SPL), and SUPERMAN (SUP), suggesting that it controls the expression of a battery of seemingly flower-specific factors. Carpel formation also involves the MADS-box gene FRUITFULL (FUL) [64, 66]. Some of these genes have additional activities in addition to those that aid in the development of floral organs. For instance, AG regulates the floral meristem determinacy, while FUL and AP1 are both essential in the transition from inflorescence meristem to floral meristem identity. Before they may access the nucleus, several MADS domain proteins must be in a dimeric state, either homo- or hetero-dimeric [67]. Proteins with the consensus CC(A/T)6GG sequence, also referred to as the CArG box, bind to DNA sequences of the target gene in the nucleus [44]. Either a dimer or a multimeric structure, such as a tetrameric form as suggested by the "quartet" hypothesis, can bind to the DNA. A study that integrated interaction data from a large-scale microarray co-expression investigation of the relevant genes with systematic yeast two-hybrid interaction data considerably increased our understanding of the interactions of MADS domain-containing proteins. Microarray studies only provide a very general overview of the spatio-temporal pattern of gene expression and lack the requisite specificity to show that the encoded proteins are co-localized. The expression patterns are further resolved by in situ mRNA hybridization studies and promoter-reporter experiments, such as those documented for AG, SEP3, AP1, and FUL [24, 64]. A thorough investigation demonstrates that FUL is present in the early stages of the flower bud, suggesting that it may operate similarly to AP1 in controlling these flowering period genes. This is also in agreement with the previous genetic studies that revealed synergy between AP1 and FUL during the floral transition [21] and with the idea that both AP1 and FUL act as links between the floral organ protein network and the flower induction protein network [68]. The reported mRNA patterns also failed to explain the asymmetric localization of SEP3 in the epidermis of both sepals and petals and its potential significance in the adaxial/abaxial patterning of these organs, as well as the presence of FUL protein in the replum of developing pistils and AG in the nucellus of the developing ovule during the down-regulation of WUS. In shoot and floral meristems, the WUSCHEL (WUS) protein is essential for maintaining stem cell populations [69].

The Arabidopsis AGL11 gene and the Petunia FBP7/11 gene are significantly similar to TAGL11 [70]. According to the findings, the expression patterns of TAGL11 and FBP7/11 are comparable.

These genes begin to be expressed in the inner integument of ovules at a late stage of floral development, and after anthesis, in the integuments of developing seeds [70]. It has been postulated that FBP11 and FBP7 may be necessary for post-fertilization activities as well as ovule development [70]. Since both of these proteins are present in the pericarp at anthesis, the TAG1-TDR4 dimer may be crucial for regulation in the pericarp. As a result, both the genes, viz., *FBP7/11* and *TAGL11*, represent class D genes in the modified ABC model and are necessary for appropriate ovule and seed development. Two new tomato MADS-box genes were recently discovered. The gene *JOINTLESS* is in a different evolutionary group from those responsible for the establishment of abscission zones in flowers and fruits. At the abscission zones, *JOINTLESS* gene expresses itself specifically [71].

The protein AMS EN 48, which is encoded by the transcription factor ABORTED MICRO-SPORES (AMS), controls male fertility and pollen differentiation [72, 73]. Exine production is necessary for normal pollen growth and is regulated by genes like *LESS ADHERENT POLLEN3* (*LAP3*), which codes for the protein strictosidine synthase-like 13. The genes *LESS ADHESIVE POLLEN5* (*LAP5*) and *LESS ADHESIVE POLLEN6* (*LAP6*) encode the Type III polyketide synthase B protein, which is necessary for pollen development, and the Type III polyketide synthase A protein, which aids in the biosynthesis of pollen fatty acids [74, 75]. These genes, often known as male candidate genes, encode anther-specific proteins necessary for pollen exine formation [76].

MALE SPECIFIC EXPRESSION 1 (MSE1), a putatively duplicated homeodomain-like superfamily protein that is only produced in male flowers, has a close link to the Y chromosome, exhibits a distinctive expression pattern throughout the early stages of anther development, and has a loss of function on the X chromosome. It is essential for the early development of anther [77]. POLYGALACTURONASES, such as PG1 Polygalacturonase 1 beta-like protein 2 and PG2 Polygalacturonase 1 beta-like protein 3, are involved in determining cell size and may also function as chaperones. The enzyme PECTIN METHYLESTERASE (PME), which codes for pectin esterase, is crucial for the production of pollen tubes as well as other processes. After the mitosis in the microspore, the tissues of mature pollen grains express the proteins POLYGALACTURONASES and PME, which are important for the development of tapetum and pollen and are specific [54].

Female potential genes involved in the identification of pistillate tissues and the detection of pollen include the filamentous floral proteins YABBY1 and YABBY4 as well as the transcription factor ANT. The gynoecium gene *YAB1 Axial regulator* produces the YABBY1 protein, which is expressed in the abaxial cell layers and aids in the development of the valve meristem. The gene also produces the YAB4 protein, which is necessary for the formation and asymmetric growth of the ovule outer the integument along the abaxial-adaxial axis [78–80]. The *ANT* gene encodes an AP2-like ethylene-responsive transcription factor, which activates ANT transcription and is necessary for the formation of the integument of the ovule as well as the development of the female gametophyte [38, 40, 81]. The gene *STIGMA SPECIFIC1* (*STIG1*) encodes a protein that is specific to stigmas and is involved in the timing of exudate secretion onto the stigma. In the *Quercus suber* species, this protein was also discovered as a female candidate [76]. Another female potential gene, *SUP*, encodes for the transcriptional regulator SUPERMAN, which controls transcription [53]. It is assumed to be a cadastral protein that indirectly inhibits the actions of the B-class homeotic proteins APETALA3 and maybe PISTILLATA in the gynoecium whorl [41, 78]. Axillary bud outgrowth is inhibited by the transcription factor Protein BRANCHED 2, which is encoded by the gene *TCP12* which is another female candidate gene [82, 83]. A female candidate gene called *MYB61* codes for a transcription factor MYB61, which regulates a compact network of downstream target genes necessary for various aspects of plant development and growth [36, 84, 85].

4.4 PROTEINS EXPRESSED DURING FLORAL ANTHESIS

The stage when a flower is fully open and functioning is known as anthesis. It might also indicate the onset of the period. The flower becomes useful in this way. The style of a flower in anthesis extends above its upper perianth, and it further helps facilitate pollination. When pollinators are attracted to facilitate cross-pollination, that could improve reproductive success and species con- servation; flower blossoming to reveal stamens and gynoecia is a crucial event. The coordinated movement of different floral parts, in particular petals, occurs in tandem with the process of flower opening [86]. The activity of the enzyme BEAT changes along with the maturation of the bud, and levels of BEAT mRNA in the petals peak at anthesis. BEAT (acetyl-CoA:benzylalcohol acetyltrans- ferase) is an enzyme that catalyses the synthesis of benzylacetate [87, 88].

In the formation of petals, there are two distinct developmental phases: the first includes cell division, and the second, cell expansion [66, 89]. In Gerbera, GEG [Gerbera hybrida homolog of the gibberellin [GA]-stimulated transcript 1 [GAST1] from tomato] has been suggested to have a significance in the regulation of corolla morphology, and NAP gene has been demonstrated to func- tion in the transition from active cell division to cell expansion in Arabidopsis [90, 91]. Moreover, less is understood about the genetic mechanism responsible for determining corolla structure in general and phase transition in particular [46]. When mitotic activity has essentially stopped during the later stages of flower formation, the significant rise in petal size occurs [89]. Different metabolic pathways are active during the cell expansion phase to produce distinctive secondary metabolites, such as flavonoids, carotenoids, and many aroma substances [92]. The growing petals contain many stress-related proteins, such as low- and high-molecular-weight HSPs, peroxidase, superoxide dis- mutase, and catalase. The homologue of a protein associated with chromoplast carotenoid is another member of this group [93]. However, it has been demonstrated that chromoplast and carotenoid proteins also function in protecting against biotic and abiotic stressors in addition to being involved in chromoplastogenesis during flower development [94]. Stress-related genes make up a significant fraction of the petal ESTs in both Arabidopsis and *Antirrhinum*, and they are controlled by a class B floral homeotic gene [47, 95, 96]. Jasmonic acid, a hormone associated with stress, has also been demonstrated to build up to high amounts in growing flowers [97]. The build-up of stress-associated proteins in petals may be a sign that the cell needs to be protected from either intracellular stress (such as an oxidative environment) or external stress during floral development. Petal tissues in par- ticular and the flower as a whole are particularly sensitive tissues, and even low levels of biotic or abiotic stress can cause organ death. Even though processes associated to secondary metabolism are quite active throughout the later stages of petal development, only one discovered protein, dihydro- flavonol reductase, could be directly linked to secondary metabolism [98, 99]. Recent studies have shown that the homeodomain transcription factor (TF) gene PETAL MOVEMENT-RELATED PROTEIN1 (RhPMP1) is a direct target of the TF ETHYLENE INSENSITIVE3, which works downstream of ethylene signalling. RhAPC3b, a gene that encodes the core subunit of the anaphase- promoting complex, was discovered to be induced by ethylene, and this selectively activated the endoreduplication of parenchyma cells on the adaxial side of the petal (ADSP) base [100, 101]. Following an increase in cell proliferation of the parenchyma on the ADSP base, the petal base grew asymmetrically, resulting in the usual epinastic movement of petals and flower opening [102]. In roses, ethylene speeds up flower opening, and production rises during flower opening before reach- ing a peak at the fully opened stage [103–106].

4.5 PROTEINS EXPRESSED DURING FLORAL PRE-POLLINATION

Phosphoglycerate kinase and CYB5R were found to be up-regulated in the pre-pollination phase (fully opened with the keels still closed) but down-regulated in the pollination phase (2 hours after pollination) and post-pollination senesced stage (24 hours after pollination stage). Phosphoglycerate kinase is involved in glycolysis/gluconeogenesis, whereas CYB5R is associated with nucleotide sugar and amino sugar metabolism [107]. The metabolism of the organism could become disordered if this enzyme is lacking. In addition to being necessary for germination, flowering, ageing, and the stress response, sugars play a vital regulatory role in plant metabo- lism and growth [108]. Studies have shown that in the male sterile hybrid pummelo, early in the development of the floral bud, high quantities of the most abundant sugars are stored, and during the mature pollen stage, they are down-regulated, indicating that the accumulated sugars provide nutrition for the flower nutrition and pollen development [109–111]. In *Pinus strobus*, sugars were only found during the development of the pollen tube [112]. Similar findings were found in alfalfa flower study; proteins related to the high accumulation of carbohydrates metabolism were up-regulated prior to pollination. These proteins would provide the nutrition and energy for pollen formation in alfalfa flowers. The biochemical reactions that ethylene regulates include those that affect plant growth [101]. The secondary metabolite polyamine enhances plant resistance, controls plant growth and development, delays senescence, and regulates morphology and architecture. It had been proposed that polyamine could control the process of flowering and thereby encourage fertilization [113]. Research shows that SAMS are up-regulated in the pre-pollinated stage, down-regulated in the pollinated stage, and senesced post-pollinated. SAMS were up-regulated before pollination and played role in amino acid metabolism, according to research on the stenospermocarpic table grape (*Vitis vinifera* L.) [114] and petunia [115]. Following corolla senescence, they were subsequently down-regulated. A crucial component of eucaryon development is the PTK kinase splALs, which takes part in cell growth, proliferation, differentiation, and signal transduction [116]. Kinase splALs were found to be up-regulated in the pre-pollinated stage but down-regulated in the pollination and post-pollination stages. They were also involved in other signal transduction pathways, including the MAPK pathway. These findings suggest that kinase splALs has a significant signalling role in alfalfa pollination [37, 82].

The metalloenzyme CA, which contains zinc, has been recognized as a crucial photosynthetic enzyme because it can catalyse the reversible hydration reaction of CO_2 [117]. Light stimulus can control changes in plant shape and structure, including the flowering onset, the growth of new leaves, the lengthening of stems, and the germination of seeds [108]. CA may control a number of light-dependent processes in alfalfa flowers. NQOL mRNA levels were reported to be up-regulated during pollination and post-pollination stages, and NQOLs were shown to be up-regulated at both the stages as well [118].

4.6   PROTEINS EXPRESSED DURING FLORAL POLLINATION

Pollination is influenced by numerous genes, proteins, and environmental factors that interact with one another. Only a limited number of flowers and ovules turn into fruits or seeds due to the complexity of the pollen-stigma interaction [5]. In rice, pollination and flower opening occur almost simultaneously, and lemma and palea open simultaneously when the pollens land on the stigma, exposing the pistil to intense light and high temperatures. In Arabidopsis, pollination occurs before flower opening.

Proteomics has made significant advancements with regard to the study of several proteins involved in the pollen-stigma interaction. Pollen coat structure, lipids, and the pollen coat proteins play critical roles during pollen-stigma identification, according to several researches focusing on pollens [9, 119–123]. Studies revealed that a successful pollination also depends on the stigma. It has been discovered that during pollination, the proteinaceous stigmatic pellicle, the outermost layer of the stigmatic papilla, communicates with the pollen grain. The failure of pollen tubes to access the stigma may be caused by pellicle ablation [124]. The interactions between stigma-specific S receptor kinase (SRK) and the pollen coat protein-S-locus Cys-rich/S-locus protein 11 (SCR/SP11) in *Brassica* stigma led to the self-incompatibility reactions [125–127]. Exo70A1 and ARM-repeat containing 1 (ARC1), two additional components, were found to be involved in the rejection of self-pollen [77, 128–130]. It was discovered that through modulating the synthesis of stigmatic estolide in the Petunia stigma, the cytochrome P450 enzyme CYP86A22 could also have an impact on pollination [131]. Pollentube growth is an actin-dependent process that necessitates dynamic remodelling of the actin cytoskeleton. Few protein-expression patterns in pollen tubes have been identified, in contrast to the vast majority of morphological research that examined pollen germination and tube expansion. Gymnosperm pollen tubes differ significantly from angiosperm pollen tubes in that they grow more slowly, spend longer periods of time growing, and have delayed gametogenesis. Additionally, there are differences in the cytoskeleton of the two groups and vesicular trafficking. These variations might suggest that the mechanism by which pollen tubes in gymnosperms and angiosperms grow and lengthen differs. During the development of soybean flowers, metabolism-related proteins make up 41.67% of all identified proteins [132], while related proteins and genes make up 31% and 22% during the *Agapanthus* flowering transition [133]. Because there are only low levels of these nutrients in the vegetative tissues of many plants, cysteine and methionine are crucial nutrients for flower formation [134]. The majority of the proteins that were revealed to be differently expressed in alfalfa flower development were found to be involved in metabolism [118]. The majority of them are involved in the metabolism of carbohydrates and amino acids [135]. They were found to be up-regulated during pre-pollination and down-regulated following pollination, indicating that the primary metabolism is improved to aid in pollination. It also implies that improving the major metabolisms in the pistil may improve plant pollination success [136]. The ubiquitous signalling module known as MAPK cascade controls every stage of plant growth and development, including ovule development, anther development, pollen development, pollen tube direction, morphogenesis, fertilization, gametogenesis, embryogenesis, senescence, and seed formation [137]. It has been established that Arabidopsis MPK3/MPK6 governs the structure of inflorescences and the direction of pollen tubes. The MPK3/MPK6 proteins play a unique role in signal recognition between pollen tubes and the micropyle [138]. Water passage into pollen grains is reportedly regulated by an aquaporin-like protein in the pistil. Pollen germination and seed set may be affected by lowering the expression of the pistil gene *pis63*, which encodes a protein with an unidentified function. A few additional proteins, including an ATP-binding cassette (ABC) transporter from *Nicotiana tabacum* and a peroxidase specific to the pistil from *Senecio squalidus*, may also play a role in pollination. As a crucial signal receptor, the receptor kinase has been linked to the detection of stigma and pollen. In *Brassica*, incompatible pollination begins when a receptor-like kinase recognizes pollen S-proteins and encourages phosphorylation and ubiquitination in the pistil. According to a report, the key signals used in the pollination process are auxin and ethylene, both of which can be induced. The ubiquitination-mediated degradation of EXO 70A1, a component of the vesicle complex controlling cell exocytosis, is one of the crucial steps in Brassica self-incompatible processes. This could imply that the trafficking of vesicles plays a beneficial role in compatible pollination. Major proteins in the pollen coat of maize include beta-1,4-xylanase and beta-glucanase. As model plants for the study of pollen-pistil interaction, crucifers are the typical family with dry stigma. The extracellular matrix components are controlled by the proteins expressed by the genes through a signal transduction pathway during pollination. SRK, M-locus protein kinase (MLPK), and arm repeat containing (ARC1) are three of these proteins that contribute to self-incompatibility. The largest group of proteins, as determined by the MALDI-TOF-MS results based on the 2D-gel and functional classification, belonged to the metabolism-related protein category. Among them, the expression of sucrose-phosphate synthase (U18) increased, whereas other glutamine synthetase isoforms declined. These showed that the primary metabolism is improved to aid in pollination and the subsequent expansion of the pollen tube. *Solanaceae*, *Rosaceae*, and *Liliaceae* are the principal plant families having wet stigmas. For the suitable pollen grains to be ingested, the stigmatic secretion (SE) is crucial. In addition to lipids and carbohydrates, SE also includes a variety of proteins with important roles, such as the sigma-specific protein 1 (STIG1), which functions when SE accumulates in plants like tobacco and petunia [136, 139].

4.7 PROTEINS EXPRESSED DURING FLORAL SENESCENCE

A period of steady growth is first visible in flower buds, and then, as the petals unfold and the flower opens, they frequently go through a brief period of rapid cell elongation. Biological and abiotic stressors have the ability to trigger senescence. As the most vibrant and vulnerable component of plants, flowers are quickly impacted by a variety of environmental conditions, including insect-mediated pollination, seasonal changes, a shortage of water, and various stresses like pathogen invasion or predator attack. Depending on the species, after floral opening, the petals or tepals in flowers like iris may thereafter be shed within a few hours or persist for as long as three quarters of a year [11]. Asparagine synthetase, proteases, and genes involved in lipid degradation including malate synthase and isocitrate lyase are only a few of the genes that are up-regulated prior to the appearance of apparent flower senescence. Senescence becomes apparent as an end outcome, which typically involves significant cell death. Plant senescence is tightly controlled and may already be progressing before it is apparent to the unaided eyes. A scheme of the control mechanisms involved in flower senescence is depicted in Figure 4.3. The PCD signal is produced by both internal and external stimuli, and it is then translated into other signals that cause an imbalance in the hormone production. This changed hormone level further activates a number of cascades and transcriptional controls. The expression of various senescence-associated genes (SAGs), such as proteases, nucleases, wall-degrading, and oxidative enzymes, signifies the onset of senescence. Now with all of these enzymes working together, the senescence process is accelerating and becoming irreversible. Senescent signs begin to appear at a later stage, which ultimately causes

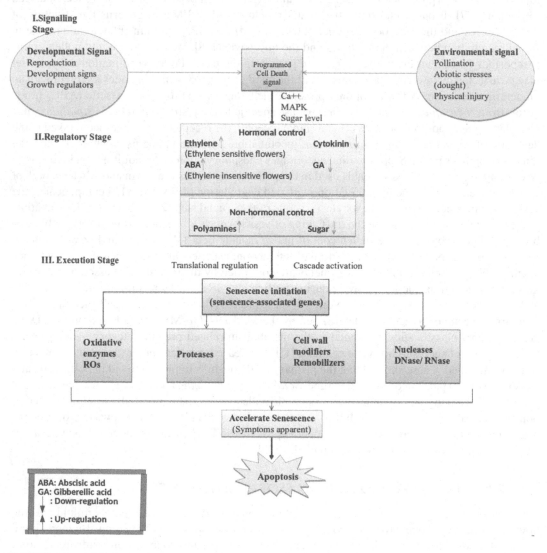

FIGURE 4.3   A brief overview of floral senescence and the associated signals [142].

flower cell death or apoptosis. Numerous biological and physiological processes undergo multiple regulated modifications during the senescence of petals [89]. An important part of protein degradation, which is a sign of organ senescence, is ubiquitination. Proteins associated to peptidases in proteolysis and autophagy pathways were found to be abundant in the proteome by Gene Ontology (GO) and Kyoto Encyclopedia of Genes and Genomes (KEGG) analysis, indicating that protein degradation and autophagy are significant contributors to petal senescence. Additionally, numerous transporter proteins accumulated in senescent petals, and a number of transport mechanisms were concentrated in the ubiquitome, showing that ubiquitination regulates substance transfer and is linked to petal senescence. Van Doorn and Woltering (2005) attempted to classify plant PCD into three categories based on the PCD mechanism identified in morphological kinds of metazoan cell death: apoptotic-like, autophagic, and neither apoptotic nor autophagic [11]. Similar to animal apoptosis, tapetum and pollen tubes provided convincing evidence for the critical role of the mitochondrion and the involvement of enzyme caspases. A study demonstrated a functional relationship between actin polymerization and PCD in pollen by demonstrating that changes in actin filament levels or dynamics serve a functional role in activating PCD in *Papaver* pollen [15]. The fact that various constituents of the brassinosteroid (BR) production and signalling pathway underwent considerable protein and ubiquitination level changes further suggests that BR is crucial to petal senescence. Petunia (*Petunia hybrida*) petals lose a significant amount of protein during petal senescence, while protease activity in petunia, daylily, and iris has been found to significantly increase. During senescence, the phospholipid content of rose, carnation, and tradescantia petals also drops, which results in impaired membrane integrity. Small molecules, including sucrose, amino acids, and mineral ions, are transferred from senescing petals to growing tissues via a variety of transporters in the phloem to recycle nutrients. Furthermore, phytohormones play a crucial part in controlling the advancement of senescence, with ethylene and abscisic acid (ABA) speeding up petal senescence and cytokinin retarding it [140]. Through controlling the PCD of the petals, ethylene is a key factor in the onset of floral senescence. Plant development includes PCD, which is essential. Numerous proteins that were up-regulated during senescence were reportedly associated to catabolism of macromolecules or stress and defence responses. The GO enrichment experiments for these differentially abundant proteins (DAPs) to assess their possible roles revealed that DAPs were particularly high in the molecular functions of hydrolase catalysis, ribonucleoside binding, transferase, guanosine triphosphate (GTP) binding, carboxypeptidase, phosphatase, oxidoreductase, serine-type carboxypeptidase, and serine-type exopeptidase [141]. While hydrolase activity, phosphoric ester hydrolase activity, phosphatase activity, and carboxypeptidase activity were all enhanced in down-regulated DAPs, activities of carboxylic ester hydrolase, hydrolase, and S-acyltransferase were all enhanced in up-regulated DAPs. Studies indicate that protein modification and catabolic mechanisms play significant roles in petal senescence [140]. Additionally, proteins with peptidase activity connected to proteolysis were also reported to be markedly enriched. Research shows that enhanced protein proteolysis occurs alongside organ senescence in species like petunias and iris flowers. During petal senescence, biological processes were affected. Protein kinases (PKs) are crucial senescence signalling regulators. Receptor-like kinases or PKs were abundant in the rose according to a proteomic investigation (RLKs).

4.8 IMPORTANT PROTEINS EXPRESSED DURING FLOWER DEVELOPMENT IN RICE PLANT

Rice (*O. sativa* L.) is one of the most fascinating model species to study flower development since it has a unique inflorescence and floral structure that differs significantly from those of other model eudicots, including *A. thaliana*, *Antirrhinum majus*, and *Petunia*. OsMADS1, one of the SEPALLATA-like MADS-box proteins, has been discovered to be essential for determining the floral meristem and specifying the identity of the rice floral organ [143].

Heading date 3a (Hd3a), which is hypothesized to encode a mobile flowering signal and facilitate floral transition under short-day (SD) circumstances, is most closely related to RICE FLOWERING LOCUS T 1 (RFT1/FT-L3). On chromosome 6, RFT1 is only 11.5 kb away from Hd3a. The Hd3a and RFT1 proteins spatiotemporally coordinated transitions among these reproductive meristems, which start with inflorescence meristem (IM) initiation produced by the buildup of rice florigen, thereby governing the morphology of the rice panicle [143]. The interaction between pollen and stigma is a multi-step, intricate physiological process that involves numerous signalling and metabolic pathways. Comparative proteins from the several pollen phases in *O. sativa* such as pollen mother cell, microspore, binucleate, heading stage were examined using proteomic analysis. The elongation and growth of cells, as well as the participation of the 33 distinct proteins with the same varying trend in these processes, were crucial for pollen germination [73]. Similar to this, pollen grains from mature and germinated rice plants were selected before being applied to 2D gels to produce protein spots. A total of 186 proteins out of about 2300 proteins were differentially expressed when proteins from the two distinct growth phases were compared [143, 144]. These proteins play a role in metabolic and regulatory activities, including protein and cytoskeleton dynamics. The up-regulated proteins serve functions in carbohydrate, nucleotide and protein metabolism, signal transduction, and stress responses when comparing the protein expression of the canola (*Brassica napus*) germinating pollen and the mature pollen through DIGE coupled to MALDI-TOF/TOF. In contrast, it was discovered that certain LEA proteins and catalase were down-regulated. These demonstrated that the pollen germination was dependent on proteins involved in the metabolism of macromolecules and enzymes in signalling pathways. A member of the CBL (calcineurin B-like) interacting protein kinase (CIPK) family, OsCIPK23, was reported to play a role in pollination and stress responses in rice. The differentially expressed proteins in pollen tubes compared to non-germinated pollens may show the unique physiological processes that occur during the development of the pollen tubes [145]. According to molecular studies, OsCIPK23 is mostly expressed in the pistil and anther but is also activated by pollination, other abiotic stressors and phytohormones. OsCIPK23 may have a role in both pollination and drought stress as evident from the drastically reduced seed set and hypersensitivity response to drought stress caused by RNA interference-mediated suppression of OsCIPK23 expression. Numerous genes linked to drought tolerance were stimulated to express when OsCIPK23 was overexpressed [146]. When considered as a whole, these findings showed that OsCIPK23 is a multistress-triggered gene and probably regulates a signalling pathway that is frequently shared by rice pollination stress responses and drought stress responses. In developing rice flowers, the *OsPRP3* (*O. sativa Proline-Rich Protein 3*) gene is regulated. A cell wall protein called OsPRP3 is essential for determining the extracellular matrix structure of floral organs. This rice gene contributes to the development of rice flower cell wall [147].

4.9 CONCLUSION

Flowering, which marks the transition from vegetative to reproductive development, is an essential biological process for flowering plants to successfully produce seeds. Plants have developed a number of major genetic systems that are controlled by both internal and external stimuli to ensure timely flowering in a variety of conditions. Presently, it is documented that six main pathways—photoperiod, vernalization, ambient temperature, gibberellin, the autonomous pathway, and age—control flowering. In many crops, the subject of fertility is particularly crucial to both genetics and breeding efforts. The discovery of the genetic regulatory mechanisms governing the flowering process may be aided by the identification and functional characterization of key regulatory TFs implicated in these pathways. Combining data from morphophysiological studies and analysis of transcriptome and proteome profiles, it is possible to identify the precise mechanisms and genes involved in gametogenesis as well as changes in global gene expression. Future research on these genes will contribute to a better understanding of the early response to flowering and how it affects subsequent events in the cascade.

REFERENCES

1. Tan, F.C. and Swain, S.M., 2006. Genetics of flower initiation and development in annual and perennial plants. *Physiologia Plantarum*, *128*(1), pp. 8–17.
2. Davis, F.S. and Albrigo, L.J., 1994. Description of Mayor Rootstocks. *Citrus*. CAB International, UK.
3. Lawson, E.J. and Poethig, R.S., 1995. Shoot development in plants: time for a change. *Trends in Genetics*, *11*(7), pp. 263–268.
4. Li, M., Sha, A., Zhou, X. and Yang, P., 2012. Comparative proteomic analyses reveal the changes of metabolic features in soybean (*Glycine max*) pistils upon pollination. *Sexual Plant Reproduction*, *25*(4), pp. 281–291.
5. Arathi, H.S., Ganeshaiah, K.N., Shaanker, R.U. and Hegde, S.G., 1999. Seed abortion in *Pongamia pinnata* (Fabaceae). *American Journal of Botany*, *86*(5), pp. 659–662.
6. Ohyanagi, H., Sakata, K. and Komatsu, S., 2012. Soybean proteomedatabase 2012: update on the comprehensive data repository for soybean proteomics. *Frontiers in Plant Science*, *3*, p. 110.
7. Goldberg, R.B., 1988. Plants: novel developmental processes. *Science*, *240*(4858), pp. 1460–1467.
8. Hill, J.P. and Lord, E.M., 1989. Floral development in *Arabidopsis thaliana*: a comparison of the wild type and the homeotic pistillata mutant. *Canadian Journal of Botany*, *67*(10), pp. 2922–2936.
9. Suen, D.F., Wu, S.S., Chang, H.C., Dhugga, K.S. and Huang, A.H., 2003. Cell wall reactive proteins in the coat and wall of maize pollen: potential role in pollen tube growth on the stigma and through the style. *Journal of Biological Chemistry*, *278*(44), pp. 43672–43681.
10. Samach, A. and Smith, H.M., 2013. Constraints to obtaining consistent annual yields in perennials. II: environment and fruit load affect induction of flowering. *Plant Science*, *207*, pp. 168–176.
11. van Doorn, W.G. and Woltering, E.J., 2008. Physiology and molecular biology of petal senescence. *Journal of Experimental Botany*, *59*(3), pp. 453–480.
12. Orzaez, D., Blay, R. and Granell, A., 1999. Programme of senescence in petals and carpels of *Pisum sativum* L. flowers and its control by ethylene. *Planta*, *208*(2), pp. 220–226.
13. Bernier, G. and Périlleux, C., 2005. A physiological overview of the genetics of flowering time control. *Plant Biotechnology Journal*, *3*(1), pp. 3–16.
14. Garner, W.W. and Allard, H.A., 1920. Effect of the relative length of day and night and other factors of the environment on growth and reproduction in plants. *Monthly Weather Review*, *48*(7), pp. 415–415.
15. Thomas, B. and Vince-Prue, D., 1997. Photoperiodism in plants.
16. Simpson, G.G. and Dean, C., 2002. Arabidopsis, the Rosetta stone of flowering time? *Science*, *296*(5566), pp. 285–289.
17. Alejandra Mandel, M., Gustafson-Brown, C., Savidge, B. and Yanofsky, M.F., 1992. Molecular characterization of the Arabidopsis floral homeotic gene APETALA1. *Nature*, *360*(6401), pp. 273–277.
18. Bowman, J.L., Alvarez, J., Weigel, D., Meyerowitz, E.M. and Smyth, D.R., 1993. Control of flower development in *Arabidopsis thaliana* by APETALA1 and interacting genes. *Development*, *119*(3), pp. 721–743.
19. Ditta, G., Pinyopich, A., Robles, P., Pelaz, S. and Yanofsky, M.F., 2004. The SEP4 gene of *Arabidopsis thaliana* functions in floral organ and meristem identity. *Current Biology*, *14*(21), pp. 1935–1940.
20. Ferrándiz, C., Gu, Q., Martienssen, R. and Yanofsky, M.F., 2000. Redundant regulation of meristem identity and plant architecture by FRUITFULL, APETALA1 and CAULIFLOWER. *Development*, *127*(4), pp. 725–734.
21. Gu, Q., Ferrándiz, C., Yanofsky, M.F. and Martienssen, R., 1998. The FRUITFULL MADS-box gene mediates cell differentiation during Arabidopsis fruit development. *Development*, *125*(8), pp. 1509–1517.
22. de Folter, S., Immink, R.G., Kieffer, M., Parenicova, L., Henz, S.R., Weigel, D., Busscher, M., Kooiker, M., Colombo, L., Kater, M.M. and Davies, B., 2005. Comprehensive interaction map of the Arabidopsis MADS box transcription factors. *The Plant Cell*, *17*(5), pp. 1424–1433.
23. Sommer, H., Beltrán, J.-P., Huijser, P., Pape, H., Lönnig, W.-E., Saedler, H. and Schwarz-Sommer, Z., 1990. Deficiens, a homeotic gene involved in the control of flower morphogenesis in *Antirrhinum majus*: the protein shows homology to transcription factors. *The EMBO Journal*, *9*, pp. 605–613.
24. Yanofsky, M.F., Ma, H., Bowman, J.L., Drews, G.N., Feldmann, K.A. and Meyerowitz, E.M., 1990. The protein encoded by the Arabidopsis homeotic gene agamous resembles transcription factors. *Nature*, *346*(6279), pp. 35–39.
25. Okada, K. and Shimura, Y., 1994. Genetic analyses of signalling in flower development using Arabidopsis. *Plant Molecular Biology*, *26*(5), pp. 1357–1377.
26. Theissen, G. and Saedler, H., 2001. Floral quartets. *Nature*, *409*(6819), pp. 469–471.

27. Weigel, D. and Meyerowitz, E.M., 1994. The ABCs of floral homeotic genes. *Cell*, *78*(2), pp. 203–209.
28. Passmore, S., Maine, G.T., Elble, R., Christ, C. and Tye, B.K., 1988. Saccharomyces cerevisiae protein involved in plasmid maintenance is necessary for mating of MATα cells. *Journal of Molecular Biology*, *204*(3), pp. 593–606.
29. Egea-Cortines, M., Saedler, H. and Sommer, H., 1999. Ternary complex formation between the MADS-box proteins SQUAMOSA, DEFICIENS and GLOBOSA is involved in the control of floral architecture in *Antirrhinum majus*. *The EMBO Journal*, *18*(19), pp. 5370–5379.
30. Norman, C., Runswick, M., Pollock, R. and Treisman, R., 1988. Isolation and properties of cDNA clones encoding SRF, a transcription factor that binds to the c-fos serum response element. *Cell*, *55*(6), pp. 989–1003.
31. Goto, K., Kyozuka, J. and Bowman, J.L., 2001. Turning floral organs into leaves, leaves into floral organs. *Current Opinion in Genetics & Development*, *11*(4), pp. 449–456.
32. Mandel, M.A. and Yanofsky, M.F., 1995. A gene triggering flower formation in Arabidopsis. *Nature*, *377*(6549), pp. 522–524.
33. Ferrario, S., Immink, R.G. and Angenent, G.C., 2004. Conservation and diversity in flower land. *Current Opinion in Plant Biology*, *7*(1), pp. 84–91.
34. Canovas, F.M., Dumas-Gaudot, E., Recorbet, G., Jorrin, J., Mock, H.P. and Rossignol, M., 2004. Plant proteome analysis. *Proteomics*, *4*(2), pp. 285–298.
35. Sheoran, I.S. and Sawhney, V.K., 2010. Proteome analysis of the normal and Ogura (ogu) CMS anthers of *Brassica napus* to identify proteins associated with male sterility. *Botany*, *88*(3), pp. 217–230.
36. Liang, Y.K., Dubos, C., Dodd, I.C., Holroyd, G.H., Hetherington, A.M. and Campbell, M.M., 2005. AtMYB61, an R2R3-MYB transcription factor controlling stomatal aperture in *Arabidopsis thaliana*. *Current Biology*, *15*(13), pp. 1201–1206.
37. Chen, X., 2004. A microRNA as a translational repressor of APETALA2 in Arabidopsis flower development. *Science*, *303*(5666), pp. 2022–2025.
38. Elliott, R.C., Betzner, A.S., Huttner, E., Oakes, M.P., Tucker, W.Q., Gerentes, D., Perez, P. and Smyth, D.R., 1996. AINTEGUMENTA, an APETALA2-like gene of Arabidopsis with pleiotropic roles in ovule development and floral organ growth. *The Plant Cell*, *8*(2), pp. 155–168.
39. Jofuku, K.D., Den Boer, B.G., Van Montagu, M. and Okamuro, J.K., 1994. Control of Arabidopsis flower and seed development by the homeotic gene APETALA2. *The Plant Cell*, *6*(9), pp. 1211–1225.
40. Klucher, K.M., Chow, H., Reiser, L. and Fischer, R.L., 1996. The AINTEGUMENTA gene of Arabidopsis required for ovule and female gametophyte development is related to the floral homeotic gene APETALA2. *The Plant Cell*, *8*(2), pp. 137–153.
41. Goto, K. and Meyerowitz, E.M., 1994. Function and regulation of the Arabidopsis floral homeotic gene PISTILLATA. *Genes & Development*, *8*(13), pp. 1548–1560.
42. Jack, T., Brockman, L.L. and Meyerowitz, E.M., 1992. The homeotic gene APETALA3 of *Arabidopsis thaliana* encodes a MADS box and is expressed in petals and stamens. *Cell*, *68*(4), pp. 683–697.
43. McGonigle, B., Bouhidel, K. and Irish, V.F., 1996. Nuclear localization of the Arabidopsis APETALA3 and PISTILLATA homeotic gene products depends on their simultaneous expression. *Genes & Development*, *10*(14), pp. 1812–1821.
44. Bemer, M., Wolters-Arts, M., Grossniklaus, U. and Angenent, G.C., 2008. The MADS domain protein DIANA acts together with AGAMOUS-LIKE80 to specify the central cell in Arabidopsis ovules. *The Plant Cell*, *20*(8), pp. 2088–2101.
45. Pelaz, S., Ditta, G.S., Baumann, E., Wisman, E. and Yanofsky, M.F., 2000. B and C floral organ identity functions require SEPALLATA MADS-box genes. *Nature*, *405*(6783), pp. 200–203.
46. Sablowski, R.W. and Meyerowitz, E.M., 1998. A homolog of NO APICAL MERISTEM is an immediate target of the floral homeotic genes APETALA3/PISTILLATA. *Cell*, *92*(1), pp. 93–103.
47. Zik, M. and Irish, V.F., 2003. Global identification of target genes regulated by APETALA3 and PISTILLATA floral homeotic gene action. *The Plant Cell*, *15*(1), pp. 207–222.
48. Bowman, J.L., Drews, G.N. and Meyerowitz, E.M., 1991. Expression of the Arabidopsis floral homeotic gene AGAMOUS is restricted to specific cell types late in flower development. *The Plant Cell*, *3*(8), pp. 749–758.
49. Yang, X., Pang, H.B., Liu, B.L., Qiu, Z.J., Gao, Q., Wei, L., Dong, Y. and Wang, Y.Z., 2012. Evolution of double positive autoregulatory feedback loops in CYCLOIDEA2 clade genes is associated with the origin of floral zygomorphy. *The Plant Cell*, *24*(5), pp. 1834–1847.
50. Mizukami, Y. and Ma, H., 1995. Separation of AG function in floral meristem determinacy from that in reproductive organ identity by expressing antisense AG RNA. *Plant Molecular Biology*, *28*(5), pp. 767–784.

51. Gao, X., Liang, W., Yin, C., Ji, S., Wang, H., Su, X., Guo, C., Kong, H., Xue, H. and Zhang, D., 2010. The SEPALLATA-like gene OsMADS34 is required for rice inflorescence and spikelet development. *Plant Physiology*, *153*, pp. 728–740.

52. Magnani, E., Sjölander, K. and Hake, S., 2004. From endonucleases to transcription factors: evolution of the AP2 DNA binding domain in plants. *The Plant Cell*, *16*(9), pp. 2265–2277.

53. Li, H., Liang, W., Hu, Y., Zhu, L., Yin, C., Xu, J., Dreni, L., Kater, M.M. and Zhang,D., 2011. Rice *MADS6* interacts with the floralhomeoticgenes *SUPERWOMAN1*, *MADS3*, *MADS58*, *MADS13*, and *DROOPING LEAF* in specifying floral organ identities and meristemfate. *The Plant Cell*, *23*(7), pp. 2536–2552.

54. Futamura, N., Mori, H., Kouchi, H. and Shinohara, K., 2000. Male flower-specific expression of genes for polygalacturonase, pectin methylesterase and β-1,3-glucanase in a dioecious willow (*Salix gilgianaseemen*). *Plant and Cell Physiology*, *41*(1), pp. 16–26.

55. Pinyopich, A., Ditta, G.S., Savidge, B., Liljegren, S.J., Baumann, E., Wisman, E. and Yanofsky, M.F., 2003. Assessing the redundancy of MADS-box genes during carpel and ovule development. *Nature*, *424*(6944), pp. 85–88.

56. Immink, R.G., Gadella, T.W. Jr, Ferrario, S., Busscher, M. and Angenent, G.C., 2002. Analysis of MADS box protein–protein interactions in living plant cells. *Proceedings of the National Academy of Sciences*, *99*(4), pp. 2416–2421.

57. Corbesier, L., Vincent, C., Jang, S., Fornara, F., Fan, Q., Searle, I., Giakountis, A., Farrona, S., Gissot, L., Turnbull, C. and Coupland, G., 2007. FT protein movement contributes to long-distance signaling in floral induction of Arabidopsis. *Science*, *316*(5827), pp. 1030–1033.

58. Samach, A., Onouchi, H., Gold, S.E., Ditta, G.S., Schwarz-Sommer, Z., Yanofsky, M.F. and Coupland, G., 2000. Distinct roles of CONSTANS target genes in reproductive development of Arabidopsis. *Science*, *288*(5471), pp. 1613–1616.

59. Banfield, M.J. and Brady, R.L., 2000. The structure of Antirrhinum centroradialis protein (CEN) suggests a role as a kinase regulator. *Journal of Molecular Biology*, *297*(5), pp. 1159–1170.

60. Kardailsky, I., Shukla, V.K., Ahn, J.H., Dagenais, N., Christensen, S.K., Nguyen, J.T., Chory, J., Harrison, M.J. and Weigel, D., 1999. Activation tagging of the floral inducer FT. *Science*, *286*(5446), pp. 1962–1965.

61. Kobayashi, Y., Kaya, H., Goto, K., Iwabuchi, M. and Araki, T., 1999. A pair of related genes with antagonistic roles in mediating flowering signals. *Science*, *286*(5446), pp. 1960–1962.

62. Lemon, W.J., Liyanarachchi, S. and You, M., 2003. A high performance test of differential gene expression for oligonucleotide arrays. *Genome Biology*, *4*(10), pp. 1–11.

63. Abe, M., Kobayashi, Y., Yamamoto, S., Daimon, Y., Yamaguchi, A., Ikeda, Y., Ichinoki, H., Notaguchi, M., Goto, K. and Araki, T., 2005. FD, a bZIP protein mediating signals from the floral pathway integrator FT at the shoot apex. *Science*, *309*(5737), pp. 1052–1056.

64. O'Maoileidigh, D.S., Graciet, E. and Wellmer, F., 2014. Genetic control of Arabidopsis flower development. In *Advances in botanical research* (Vol. 72, pp. 159–190). Academic Press.

65. Mandel, M.A. and Yanofsky, M.F., 1998. The Arabidopsis AGL9 MADS box gene is expressed in young flower primordia. *Sexual Plant Reproduction*, *11*(1), pp. 22–28.

66. Yu, H., Ito, T., Zhao, Y., Peng, J., Kumar, P. and Meyerowitz, E.M., 2004. Floral homeotic genes are targets of gibberellin signaling in flower development. *Proceedings of the National Academy of Sciences*, *101*(20), pp. 7827–7832.

67. de Folter, S., Urbanus, S.L., van Zuijlen, L.G., Kaufmann, K. and Angenent, G.C., 2007. Tagging of MADS domain proteins for chromatin immunoprecipitation. *BMC Plant Biology*, *7*(1), pp. 1–11.

68. Bowman, J.L., Smyth, D.R. and Meyerowitz, E.M., 1991. Genetic interactions among floral homeotic genes of Arabidopsis. *Development*, *112*(1), pp. 1–20.

69. Lenhard, M., Bohnert, A., Jürgens, G. and Laux, T., 2001. Termination of stem cell maintenance in Arabidopsis floral meristems by interactions between WUSCHEL and AGAMOUS. *Cell*, *105*(6), pp. 805–814.

70. Colombo, L., Franken, J., Koetje, E., van Went, J., Dons, H.J., Angenent, G.C. and van Tunen, A.J., 1995. The petunia MADS box gene FBP11 determines ovule identity. *The Plant Cell*, *7*(11), pp. 1859–1868.

71. Busi, M.V., Bustamante, C., D'angelo, C., Hidalgo-Cuevas, M., Boggio, S.B., Valle, E.M. and Zabaleta, E., 2003. MADS-box genes expressed during tomato seed and fruit development. *Plant Molecular Biology*, *52*(4), pp. 801–815.

72. Lou, Y., Xu, X.F., Zhu, J., Gu, J.N., Blackmore, S. and Yang, Z.N., 2014. The tapetal AHL family protein TEK determines nexine formation in the pollen wall. *Nature Communications*, *5*(1), pp. 1–9.

73. Xu, J., Yang, C., Yuan, Z., Zhang, D., Gondwe, M.Y., Ding, Z., Liang, W., Zhang, D. and Wilson, Z.A., 2010. The ABORTED MICROSPORES regulatory network is required for postmeiotic male reproductive development in *Arabidopsis thaliana. The Plant Cell*, 22(1), pp. 91–107.

74. Dobritsa, A.A., Lei, Z., Nishikawa, S.I., Urbanczyk-Wochniak, E., Huhman, D.V., Preuss, D. and Sumner, L.W., 2010. LAP5 and LAP6 encode anther-specific proteins with similarity to chalcone synthase essential for pollen exine development in Arabidopsis. *Plant Physiology*, 153(3), pp. 937–955.

75. Dobritsa, A.A., Nishikawa, S.I., Preuss, D., Urbanczyk-Wochniak, E., Sumner, L.W., Hammond, A., Carlson, A.L. and Swanson, R.J., 2009. LAP3, a novel plant protein required for pollen development, is essential for proper exine formation. *Sexual Plant Reproduction*, 22(3), pp. 167–177.

76. Rocheta, M., Sobral, R., Magalhães, J., Amorim, M.I., Ribeiro, T., Pinheiro, M., Egas, C., Morais-Cecílio, L. and Costa, M.M., 2014. Comparative transcriptomic analysis of male and female flowers of monoecious *Quercus suber. Frontiers in Plant Science*, 5, p. 599.

77. Murase, K., Shiba, H., Iwano, M., Che, F.S., Watanabe, M., Isogai, A. and Takayama, S., 2004. A membrane-anchored protein kinase involved in Brassica self-incompatibility signaling. *Science*, 303(5663), pp. 1516–1519.

78. Meister, R.J., Kotow, L.M. and Gasser, C.S., 2002. SUPERMAN attenuates positive INNER NO OUTER autoregulation to maintain polar development of Arabidopsis ovule outer integuments.

79. Siegfried, K.R., Eshed, Y., Baum, S.F., Otsuga, D., Drews, G.N. and Bowman, J.L., 1999. Members of the YABBY gene family specify abaxial cell fate in Arabidopsis. *Development*, 126(18), pp. 4117–4128.

80. Villanueva, J.M., Broadhvest, J., Hauser, B.A., Meister, R.J., Schneitz, K. and Gasser, C.S., 1999. INNER NO OUTER regulates abaxial–adaxial patterning in Arabidopsis ovules. *Genes & Development*, 13(23), pp. 3160–3169.

81. Long, J.A. and Barton, M.K., 1998. The development of apical embryonic pattern in Arabidopsis. *Development*, 125(16), pp. 3027–3035.

82. Busch, A., Horn, S., Mühlhausen, A., Mummenhoff, K. and Zachgo, S., 2011. Corolla monosymmetry: evolution of a morphological novelty in the Brassicaceae family. *Molecular Biology and Evolution*, 29(4), pp. 1241–1254.

83. Zhu, J.Y., Sae-Seaw, J. and Wang, Z.Y., 2013. Brassinosteroid signalling. *Development*, 140(8), pp. 1615–1620.

84. Murase, K., Shigenobu, S., Fujii, S., Ueda, K., Murata, T., Sakamoto, A., Wada, Y., Yamaguchi, K., Osakabe, Y., Osakabe, K. and Kanno, A., 2017. MYB transcription factor gene involved in sex determination in *Asparagus officinalis. Genes to Cells*, 22(1), pp. 115–123.

85. Romano, J.M., Dubos, C., Prouse, M.B., Wilkins, O., Hong, H., Poole, M., Kang, K.Y., Li, E., Douglas, C.J., Western, T.L. and Mansfield, S.D., 2012. AtMYB61, an R2R3-MYB transcription factor, functions as a pleiotropic regulator via a small gene network. *New Phytologist*, 195(4), pp. 774–786.

86. Jack, T., 2004. Molecular and genetic mechanisms of floral control. *The Plant Cell*, 16(suppl_1), pp. S1–S17.

87. Dudareva, N., D'auria, J.C., Nam, K.H., Raguso, R.A. and Pichersky, E., 1998. Acetyl-CoA: Benzylalcohol acetyltransferase—An enzyme involved in floral scent production in *Clarkia breweri. The Plant Journal*, 14(3), pp. 297–304.

88. Dudareva, N., Pichersky, E. and Gershenzon, J., 2004. Biochemistry of plant volatiles. *Plant Physiology*, 135(4), pp. 1893–1902.

89. Martin, C. and Gerats, T., 1993. Control of pigment biosynthesis genes during petal development. *The Plant Cell*, 5(10), p. 1253.

90. Carrari, F. and Fernie, A.R., 2006. Metabolic regulation underlying tomato fruit development. *Journal of Experimental Botany*, 57(9), pp. 1883–1897.

91. Kotilainen, M., Helariutta, Y., Mehto, M., Pöllänen, E., Albert, V.A., Elomaa, P. and Teeri, T.H., 1999. GEG participates in the regulation of cell and organ shape during corolla and carpel development in *Gerbera hybrida. The Plant Cell*, 11(6), pp. 1093–1104.

92. Ben-Meir, H., Zuker, A., Weiss, D. and Vainstein, A., 2002. Molecular control of floral pigmentation: anthocyanins. In *Breeding for ornamentals: Classical and molecular approaches* (pp. 253–272). Springer, Dordrecht.

93. Dafny-Yelin, M., Guterman, I., Menda, N., Ovadis, M., Shalit, M., Pichersky, E., Zamir, D., Lewinsohn, E., Adam, Z., Weiss, D. and Vainstein, A., 2005. Flower proteome: changes in protein spectrum during the advanced stages of rose petal development. *Planta*, 222(1), pp. 37–46.

94. Langenkämper, G., Manac'h, N., Broin, M., Cuiné, S., Becuwe, N., Kuntz, M. and Rey, P., 2001. Accumulation of plastid lipid-associated proteins (fibrillin/CDSP34) upon oxidative stress, ageing and biotic stress in Solanaceae and in response to drought in other species. *Journal of Experimental Botany*, *52*(360), pp. 1545–1554.

95. Bey, M., Stüber, K., Fellenberg, K., Schwarz-Sommer, Z., Sommer, H., Saedler, H. and Zachgo, S., 2004. Characterization of Antirrhinum petal development and identification of target genes of the class B MADS box gene DEFICIENS. *The Plant Cell*, *16*(12), pp. 3197–3215.

96. Schwarz-Sommer, Z., Hue, I., Huijser, P., Flor, P., Hansen, H., Tetens, F., Lo nnig, W.E., Saedler, H. and Sommer, H., 1992. Characterization of the Antirrhinum floral homeotic MADS-box gene deficiens: evidence for DNA-binding and autoregulation of its persistent expression throughout flower development. *The EMBO Journal*, *11*, pp. 251–263.

97. Meyer, A., Miersch, O., Büttner, C., Dathe, W. and Sembdner, G., 1984. Occurrence of the plant growth regulator jasmonic acid in plants. *Journal of Plant Growth Regulation*, *3*(1), pp. 1–8.

98. Francis, M.J.O. and Allcock, C., 1969. Geraniol β-D-glucoside; occurrence and synthesis in rose flowers. *Phytochemistry*, *8*(8), pp. 1339–1347.

99. Suzuki, K.I., Tsuda, S., Fukui, Y., Fukuchi-Mizutani, M., Yonekura-Sakakibara, K., Tanaka, Y. and Kusumi, T., 2000. Molecular characterization of rose flavonoid biosynthesis genes and their application in petunia. *Biotechnology & Biotechnological Equipment*, *14*(2), pp. 56–62.

100. Ma, N., Tan, H., Liu, X., Xue, J., Li, Y. and Gao, J., 2006. Transcriptional regulation of ethylene receptor and CTR genes involved in ethylene-induced flower opening in cut rose (*Rosa hybrida*) cv. Samantha. *Journal of Experimental Botany*, *57*(11), pp. 2763–2773.

101. Ma, N., Cai, L., Lu, W., Tan, H. and Gao, J., 2005. Exogenous ethylene influences flower opening of cut roses (*Rosa hybrida*) by regulating the genes encoding ethylene biosynthesis enzymes. *Science in China Series C: Life Sciences*, *48*(5), pp. 434–444.

102. Cheng, C., Yu, Q., Wang, Y., Wang, H., Dong, Y., Ji, Y., Zhou, X., Li, Y., Jiang, C.Z., Gan, S.S. and Zhao, L., 2021. Ethylene-regulated asymmetric growth of the petal base promotes flower opening in rose (*Rosa hybrida*). *The Plant Cell*, *33*(4), pp. 1229–1251.

103. Ma, N., Xue, J., Li, Y., Liu, X., Dai, F., Jia, W., Luo, Y. and Gao, J., 2008. Rh-PIP2; 1, a rose aquaporin gene, is involved in ethylene-regulated petal expansion. *Plant Physiology*, *148*(2), pp. 894–907.

104. Pei, H., Ma, N., Tian, J., Luo, J., Chen, J., Li, J., Zheng, Y., Chen, X., Fei, Z. and Gao, J., 2013. An NAC transcription factor controls ethylene-regulated cell expansion in flower petals. *Plant Physiology*, *163*(2), pp. 775–791.

105. Reid, M.S., Evans, R.Y., Dodge, L.L. and Mor, Y., 1989. Ethylene and silver thiosulfate influence opening of cut rose flowers. *Journal of the American Society for Horticultural Science*, *114*(3), pp. 436–440.

106. Xue, J., Li, Y., Tan, H., Yang, F., Ma, N. and Gao, J., 2008. Expression of ethylene biosynthetic and receptor genes in rose floral tissues during ethylene-enhanced flower opening. *Journal of Experimental Botany*, *59*(8), pp. 2161–2169.

107. Brice, D.C., Bryant, J.A., Dambrauskas, G., Drury, S.C. and Littlechild, J.A., 2004. Cloning and expression of cytosolic phosphoglycerate kinase from pea (*Pisum sativum* L.). *Journal of Experimental Botany*, *55*(398), pp. 955–956.

108. Fait, A., Angelovici, R., Less, H., Ohad, I., Urbanczyk-Wochniak, E., Fernie, A.R. and Galili, G., 2006. Arabidopsis seed development and germination is associated with temporally distinct metabolic switches. *Plant Physiology*, *142*(3), pp. 839–854.

109. Lebon, G., Wojnarowiez, G., Holzapfel, B., Fontaine, F., Vaillant-Gaveau, N. and Clément, C., 2008. Sugars and flowering in the grapevine (*Vitis vinifera* L.). *Journal of Experimental Botany*, *59*(10), pp. 2565–2578.

110. Zeng, G.J., Li, C.M., Zhang, X.Z., Han, Z.H., Yang, F.Q., Gao, Y., Chen, D.M., Zhao, Y.B., Wang, Y., Teng, Y.L. and Dong, W.X., 2010. Differential proteomic analysis during the vegetative phase change and the floral transition in *Malus domestica*. *Development, Growth & Differentiation*, *52*(7), pp. 635–644.

111. Zheng, B.B., Fang, Y.N., Pan, Z.Y., Sun, L., Deng, X.X., Grosser, J.W. and Guo, W.W., 2014. iTRAQ-based quantitative proteomics analysis revealed alterations of carbohydrate metabolism pathways and mitochondrial proteins in a male sterile cybrid pummelo. *Journal of Proteome Research*, *13*(6), pp. 2998–3015.

112. Fernando, D.D., 2005. Characterization of pollen tube development in *Pinus strobus* (Eastern white pine) through proteomic analysis of differentially expressed proteins. *Proteomics*, *5*(18), pp. 4917–4926.

113. Aloisi, I., Cai, G., Serafini-Fracassini, D. and Del Duca, S., 2016. Polyamines in pollen: from microsporogenesis to fertilization. *Frontiers in Plant Science*, *7*, p. 155.

114. Domingos, S., Fino, J., Paulo, O.S., Oliveira, C.M. and Goulao, L.F., 2016. Molecular candidates for early-stage flower-to-fruit transition in stenospermocarpic table grape (*Vitis vinifera* L.) inflorescences ascribed by differential transcriptome and metabolome profiles. *Plant Science, 244*, pp. 40–56.

115. Bai, S., Willard, B., Chapin, L.J., Kinter, M.T., Francis, D.M., Stead, A.D. and Jones, M.L., 2010. Proteomic analysis of pollination-induced corolla senescence in petunia. *Journal of Experimental Botany, 61*(4), pp. 1089–1109.

116. Nuckolls, G.H., Osherov, N., Loomis, W.F. and Spudich, J.A., 1996. The *Dictyostelium* dual-specificity kinase splA is essential for spore differentiation. *Development, 122*(10), pp. 3295–3305.

117. Atkins, C.A., Patterson, B.D. and Graham, D., 1972. Plant carbonic anhydrases: I. Distribution of types among species. *Plant Physiology, 50*(2), pp. 214–217.

118. Chen, L., Chen, Q., Zhu, Y., Hou, L. and Mao, P., 2016. Proteomic identification of differentially expressed proteins during alfalfa (*Medicago sativa* L.) flower development. *Frontiers in Plant Science, 7*, p. 1502.

119. Fiebig, A., Mayfield, J.A., Miley, N.L., Chau, S., Fischer, R.L. and Preuss, D., 2000. Alterations in CER6, a gene identical to CUT1, differentially affect long-chain lipid content on the surface of pollen and stems. *The Plant Cell, 12*(10), pp. 2001–2008.

120. Mayfield, J.A., Fiebig, A., Johnstone, S.E. and Preuss, D., 2001. Gene families from the *Arabidopsis thaliana* pollen coat proteome. *Science, 292*(5526), pp. 2482–2485.

121. Swanson, R., Edlund, A.F. and Preuss, D., 2004. Species specificity in pollen-pistil interactions. *Annual Review of Genetics, 38*, pp. 793–818.

122. Zinkl, G.M. and Preuss, D., 2000. Dissecting Arabidopsis pollen–stigma interactions reveals novel mechanisms that confer mating specificity. *Annals of Botany, 85*, pp.15–21.

123. Zinkl, G.M., Zwiebel, B.I., Grier, D.G. and Preuss, D., 1999. Pollen-stigma adhesion in Arabidopsis: A species-specific interaction mediated by lipophilic molecules in the pollen exine. *Development, 126*(23), pp. 5431–5440.

124. Roberts, I.N., Harrod, G. and Dickinson, H.G., 1984. Pollen-stigma interactions in Brassica oleracea. I. Ultrastructure and physiology of the stigmatic papillar cells. *Journal of Cell Science, 66*(1), pp. 241–253.

125. Schopfer, C.R., Nasrallah, M.E. and Nasrallah, J.B., 1999. The male determinant of self-incompatibility in Brassica. *Science, 286*(5445), pp. 1697–1700.

126. Silva, N.F., Stone, S.L., Christie, L.N., Sulaman, W., Nazarian, K.A.P., Burnett, L.A., Arnoldo, M.A., Rothstein, S.J. and Goring, D.R., 2001. Expression of the S receptor kinase in self-compatible *Brassica napus* cv. Westar leads to the allele-specific rejection of self-incompatible *Brassica napus* pollen. *Molecular Genetics and Genomics, 265*(3), pp. 552–559.

127. Takayama, S., Shiba, H., Iwano, M., Shimosato, H., Che, F.S., Kai, N., Watanabe, M., Suzuki, G., Hinata, K. and Isogai, A., 2000. The pollen determinant of self-incompatibility in Brassica campestris. *Proceedings of the National Academy of Sciences, 97*(4), pp. 1920–1925.

128. Goring, D.R. and Walker, J.C., 2004. Self-rejection—a new kinase connection. *Science, 303*(5663), pp. 1474–1475.

129. Kakita, M., Murase, K., Iwano, M., Matsumoto, T., Watanabe, M., Shiba, H., Isogai, A. and Takayama, S., 2007. Two distinct forms of M-locus protein kinase localize to the plasma membrane and interact directly with S-locus receptor kinase to transduce self-incompatibility signaling in *Brassica rapa*. *The Plant Cell, 19*(12), pp. 3961–3973.

130. Samuel, M.A., Chong, Y.T., Haasen, K.E., Aldea-Brydges, M.G., Stone, S.L. and Goring, D.R., 2009. Cellular pathways regulating responses to compatible and self-incompatible pollen in Brassica and Arabidopsis stigmas intersect at Exo70A1, a putative component of the exocyst complex. *The Plant Cell, 21*(9), pp. 2655–2671.

131. Schmutz, J., Cannon, S.B., Schlueter, J., Ma, J., Mitros, T., Nelson, W., Hyten, D.L., Song, Q., Thelen, J.J., Cheng, J. and Xu, D., 2010. Genome sequence of the palaeopolyploid soybean. *Nature, 463*(7278), pp. 178–183.

132. Ahsan, N. and Komatsu, S., 2009. Comparative analyses of the proteomes of leaves and flowers at various stages of development reveal organ-specific functional differentiation of proteins in soybean. *Proteomics, 9*(21), pp. 4889–4907.

133. Zhang, D., Ren, L., Yue, J.H., Wang, L., Zhuo, L.H. and Shen, X.H., 2013. A comprehensive analysis of flowering transition in *Agapanthus praecox* ssp. orientalis (Leighton) Leighton by using transcriptomic and proteomic techniques. *Journal of Proteomics, 80*, pp. 1–25.

134. Azevedo, R.A., Lancien, M. and Lea, P.J., 2006. The aspartic acid metabolic pathway, an exciting and essential pathway in plants. *Amino Acids, 30*(2), pp. 143–162.

135. Badger, M.R. and Price, G.D., 1994. The role of carbonic anhydrase in photosynthesis. *Annual Review of Plant Physiology and Plant Molecular Biology*, *45*(1), pp. 369–392.
136. Tung, C.W., Dwyer, K.G., Nasrallah, M.E. and Nasrallah, J.B., 2005. Genome-wide identification of genes expressed in Arabidopsis pistils specifically along the path of pollen tube growth. *Plant Physiology*, *138*(2), pp. 977–989.
137. Xu, J. and Zhang, S., 2015. Mitogen-activated protein kinase cascades in signaling plant growth and development. *Trends in Plant Science*, *20*(1), pp. 56–64.
138. Guan, Y., Lu, J., Xu, J., McClure, B. and Zhang, S., 2014. Two mitogen-activated protein kinases, MPK3 and MPK6, are required for funicular guidance of pollen tubes in Arabidopsis. *Plant Physiology*, *165*(2), pp. 528–533.
139. Verhoeven, T., Feron, R., Wolters-Arts, M., Edqvist, J., Gerats, T., Derksen, J. and Mariani, C., 2005. STIG1 controls exudate secretion in the pistil of petunia and tobacco. *Plant Physiology*, *138*(1), pp. 153–160.
140. Channelière, S., Rivière, S., Scalliet, G., Szecsi, J., Jullien, F., Dolle, C., Vergne, P., Dumas, C., Bendahmane, M., Hugueney, P. and Cock, J.M., 2002. Analysis of gene expression in rose petals using expressed sequence tags. *FEBS Letters*, *515*(1–3), pp. 35–38.
141. Krizek, B.A. and Fletcher, J.C., 2005. Molecular mechanisms of flower development: an armchair guide. *Nature Reviews Genetics*, *6*(9), pp. 688–698.
142. Tripathi, S.K. and Tuteja, N., 2007.Integrated signaling in flower senescence: An overview. *Plant Signaling & Behavior*, *2*(6), pp. 437–45. doi: 10.4161/psb.2.6.4991. PMID: 19517004; PMCID: PMC2634333.
143. Hu, Y., Wang, L., Jia, R., Liang, W., Zhang, X., Xu, J., Chen, X., Lu, D., Chen, M., Luo, Z., Xie, J., Cao, L., Xu, B., Yu, Y., Persson, S., Zhang, D. and Yuan, Z., 2021. Rice transcription factor MADS32 regulates floral patterning through interactions with multiple floral homeotic genes. *Journal of Experimental Botany*, *72*(7), pp. 2434–2449.
144. Hu, Y., Liang, W., Yin, C., Yang, X., Ping, B., Li, A., Jia, R., Chen, M., Luo, Z., Cai, Q., Zhao, X., (2015). Interactions of OsMADS1 with Floral Homeotic Genes in Rice Flower Development. *Molecular Plant*, *8*(9), pp. 1366–1384.
145. Ouyang, S.Q., Liu, Y.F., Liu, P., Lei, G., He, S.J., Ma, B., Zhang, W.K., Zhang, J.S. and Chen, S.Y., 2010. Receptor-like kinase OsSIK1 improves drought and salt stress tolerance in rice (*Oryza sativa*) plants. *The Plant Journal*, *62*(2), pp. 316–29.
146. Arsovski, A.A., Villota, M.M., Rowland, O., Subramaniam, R. and Western, T.L., 2009. MUM ENHANCERS are important for seed coat mucilage production and mucilage secretory cell differentiation in *Arabidopsis thaliana*. *Journal of Experimental Botany*, *60*(9), pp. 2601–2612.
147. Shin, J.S., Gothandam, K.M., Nalini, E. and Karthikeyan, S., 2010. OsPRP3, a flower specific proline-rich protein of rice, determines extracellular matrix structure of floral organs and its overexpression confers cold-tolerance. *Plant Molecular Biology*, *72*(1–2), pp. 125–35.

5 Proteomic Responses to Salinity and Drought Stress in Plants

Pradeep Krishnan S., Suriyaprakash Rajendran, Balaji Balamurugan, Dharani Elangovan, Sivapragasam Ezhumalai, Harika Amooru, Shricharan Senthilkumar, Manduparambil Subramanian Nimmy, Shailendra K. Jha, and Lekshmy Sathee

5.1 INTRODUCTION

Plant science research has been influenced by breakthroughs in high-throughput "Omics" solutions. Proteomics is one of the greatest alternatives for analysing the genome functionality and producing comprehensive conclusions. When supplemented with findings by other conventional and "Omics" approaches, this would provide a better comprehension of the plant developmental processes. Plants are sessile in nature and are consistently exposed to environmental hazards, which significantly reduce economic production (Komatsu & Jorrin-Novo, 2021). Abiotic stresses like salinity, water logging or deficit, high or low temperature and heavy metal toxicity pose a serious risk to agriculture and the ecosystem which are responsible for a significant loss in agricultural output and food security. The crucial technology of mass spectrometry (MS) is utilized to profile proteins in the cell. However, because of its intricacy and dynamic nature, biomarker identification continues to be the key difficulty in proteomics. The insights on plant systems will thus be understood by fusing the proteomics technique with genomics and bioinformatics (Al-Amrani et al., 2021).

Proteomics types include expression proteomics, which is the quantitative analysis of protein expression in samples that differ in some way. This allows for the comparison of protein expression across samples, whether it is throughout the full proteome or across sub-proteomes (Graves & Haystead, 2002). Utilizing 2-DE and MS methods, it is possible to identify protein expression and changes in stressed and non-stressed plants (Ahmad et al., 2016). The structure and nature of protein complexes present particularly in a given cellular organelle are mapped out by structural proteomics. The objective is to identify every protein in a complex and describe protein-protein interaction (Jung et al., 2000). For example, candidate plasmodesmata proteins for the legume *Medicago truncatula* were studied using plasmodesmata in silico proteome 1 (PIP1) and transcriptome meta-analyses. This resulted in the identification of Medtr1g073320, a new receptor-like protein that localizes to plasmodesmata (Kirk et al., 2022). Molecular definitions of cellular systems and the clarification of the biological function of unidentified proteins are the targets of the newly emerging field of functional proteomics (Monti et al., 2005). Proteome analysis techniques include advanced protein microarrays, generally called protein chips which are parallel assay devices that are minute and include high-density arrays of tiny quantities of pure proteins. They make it possible to determine several analytes simultaneously from tiny samples in a single experiment (Chen & Zhu, 2006). Proteomic analysis is accomplished by separating and identifying proteins using 2-DE or coupled gel-free shotgun liquid chromatography-tandem mass spectrometry (LC-MS/MS) platforms followed by an investigation of protein mapping, post-translational modification (PTM) characterization and protein-protein interactions to clarify protein activities and protein functional networks in plant metabolic and signalling processes and finally by using

DOI: 10.1201/b23255-5

of databases and bioinformatic techniques for model and non-model plant species (Holman et al., 2013; Hu et al., 2015).

The most popular technique for separating and calculating the global protein content is gel-based proteomics. In comparison to gel-free proteomics, this is a more established method for screening protein expression on a wide scale (Chevalier, 2010). Mass spectrometric or tandem MS (MS/MS) analysis in conjunction with high-resolution (isoelectric focusing/SDS-PAGE) two-dimensional gel electrophoresis (2DE) protein separation is the standard procedure for quantitative proteome analysis (Lopez, 2007). Protein solubilization is the initial step in gel-based proteome analysis. IEF is an electrophoretic separation, wherein a strip made by the polymerization of acrylamide monomers, connected by bis-acrylamide with molecules of covalently attached immobilien, results in the formation of a dry gel used to produce the first dimension of the 2DE (Carrette et al., 2006). The SDS-PAGE forms the second dimension. The molecular weight of polypeptides governs migration in SDS polyacrylamide gel electrophoresis. In contrast to a molecular weight marker, the SDS-denatured and reduced proteins are divided based on an apparent molecular weight. After running the gel, the second dimension is complete, and the acrylamide gel may be removed from the glass plates after the bromophenol blue dye migration has reached the bottom of the gel (Kendrick et al., 2019). Depending on the staining process chosen, the gel must first be submerged in a fixation solution that contains an acid (phosphoric acid or acetic acid) and an alcohol (ethanol or methanol). Coomassie Blue, colloidal Coomassie Blue and silver nitrate are commonly used visible dyes with distinct sensitivities (50, 10 and 0.5 ng of detectable protein/spot, respectively) (Lanne & Panfilov, 2005; Smejkal, 2004). Depending on the staining methodology used, stained gels are scanned on either a visible or a fluorescent scanner. The image can then be loaded into suitable image analysis applications for comparison and analysis. Gels may be utilized for identification and other purposes by MS, and software such as Image Master, Progenesis, PDQuest and SameSpots can be used to identify spots and compare the spot intensity between samples (Kang et al., 2009; Rosengren et al., 2003). In recent years, the descriptive and comparative proteomic studies of plant development and metabolic strategies have made extensive use of gel-free protein separation methods and second-generation proteomic techniques like multidimensional protein identification technology (MudPIT), isotope-coded affinity tags (ICATs), targeted mass tags (TMTs) and isobaric tags for relative and absolute quantitation (iTRAQ) (Hu et al., 2015).

The MudPIT combines the separation of peptides using two-dimensional liquid chromatography (2D-LC) on a microcapillary column with detection using a tandem mass spectrometer. An initial reduction, alkylation and digestion of a protein or combination of proteins result in a complicated peptide mixture. It is put onto a microcapillary column together with the digested sample. The column is loaded and placed on a platform connected to the mass spectrometer, and the column is interfaced with an HPLC pump and then positioned such that the intake is in line. Peptides with a comparable isoelectric point are successively shifted from the Strong Cation exchange resin to the C18, where they are separated based on their size and hydrophobicity (Delahunty & Yates, 2007). Maor et al. (2007) investigated ubiquitin-dependent regulatory pathways in plants using the MudPIT system. High molecular weight ubiquitinated proteins were separated by SDS-PAGE and the trypsin-digested samples were examined using the MudPIT system. There were 294 distinct proteins that the GST-tagged ubiquitin-binding domains selectively bound. From these 56 proteins, 85 ubiquitinated lysine residues were found.

The ICAT method has been used to evaluate pairwise changes in protein expression by differentially labelling proteins or peptides with stable isotopes and then identifying and quantifying the labelled proteins or peptides using a mass spectrometer. With this method, it is possible to compare the expression of several proteins in two different biological samples at the same time (Chan et al., 2015). Isobaric tags can be used to measure protein expression between samples in both a relative and an absolute manner. The accuracy of protein identification and quantification may be improved by using this potent new proteomics method, which can lower the possibility of variance in several MS runs. Isobaric tags for relative and absolute quantification (iTRAQ) are capable of labelling all enzymatic peptides, certain hydrophobins and membrane proteins, in contrast to conventional proteomics technologies (Tian et al., 2019). The involvement of Phytochrome A in controlling central metabolism in tomato seedlings grown under far red (FR) light was revealed by iTRAQ-based quantitative proteomics and metabolomics. The findings of Thomas et al. (2021) suggest that Phytochrome A plays a key role in the FR-mediated maturation of the seedling proteome, and additional proteomics studies revealed a lower abundance of photosynthesis- and carbon-fixing-related enzymes in the mutant.

A variety of diverse eukaryotic species have recently been studied using the potent comparative quantitative proteomics approach known as stable isotope labelling by amino acids in cell culture (SILAC) (Lewandowska et al., 2013). Due to poor metabolic labelling efficiency, which has an impact on the precision of peptide ratio measurement, SILAC has traditionally been seen as being inappropriate for plant systems. Quantitation is difficult since there have been only two studies using SILAC labelling in plant systems, both of which used Arabidopsis cell cultures and were labelled with around 80% and 83–91% efficiency, respectively (Schütz et al., 2011). Lewandowska et al. (2013) looked at the reaction in the shoots of seedlings subjected to mild salt stress to show the value of SILAC labelling of entire seedlings to address dynamic changes in protein composition. In comparison to plants not treated with salt, 92 and 123 proteins were significantly up- or downregulated in the shoot. Chloroplast structural proteins, photosynthetic and light-responsive proteins, as well as some abiotic stress response proteins, were enriched among upregulated proteins in shoots treated for eight days with 80-mM NaCl, while ROS-inactivating proteins and other biotic and abiotic stress response proteins, such as salt and osmotic stress were downregulated.

5.2 PROTEOMIC ANALYSIS OF DROUGHT STRESS IN LEGUME CROPS

When exposed to drought stress, 373 seed proteins in pigeon peas displayed abundant expression, according to MALDI-TOF-MS/MS analyses; among them, superoxide dismutases, peroxiredoxins (Prx) and late-embryogenesis-abundant (LEA) proteins were found in seed embryo and aleurone layer (Krishnan et al., 2017). Differential expression of salt-stress-induced proteins such as photosynthesis-related proteins (ATP synthase beta subunit), oxidative stress (cysteine synthase, MLP-like protein 43), carbohydrate metabolism (triosephosphate isomerase, fructose-bisphosphate aldolase, phosphoribulokinase), protein metabolism (glutamine synthetase leaf isoenzyme ubiquitin-binding protein) and lignan biosynthesis (isoflavone reductase) which were contributed in all the major metabolic pathways of plant growth and development (Awana et al., 2020).

The development of parenchyma cells in the soyabean seed coat is greatly aided by the presence of the BURP domain protein 1 (SCB1) in the seed coat (Ae et al., 2002). The disease resistance response protein 1 and the BURP (BNM2, USP, RD22 and PG1b) proteins are both expressed in soyabean under drought stress, according to a proteomic study (Yu et al., 2016). gamma-aminobutyric acid (GABA) reduces damage caused by ROS and improves plant tolerance to oxidative stress caused by various abiotic stresses (Nayyar et al., 2014). The germination of embryonic axes of soyabean under salt stress showed an increased accumulation of LEA proteins and aquaporins (Fercha et al., 2016). The iTRAQ-based proteomic analysis of dark-germinated soyabeans under salt stress revealed an increased accumulation of gamma-aminobutyric acid (GABA), ROS-scavenging enzymes, stress-responsive α–β barrel domain-containing protein, LEA proteins, dehydrins and ribosomal proteins (Yin et al., 2018).

Nucleoside diphosphate kinase (NDPK), superoxide dismutase (SOD) and pathogenesis-related (PR)10 proteins were all accumulated in greater amounts in pea (*Pisum sativum* L.) as a result of salt-induced root proteome alterations (Kav et al., 2004). The overexpression of AtNDPK2 and its coordination with MAPK-mediated H_2O_2 signalling caused an increased accumulation of multiple antioxidants and enhanced oxidative stress resistance and multiple stress tolerance (Moon et al., 2003). SOD improved stress resistance in plants by scavenging ROS and reducing the salt-induced oxidative damage to plant cells (Lu et al., 2020).

In common bean, many types of proteins including formate dehydrogenase, quinine oxidoreductase-like protein and NDPK have been identified along with an increased production of proteins involved in secondary metabolism; the signal transduction pathway (2-C-methyl-D-erythritol 2,4-cyclodiphosphate synthase, 1-deoxy-D-xylulose5-phosphate reducto-isomerase) and protein biosynthesis (ribosomal proteins, cysteine synthetase, glutamine synthetase and acetohydroxy acid synthase) were upregulated under drought stress (Zadražnik et al., 2013). Seven proteins that were differently expressed during drought stress in a common bean were found using SDS-PAGE and MALDI-MS analyses. Among these identified proteins, five proteins like Glucan endo-1,3-betaglucosidase, endochitinase, endochitinase CH5B, auxin-responsive protein and L-type lectin domain-containing receptor kinase IV.3 are upregulated, whereas two proteins including (ribulose bisphosphate carboxylase/oxygenase activase and ADP-ribosylation factor-like protein 8a) were found to be downregulated; all these proteins play a vital role in photosynthesis, defence and regulatory networks of metabolic pathways (Gupta et al., 2019).

The proteins and genes linked to plant stress response can be found using high-throughput proteomics technology. These proteins and genes can then be used in breeding programmes to find specific biomarkers and isolate candidate genes, which can then be integrated using marker-based gene pyramiding and proteomic-based marker-assisted selection (Jan et al., 2022).

Chickpea (*Cicer arietinum* L.), which is cultivated over an area of 11.5 million hectares and yields about 9.7 million tons, is the second most widely grown legume in the world (Goa & Gezahagn, 2018). It is mainly grown in arid and semi-arid areas and may withstand drought stress to a certain extent due to the structure of its roots; however, under moderate-to-severe drought stress, the overall yield drops by 40–50% (Gaur et al., 2012; Sinha et al., 2016). Drought stress is still the major abiotic factor that affects crop yields globally. Drought stress has a negative impact on blooming and seed production. Globally, the effects of terminal drought stress are causing a 40–50% annual decline in chickpea yield (Ramamoorthy et al., 2017). Under terminal drought stress, flower abortion and empty pod formation had a significant impact on production decrease since lower soil moisture levels reduced pollen viability and inhibited pollen development, thus increasing the number of sterile pods. Ultimately, the yield is reduced by the increased number of empty pods and reduced seed size under drought stress. The reproductive phase of flowering, pod formation and seed-set are the three more sensitive stages in chickpeas during drought stress (Pang et al., 2017; Ramamoorthy et al., 2017). Therefore, understanding the response mechanisms of plants to drought stress is important to develop strategies that improve drought tolerance in crops (Roy, 2014). The majority of the proteins produced in response to these various stresses are almost similar and belong to metabolic pathways that are shared by processes such as photosynthesis, signal transmission, protein metabolism, antioxidant defence and cell defence mechanisms (Parankusam et al., 2017). The drought tolerance traits may be affected by the alteration of protein synthesis or degradation (Chandler & Robertson, 1994; Ouvrard et al., 1996). Both quantitative and qualitative changes in proteins have been detected during dehydration stress (Riccardi et al., 1998). Increasing evidence indicates a relationship between the accumulation of dehydration-induced proteins and physiological adaptations to water limitation (Tables 5.1 and 5.2).

By the comparative study of drought-tolerant (*Cicer reticulatum*) and drought-sensitive (*C. arietinum*) chickpea genotypes for leaf proteome analysis, a total of 24 differently expressed proteins were detected by using MALDI-TOF/TOF-MS/MS in response to drought (Çevik et al., 2019). Among them (Table 5.1) light-harvesting complex protein and chlorophyll-a/b-binding protein, ferredoxin NADP reductase (FNR), oxygen-evolving enhancer protein 1 (OEE1) and oxygen-evolving enhancer protein 2 (OEE2) during drought stress were found to be increased (Çevik et al., 2019). Oxygen-evolving enhancer proteins contributed to optimizing the manganese cluster during water photolysis (Heide et al., 2004) and protect reaction centre proteins from ROS as an antioxidant (Kim et al., 2015) for PSII core stability. Drought stress decreased the expression of ferredoxin-$NADP^+$ oxidoreductase (FNR) in drought-sensitive plants, while in drought-tolerant plants, there were no changes (Çevik et al., 2019). FNR activity positively affected the photosynthetic activity (Tebini et al., 2022). Thus, the FNR enzyme reduction might decrease the photosynthesis rate in drought susceptible plants. In plants, chlorophylls and carotenoids harvest solar energy efficiently

TABLE 5.1

Differentially Expressed Proteins in Response to Drought Stress Identified by Comparative Proteomics of Tolerant and Sensitive Chickpea Cultivars

Proteins Affected Under Drought Stress	Function	Remarks
Carbonic anhydrase, a zinc-containing metalloenzyme	Increases the availability of CO_2 to RuBisCo and promotes CO_2 transport in chloroplasts (Budak et al., 2013)	A high concentration of this protein might improve the use of scarce resources during drought-stress circumstances
Glutamine synthetase (GS)	Plays a critical function in the synthesis of proline and the metabolism of nitrogen (Wang et al., 2015)	Under stressful circumstances, proline build-up may result from glutamine synthase modifications
Fructose-bisphosphate aldolase	An essential enzyme in the production of sucrose	Under drought stress, accumulating solutes like sucrose is a key mechanism for plants to conserve water
GDP-mannose-epimerase (GME)	By catalysing the conversion of GDP-D mannose to GDP-L-galactose, ascorbic acid (ASC) can be produced (Ma et al., 2014)	ASC is a small, water-soluble antioxidant molecule that takes part in a cyclic process for the enzymatic detoxification of hydrogen peroxide (Dolatabadian et al., 2008)
Enolase (energy metabolism associated protein) (Zadražnik et al., 2013)	Catalyse the conversion of 2-phosphoglycerate to phosphoenolpyruvate, which is involved in glycolysis	The high expression of the proteins involved in energy metabolism may boost the energy supply to protect chickpea from drought stress
CSP41 proteins/chloroplast RNA–binding proteins (CRB)	Regulate transcription and translation of chloroplast-encoded RNAs through binding and stabilization of distinct plastid transcripts (Bollenbach et al., 2003; Qi et al., 2012)	Involvement of CRB in chloroplast stabilization in synergy with CaPDZ1; PSI-F is a light-harvesting complex I protein acting as dehydration-responsive proteins (Lande et al., 2022). The CRB proteins contribute to a high photosynthetic rate (Ariga et al., 2015)

by LHC protein complexes (Yang et al., 2001). This light-harvesting chlorophyll-a/b-binding protein increased under drought in susceptible cultivar; at the same time, drought stress increased the expression of phosphoglycerate kinase (PGK) and chloroplastic glyceraldehyde 3-phosphate dehydrogenase (G3PDH) enzymes in the tolerant cultivar but was not changed in the susceptible cultivars. It has been stated that the stress treatment may have resulted in an increase in photosynthesis-related proteins, which may have helped to maintain some photosynthesis during the early stages of water deficit by partially offsetting the decrease in internal CO_2 concentration (Bogeat-Triboulot et al., 2007).

Roots function as a first signal transducer of drought stress since they directly contact with soil moisture (Agrawal et al., 2016). The most potential source of drought tolerance was identified as the chickpea germplasm ICC 4958 which revealed that a large root system appeared to be effective in greater extraction of available soil moisture (Saxena et al., 2002). Marker-assisted selection techniques improve the effectiveness of breeding for the crucial trait of drought tolerance (Saxena et al., 2002). Differential expression of genes for root and shoot of chickpeas under drought stress has been investigated (Garg et al., 2016; Mashaki et al., 2018). Due to numerous post-transcriptional and PTMs, localization and final product efficacy, these changes in gene expression were unable to reveal the changes in their functional gene products (Agrawal et al., 2016). Among 75 drought-responsive proteins (DRPs) in roots, 46 showed an increase in abundance (Table 5.2). Of 75 differentially expressed proteins (DEPs), eight groups were divided as carbon and energy metabolism (71%), N and amino acid metabolism (80%), stress-responsive protein (75%), ROS metabolism (60%), protein metabolism (64%) and signal transduction (57%), whereas secondary metabolism

TABLE 5.2

Different Functional Classes of Differentially Expressed Proteins Identified from Drought Stressed Chickpea Roots

Classes of Differentially Expressed Proteins	Differentially Expressed Proteins
Carbon and energy metabolism	1. 6-phosphogluconate dehydrogenase 2. 2,3-bisphosphoglycerate-independent phosphoglycerate mutase 3. Fructose-bisphosphate aldolase 4. Cytoplasmic isozyme of malate dehydrogenase 5. Dihydrolipoyl dehydrogenase 6. ATP synthase subunit beta 7. Mitochondrial V-type proton ATPase subunit E-like
N and amino acid metabolism	1. Methyl tetra hydroperoxyl glutamate homocysteine methyl transferase 2. Dihydrolipoyl lysine residue 3. Succinyl transferase mitochondrial 4. Glutamine synthetase leaf isozyme 5. Aminoacylase-1 isoform 6. Cysteine synthase, chloroplastic/chromoplastic
Stress-responsive protein	1. Betaine aldehyde dehydrogenase 1 2. 2-methylene-furan-3-one reductase 3. NAD(P)H dehydrogenase (quinone) 4. MLP-like protein 34, MLP-like protein 4 5. ABA-responsive protein ABR18-like
ROS metabolism	6. Glutathione S-transferase F9-like 7. L-ascorbate peroxidase, Kunitz proteinase inhibitor

protein (75%) and some other protein like actin and adenosine kinase, transaldolase were decreased under drought stress (Gupta et al., 2020). The differential abundance of carbon and amino acid metabolism and defence-related proteins indicates towards cross-protection of chickpea plants to survive under stress through enhancing metabolism and preventing excessive oxidative damage and stress (Gupta et al., 2020).

5.3 PROTEOMICS FOR IMPROVING SALINITY TOLERANCE IN CHICKPEAS AND OTHER LEGUMES

One of the key abiotic stresses that significantly restrict crop output is salinity stress, especially in dry and semi-arid environments (Greenway & Munns, 1980; Mano & Takeda, 1997; Shannon, 1985) where the chickpea cultivation mostly occurs. High salt concentration in plant cells alters ABA production, which causes stomata to close, photosynthesis to be less active and ROS levels to rise leading to protein damage or misfolding. Chickpea root and shoot growth is slowed down by saline concentration because high salinity may prevent root and shoot elongation by delaying the ability of the plant to absorb water (Tebini et al., 2022). A diallel cross study of both additive and dominance gene effects in chickpeas results stronger dominance effects for the three variables (seed production, pods per plant and seeds per plant) (Ashraf & Waheed, 1998). Chickpeas are well-described for their susceptibility to salt stress, and at various salinity levels, plant growth is inhibited (Flowers et al., 2010).

Through 2-DE and LC-MS/MS studies on chickpea, 65 DEPs were found and categorized based on their putative activities by the functional analysis approach through InterPro that represented proteins involved in photosynthesis and bioenergy (31%), stress responsiveness (22%), protein synthesis and degradation (17.2%), gene transcription and replication (10.9%), amino acid and nitrogen metabolism (6.2%), photorespiration (4.7%), signalling (3.1%) and other metabolisms (4.7%). Among them, five DEPs (49, 52, 74, 68 and 69) are involved in the light-harvesting reaction and photosystems (Arefian et al., 2019).

Furthermore, SOTA analysis for the expression pattern of the DEPs showed that 14 proteins were upregulated in salinity-tolerant plants at a later stage of stress treatment. The upregulated proteins were chlorophyll-a/b-binding (CAB) protein 3 and psbP domain-containing protein 1 (PPD1) (Arefian et al., 2019) involved in the transfer of excitation energy to the reaction centre, where the accumulating plastocyanin could provide more electrons to photosystem I (PSI) to convert $NADP^+$ to NADPH (Benešová et al., 2012), chloroplastic oxygen-evolving enhancer protein (OEE) 1, ATP synthase which could have chaperone-like activity (Suzuki et al., 1997) which is a crucial coping strategy for plants under salt stress (Liu et al., 2014) and carbon assimilating protein such as carbonic anhydrase, phosphoribulokinase (PRK, spot 26), ribulose 1,5-bisphosphate carboxylase/oxygenase (RuBisCO), phosphoglycerate kinase (PGK), fructose-bisphosphate aldolase and transketolase (Arefian et al., 2019). Magnesium chelatase, the first committed enzyme in the chlorophyll biosynthesis pathway (Bandehagh et al., 2011), was also upregulated under salt stress conditions (Arefian et al., 2019). Antioxidant and detoxifying enzymes (superoxide dismutase, ascorbate peroxidase and glutathione S-transferase) and defensive proteins (apolipoprotein D) were also upregulated under stress treatment in the tolerant plant. These results demonstrate the significance of proteins involved in photosynthesis and those that respond to stress in the adaptation of chickpeas to salinity stress (Arefian et al., 2019).

5.4 PROTEOMICS FOR IMPROVING SALINITY AND DROUGHT TOLERANCE IN CEREALS

Drought and salinity are the two major abiotic stresses that have a huge impact on the physiological process of plants. Plants respond to these stresses in different ways including stomatal closure, osmoregulation, photosynthetic rates, stress-responsive metabolite synthesis, check in water absorption in roots, nutrient accumulation and transport, regulation of protein synthesis and morphological changes (Hussain et al., 2017; Singh & Roychoudhury, 2021). Since proteins are the ultimate functional unit of a biological system, understanding them through the level of expression, modification and regulation of metabolic pathways and responses is significant for developing crop tolerance to drought and salinity stress (Rabara et al., 2021). Proteomics analyses at the translational and post-translational levels are used to study the effects on protein biosynthesis and the role of several proteins of plants during stress. Apart from that, these can be used as biochemical markers for drought and salt stress (Roychoudhury et al., 2011).

5.5 PROTEOMICS FOR IMPROVING SALINITY AND DROUGHT TOLERANCE OF RICE

Rice is highly prone to salinity and soil salinization among cereal crops affecting rice cultivation globally. Rice has a minimum salinity threshold level of 3.0 dS/m (Roychoudhury et al. 2008). In the period between 1980 and 2018, an area of about 11.37 km^2 was salt affected, and it has been estimated that 16.49 Mha of salinity-affected soil globally fall under croplands (Hassani et al., 2020). An increase in the soluble salt in the soil around the roots causes saline stress in plants (Abrol et al., 1988). Proteomics analysis of salinity-tolerating rice plants identifies novel proteins playing a role, especially in salinity tolerance. OsRPK1, ABA-responsive and membrane-stabilizing protein under salt stress have been identified along with 18 members associated with salt-responsive protein through analysing proteomic changes in rice plasma membrane under salt stress (Cheng et al., 2009; Basu and Roychoudhury 2014). The upregulation of several proteins during salt stress has a role in stabilizing the damaged plasma membrane. Cytoskeleton protein upregulation is expected to sense salt stress signals through the cell wall and induce the callose synthase expression, thus significantly maintaining plasma membrane stability under salt stress (Cheng et al., 2009). Salt stress also influences the channel proteins in the plasma membrane. For instance, aquaporin plasma membrane intrinsic protein and tonoplast intrinsic protein decreased under salt stress (Habibpourmehraban et al., 2022). The salt-tolerant rice variety Cheriviruppu showed upregulation in fructokinase-2 and dirigent-like protein, while Cu/Zn superoxide dismutase and putative pollen allergen ph1 p11 were downregulated under salt stress (Sarhadi et al., 2012). These proteins have a role in pollen metabolism by reconstructing the anther walls significant for salt tolerance at the reproductive stage. Among the plant parts, roots have the major importance on the salinity response. Comparative proteome analysis in roots of OSRK1 transgenic rice identified OSRK1 – rice sucrose non-fermenting-1-related kinase2 (SnRK2) having a major role in activating the salt stress-responsive metabolic pathways in rice (Nam et al., 2012). Plants do not have any specialized receptors for sodium ions and sensing this abiotic stress is more challenging. They respond to salinity through the Salt Overly Sensitive (SOS) signalling pathway. Rice Salt Overlay Sensitive Proteins (OsSOS) and Na^+/H^+ ion exchanger proteins have a crucial role in salt sensing and tolerance in rice crops (Martínez-Atienza et al., 2007). Furthermore, OsSOS2 and OsSOS3 also influence the salinity tolerance of rice (Kumar et al., 2022; Roychoudhury et al. 2013). Salt stress has a profound effect on photosynthesis, and this has been evident with the proteomic analysis of enzymes and other proteins participating in photosynthesis. The work of Xu et al. (2015) on proteomic analysis of the salt response in rice showed a reduction in the level of light-harvesting complex 1 subunit 1 (Lhca1), putative Lhca2 and Lhca4 precursors by 41.9, 36.5 and 45.1%, respectively. The expression of the protein RSS1 is highly significant under stress conditions and ensures cell division, especially in meristematic tissues by interacting with protein phosphatase 1 (PP1) under salinity stress (Ogawa et al., 2011.). Several proteins negatively regulate the salt tolerance in rice including OsRMC (Guo et al., 2009). Lee et al., (2011) found that 23 proteins were upregulated during salt stress of two different salt-sensitive rice genotypes Dalseongaengmi-44 and Dongjin. Out of the 23 proteins, six proteins were upregulated in salt-tolerant Dongjin which included ATP synthase F1 beta chain, 4 fragments of RuBisCo large chains, putative transcription factor X2 (TF X2), dehydroascorbate reductase, 2-Cys peroxiredoxin and RuBisCo activase isoform. Seven proteins containing ATP synthase beta subunit, Class III peroxidase 29 precursor, FBP aldolase class-1, proteasome subunit alpha type V, putative ribosomal protein L12, putative triose phosphate isomerase chloroplast precursor and drought-induced S-like ribonuclease were upregulated in salt-sensitive Dalseongasengmi-44.

In salt-sensitive rice plants, the antioxidant enzyme peroxidase is downregulated after several hours of exposure to salt stress and is not able to scavenge the hydrogen peroxide (Cheng et al., 2009). Several enzymes with antioxidant function are also downregulated under salt stress such as thioredoxin X (TRX), putative glutathione S-transferase 3 (GST3), thioredoxin M-like, putative thioredoxin peroxidase (TPx) by 52.3, 36.5, 39.2 and 34%, respectively (Xu et al., 2015). OsCYP2 cyclophilin protein has a role in ROS scavenging through modulating at the translational level of antioxidant enzymes (Ruan et al., 2011). This can be further confirmed through knock-out or overexpression studies of the gene of the corresponding proteins. For instance, ascorbate peroxidase 2 (OsAPX2) has the potential to regulate the cytosolic H_2O_2 concentration and overexpression of it improves the salinity tolerance of rice (Zhang et al., 2013). Under salt stress, rice dehydroascorbate reductase 1 (OsDHAR1) reduces the hydroperoxide and malondialdehyde levels which reduce the adverse effects of stress (Kim et al., 2014). Mitochondrial proteome analysis of salt-stressed rice showed an increase in mitochondrial heat shock protein 70 (HSP70) and glycoside hydrolase (GH) which may regulate the programmed cell death (PCD) and electron transport chain (ETC), respectively (Chen et al., 2009).

There are several rice proteome databases available that have a list of identified proteins and several functional annotated proteins. These include the Global Proteome Machine Database, PeptideAtlas, Manually Curated Database of Rice Protein (MCRDP), Proteomics ID Entifications

(PRIDE), P³DB and ARAMENNON (Gao et al., 2009; Martens et al., 2005; Samaras et al., 2020; Vizcaíno et al., 2013). In UniProt under the proteome ID: UP000059680, around 48,900 protein data from rice is available (Schneider & Poux, 2012). Rice SRTFDB is a database containing the list of transcription factors especially responsive to drought and salinity stress conditions (Priya et al., 2013). These proteomics databases are highly useful in identifying target proteins of interest for salinity-tolerant and drought-tolerant analysis in rice.

Changes in the expression of several proteins accompany the responses of plants to drought stress situations. Therefore, a proteomics method is an effective tool for identifying and characterizing the proteins that are altered in response to stress conditions and their function in drought tolerance (Wang et al., 2016). The identification of several DRPs in rice has been made easier by technical developments in proteomics during the past ten years. With the help of both bottom-up and top-down proteomics techniques, the proteomes of various tissues such as leaves, roots, spikes, spikelets and seeds have been examined to determine which proteins are responsive to drought stress (Kim et al., 2014). Besides this, proteome analyses of the nucleus (Choudhary et al., 2019; Jaiswal et al., 2013), extra-cellular matrix (Pandey et al., 2010) and chloroplast (Gayen et al., 2019) were carried out primarily utilizing a gel-based proteomics approach, except the chloroplast proteome analysis which was an iTRAQ-based. The identified DRPs have been mapped to several pathways, including ROS detoxification, primary and secondary metabolism, protein-folding (chaperone activity) and stress/defence response. Beta-expansin, an actin-binding protein, glyceraldehyde-3-phosphate dehydrogenase and pectin esterase inhibitor domain-containing proteins were found in the proteome analysis of two rice genotypes, IR64 (drought-sensitive) and Moroberekan (drought-resistant), without affecting starch accumulation in the tolerant genotype, while no such effects were seen in the sensitive genotype (Liu & Bennett, 2011).

Comparative proteomics is the foundation stone for the analysis of drought stress and other abiotic stresses in plants (Gupta et al., 2015). Rice stress-responsive proteins have been identified using two-dimensional gel electrophoresis (2-DGE), but efforts have also been made to use high-throughput shotgun proteomics (Meng et al., 2018, 2019) including label-free quantification and/or tandem mass tags/iTRAQ (TMT/iTRAQ)-based quantification (Gupta et al., 2019). To identify the distinctive proteins under drought stress, comparative proteomics studies have utilized a variety of proteomic methods including 2-DE, DIGE, iTRAQ and LC-MS/MS (Singh et al., 2022). Interestingly, multiple studies based on root proteome analysis (Mirzaei et al., 2012) and extracellular matrix proteome analysis have indicated higher abundances of heat shock proteins (HSPs) and other chaperones in rice during drought stress, suggesting their significance in drought stress tolerance (Shu et al., 2011). The accumulation of HSP18.6 and four LEA proteins was the highest in the drought-stress-tolerant N22 genotype, demonstrating the critical functions of these proteins in drought stress tolerance (Hamzelou et al., 2020). HSPs are stress-responsive proteins that are induced as an acclimation response of plants to dehydration stress to sustain their growth and survival (UlHaq et al., 2019). Additionally, the proteomes of two rice cultivars, IAC1131 (drought-tolerant) and drought-sensitive Nipponbare, were compared using a combination of label-free and TMT-based quantitative proteome analysis (Wu et al., 2016). When exposed to extreme drought stress, it was noted that both the cultivars displayed increased abundances of the chaperone protein ClpB1, 17.9-kDa class I heat shock protein (Hsp17.9) and 18.6-kDa class III heat shock protein (Hsp18.6), indicating that both these cultivars use a similar mechanism to combat the drought stress. When taken as a whole, these results indicate that HSPs and other chaperonins play a significant role in the ability to withstand drought stress. Numerous proteins are linked to the production of plant hormones, their signalling systems and drought resistance (Rabello et al., 2014). The low quantity of hormone-responsive proteins makes it challenging to use a gel-based proteomics technique to identify the proteins involved in hormone production and signalling. As a result, substantially fewer proteins susceptible to drought-stress-modulating hormones have been found so far. However, ABA was positively regulated whereas JA was negatively regulated with regard to withstanding drought stress (Dhakarey et al., 2017). However, it has also been shown that JA is positively regulated in the signalling of drought stress (Wu et al., 2016). The proteomes of transgenic and wild-type rice were compared using a label-free quantitative proteomics approach, and OsPP2C was shown to be a downregulated protein in transgenic rice. OsPP2C is a negative regulator of ABA signalling in plants (Gupta et al., 2018); therefore, the fact that this protein is downregulated in response to drought stress shows that ABA is positively regulated in rice, which can withstand drought stress (Gupta et al., 2019). The use of a 2-DGE-based proteomic method allowed researchers to better understand the impact of periodic drought stress on rice leaves. A total of 15 differential proteins were found in this investigation, and 7 of them, including an APX- and a GSH-dependent dehydro-ascorbate reductase 1 (Rabello et al., 2014), exhibited higher abundance. In response to drought stress, the number of proteins involved in energy metabolism and anabolic activities increased (Jaiswal et al., 2013; Shu et al., 2011). The PR proteins (pathogenesis-related proteins) may be essential in modulating plant metabolism and development as well as enhancing drought tolerance under stress conditions (Lee et al., 2016) Following drought stress, shotgun proteome analysis revealed an increase of six PR proteins in rice roots (Mirzaei et al., 2012).

Alpha-tubulin, putative elongation factor-2, thiamine biosynthesis protein, putative beta-alanine synthase and cysteine synthase were among the proteins that were abundantly detected in rice under drought stress. Other proteins, such as tricin synthase-1, triose phosphate isomerase, proteasome subunit alpha type-1, homocysteine methyltransferase, elongation factor 1-beta, proteasome subunit beta type 3 and ubiquitin-activating enzyme E1-2 had variable expression patterns (Agrawal et al., 2016). During drought stress, there was a difference in the expression of glycine-rich RNA-binding proteins (GR-RBPs) (Yang et al., 2014). In response to drought stress, particular gene expression in rice was controlled by an increase in glycine-rich RNA-binding proteins. Additionally, the proteomic investigation indicated that S-like ribonucleases (RNases) were much more prevalent and phosphorylated in rice under drought constraints (Wang et al., 2016). Five DAPs, including chitinase, were shown to be abundant during drought stress by conducting the comparative morphological and proteomic investigation of two rice genotypes (Anupama et al., 2019). By performing physiological assessments and leaf proteome analyses on thylakoidal APX-knockdown rice plants (apx8) and unaltered control plants, researchers identified APX in rice thylakoids in response to mild drought stress (Cunha et al., 2019). Systemic proteome analysis using a variety of proteomic techniques is a potent tool for examining the entire proteomes of rice plants at various levels, including the organelle, cell, tissue, organ and organism levels. It also offers insights into evaluating proteomes after plants are exposed to stressful conditions.

5.6 PROTEOMICS FOR SALINITY AND DROUGHT TOLERANCE IMPROVEMENT IN WHEAT

Wheat (*Triticum aestivum*) is a crop that is modest in tolerance to salinity stress. Bread wheat, corn (*Zea mays*) and sorghum (*Sorghum bicolor*) have a higher salt tolerance than durum (*Triticum turgidum* ssp. *durum*) (Maas & Hoffman, 1977). Proteome analyses of Poaceae species have shown that these plants use a diversity of mechanisms to withstand salt stress. When wheat plants are exposed to NaCl, the concentrations of the most prevalent amino acids dropped, but the concentrations of valine, isoleucine, aspartic acid, proline and total free amino acids were increased (El-Shintinawy & El-Shourbagy, 2001). Quantitative proteomics showed that in genotypes with different salt tolerance, salt-responsive proteins were differentially upregulated. These proteins are primarily involved in the regulation of salt stress responses, including oxidation-reduction responses, photosynthesis, carbohydrate and energy metabolism, ion homeostasis, compatible solute production, hormone modulation, cytoskeleton stability, cellular detoxification, membrane stabilization and signal transduction (Chen et al., 2020; Kosová et al., 2013; Li et al., 2017; Yang et al., 2018).

Kamal et al. (2012) first evaluated the proteome in "Keumgang" leaf chloroplasts and revealed 65 distinct proteins. Proteins such as cytochrome b6–f, germin-like protein, the c-subunit of ATP synthase, glutamine synthetase, fructose-bisphosphate aldolase and S-adenosylmethionine synthase, used as stress-responsive marker proteins, were upregulated. The first proteome analysis, in the mitochondria of wheat shoot and root tissues under salinity, was conducted by Jacoby et al. (2010). Enzymes such as Mn-SOD, serine hydroxymethyl transferase, aconitase, malate dehydrogenase and beta (β)-cyanoalanine synthase were the key DEPs exhibiting differential abundance between varieties under salinity stress relative to the control. Aspartate aminotransferase and glutamate dehydrogenase expression were elevated in shoots but downregulated in roots (Jacoby et al., 2013).

The seedling stage is assumed to be most sensitive to salt stress, and it serves as an excellent determinant of adult stage tolerance to salinity. Consequently, a number of proteome research on wheat has been undertaken during the seedling stage under salinity stress (Halder et al., 2022). Upregulation of ribulose-1,5-bisphosphate carboxylase oxygenase (RuBisCo) small subunit chaperone proteins, such as peptidyl-prolyl cis-trans isomerase and calnexin (a calcium-binding protein), was observed when wheat seedlings were subjected to the severe salt stress (Caruso et al., 2008). Proteins such as dehydrin WCOR410, D-11 protein, chaperones, chitinase and trypsin inhibitor (defence-related protein), mitochondrial protein CBSX3 (cell redox-homeostasis) and chloroplastic UDP-sulfoquinovose synthase were upregulated in wheat seedlings under high-salinity conditions. In the same manner, JA biosynthesis enzymes like allene oxide synthase, 4-coumarate-CoA ligase-like 4 and lipoxygenase 2.2 were upregulated; P5CS overproduction was also observed in salt-tolerant durum wheat at high salinity. Under salt stress, (9S)-lipoxygenase (LOX) 1 in the chloroplast membrane was upregulated, contributing to the biosynthesis of oxylipins (Capriotti et al., 2014). In durum wheat (*T. turgidum*), it was shown that ascorbate treatment breaks salinity-induced dormancy, helping the germinating embryos survive during early salinity stress. The metabolic proteome of germinating seeds identified 167 DEPs and 69 DEPs in embryos and embryo-surrounding tissues, respectively. Of these DEPs, 129 proteins (45 upregulated and 84 downregulated) in embryos and 53 (26 upregulated and 27 downregulated) in embryo-surrounding tissues were differentially accumulated in an unprimed salt-stressed condition, proving that salinity-induced germination reduction or dormancy results from an altered proteome in seeds. The majority of DEPs belonged to the categories of metabolism, energy, disease/defence, protein destination and storage. Proteome profiling in germinating wheat seeds under salinity stress identified 397 DEPs mainly related to small molecule metabolic processes, fatty acid degradation and phenylpropanoid biosynthesis pathway; 207 DEPs were involved in protein, amino acid and organic acid metabolic processes (Yan et al., 2020, 2021). In the leaves of wheat seedlings, the upregulation of defence-related proteins of chloroplasts, such as chloroplast ABC1-like family protein and type 2 phosphatidic acid phosphatase family protein, was noted for salinity and drought stress tolerance (Zhu et al., 2021).

Under various environmental conditions, root system architecture (RSA) plays an important role in ensuring sustainable yield. Improved RSA is important for salt tolerance improvement. Dissanayake et al. (2022) found that root tips are more sensitive to salinity stress than mature roots in wheat. A total of 50 and 172 DAPs were identified in mature roots and root tips, respectively. Proteins such as alcohol dehydrogenase, phenylalanine ammonia-lyase and O-methyltransferase, which produce secondary metabolites, GST and peroxidase (redox-reaction-associated proteins) and endo-1,3(4)-β-glucanase 1 (stress-associated protein), were upregulated in root tips, and secondary-metabolite-associated proteins and peroxidases were upregulated in mature roots. Ethylene-dependent salt tolerance occurs through the activation of ribosomal proteins (RPs), which decreases ROS accumulation, chaperone synthesis, ROS scavenging and changes in carbohydrate metabolism; 1140 proteins were identified in a study with significant differential expression of proteins including RPs, CDPKs, transaldolases, β-glucosidases, phosphoenolpyruvate carboxylases, SODs and 6-phosphogluconate dehydrogenases. A total of 8 and 49 DEPs were noticed in roots and shoots, respectively, and 48 RPs in roots (Ma et al., 2020). Three salt-tolerance-associated proteins, a pyruvate orthophosphate dikinase (PPDK) and two LEA proteins encoded by TaPPDK, TaLEA1 and TaLEA2, respectively, were reported in the roots of wheat seedling (Jiang et al., 2017). Salinity stress tolerance is also greatly influenced by root proteins such as speckle-type POZ protein, coronatine insensitive 1, F-box proteins and ubiquitin-like proteins (Jiang et al., 2017). Under salinity stress, upregulation of GSTs, dehydrin (DHNs) and V-ATPase in roots, LEA and DHN in shoots and Cu/Zn SODs in both the tissues of wheat were found. Furthermore, salinity-stress-responsive biomarkers, such as betaine-aldehyde dehydrogenase, cp31BHv, Cu/Zn SOD, leucine aminopeptidase 2 and cytosolic and 2-Cys peroxiredoxin BAS1, were identified through phosphoproteomics (Lv et al., 2016). The upregulation of cell-wall-strengthening and cell-structure-protecting proteins, such as tubulin, profilin, retinoblastoma, Casparian strip membrane protein and xyloglucan endo trans-glycosylase, and ion transporter proteins (e.g., malate transporter), metabolic pathway and protein synthesis during salinity stress in wheat was evidently beneficial to salt tolerance (Singh et al., 2017). In a subsequent salinity stress, Han et al. (2019) also found upregulated SODs, malate dehydrogenases and dehydrin proteins, V-ATPase protein in roots, Cu/Zn, LEA protein and DHN proteins in leaves.

Two important molecular and cellular mechanisms help plants adapt to drought stress: (i) accumulation of several osmolytes, including Pro, glutamate, GB and sugars (mannitol, sorbitol and trehalose), which are essential for preventing membrane deterioration and the activation of enzymes induced by drought stress (Bhushan et al., 2007; Mahajan et al., 2005), and (ii) large-scale drought-responsive gene activation and specialized protective protein expression under drought tolerance (Reddy et al., 2004; Zang et al., 2007). The multidimensional impacts of drought-mediated stress sensitivity on transcriptional and proteomic alterations change the morphological, physiological, metabolic and hormonal responses. Additionally, proteomic modulation is associated with anti-oxidant defence, photosynthesis, respiration, stomatal conductance, cell signalling and PTMs of proteins. These elements exhibit effective plant mitigation methods in response to water shortages (Singh et al., 2022).

Based on the research on drought proteomics coupled with temperature stress, albumins-globulins and amphiphiles (structural proteins) increased during early grain filling, and gliadins and glutenins (storage proteins) accumulated during early and late grain filling, respectively (Triboï et al., 2003). Similar to this, proteome study of the CS-1Sl (1B) Chinese Spring wheat, *Aegilops longissima* chromosome substitution line showed that albumins and globulins have a role in drought stress and further supported the utilization of the 1Sl chromosome as a potential gene resource for improvement of drought tolerance in wheat (Zhou et al., 2016). Higher levels of mannitol 1-phosphate dehydrogenase (mtlD) increased wheat tolerance to salinity and water stress (Abebe et al., 2003). Comparative proteomics research identified 98 and 85 proteins that are differently expressed in the leaves and roots of drought-tolerant wild wheat (*Triticum boeoticum*), respectively, which were involved in carbon metabolism, amino acid/nitrogen/protein metabolism, nucleotide metabolism, signal transduction and photosynthesis. The impact of ABA on the root proteome under drought stress was investigated using LC-based quantitative proteomics (iTRAQ) analysis of drought-tolerant (Nesser) and sensitive (Opata) wheat varieties (Alvarez et al., 2014). In addition, drought stress tolerance in wheat is also influenced by RuBisCo- and ABA-responsive proteins such GST, helicase, LEA and proline (Nezhadahmadi et al., 2013). In a combined metabolomics and proteomics study of wheat, ADP-glucose pyrophosphatase (AGPPase) overexpression, sustained ribulose biphosphate (RuBP) synthesis, organic osmolyte accumulation and downregulation of auxin production were also suggested as engineering strategies for the enhancement of drought tolerance (Michaletti et al., 2018). The C1–2i subclass of *TaZFP* (zinc finger protein) is also responsive to drought tolerance in wheat (Cheuk & Houde, 2016). A drought-stress-tolerant wheat genotype (Khazar-1) exposed to drought stress showed elevation of ROS-associated proteins (GST and Trxs h), defence-associated proteins (α-amylase inhibitor), metabolism-related proteins (mitochondrial aldehyde dehydrogenase) and seed storage proteins (gliadin) (Hajheidari et al., 2007). The effect of gliadin on grain development and quality enhancement under drought stress has been reported in several studies (Jiang et al., 2012; Labuschagne et al., 2020).

According to leaf proteomics, flag leaves are more vulnerable to drought than grain, which proves that drought stress has a considerable impact on photosynthesis-related proteins that are localized in flag leaf chloroplasts (Deng et al., 2018). Under drought stress, L-ascorbate peroxidase-1 was significantly upregulated in grain and improved stress tolerance by maintaining the balance in the ascorbate-glutathione cycle and improving H_2O_2 removal efficiency (Deng et al., 2018). Upregulation of other carbohydrate metabolism proteins, such as pyrroline-5-carboxylate dehydrogenase and pyrroline-5-carboxylate synthetase, in wheat leaves was observed under drought stress (Cui et al., 2019). Moreover, proteins involved in photosynthesis (fructose-bisphosphate aldolase), stress defence (HPs) and detoxification (superoxide dismutase, peroxidase), transporter proteins (ATP-binding cassette) and proteins involved in protein synthesis (60S ribosomal protein L31-1) also play a crucial role in reducing the effects of drought stress and increasing yield. Click or tap here to enter text. In "Hanxuan 10" (drought-tolerant) variety of wheat, calreticulin-like protein (signal transduction protein), peroxidases (defence-related proteins), RuBisCo large subunit, oxygen-evolving enhancer protein 2 (OEE2) and RuBisCo activase (photosynthesis) were upregulated during stress but downregulated during recovery. Upregulation of HSPs during drought stress, viz., HSP60, HSP70 and HOP (Hsp70-Hsp90-organizing protein) in roots, HSP104 and HSP70 in leaves and 23.5-kDa HSP in both roots and leaves, showed increased phosphorylation. Dehydroascorbate reductase, associated with photosynthesis, transpiration and antioxidant activity of catalase (Osipova et al., 2011), was upregulated in leaves during drought stress (Hao et al., 2015).

In a comparative leaf and root proteomics study, S-adenosylmethionine synthase, ferredoxin-NADP(H) oxidoreductase and hairpin-binding protein 1 were upregulated in leaves, while germin-like proteins were significantly upregulated in the roots of the tolerant genotype (Faghani et al., 2015). Intriguingly, the authors found the enzymes 3-ketoacyl-CoA synthase and ATP-binding cassette transporter in leaf tissue, which help plants avoid drought by producing wax (Ghatak et al., 2021). A recent study proposed retaining RuBP synthesis, limiting starch biosynthesis by overexpressing ADP-glucose pyrophosphatase, raising the glutathione response, storing organic osmolytes and decreasing auxin production to produce drought stress-tolerant wheat via omics-assisted breeding (Michaletti et al., 2018).

5.7 PROTEOMICS FOR SALINITY AND DROUGHT STRESS TOLERANCE IN MAIZE

Maize (*Z. mays* L.) is a cross-pollinated crop with high variability in salinity tolerance. Proteomics can aid in the discovery of positional, functional and expressional candidate genes. Proteomic investigations can eventually help to analyse the probable links between protein alterations and plant stress tolerance, because proteins are directly implicated in plant stress responses (Zenda et al., 2018). In response to salinity, five different proteins from the LEA protein family (B4F9K0, A3KLI0, A3KLI1, B7U627 and C4J477) help the plants adapt to dehydration-related stresses and are predicted to protect molecules and structures (Hong-Bo et al., 2005; Soares et al., 2018). A number of phosphoproteins were found in maize, but only ten of them, including fructokinase, UDP-glucosyl transferase BX9 and 2-Cys-peroxyredoxin, were enhanced and phosphorylated, while the remaining six, including isocitrate-dehydrogenase, calmodulin, maturase and a 40S ribosomal protein, were dephosphorylated after the transition to saline conditions (Zörb et al., 2010). Salt tolerance in maize seedlings is connected with the ability to eliminate Na^+ while retaining potassium (K^+) and the growth reduction could be attributable to a decrease in K^+ levels in the leaves or roots. Salt tolerance in maize increased in response to Pi treatment via enhancing Na^+ exclusion. High Na^+ and low potassium (K^+) efflux were observed in the roots, indicating that high K^+ retention directly triggered Na^+ exclusion rather than Pi absorption (Sun et al., 2018).

The DRPs (differentially regulated proteins) in salt-stressed maize were predominantly related to the pentose phosphate pathway, glutathione metabolism and nitrogen metabolism. Furthermore, salt-responsive proteins in maize seedlings have been linked to redox homeostasis maintenance, osmotic homeostasis regulation, energy management, ammonia detoxification, biotic cross-tolerance, stress defence and adaptation and gene expression regulation (Luo et al., 2018). Nitric oxide (NO) improves salt tolerance in maize seedlings by elevating antioxidant enzyme activity and regulating H_2O_2 levels, and these effects are complemented by a variety of downstream defence responses (Bai et al., 2011). The iTRAQ method was used to identify 28 salt-responsive proteins from protein analysis of the maize genotypes "F63" (tolerant) and "F35" (sensitive), of which 22 were expressed explicitly in "F63" (Cui et al., 2015). According to Soares et al. (2018), 1747 proteins were identified, of which 209 were abundant under salt stress and primarily linked to oxidative stress, dehydration, respiration and translation. Therefore, maize increases its ability to tolerate salt by improving its lipid metabolism (upregulation of lipid transporters and the lipid transfer-like protein VAS), promoting lignin biosynthesis and activating the abscisic acid signalling pathway (increased expression of CAD6 as well as PRP1 and PRP10), maintaining the dynamic energy balance of the maize cells (upregulation of ADK2 and adenylate kinase expression) and improving ROS clearance and protection mechanisms (upregulation of peroxidase 12, peroxidase 67, glutathione transferase 9 and the putative laccase family protein and the downregulation of peroxidase 72) (Chen et al., 2022). In the roots of salt-treated maize seedlings, 57 protein spots were found to be differentially expressed. These proteins are involved in the antioxidant system, secondary metabolite biosynthesis, glucose metabolism, protein translation, protein refolding, proteolysis, energy metabolism and transcriptional control (Cheng et al., 2014).

In maize, drought stress can alter protein content via changes in gene expression or altered protein stability, degradation or modifications associated with various cellular processes reflecting both failure and adjustment to drought-induced damage/metabolism, adaptation and homeostasis maintenance. Water deficiency also causes the expression of proteins that are not specifically related to this stress but rather to cell-damage reactions. The most common functional category of proteins responding to drought included chaperones, chaperonins, HSPs and other proteins involved in protein folding (Benešová et al., 2012). The drought tolerance of maize (Z. mays) as influenced by rhizobiota cross-inoculation was associated with the upregulation of proteins involved in glutathione metabolism and the endoplasmic reticulum-associated degradation process at the proteome level (Zhang et al., 2022). In maize seedlings, more than 200 DAPS (differentially accumulated protein species) were drought-responsive. These DAPS were involved in protein synthesis and turnover, ROS scavenging, osmotic regulation, specific gene expression regulation and drought signal transduction (Jiang et al. 2019).

According to proteomic analysis of maize seedling root proteins, drought tolerance is attributed to the higher water retention capacity, the removal of damaging free radicals and the reduction of oxidative stress through the synergistic action of antioxidant enzymes, such as POD, the strengthening of the cell wall by xyloglucan endotransglucosylase/hydrolase, the reduction in the extravasation of cellular contents, the maintenance of osmotic balance and the stabilization of the membrane system (Zeng et al., 2019). Exogenous ABA and water stress induce RAB17 (s1306) in immature embryos and plantlet leaves, which was first identified in maize embryos during the maturation phase. This protein is present in both the nucleus and the cytosol, which may contribute to nuclear protein transport by interacting with nuclear-localization signal peptides. In maize, several enzymes involved in fundamental metabolic cellular pathways like glycolysis and the Krebs cycle (e.g., enolase and triose phosphate isomerase) as well as a number of others such as caffeate O-methyltransferase, whose induction may be connected to lignification, were discovered in addition to proteins already known to be involved in the response to water stress [for example, RAB17 Responsive to ABA]. In maize, water deficit led to the induction of ASR1 protein not only in the young growing leaves but also in the roots of seedling and field-grown plant leaves (Riccardi et al., 2004). Proteomic analysis of maize revealed significantly higher drought tolerance of NongDan 476 (ND476), which may be due to reduced synthesis of excess proteins to help plants conserve energy to combat drought stress, elevated expression of stress defence proteins and upregulated expression of some unknown-function proteins (Dong et al., 2020).

REFERENCES

Abebe, T., Guenzi, A. C., Martin, B., & Cushman, J. C. (2003). Tolerance of mannitol-accumulating transgenic wheat to water stress and salinity. Plant Physiology, 131(4), 1748–1755. https://doi.org/10.1104/PP.102.003616

Abrol, I., Yadav, J., & Massoud, F. (1988). Salt-affected soils and their management. https://books.google.com/books?hl=en&lr=&id=II1HAYDSGsC&oi=fnd&pg=PR3&dq=Abrol,+I.+P.,+Yadav,+J.+S.+P.,+%26+Massoud,+F.+I.+(1988).+Salt-affected+soils+and+their+management+(No.+39).+Food+%26+Agriculture+Org.&ots=8NhF_yrbmb&sig=hWlKghheakGZK171Bd3dy5rEyJs

Ae, A. K. B., Boutilier, K., Miller, A. S. S., Hattori, J., Lu, A. E., Ae, A. B., Hu, M., Lantin, S., Douglas, A. E., Ae, A. J., & Miki, B. L. A. (2002). SCB1, a BURP-domain protein gene, from developing soybean seed coats. Planta, 215(4), 523–532. https://doi.org/10.1007/s00425-002-0798-1

Agrawal, L., Gupta, S., Mishra, S. K., Pandey, G., Kumar, S., Chauhan, P. S., Chakrabarty, D., & Nautiyal, C. S. (2016, September). Elucidation of complex nature of PEG induced drought-stress response in rice root using comparative proteomics approach. Frontiers in Plant Science, 7. https://doi.org/10.3389/fpls.2016.01466

Ahmad, P., Abdel Latef, A. A., Rasool, S., Akram, N. A., Ashraf, M., & Gucel, S. (2016). Role of proteomics in crop stress tolerance. Frontiers in Plant Science, 7, 1336. https://doi.org/10.3389/fpls.2016.01336

Al-Amrani, S., Al-Jabri, Z., Al-Zaabi, A., Alshekaili, J., & Al-Khabori, M. (2021). Proteomics: Concepts and applications in human medicine. World Journal of Biological Chemistry, 12(5), 57–69. https://doi.org/10.4331/wjbc.v12.i5.57

Alvarez, S., Roy Choudhury, S., & Pandey, S. (2014). Comparative quantitative proteomics analysis of the ABA response of roots of drought-sensitive and drought-tolerant wheat varieties identifies proteomic signatures of drought adaptability. Journal of Proteome Research, 13(3), 1688–1701. https://doi.org/10.1021/PR401165B

Anupama, A., Bhugra, S., Lall, B., Chaudhury, S., & Chugh, A. (2019). Morphological, transcriptomic and proteomic responses of contrasting rice genotypes towards drought stress. Environmental and Experimental Botany, 166, 103795.

Arefian, M., Vessal, S., Malekzadeh-Shafaroudi, S., Siddique, K. H. M., & Bagheri, A. (2019). Comparative proteomics and gene expression analyses revealed responsive proteins and mechanisms for salt tolerance in chickpea genotypes. BMC Plant Biology, 19(1). https://doi.org/10.1186/s12870-019-1793-z

Ariga, H., Tanaka, T., Ono, H., Sakata, Y., Hayashi, T., & Taji, T. (2015). CSP41b, a protein identified via FOX hunting using *Eutrema salsugineum* cDNAs, improves heat and salinity stress tolerance in transgenic Arabidopsis thaliana. Biochemical and Biophysical Research Communications, 464(1), 318–323. https://doi.org/10.1016/j.bbrc.2015.06.151

Ashraf, M., & Waheed, A. (1998). Components of genetic variation of salt tolerance in chick pea (*Cicer arietinum* L.). Archives of Agronomy and Soil Science, 42(6), 415–424.

Awana, M., Jain, N., Samota, M. K., Rani, K., Kumar, A., Ray, M., Gaikwad, K., Praveen, S., Singh, N. K., & Singh, A. (2020). Protein and gene integration analysis through proteome and transcriptome brings new insight into salt stress tolerance in pigeonpea (*Cajanus cajan* L.). International Journal of Biological Macromolecules, 164, 3589–3602. https://doi.org/10.1016/j.ijbiomac.2020.08.223

Bai, X., Yang, L., Yang, Y., Ahmad, P., Yang, Y., & Hu, X. (2011). Deciphering the protective role of nitric oxide against salt stress at the physiological and proteomic levels in maize. Journal of Proteome Research, 10(10), 4349–4364. https://doi.org/10.1021/PR200333F

Bandehagh, A., Salekdeh, G. H., Toorchi, M., Mohammadi, A., & Komatsu, S. (2011). Comparative proteomic analysis of canola leaves under salinity stress. Proteomics, 11(10), 1965–1975. https://doi.org/10.1002/PMIC.201000564

Basu, S., & Roychoudhury, A. (2014). Expression profiling of abiotic stress-inducible genes in response to multiple stresses in rice (*Oryza sativa* L.) varieties with contrasting level of stress tolerance. BioMed Research International, 2014,706890.https://doi.org/10.1155/2014/706890

Benešová, M., Holá, D., Fischer, L., Jedelský, P. L., Hnilička, F., Wilhelmová, N., Rothová, O., Kočová, M., Procházková, D., Honnerová, J., Fridrichová, L., & Hnilićková, H. (2012). The physiology and proteomics of drought tolerance in maize: Early stomatal closure as a cause of lower tolerance to short-term dehydration? PLoS ONE, 7(6). https://doi.org/10.1371/journal.pone.0038017

Bhushan, D., Pandey, A., Choudhary, M. K., Datta, A., Chakraborty, S., & Chakraborty, N. (2007). Comparative proteomics analysis of differentially expressed proteins in chickpea extracellular matrix during dehydration stress. Molecular and Cellular Proteomics, 6(11), 1868–1884. https://doi.org/10.1074/mcp.M700015-MCP200

Bogeat-Triboulot, M. B., Brosché, M., Renaut, J., Jouve, L., Le Thiec, D., Fayyaz, P., Vinocur, B., Witters, E., Laukens, K., Teichmann, T., Altman, A., Hausman, J. F., Polle, A., Kangasjärvi, J., & Dreyer, E. (2007). Gradual soil water depletion results in reversible changes of gene expression, protein profiles, ecophysiology, and growth performance in *Populus euphratica*, a poplar growing in arid regions. Plant Physiology, 143(2), 876–892. https://doi.org/10.1104/pp.106.088708

Bollenbach, T. J., Tatman, D. A., & Stern, D. B. (2003). CSP41a, a multifunctional RNA-binding protein, initiates mRNA turnover in tobacco chloroplasts. Plant Journal, 36(6), 842–852. https://doi.org/10.1046/j.1365-313X.2003.01935.x

Budak, H., Akpinar, B. A., Unver, T., & Turktas, M. (2013). Proteome changes in wild and modern wheat leaves upon drought stress by two-dimensional electrophoresis and nanoLC-ESI-MS/MS. Plant Molecular Biology, 83(1–2), 89–103. https://doi.org/10.1007/s11103-013-0024-5

Capriotti, A. L., Borrelli, G. M., Colapicchioni, V., Papa, R., Piovesana, S., Samperi, R., Stampachiacchiere, S., & Laganà, A. (2014). Proteomic study of a tolerant genotype of durum wheat under salt-stress conditions. Analytical and Bioanalytical Chemistry, 406(5), 1423–1435. https://doi.org/10.1007/s00216-013-7549-y

Carrette, O., Burkhard, P. R., Sanchez, J. C., & Hochstrasser, D. F. (2006). State-of-the-art two-dimensional gel electrophoresis: A key tool of proteomics research. Nature Protocols, 1(2), 812–823.

Caruso, G., Cavaliere, C., Guarino, C., Gubbiotti, R., Foglia, P., & Laganà, A. (2008). Identification of changes in *Triticum durum* L. leaf proteome in response to salt stress by two-dimensional electrophoresis and MALDI-TOF mass spectrometry. Analytical and Bioanalytical Chemistry, 391(1), 381–390. https://doi.org/10.1007/S00216-008-2008-X

Çevik, S., Akpinar, G., Yildizli, A., Kasap, M., Karaosmanoğlu, K., & Ünyayar, S. (2019). Comparative physiological and leaf proteome analysis between drought-tolerant chickpea *Cicer reticulatum* and drought-sensitive chickpea *C. arietinum*. Journal of Biosciences, 44(1). https://doi.org/10.1007/s12038-018-9836-4

Chan, J., Zhou, C. Y., & Chan, L., E. C. Y. (2015). The isotope-coded affinity tag method for quantitative protein profile comparison and relative quantitation of cysteine redox modifications. Current Protocols in Protein Science, 82, 23.2.1–23.2.19. https://doi.org/10.1002/0471140864.ps2302s82

Chandler, P. M., & Robertson, M. (1994). Gene expression regulated by abscisic acid and its relation to stress tolerance. Annual Review of Plant Biology, 45(1), 113–141.

Chen, C. S., & Zhu, H. (2006). Protein microarrays. Biotechniques, 40(4), 423–429.

Chen, F., Fang, P., Zeng, W., Ding, Y., Zhuang, Z., & Peng, Y. (2020). Comparing transcriptome expression profiles to reveal the mechanisms of salt tolerance and exogenous glycine betaine mitigation in maize seedlings. PLoS ONE, 15(5). https://doi.org/10.1371/JOURNAL.PONE.0233616

Chen, F., Ji, X., Zhuang, Z., & Peng, Y. (2022). Integrated transcriptome and proteome analyses of maize inbred lines in response to salt stress. Agronomy, 12(5), 1053.

Chen, X., Wang, Y., Li, J., Jiang, A., Cheng, Y., & Zhang, W. (2009). Mitochondrial proteome during salt stress-induced programmed cell death in rice. Plant Physiology and Biochemistry, 47(5), 407–415.

Cheng, Y., Chen, G., Hao, D., Lu, H., Shi, M., Mao, Y., Huang, X., Zhang, Z., & Xue, L. (2014). Salt-induced root protein profile changes in seedlings of maize inbred lines with differing salt tolerances. Chilean Journal of Agricultural Research, 74(4), 468–476.

Cheng, Y., Qi, Y., Zhu, Q., Chen, X., Wang, N., Zhao, X., Chen, H., Cui, X., Xu, L., & Zhang, W. (2009). New changes in the plasma-membrane-associated proteome of rice roots under salt stress. Proteomics, 9(11), 3100–3114. https://doi.org/10.1002/PMIC.200800340

Cheuk, A., & Houde, M. (2016). Genome wide identification of C1 2i zinc finger proteins and their response to abiotic stress in hexaploid wheat. Molecular Genetics and Genomics, 291(2), 73–90. https://doi.org/10.1007/S00438-015-1152-1

Chevalier, F. (2010). Highlights on the capacities of "gel-based" proteomics. Proteome Science, 8, 23. https://doi.org/10.1186/1477-5956-8-23

Choudhary, M. K., Basu, D., Datta, A., Chakraborty, N. and Chakraborty, S. (2009). Dehydration-responsive nuclear proteome of rice (*Oryza sativa* L.) illustrates protein network, novel regulators of cellular adaptation, and evolutionary perspective. Molecular & Cellular Proteomics, 8(7), 1579–1598.

Cui, D., Wu, D., Liu, J., Li, D., Xu, C., Li, S., Li, P., Zhang, H., Liu, X., Jiang, C., Wang, L., Chen, T., Chen, H., & Zhao, L. (2015). Proteomic analysis of seedling roots of two maize inbred lines that differ significantly in the salt stress response. PLoS ONE, 10(2). https://doi.org/10.1371/JOURNAL.PONE.0116697

Cui, G., Zhao, Y., Zhang, J., Chao, M., Xie, K., Zhang, C., Sun, F., Liu, S., & Xi, Y. (2019). Proteomic analysis of the similarities and differences of soil drought and polyethylene glycol stress responses in wheat (*Triticum aestivum* L.). Plant Molecular Biology, 100(4–5), 391–410. https://doi.org/10.1007/S11103-019-00866-2

Cunha, A.P.M.A., Zeri, M., Deusdará Leal, K., Costa, L., Cuartas, L. A., Marengo, J. A., Tomasella, J., Vieira, R. M., Barbosa, A. A., Cunningham, C., et al. (2019). Extreme drought events over Brazil from 2011 to 2019. Atmosphere, 10, 642. https://doi.org/10.3390/atmos10110642

Delahunty, C. M., & Yates, J. R, 3rd. (2007). MudPIT: Multidimensional protein identification technology. Biotechniques, 43(5), 563–565.

Deng, X., Liu, Y., Xu, X., Liu, D., Zhu, G., Yan, X., Wang, Z., & Yan, Y. (2018). Comparative proteome analysis of wheat flag leaves and developing grains under water deficit. Frontiers in Plant Science, 9. https://doi.org/10.3389/fpls.2018.00425

Dhakarey, R., Raorane, M. L., Treumann, A., Peethambaran, P. K., Schendel, R. R., Sahi, V. P., Hause, B., Bunzel, M., Henry, A., Kohli, A., & Riemann, M. (2017). Physiological and proteomic analysis of the rice mutant cpm2 suggests a negative regulatory role of jasmonic acid in drought tolerance. Frontiers in Plant Science, 8. https://doi.org/10.3389/fpls.2017.01903

Dissanayake, B. M., Staudinger, C., Munns, R., Taylor, N. L., & Millar, A.H. (2022). Distinct salinity-induced changes in wheat metabolic machinery in different root tissue types. Journal of Proteomics, 256, 104502. https://doi.org/10.1016/j.jprot.2022.104502

Dolatabadian, A., Sanavy, S. A. M. M., &Chashmi, N. A. (2008). The effects of foliar application of ascorbic acid (vitamin C) on antioxidant enzymes activities, lipid peroxidation and proline accumulation of canola (Brassica napus L.) under conditions of salt stress. Journal of Agronomy and Crop Science, 194(3), 206–213. https://doi.org/10.1111/j.1439-037X.2008.00301.x

Dong, A., Yang, Y., Liu, S., Zenda, T., Liu, X., Wang, Y., Li, J., & Duan, H. (2020). Comparative proteomics analysis of two maize hybrids revealed drought-stress tolerance mechanisms. Biotechnology and Biotechnological Equipment, 34(1), 763–780. https://doi.org/10.1080/13102818.2020.1805015

El-Shintinawy, F., & El-Shourbagy, M. N. (2001). Alleviation of changes in protein metabolism in NaCl-stressed wheat seedlings by thiamine. Biologia Plantarum, 44(4), 541–545. https://doi.org/10.1023/A:1013738603020

Fercha, A., Caruso, G., Cavaliere, C., Chiozzi, R. Z., Capriotti, A. L., Stampachiacchiere, S., Laganà, A., & Laganà, L. (2016). Shotgun proteomic analysis of soybean embryonic axes during germination under salt stress. Proteomics, 16(10), 1537–1546. https://doi.org/10.1002/pmic.201500283

Flowers, T. J., Gaur, P. M., Gowda, C. L. L., Krishnamurthy, L., Samineni, S., Siddique, K. H. M., Turner, N. C., Vadez, V., Varshney, R. K., & Colmer, T. D. (2010). Salt sensitivity in chickpea. Plant, Cell and Environment, 33(4), 490–509. https://doi.org/10.1111/j.1365-3040.2009.02051.x

Gao, J., Agrawal, G. K., Thelen, J. J., and Xu, D. (2009). P3DB: A plant protein phosphorylation database. Nucleic Acids Research, 37(suppl_1), D960–D962.

Garg, R., Shankar, R., Thakkar, B., Kudapa, H., Krishnamurthy, L., Mantri, N., Varshney, R. K., Bhatia, S., & Jain, M. (2016). Transcriptome analyses reveal genotype- and developmental stage-specific molecular responses to drought and salinity stresses in chickpea. Scientific Reports, 6. https://doi.org/10.1038/srep19228

Gaur, P. M., Jukanti, A. K., & Varshney, R. K. (2012). Impact of genomic technologies on chickpea breeding strategies. Agronomy, 2(3), 199–221.

Gayen, D., Barua, P., Lande, N.V., Varshney, S., Sengupta, S., Chakraborty, S., & Chakraborty, N. (2019). Dehydration-responsive alterations in the chloroplast proteome and cell metabolomic profile of rice reveals key stress adaptation responses. Environmental and Experimental Botany, 160, 12–24.

Ghatak, A, Chaturvedi, P., Bachmann, G., Valledor, L., Ramšak, Ž., Bazargani, M. M., Bajaj, P., Jegadeesan, S., Li, W., Sun, X., Gruden, K., Varshney, R. K., & Weckwerth, W. (2021). Physiological and proteomic signatures reveal mechanisms of superior drought resilience in pearl millet compared to wheat. Front Plant Sci., 11, 600278. https://doi.org/10.3389/fpls.2020.600278

Goa, Y., & Gezahagn, G. (2018). Introduction of desi chickpea (Cicer arietinum L.) varieties through participatory variety selection: A case for Konta and Tocha Districts in Southern Ethiopia. Journal of Genetics and Genomes, 2(1), 3–4.

Graves, P. R., & Haystead, T. A. (2002). Molecular biologist's guide to proteomics. Microbiology and Molecular Biology Reviews: MMBR, 66(1), 39–63. https://doi.org/10.1128/MMBR.66.1.39-63.2002

Greenway, H., & Munns, R. (1980). Mechanisms of salt tolerance in nonhalophytes. Annual Review of Plant Physiology, 31(1), 149–190. https://doi.org/10.1146/annurev.pp.31.060180.001053

Guo, Y., & Song, Y. (2009). Differential proteomic analysis of apoplastic proteins during initial phase of salt stress in rice. Plant Signaling & Behavior, 4(2), 121–122.

Gupta, R., Min, C. W., Kim, Y. J., & Kim, S. T. (2019). Identification of Msp1-induced signaling components in rice leaves by integrated proteomic and phosphoproteomic analysis. Int J Mol Sci., 20(17), 4135. https://doi.org/10.3390/ijms20174135

Gupta, R., Woo Min, C., Kumar Agrawal, G., Rakwal, R., Kramer, K., Park, K.-H., Wang, Y., Finkemeier, I., Tae Kim, S., Gupta, R., Min, C. W., Kim, S. T., Kramer, K., Finkemeier, I., Agrawal, G. K., & Rakwal, R. (2018). A multi-omics analysis of *Glycine max* leaves reveals alteration in flavonoid and isoflavonoid metabolism upon ethylene and abscisic acid treatment. Proteomics, 18(7). https://doi.org/10.1002/pmic.201700366

Gupta, R., Woo Min, C., Wun Kim, S., Wang, Y., Kumar Agrawal, G., Rakwal, R., Gon Kim, S., Won Lee, B., Min Ko, J., Yeol Baek, I., Won Bae, D., & Tae Kim, S. (2015). Comparative investigation of seed coats of brown-versus yellow-colored soybean seeds using an integrated proteomics and metabolomics approach. Proteomics, 15(10), 1706–1716. https://doi.org/10.1002/pmic.201400453

Gupta, S., Mishra, S. K., Misra, S., Pandey, V., Agrawal, L., Nautiyal, C. S., & Chauhan, P. S. (2020). Revealing the complexity of protein abundance in chickpea root under drought-stress using a comparative proteomics approach. Plant Physiology and Biochemistry, 151, 88–102. https://doi.org/10.1016/j.plaphy.2020.03.005

Habibpourmehraban, F., Atwell, B. J., & Haynes, P. A. (2022). Unique and shared proteome responses of rice plants (*Oryza sativa*) to individual abiotic stresses. Int J Mol Sci., 23(24), 15552. https://doi.org/10.3390/ijms232415552

Hajheidari, M., Eivazi, A., Buchanan, B. B., Wong, J. H., Majidi, I., & Salekdeh, G. H. (2007). Proteomics uncovers a role for redox in drought tolerance in wheat. Journal of Proteome Research, 6(4), 1451–1460. https://doi.org/10.1021/PR060570J

Halder, T., Choudhary, M., Liu, H., Chen, Y., Yan, G., & Siddique, K. H. M. (2022). Wheat proteomics for abiotic stress tolerance and root system architecture: Current status and future prospects. Mdpi.Com, 10(2). https://doi.org/10.3390/proteomes10020017

Hamzelou, S., Pascovici, D., Kamath, K. S., Amirkhani, A., McKay, M., Mirzaei, M., Atwell, B. J., Haynes, P. A. (2020). Proteomic responses to drought vary widely among eight diverse genotypes of rice (*Oryza sativa*). Int J Mol Sci., 21(1), 363. https://doi.org/10.3390/ijms21010363

Han, L., Xiao, C., Xiao, B., Wang, M., Liu, J., Bhanbhro, N., Khan, A., Wang, H., Wang, H., & Yang, C. (2019). Proteomic profiling sheds light on alkali tolerance of common wheat (*Triticum aestivum* L.). Plant Physiol Biochem., 138, 58–64. https://doi.org/10.1016/j.plaphy.2019.02.024.

Hao, P., Zhu, J., Gu, A., Lv, D., Ge, P., Chen, G., Li, X., & Yan, Y. (2015). An integrative proteome analysis of different seedling organs in tolerant and sensitive wheat cultivars under drought stress and recovery. Proteomics, 15(9), 1544–1563. https://doi.org/10.1002/PMIC.201400179

Hassani, A., Azapagic, A., & Shokri, N. (2020). Predicting long-term dynamics of soil salinity and sodicity on a global scale. Proceedings of the National Academy of Sciences of the United States of America, 117(52), 33017–33027. https://doi.org/10.1073/PNAS.2013771117

Heide, H., Kalisz, H. M., & Follmann, H. (2004). The oxygen evolving enhancer protein 1 (OEE) of photosystem II in green algae exhibits thioredoxin activity. International Journal of Plant Physiology, 161. http://www.elsevier-deutschland.de/jplhp

Holman, J. D., Dasari, S., & Tabb, D. L. (2013). Informatics of protein and post-translational modification detection via shotgun proteomics. Methods in Molecular Biology, 1002, 167–179. https://doi.org/10.1007/978-1-62703-360-2_14

Hong-Bo, S., Zong-Suo, L., & Ming-An, S. (2005). LEA proteins in higher plants: Structure, function, gene expression and regulation. Colloids Surf B Biointerfaces, 45(3-4), 131–135. https://doi.org/10.1016/j.colsurfb.2005.07.017

Hu, J., Rampitsch, C., & Bykova, N. V. (2015) Advances in plant proteomics toward improvement of crop productivity and stress resistance. Frontiers in Plant Science, 6, 209. https://doi.org/10.3389/fpls.2015.00209

Hussain, N., Sarwar, G., Schmeisky, H., Al-Rawahy, S., & Ahmad, M. (2010). Salinity and drought management in legume crops. In: Climate Change and Management of Cool Season Grain Legume Crops. Springer,, pp. 171–191. ISBN 9789048137. https://doi.org/10.1007/978-90-481-3709-1_10

Hussain, S., Zhang, J., Zhong, C., Zhu, L., Cao, X., Yu, S., Bohr, J. A., Hu, J., & Jin Q. (2017). Effects of salt stress on rice growth, development characteristics, and the regulating ways: A review. Journal of Integrative Agriculture,, 16(11), 2357–2374. Retrieved December 27, 2022, from https://www.sciencedirect.com/science/article/pii/S2095311916616088

Jacoby, R. P., Millar, A. H., & Taylor, N. L. (2010). Wheat mitochondrial proteomes provide new links between antioxidant defense and plant salinity tolerance. Journal of Proteome Research, 9(12), 6595–6604. https://doi.org/10.1021/PR1007834

Jacoby, R. P., Millar, A. H., & Taylor, N. L. (2013). Investigating the role of respiration in plant salinity tolerance by analyzing mitochondrial proteomes from wheat and a salinity-tolerant amphiploid (wheat × lophopyrumelongatum). Journal of Proteome Research, 12(11), 4807–4829. https://doi.org/10.1021/PR400504A

Jaiswal, D. K., Ray, D., Choudhary, M. K., Subba, P., Kumar, A., Verma, J., Kumar, R., Datta, A., Chakraborty, S., & Chakraborty, N. (2013). Comparative proteomics of dehydration response in the rice nucleus: New insights into the molecular basis of genotype-specific adaptation. Proteomics, 13(23–24), 3478–3497. https://doi.org/10.1002/PMIC.201300284

Jan, N., Rather, A. M. U. D., John, R., Chaturvedi, P., Ghatak, A., Weckwerth, W., Zargar, S. M., Mir, R. A., Khan, M. A., & Mir, R. R. (2022). Proteomics for abiotic stresses in legumes: Present status and future directions. Critical Reviews in Biotechnology. https://doi.org/10.1080/07388551.2021.2025033

Jiang, Q., Li, X., Niu, F., Sun, X., Hu, Z., & Zhang, H. (2017). iTRAQ-based quantitative proteomic analysis of wheat roots in response to salt stress. Proteomics, 17(8), 1600265. https://doi.org/10.1002/PMIC.201600265

Jiang, S.-S., Liang, X.-N., Li, X., Wang, S.-L., Lv, D.-W., Ma, C.-Y., Li, X.-H., Ma, W.-J., & Yan, Y.-M. (2012). Wheat drought-responsive grain proteome analysis by linear and nonlinear 2-DE and MALDI-TOF mass spectrometry. Mdpi.Com, 13, 16065–16083. https://doi.org/10.3390/ijms131216065

Jiang, Z., Jin, F., Shan, X., & Li, Y. (2019). iTRAQ-based proteomic analysis reveals several strategies to cope with drought stress in maize seedlings. Int J Mol Sci., 20(23), 5956. https://doi.org/10.3390/ijms20235956

Jung, E., Heller, M., Sanchez, J. C., & Hochstrasser, D. F. (2000). Proteomics meets cell biology: The establishment of subcellular proteomes. Electrophoresis: An International Journal, 21(16), 3369–3377.

Kamal, A. H. M., Cho, K., Kim, D. E., Uozumi, N., Chung, K. Y., Lee, S. Y., Choi, J. S., Cho, S. W., Shin, C. S., & Woo, S. H. (2012). Changes in physiology and protein abundance in salt-stressed wheat chloroplasts. Molecular Biology Reports, 39(9), 9059–9074. https://doi.org/10.1007/S11033-012-1777-7

Kang, Y., Techanukul, T., Mantalaris, A., & Nagy, J. M. (2009). Comparison of three commercially available DIGE analysis software packages: Minimal user intervention in gel-based proteomics. Journal of Proteome Research, 8(2), 1077–1084.

Kav, N. N. V., Srivastava, S., Goonewardene, L., & Blade, S. F. (2004). Proteome-level changes in the roots of *Pisum sativum* in response to salinity. Annals of Applied Biology, 145(2), 217–230. https://doi.org/10.1111/J.1744-7348.2004.TB00378.X

Kendrick, N., Darie, C. C., Hoelter, M., Powers, G., & Johansen, J. (2019). 2D SDS PAGE in combination with Western blotting and mass spectrometry is a robust method for protein analysis with many applications. Advances in Experimental Medicine and Biology, 1140, 563–574. https://doi.org/10.1007/978-3-030-15950-4_33

Kim, E. Y., Choi, Y. H., Lee, J. I., Kim, I. H., & Nam, T. J. (2015). Antioxidant activity of oxygen evolving enhancer protein 1 purified from *Capsosiphon fulvescens*. Journal of Food Science, 80(6), H1412–H1417. https://doi.org/10.1111/1750-3841.12883

Kim, S. T., Kim, S. G., Agrawal, G. K., Kikuchi, S., & Rakwal, R. (2014). Rice proteomics: A model system for crop improvement and food security. Proteomics, 14(4–5), 593–610. https://doi.org/10.1002/PMIC.201300388

Kim, Y. S., Kim, I. S., Shin, S. Y., Park, T. H., Park, H. M., Kim, Y. H., Lee, G. S., Kang, H. G., Lee, S. H., & Yoon, H. S. (2014). Overexpression of dehydroascorbate reductase confers enhanced tolerance to salt stress in rice plants (*Oryza sativa* L. japonica). Journal of Agronomy and Crop Science, 200(6), 444–456. https://doi.org/10.1111/jac.12078

Kirk, P., Amsbury, S., German, L., Gaudioso-Pedraza, R., & Benitez-Alfonso, Y. (2022). A comparative metaproteomic pipeline for the identification of plasmodesmata proteins and regulatory conditions in diverse plant species. BMC Biology, 20(1), 1–21.

Komatsu, S., & Jorrin-Novo, J. V. (2021). Plant proteomic research 3.0: Challenges and perspectives. International Journal of Molecular Sciences, 22(2), 766.

Kosová, K., Vítámvás, P., Planchon, S., Renaut, J., Vanková, R., & Prášil, I. T. (2013). Proteome analysis of cold response in spring and winter wheat (*Triticum aestivum*) crowns reveals similarities in stress adaptation and differences in regulatory processes between the growth habits. Journal of Proteome Research, 12(11), 4830–4845. https://doi.org/10.1021/PR400600G

Krishnan, H. B., Natarajan, S. S., Oehrle, N. W., Garrett, W. M., & Darwish, O. (2017). Proteomic analysis of pigeonpea (*Cajanus cajan*) seeds reveals the accumulation of numerous stress-related proteins. Journal of Agricultural and Food Chemistry, 65(23), 4572–4581. https://doi.org/10.1021/acs.jafc.7b00998

Kumar, G., Basu, S., Singla-Pareek, S. L., & Pareek, A. (2022). Unraveling the contribution of OsSOS2 in conferring salinity and drought tolerance in a high-yielding rice. Physiol Plant., 174(1), e13638.

Labuschagne, M., Masci, S., Tundo, S., Muccilli, V., Saletti, R., & van Biljon, A. (2020). Proteomic analysis of proteins responsive to drought and low temperature stress in a hard red spring wheat cultivar. Mdpi. Com. https://doi.org/10.3390/molecules25061366

Lande, N. V., Barua, P., Gayen, D., Wardhan, V., Jeevaraj, T., Kumar, S., Chakraborty, S., & Chakraborty, N. (2022). Dehydration-responsive chickpea chloroplast protein, CaPDZ1, confers dehydration tolerance by improving photosynthesis. Physiologia Plantarum, 174(1). https://doi.org/10.1111/ppl.13613

Lanne, B., & Panfilov, O. (2005). Protein staining influences the quality of mass spectra obtained by peptide mass fingerprinting after separation on 2-d gels. A comparison of staining with coomassie brilliant blue and sypro ruby. Journal of Proteome Research, 4(1), 175–179.

Lee, D. K., Jung, H., Jang, G., Jeong, J. S., Kim, Y. S., Ha, S. H., Do Choi, Y., Kim, J. K. (2016). Overexpression of the OsERF71 transcription factor alters rice root structure and drought resistance. Plant Physiol., 172(1), 575–588. https://doi.org/10.1104/pp.16.00379

Lee, D.-G., Woong Park, K., Young An, J., Geol Sohn, Y., Ki Ha, J., Yoon Kim, H., Won Bae, D., Hee Lee, K., Jun Kang, N., Lee, B.-H., Young Kang, K., & Joo Lee, J. (2011). Proteomics analysis of salt-induced leaf proteins in two rice germplasms with different salt sensitivity. Cdnsciencepub.Com, 91(2), 337–349. https://doi.org/10.4141/CJPS10022

Lewandowska, D., ten Have, S., Hodge, K., Tillemans, V., Lamond, A. I., & Brown, J. W. (2013). Plant SILAC: Stable-isotope labelling with amino acids of Arabidopsis seedlings for quantitative proteomics. PLoS ONE, 8(8), e72207.

Li, P., Cao, W., Fang, H., Xu, S., Yin, S., Zhang, Y., Lin, D., Wang, J., Chen, Y., Xu, C., & Yang, Z. (2017). Stranscriptomic profiling of the maize (Zea mays L.) leaf response to abiotic stresses at the seedling stage. Frontiers in Plant Science, 8. https://doi.org/10.3389/FPLS.2017.00290/FULL

Liu, C. W., Chang, T. S., Hsu, Y. K., Wang, A. Z., Yen, H. C., Wu, Y. P., Wang, C. S., & Lai, C. C. (2014). Comparative proteomic analysis of early salt stress responsive proteins in roots and leaves of rice. Proteomics, 14(15), 1759–1775. https://doi.org/10.1002/PMIC.201300276

Liu, J. X., & Bennett, J. (2011). Reversible and irreversible drought-induced changes in the anther proteome of rice (Oryza sativa L.) genotypes IR64 and Moroberekan. Molecular Plant, 4(1), 59–69. https://doi.org/10.1093/mp/ssq039

Lopez, J. L. (2007). Two-dimensional electrophoresis in proteome expression analysis. Journal of Chromatography B, 849(1–2), 190–202.

Lu, W., Duanmu, H., Qiao, Y., Jin, X., Yu, Y., Yu, L., & Chen, C. (2020). Genome-wide identification and characterization of the soybean SOD family during alkaline stress. PeerJ., 8, e8457. https://doi.org/10.7717/peerj.8457

Luo, M., Zhao, Y., Wang, Y., Shi, Z., Zhang, P., Zhang, Y., Song, W., & Zhao, J. (2018). Comparative proteomics of contrasting maize genotypes provides insights into salt-stress tolerance mechanisms. Journal of Proteome Research, 17(1), 141–153. https://doi.org/10.1021/ACS.JPROTEOME.7B00455

Lv, D. W., Zhu, G. R., Zhu, D., Bian, Y. W., Liang, X. N., Cheng, Z.W., Deng, X., & Yan, Y. M. (2016). Proteomic and phosphoproteomic analysis reveals the response and defense mechanism in leaves of diploid wheat T. monococcum under salt stress and recovery. J Proteomics., 143, 93–105. https://doi.org/10.1016/j.jprot.2016.04.013

Ma, L., Wang, Y., Liu, W., & Liu, Z. (2014). Overexpression of an alfalfa GDP-mannose 3, 5-epimerase gene enhances acid, drought and salt tolerance in transgenic Arabidopsis by increasing ascorbate accumulation. Biotechnology Letters, 36(11), 2331–2341. https://doi.org/10.1007/s10529-014-1598-y

Ma, Q., Shi, C., Su, C., & Liu, Y. (2020). Complementary analyses of the transcriptome and iTRAQ proteome revealed mechanism of ethylene dependent salt response in bread wheat (Triticum aestivum L.). Food Chem., 325, 126866. https://doi.org/10.1016/j.foodchem.2020.126866

Maas, E. V., & Hoffman, G. J. (1977). Crop salt tolerance – Current assessment. ASCE Journal of the Irrigation and Drainage Division, 103(2), 115–134. https://doi.org/10.1061/JRCEA4.0001137

Mahajan, S., & Tuteja, N. (2005). Cold, salinity and drought stresses: An overview. Arch Biochem Biophys, 444(2), 139–158. https://doi.org/10.1016/j.abb.2005.10.018

Mano, Y., & Takeda, K. (1997). Mapping quantitative trait loci for salt tolerance at germination and the seedling stage in barley (Hordeum vulgare L.). Euphytica, 94(3), 263–272.

Maor, R., Jones, A., Nühse, T. S., Studholme, D. J., Peck, S. C., & Shirasu, K. (2007). Multidimensional protein identification technology (MudPIT) analysis of ubiquitinated proteins in plants. Molecular & Cellular Proteomics, 6(4), 601–610.

Martens, L., Hermjakob, H., Jones, P., Adamsk, M., Taylor, C., States, D., Gevaert, K., Vandekerckhove, J., & Apweiler, R. (2005). PRIDE: The proteomics identifications database. Proteomics, 5(13), 3537–3545. https://doi.org/10.1002/PMIC.200401303

Martínez-Atienza, J., Jiang, X., Garciadeblas, B., Mendoza, I., Zhu, J. K., Pardo, J. M., & Quintero, F. J. (2007). Conservation of the salt overly sensitive pathway in rice. Plant Physiol., 143(2), 1001–1012. https://doi.org/10.1104/pp.106.092635

Mashaki, K. M., Garg, V., NasrollahnezhadGhomi, A. A., Kudapa, H., Chitikineni, A., Nezhad, K. Z., Yamchi, A., Soltanloo, H., Varshney, R. K., & Thudi, M. (2018). RNA-Seq analysis revealed genes associated with drought stress response in kabuli chickpea (*Cicer arietinum* L.). PLoS ONE, 13(6). https://doi.org/10.1371/journal.pone.0199774

Meng, Q., Gupta R., Min, C. W., Kim, J., Kramer, K., Wang, Y., Park, S. R., Finkemeier, I., & Kim, S. T. (2019). A proteomic insight into the MSP1 and flg22 induced signaling in *Oryza sativa* leaves. J Proteomics., 196, 120–130. https://doi.org/10.1016/j.jprot.2018.04.015

Meng, Q., Gupta, R., Min, C. W., & Kim, S. T. (2018). Label-free quantitative proteome data associated with MSP1 and flg22 induced signaling in rice leaves. Data Brief., 20, 204–209. https://doi.org/10.1016/j.dib.2018.07.063

Michaletti, A., Naghavi, M. R., Toorchi, M., Zolla, L., & Rinalducci, S. (2018). Metabolomics and proteomics reveal drought-stress responses of leaf tissues from spring-wheat. Scientific Reports, 8(1), 1–18. https://doi.org/10.1038/s41598-018-24012-y

Mirzaei, M., Pascovici, D., Atwell, B. J., & Haynes, P. A. (2012). Differential regulation of aquaporins, small GTPases and V-ATPases proteins in rice leaves subjected to drought stress and recovery. Proteomics, 12(6), 864–877. https://doi.org/10.1002/PMIC.201100389

Monti, M., Orrù, S., Pagnozzi, D., & Pucci, P. (2005). Functional proteomics. Clinicachimicaacta; International Journal of Clinical Chemistry, 357(2), 140–150. https://doi.org/10.1016/j.cccn.2005.03.019

Moon, H., Lee, B., Choi, G., Shin, D., Prasad, D. T., Lee, O., Kwak, S. S., Kim, D. H., Nam, J., Bahk, J., Hong, J. C., Lee, S. Y., Cho, M. J., Lim, C.O., & Yun, D. J. (2003). NDP kinase 2 interacts with two oxidative stress-activated MAPKs to regulate cellular redox state and enhances multiple stress tolerance in transgenic plants. Proc Natl Acad Sci USA, 100(1), 358–363.

Nam, M. H., Huh, S. M., Kim, K. M., Park, W. J., Seo, J. B., Cho, K., Kim, D. Y., Kim, B. G., & Yoon, I. S. (2012). Comparative proteomic analysis of early salt stress-responsive proteins in roots of SnRK2 transgenic rice. Proteome Science, 10(1). https://doi.org/10.1186/1477-5956-10-25

Nayyar, H., Kaur, R., Kaur, S., & Singh, R. (2014). γ-Aminobutyric acid (GABA) imparts partial protection from heat stress injury to rice seedlings by improving leaf turgor and upregulating osmoprotectants and antioxidants. Journal of Plant Growth Regulation, 33(2), 408–419. https://doi.org/10.1007/S00344-013-9389-6

Nezhadahmadi, A., Prodhan, Z. H., & Faruq, G. (2013). Drought tolerance in wheat. The Scientific World Journal, 2013. https://doi.org/10.1155/2013/610721

Ogawa, D., Abe, K., Miyao, A., Kojima, M., Sakakibara, H., Mizutani, M., Morita, H., Toda, Y., Hobo, T., Sato, Y., Hattori, T., Hirochika, H., & Takeda, S. (2011). RSS1 regulates the cell cycle and maintains meristematic activity under stress conditions in rice. Nat Commun., 2, 278. https://doi.org/10.1038/ncomms1279

Osipova, S. V., Permyakov, A. V., Permyakova, M. D., Pshenichnikova, T. A., & Börner, A. (2011). Leaf dehydroascorbate reductase and catalase activity is associated with soil drought tolerance in bread wheat. Acta Physiologiae Plantarum, 33(6), 2169–2177. https://doi.org/10.1007/S11738-011-0756-2

Ouvrard, O., Cellier, F., Ferrare, K., Tousch, D., Lamaze, T., Dupuis, J. M., & Casse-Delbart, F. (1996). Identification and expression of water stress- and abscisic acid-regulated genes in a drought-tolerant sunflower genotype. Plant Molecular Biology, 31(4), 819–829. https://doi.org/10.1007/BF00019469

Pandey, A., Rajamani, U., Verma, J., Subba, P., Chakraborty, N., Datta, A., Chakraborty, S., & Chakraborty, N. (2010). Identification of extracellular matrix proteins of rice (*Oryza sativa* L.) involved in dehydration-responsive network: A proteomic approach. J Proteome Res., 9(7), 3443–3464. https://doi.org/10.1021/pr901098p

Pang, J., Turner, N. C., Khan, T., Du, Y. L., Xiong, J. L., Colmer, T. D., Devilla, R., Stefanova, K., & Siddique, K. H. M. (2017). Response of chickpea (*Cicer arietinum* L.) to terminal drought: Leaf stomatal conductance, pod abscisic acid concentration, and seed set. Journal of Experimental Botany, 68(8), 1973–1985. https://doi.org/10.1093/jxb/erw153

Parankusam, S., Bhatnagar-Mathur, P., & Sharma, K. K. (2017). Heat responsive proteome changes reveal molecular mechanisms underlying heat tolerance in chickpea. Environmental and Experimental Botany, 141, 132–144. https://doi.org/10.1016/j.envexpbot.2017.07.007

Priya, P., & Jain, M. (2013). RiceSRTFDB: A database of rice transcription factors containing comprehensive expression, cis-regulatory element and mutant information to facilitate gene function analysis. Database: The Journal of Biological Databases and Curation, bat027. https://doi.org/10.1093/database/bat027

Qi, Y., Armbruster, U., Schmitz-Linneweber, C., Delannoy, E., De Longevialle, A. F., Rühle, T., Small, I., Jahns, P., & Leister, D. (2012). *Arabidopsis* CSP41 proteins form multimeric complexes that bind and stabilize distinct plastid transcripts. Journal of Experimental Botany, 63(3), 1251–1270. https://doi.org/10.1093/jxb/err347

Rabara, R., Msanne, J., Basu, S., Ferrer, M., & Roychoudhury, A. (2021). Coping with inclement weather conditions due to high temperature and water deficit in rice: An insight from genetic and biochemical perspectives. Physiologia Plantarum, 172, 487–504.

Rabello, F. R., Villeth, G. R., Rabello, A. R., Rangel, P. H., Guimarães, C. M., Huergo, L. F., Souza, E. M., Pedrosa, F. O., Ferreira, M. E., & Mehta, A. (2014). Proteomic analysis of upland rice (*Oryza sativa* L.) exposed to intermittent water deficit. Protein J., 33(3), 221–230.

Ramachandra Reddy, A., Chaitanya, K.V., & Vivekanandan, M. (2004). Drought-induced responses of photosynthesis and antioxidant metabolism in higher plants. J Plant Physiol., 161(11), 1189–1202. https://doi.org/10.1016/j.jplph.2004.01.013

Ramamoorthy, P., Lakshmanan, K., Upadhyaya, H. D., Vadez, V., & Varshney, R. K. (2017). Root traits confer grain yield advantages under terminal drought in chickpea (*Cicer arietinum* L.). Field Crops Research, 201, 146–161. https://doi.org/10.1016/j.fcr.2016.11.004

Riccardi, F., Gazeau, P., de Vienne, D., & Zivy, M. (1998). Protein changes in response to progressive water deficit in maize: Quantitative variation and polypeptide identification. Plant Physiology, 117(4), 1253–1263.

Riccardi, F., Gazeau, P., Jacquemot, M., et al. (2004). Deciphering genetic variations of proteome responses to water deficit in maize leaves. Elsevier.

Riccardi, F., Gazeau, P., Vienne, D. et al. (1998). Protein changes in response to progressive water deficit in maize: Quantitative variation and polypeptide identification. Academic.Oup.Com.

Rosengren, A. T., Salmi, J. M., Aittokallio, T., Westerholm, J., Lahesmaa, R., Nyman, T. A., & Nevalainen, O. S. (2003). Comparison of PDQuest and Progenesis software packages in the analysis of two-dimensional electrophoresis gels. Proteomics, 3(10), 1936–1946.

Roy, A. (2014). Proteomic analyses of alterations in plant proteome under drought stress. In: Gaur, R. K., & Sharma, P. (Eds.). Molecular Approaches in Plant Abiotic Stress, 1st ed. Taylor & Francis Group, pp. 232–247.

Roychoudhury, A., Basu, S., Sarkar, S. N., & Sengupta, D. N. (2008). Comparative physiological and molecular responses of a common aromatic indica rice cultivar to high salinity with non-aromatic indica rice cultivars. Plant Cell Reports, 27(8), 1395–1410.

Roychoudhury, A., Datta, K., & Datta, S. K. (2011). Abiotic stress in plants: From genomics to metabolomics. In: Tuteja, N., Gill, S. S., & Tuteja, R. (Eds.). Omics and Plant Abiotic Stress Tolerance. Bentham Science Publishers, pp. 91–120.

Roychoudhury, A., Paul, S., & Basu, S. (2013). Cross-talk between abscisic acid-dependent and abscisic acid-independent pathways during abiotic stress. Plant Cell Reports, 32(7), 985–1006.

Ruan, S. L., Ma, H. S., Wang, S. H., Fu, Y. P., Xin, Y., Liu, W. Z., Wang, F., Tong, J. X., Wang, S. Z., & Chen, H. Z. (2011). Proteomic identification of OsCYP2, a rice cyclophilin that confers salt tolerance in rice (*Oryza sativa* L.) seedlings when overexpressed. BMC Plant Biology, 11. https://doi.org/10.1186/1471-2229-11-34

Samaras, P., Schmidt, T., et al. (2020). ProteomicsDB: A multi-omics and multi-organism resource for life science research. Academic.Oup.Com. Retrieved December 27, 2022, from https://academic.oup.com/nar/article-abstract/48/D1/D1153/5609531

Sarhadi, E. Bazargani, M., et al. (2012). Proteomic analysis of rice anthers under salt stress. Elsevier. Retrieved December 27, 2022, from https://www.sciencedirect.com/science/article/pii/S098194281200191X

Saxena, N. P., O'Toole, J. C., & International Crops Research Institute for the Semi-Arid Tropics. (2002). Field screening for drought tolerance in crop plants with emphasis on rice. Proceedings of an International Workshop on Field Screening for Drought Tolerance in Rice, 11–14 Dec. 2000, ICRISAT, Patancheru, India. International Crops Research Institute for the Semi-Arid Tropics.

Schneider, M., & Poux, S. (2012). UniProtKB amid the turmoil of plant proteomics research. Frontiers in Plant Science, 3(DEC). https://doi.org/10.3389/FPLS.2012.00270/FULL

Schütz, W., Hausmann, N., Krug, K., Hampp, R., & Macek, B. (2011). Extending SILAC to proteomics of plant cell lines. The Plant Cell, 23(5), 1701–1705.

Shannon, M. C. (1985). Principles and strategies in breeding for higher salt tolerance. In: Biosalinity in Action: Bioproduction with Saline Water. Springer, Pp. 227–241.

Shu, L., Lou, Q., Ma, C., Ding, W., Zhou, J., Wu, J., Feng, F., Lu, X., Luo, L., Xu, G., & Mei, H. (2011). Genetic, proteomic and metabolic analysis of the regulation of energy storage in rice seedlings in response to drought. Proteomics, 11(21), 4122–4138. https://doi.org/10.1002/PMIC.201000485

Singh, A., & Roychoudhury, A. (2021). Gene regulation at transcriptional and post transcriptional levels to combat salt stress in plants. Physiologia Plantarum, 173, 1556–1572.

Singh, P. K., Indoliya, Y., Agrawal, L., Awasthi, S., Deeba, F., Dwivedi, S., Chakrabarty, D., Shirke, P. A., Pandey, V., Singh, N., Dhankher, O. P., Barik, S. K., & Tripathi, R. D. (2022). Genomic and proteomic responses to drought stress and biotechnological interventions for enhanced drought tolerance in plants. Current Plant Biology, 29, 100239. https://doi.org/10.1016/j.cpb.2022.100239

Singh, R. P., Runthala, A., Khan, S., & Jha, P. N. (2017). Quantitative proteomics analysis reveals the tolerance of wheat to salt stress in response to enterobacter cloacae SBP-8. PLoS ONE, 12(9), e0183513. https://doi.org/10.1371/JOURNAL.PONE.0183513

Sinha, R., Gupta, A., & Senthil-Kumar, M. (2016). Understanding the impact of drought on foliar and xylem invading bacterial pathogen stress in Chickpea. Frontiers in Plant Science, 7. https://doi.org/10.3389/fpls.2016.00902

Smejkal, G. B. (2004). The coomassie chronicles: Past, present and future perspectives in polyacrylamide gel staining. Expert Review of Proteomics, 1(4), 381–387.

Soares, A. L. C., Geilfus, C. M., & Carpentier, S. C. (2018). Genotype-specific growth and proteomic responses of maize toward salt stress. Frontiers in Plant Science, 9. https://doi.org/10.3389/FPLS.2018.00661/FULL

Sun, Y., Mu, C., Zheng, H., Lu, S., Zhang, H., et al. (2018). Exogenous Pi supplementation improved the salt tolerance of maize (Zea mays L.) by promoting Na+ exclusion. Nature.Com.

Suzuki, C. K., Rep, M., Van Dijl, J. M., Suda, K., Grivell, L. A., & Schatz, G. (1997). ATP-dependent proteases that also chaperone protein biogenesis. Trends in Biochemical Sciences, 22(4), 118–123. https://doi.org/10.1016/S0968-0004(97)01020-7

Tebini, M., Rabaoui, G., M'Rah, S., Luu, D. T., Ben Ahmed, H., & Chalh, A. (2022). Effects of salinity on germination dynamics and seedling development in two amaranth genotypes. Physiol Mol Biol Plants., 28(7), 1489–1500. https://doi.org/10.1007/s12298-022-01221-4

Thomas, S., Kumar, R., Sharma, K., Barpanda, A., Sreelakshmi, Y., Sharma, R., & Srivastava, S. (2021). iTRAQ-based proteome profiling revealed the role of phytochrome A in regulating primary metabolism in tomato seedling. Scientific Reports, 11(1), 1–21.

Tian, L., You, H. Z., Wu, H., Wei, Y., Zheng, M., He, L., … & Hu, X. (2019). iTRAQ-based quantitative proteomic analysis provides insight for molecular mechanism of neuroticism. Clinical Proteomics, 16(1), 1–13.

Triboï, E., Martre, P., & Triboï-Blondel, A. M. (2003). Environmentally-induced changes in protein composition in developing grains of wheat are related to changes in total protein content. J Exp Bot., 54(388), 1731–1742. https://doi.org/10.1093/jxb/erg183.

Ul Haq, S., Khan, A., Ali, M., Khattak, A. M., Gai, W. X., Zhang, H. X., Wei, A. M., & Gong, Z. H. (2019). Heat shock proteins: Dynamic biomolecules to counter plant biotic and abiotic stresses. Int J Mol Sci., 20(21), 5321.

Vizcaíno, J. A., Côté, R. G., Csordas, A., Dianes, J. A., Fabregat, A., Foster, J. M., Griss, J., Alpi, E., Birim, M., Contell, J., O'Kelly, G., Schoenegger, A., Ovelleiro, D., Pérez-Riverol, Y., Reisinger, F., Ríos, D., Wang, R., & Hermjakob, H. (2013). The PRoteomics IDEntifications (PRIDE) database and associated tools: Status in 2013. Nucleic Acids Res., 41(Database issue), D1063–D1069. https://doi.org/10.1093/nar/gks1262

Wang, N., Zhao, J., He, X., Sun, H., Zhang, G., & Wu, F. (2015). Comparative proteomic analysis of drought tolerance in the two contrasting Tibetan wild genotypes and cultivated genotype. BMC Genomics, 16(1). https://doi.org/10.1186/s12864-015-1657-3

Wang, X., Cai, X., Xu, C., Wang, Q., & Dai, S. (2016). Drought-responsive mechanisms in plant leaves revealed by proteomics. Int J Mol Sci., 17(10), 1706. https://doi.org/10.3390/ijms17101706

Wu, Y., Mirzaei, M., Pascovici, D., Chick, J., et al. (2016). Quantitative proteomic analysis of two different rice varieties reveals that drought tolerance is correlated with reduced abundance of photosynthetic machinery. Elsevier. Retrieved December 27, 2022, from https://www.sciencedirect.com/science/article/pii/S187439191630207X

Xu, J., Lan, H., Fang, H., Huang, X., Zhang, H., & Huang, J. (2015). Quantitative proteomic analysis of the rice (Oryza sativa L.) salt response. PLoS ONE, 10(3). https://doi.org/10.1371/JOURNAL.PONE.0120978

Yan, M., Xue, C., Xiong, Y., Meng, X., Li, B., Shen, R., & Lan, P. (2020). Proteomic dissection of the similar and different responses of wheat to drought, salinity and submergence during seed germination. J Proteomics., 220, 103756. https://doi.org/10.1016/j.jprot.2020.103756

Yan, M., Zheng, L., Li, B., Shen, R., & Lan, P. (2021). Comparative proteomics reveals new insights into the endosperm responses to drought, salinity and submergence in germinating wheat seeds. Plant Molecular Biology, 105(3), 287–302. https://doi.org/10.1007/S11103-020-01087-8

Yang, D.-H., Andersson, B., Aro, E.-M., & Ohad, I. (2001). The redox state of the plastoquinone pool controls the level of the light-harvesting chlorophyll a/b binding protein complex II (LHC II) during photoacclimation. Photosynthesis Research, 68(2), 163–174.

Yang, D. H., Kwak, K. J., Kim, M. K., Park, S. J., Yang, K. Y., & Kang, H. (2014). Expression of Arabidopsis glycine-rich RNA-binding protein AtGRP2 or AtGRP7 improves grain yield of rice (Oryza sativa) under drought stress conditions. Plant Sci., 214, 106–112. https://doi.org/10.1016/j.plantsci.2013.10.006

Yang, Y., & Guo, Y. (2018). Unraveling salt stress signaling in plants. J Integr Plant Biol., 60(9), 796–804.

Yin, Y., Qi, F., Gao, L., Rao, S., Yang, Z., & Fang, W. (2018). iTRAQ-based quantitative proteomic analysis of dark-germinated soybeans in response to salt stress. RSC Advances, 8(32), 17905–17913. https://doi.org/10.1039/C8RA02996B

Yu, X., James, A. T., Yang, A., Jones, A., Mendoza-Porras, O., Bétrix, C. A., Ma, H., & Colgrave, M. L. (2016). A comparative proteomic study of drought-tolerant and drought-sensitive soybean seedlings under drought stress. Crop and Pasture Science, 67(5), 528–540. https://doi.org/10.1071/CP15314

Zadražnik, T., Hollung, K., Egge-Jacobsen, W., Meglič, V., & Šuštar-Vozlič, J. (2013). Differential proteomic analysis of drought stress response in leaves of common bean (*Phaseolus vulgaris* L.). Journal of Proteomics, 78, 254–272. https://doi.org/10.1016/J.JPROT.2012.09.021

Zang, X., & Komatsu, S. (2007). A proteomics approach for identifying osmotic-stress-related proteins in rice. Phytochemistry, 68(4), 426–437. https://doi.org/10.1016/j.phytochem.2006.11.005

Zenda, T., Liu, S., Wang, X., Jin, H., Liu, G., & Duan, H. (2018). Comparative proteomic and physiological analyses of two divergent maize inbred lines provide more insights into drought-stress tolerance mechanisms. Int J Mol Sci., 19(10), 3225.

Zeng, W., Peng, Y., Zhao, X., Wu, B., Chen, F., Ren, B., Zhuang, Z., Gao, Q., & Ding, Y. (2019). Comparative proteomics analysis of the seedling root response of drought-sensitive and drought-tolerant maize varieties to drought stress. Int J Mol Sci., 20(11), 2793. https://doi.org/10.3390/ijms20112793

Zhang, Z., Singh Jatana, B., Campbell, B. J., Gill, J., Suseela, V., & Tharayil, N. (2022). Cross-inoculation of rhizobiome from a congeneric ruderal plant imparts drought tolerance in maize (*Zea mays*) through changes in root morphology and proteome. The Plan Journal, 111(1), 54–71. https://doi.org/10.1111/tpj.15775

Zhang, Z., Zhang, Q., Wu, J., Zheng, X., Zheng, S., Sun, X., Qiu, Q., & Lu, T. (2013). Gene knockout study reveals that cytosolic ascorbate peroxidase 2(OsAPX2) plays a critical role in growth and reproduction in rice under drought, salt and cold stresses. PLoS ONE, 8(2). https://doi.org/10.1371/JOURNAL.PONE.0057472

Zhou, J., Ma, C., Zhen, S., Cao, M., Zeller, F. J., Hsam, S. L. K., & Yan, Y. (2016). Identification of drought stress related proteins from 1Sl(1B) chromosome substitution line of wheat variety Chinese Spring. Botanical Studies, 57(1). https://doi.org/10.1186/S40529-016-0134-X

Zhu, D., Luo, F., Zou, R., Liu, J., Yan, Y. (2021). Integrated physiological and chloroplast proteome analysis of wheat seedling leaves under salt and osmotic stresses. J Proteomics., 234:104097. https://doi.org/10.1016/j.jprot.2020.104097

Zörb, C., Schmitt, S., & Mühling, K. H. (2010). Proteomic changes in maize roots after short-term adjustment to saline growth conditions. Proteomics, 10(24), 4441–4449. https://doi.org/10.1002/PMIC.201000231

6 Proteomics Studies to Understand Heavy Metal Stress Response in Plants

Aryadeep Roychoudhury

6.1 INTRODUCTION

Currently, high throughput omics strategies have been largely exploited for the dissection of plant molecular mechanisms involving stress tolerance cascades that are functional under conditions of heavy metal toxicity in plants. Plants have evolved coordinated homeostatic processes regulating the uptake, intracellular concentration and mobilisation of heavy metal ions, for the alleviation of heavy metal stress-associated injuries. Proteomic studies highlight the operational metabolic circuits and protein networks, primarily associated with stress tolerance and cellular detoxification of heavy metals, since the functionally translated portion of the plant genome largely modulates the plant stress-responsive mechanisms (Hossain and Komatsu 2013a, b).

By convention, elements with a specific gravity of more than five are referred to as heavy metals. Heavy metals are particularly toxic in nature and have been reported to exert detrimental effects on overall plant development and growth (Hossain et al. 2012). Some common heavy metals include copper (Cu), cadmium (Cd), lead (Pb), chromium (Cr) and zinc (Zn), as well as certain toxic metalloids like boron (B) and arsenic (As). Many reports indicate the uptake of various heavy metals via plant root-associated generic channels or ion carriers (Bubb and Lester 1991). However, plant transporters mediating the uptake of essential metal ions, including calcium (Ca^{+2}), zinc (Zn^{+2}) and iron (Fe^{+2}) rather lack specificity, allowing the uptake of toxic lead (Pb^{+2}) and cadmium (Cd^{+2}) ions within the plant system (Perfus-Barbeoch et al. 2002; Samanta and Roychoudhury 2021). Upon entry into the cell, heavy metal ions widely affect cellular functions and metabolism, viz., by binding to protein-associated sulfhydryl groups and by replacing protein binding site-localised essential cations, leading to disruption of protein and enzymatic functions. Also, heavy metal toxicity results in the generation of detrimental reactive oxygen species (ROS), which in turn evokes cellular oxidative injuries, involving widespread damages to proteins, nucleic acids and lipids (Sharma and Dietz 2009; Roychoudhury et al. 2012a, b).

Since the past decade, elaborate research on plant heavy metal stress response has been undertaken to unravel various underlying plant tolerance response mechanisms. Upcoming genomic pipelines have been found to be quite efficient in addressing plant responses against myriads of abiotic stresses, including heavy metal toxicity (Bohnert et al. 2006). Alterations in gene expression patterns at the level of transcripts are sometimes not reflected in terms of proteins. Hence, the proper identification of protein targets actively participating in plant heavy metal detoxification demands an in-depth proteomic analysis (Roychoudhury et al. 2011).

Various metal stress-related proteins mediating the sequestration of heavy metal ions, antioxidative defence machinery, role of subcellular organelles and various metabolic cascades regulating plant heavy metal detoxification demand greater emphasis. The current chapter highlights the state of art, involving the recent research developments associated with proteomic technologies and the progressing contributions made till date for the proper understanding of plant heavy metal stress responses, particularly at the protein level.

DOI: 10.1201/b23255-6

6.2 HEAVY METAL TOLERANCE STRATEGIES IN PLANTS

Higher plants have eventually evolved, in course of time, several sophisticated mechanisms for the efficient regulation of intracellular concentration, mobilisation and uptake of heavy metal ions. The most common approach employed by plants for rendering protection to plant cells from the adversities of heavy metal toxicity is the plasma membrane exclusion method. In addition, vacuolar sequestration of toxic metal ions via thiol-containing chelating compounds and membrane transporters also forms an important part of the heavy metal detoxification machinery in plants (Hossain and Komatsu 2013a, b). For the maintenance of redox homeostasis, plants tend to shoot up the abundance of defence proteins involved in the efficient scavenging of toxic ROS during stress. Moreover, plants also upregulate the activities of molecular chaperones, which aid in the reestablishment of normal conformation of proteins subjected to heavy metal toxicity (Das and Roychoudhury 2014). Heavy metal-challenged plant cells tend to exhibit high energy demand. In order to suffice for this increased energy requirement, plants modulate the vital metabolic pathways, viz., mitochondrial respiration and photosynthesis, thereby leading to increased reducing power generation. Some of the common strategies displayed by plant systems for the efficient detoxification of heavy metals are as follows:

i. Heavy metal chelation, complexion and compartmentalisation within plant cells (Verbruggen et al. 2009; Roychoudhury et al. 2020)
ii. Alteration of cellular redox homeostasis in response to heavy metal-induced oxidative stress (Dietz et al. 2006)
iii. Upregulation of the level and activity of molecular chaperones (Wang et al. 2004)
iv. Alteration of the level of proteins associated with plant energy metabolism and photosynthesis (Kieffer et al. 2008)
v. Heavy metal-induced accumulation of pathogenesis-related (PR) proteins (Durrant and Dong 2004)

6.3 ANALYSIS OF HEAVY METAL RESPONSIVE PROTEINS BY QUANTITATIVE PROTEOMIC TECHNOLOGIES

After the successful completion of genome sequencing of some common model plant species like rice and *Arabidopsis*, researchers are occasionally facing great hardships in the proper designation of the functional role of many genes and their subsequent products. Thus, it has been quite challenging for scientists to establish how biological entities interact with each other for the proper shaping of a fully functional individual. Modern "omics" strategies seem to be quite efficient in the expansion of the existing knowledge of molecular cascades mediating plant tolerance against various environmental stressors. Traditional, physiological and biochemical techniques are rather insufficient in terms of revealing the deeper insights of plant abiotic and biotic stress responses (Ahsan et al. 2009). The most widely exploited proteomic technique for the proper understanding of heavy metal-induced changes in plant proteome is conventional 2-dimensional gel electrophoresis (2-DE) in conjunction with mass spectrometric (MS) protein identification.

As a part of the heavy metal-induced stress response, plants display widespread changes in protein expression, including an inevitable quantitative and qualitative alteration in the total protein pool. However, various genomic techniques particularly monitor the alterations at the level of mRNA, i.e., transcriptional level, which necessarily might not correlate with the alterations at the protein level. It has been found that the correlation coefficient between the total mRNA pool and the protein abundance is quite low, ranging approximately around 0.5 (Anderson and Seilhamer 1997). Hence, in reality, the ratio of mRNA to protein is largely governed by the rate of translation and the extent of stability of proteins (particularly marked by the rate of de novo synthesis and degradation). Additionally, the availability of mRNA seems to be governed by post-transcriptional events occurring during the maturation of RNA, including splicing, translational initiation, transportation and RNA degradation processes (Mazzucotelli et al. 2008). Hence, it is not correct to predict the alterations in protein expression by only using the information on the mRNA level of any biological sample (Hakeem et al. 2012). Moreover, the expression of proteins is modulated at the post-translational level by a series of events, viz., phosphorylation, glycosylation, ubiquitination and sumoylation (Cobon et al. 2002; Hirano et al. 2004). The function of a protein can be perceived and studied at various levels, viz., cellular, biochemical, physiological, developmental or even at the higher level of biological organisation, encompassing organs and sometimes the whole individual. Proteins, the functional translational product of the genome, exhibit crucial functional and structural roles, as well as employ certain effector molecules, for the efficient alleviation of the adverse effects of environmental stressors in plants (Timperio et al. 2008).

Presently, there has been a rapid expansion of knowledge involving protein structure and function. Therefore, for the sake of high throughput analysis of proteins, a relatively newer technology has evolved in the form of proteomics. It has been proven that proteins are quintessential for vital cellular processes (Swinbanks 1995), and proteomics serves as a promising tool allowing the assessment of any form of alterations occurring in the protein pool upon the inception of various environmental stress conditions. A proteome is defined as the entire protein set expressed at a given point of time by a genome. Basically, a proteome represents the complete protein set expressed under a particular condition, by a specific cell type or tissue, at a particular point of time (de Hoog and Mann 2004). Proteomics encompass protein identification, protein quantification, protein expression patterns, protein-protein interactions and post-translational modification of proteins.

Proteomics particularly focuses on the actively transcribed part of the genome, which encodes for functional proteins. Hence, the applications of proteomics are expanding at a rapid rate in the field of plant heavy metal stress response. However, such proteomic studies involving metal stress responses are rather less comprehensive, as compared to those involving other forms of environmental stressors. This is largely due to the lesser availability of information depicting the detailed mechanisms regulating the processes of oxidative stress induction by toxic metals and the uptake, distribution, accumulation and detoxification of heavy metals within plants (Cuypers et al. 2001; Aravind and Prasad 2003; Hassan et al. 2005; Horvat et al. 2007; Gratão et al. 2008; Tkalec et al. 2008; Cvjetko et al. 2010; Balen et al. 2011). Thus, the proteomic approach can aid in the elucidation of new aspects, associated with plant heavy metal stress.

For the lucid description of proteome sets varying in the quality and quantity of proteins, the differential expression approach serves as the most efficient strategy, allowing the identification of proteins, as well as their relative quantitative determination. The above-mentioned methodology has been widely exploited in studies involving plant heavy metal stress response. This method is based on the variability of different proteome compositions. Basically, it deals with the comparison of the proteomes originating from heavy metal-stressed and non-stressed plant samples (Bona et al. 2007; Kieffer et al. 2008). Additionally, an alternative strategy involves the comparison of proteomes associated with different genotypes depicting distinct tolerance traits against different heavy metals. For instance, comparison of proteomes from two contrasting cadmium accumulating soybean cultivars exhibited differential responses of proteins (Ahsan et al. 2012). Similar differential responses have been observed upon comparative proteomic analysis of cadmium-exposed contrasting flax cultivars and aluminium-stressed roots, obtained from tolerant and sensitive soybean genotypes (Hradilová et al. 2010; Duressa et al. 2011).

Currently, a number of studies investigating the heavy metal stress-responsive and differentially expressed set of proteins in plants employ a classical gel-based proteomic strategy, viz., 1-D and 2-D polyacrylamide gel electrophoresis, in order to obtain a high-quality protein separation. Post separation, the proteins are identified via database search and MS analysis (Führs et al. 2008; Lee et al. 2010). However, the extraction process and preparation of samples for proteins turn out to be the most challenging part of proteomics study, especially for those comprising 2-DE as the initiating step. For a reproducible and good 2-DE procedure, efficient extraction, purification and solubilisation of proteins from plants form the most important part, because the quality and quantity of proteins extracted from plant tissues designate the protein resolution, spot number and intensity. Therefore, the efficacy of any protein isolation method largely depends on the reproducible capture and proper protein solubilisation with least post-extraction artefacts, minimal non-protein contamination and limited proteolytic degradation (Rose et al. 2004; Cho et al. 2006). Additionally, the plant samples are often dominated by a handful of high abundance proteins, which tends to obscure the presence of less abundant proteins, thereby hampering the resolution and capacity of the employed separation method. For instance, in photosynthetic tissues, ribulose bisphosphate carboxylase (RuBisCO) hinders the availability of other proteins of low abundance (Xi et al. 2006).

The extraction of protein pools and their subsequent purification from heavy metal-stressed plant tissue samples seems to be the most critical step in the 2-DE technique. In addition, the quality and amount of protein extracted from the samples aids in the proper determination of protein resolution, intensity and spot number. Oxidative and proteolytic enzymes, terpenes, phenolics, carbohydrates, organic acids, pigments and certain inhibitory ions serve as interfering agents, commonly found in recalcitrant samples. Moreover, proteolytic degradation, charge heterogeneity and streaking commonly occur due to inferior 2-D gel separation. Some of the most efficient methods for the optimal extraction of proteins and formulation of high-quality protein maps are acetone or trichloroacetic acid precipitation approach and phenol-mediated approach, which have enabled researchers to conduct elaborate proteomic analysis involving plant heavy metal stress responses (Alves et al. 2011; Ahsan et al. 2012). However, for the proper extraction of glycoproteins and the generation of superior resolution proteome map, especially from recalcitrant samples, the phenol-based approach seems to be potentially appropriate (Komatsu and Ahsan 2009).

An advanced proteomic approach involving fluorescence-based 2-dimensional difference gel electrophoresis (2-D DIGE) has replaced classical 2-DE gel staining procedures. Such advanced techniques allow researchers to compare differentially expressed proteins of stressed and control plant samples in a single gel in an appreciably efficient manner (Kieffer et al. 2008). Basically, DIGE is a gel-based approach employing fluorescent cyanine (Cy) dyes for protein labelling prior to gel electrophoresis. For quantitative and comparative multiple protein sample analysis, technology-multiplexed isobaric tagging (iTRAQ) method of peptide analysis seems to be a quite promising advanced approach. For instance, this second-generation gel-free proteomic technique has enabled scientists to efficiently unravel the plant responses, operational under conditions of metal toxicity, viz., cadmium and boron (Patterson et al. 2007; Schneider et al. 2009). Till date, 2-D DIGE has been implemented for the identification of proteins, which are differentially regulated in response to heavy metal toxicity in cadmium-stressed poplar (Kieffer et al. 2008). Also, 2-D DIGE has led to the affirmation of proteomic alterations in soybean roots and tomato cotyledons during aluminium exposure (Ahsan et al. 2012). Hence, for the efficient identification of heavy metal-specific stress protein markers and molecular characterisation of heavy metal-induced plant responses, a combinatorial approach involving 2-D DIGE and MS seems to be an efficient tool.

6.4   RECENT TRENDS IN PROTEOMICS OF PLANT HEAVY METAL STRESS RESPONSES

With the advancement in proteomic technologies associated with protein identification and separation, extensive effort needs to be applied for the proper clarification of underlying molecular mechanisms regulating heavy metal-induced responses in plants. However, the essentiality of protein profiling needs to be properly addressed. In addition, very few proteomic studies have been carried out in plant samples exposed to various stress conditions, particularly heavy metal stress (Liao et al. 2005; Ahsan et al. 2009).

Amongst the few proteomic studies carried out in plant samples under heavy metal stress, a maximum number of reports are available based on the proteomic analysis, indicating the negative impacts of cadmium exposure on plant proteome. For instance, in cadmium-exposed poplar plant samples, proteomic studies demonstrate the deleterious impacts on the expression of proteins regulating primary carbon metabolism and involved in plant oxidative stress responses (Kieffer et al. 2008). Proteomic studies of leaf and root samples of cadmium-stressed poplar also exhibited an increase in stress-related protein accumulation, viz., chaperones, heat shock proteins, proteases, pathogenesis-related proteins and foldases. However, in case of poplar root samples, there was a significant fall in the level of proteins modulating primary metabolism in plants (Kieffer et al. 2009). In addition, a series of reports indicated the upregulation and activation of ROS scavengers for the mediation of prompt anti-oxidative responses for the curbing of cadmium-induced oxidative damages in plants (Lee et al. 2010; Pandey et al. 2012). Cadmium-induced enhancement in the levels of heat shock proteins and molecular chaperones plays a crucial role in adequate protein folding, stabilisation of proteins, efficient protein translocation and assembly (Wang et al. 2004). Also, proteins involved in the chelation of toxic cadmium have been found to be upregulated in response to cadmium toxicity in plants (Horvat et al. 2007; Kieffer et al. 2008; Semane et al. 2010). One of the most effective mechanisms adopted by plants to tackle the adversities of heavy metal toxicity is to synthesise low molecular weight chelators to prevent the attachment of toxic heavy metal ions to crucial plant proteins (Verbruggen et al. 2009).

Extensive proteomic analysis established the copper-induced upregulation of proteins, which plays important role in crucial plant metabolic processes like glycolysis, photosynthesis, pentose phosphate pathway, stress response and anti-oxidative cascades in plants (Ahsan et al. 2007). In *Elsholtzia splendens*, copper exposure resulted in upregulated expression of a series of proteins involved in ion-binding and vacuolar metal sequestration, mediating protection against toxic effects of copper ions (Li et al. 2009). Elevated arsenic levels result in partial damage to the photosynthetic machinery and also lead to the disruption of many photosynthetic proteins like RuBisCO (Duquesnoy et al. 2009). On the other hand, arsenate or arsenite exposure in plants led to overexpression of anti-oxidative enzymes and proteins involved in energy production and important metabolic cascades (Ahsan et al. 2010), in order to suffice the increased energy demand and to tackle the adversities of oxidative stress upon arsenic imposition. Proteomic studies also revealed the importance of glutathione and cysteine synthase in the adaptation of rice and soybean plants against aluminium toxicity (Yang et al. 2007; Zhen et al. 2007). Moreover, aluminium stress imposition resulted in the enhanced expression of enzymes mediating detoxification of ROS and different molecular chaperone proteins required for optimal folding of disrupted proteins. In tomato root samples, aluminium exposure led to overexpression of enzymes involved in detoxification of toxic metal ions and anti-oxidative cascades (Zhou et al. 2009). Upon comparison between the proteomes of aluminium-sensitive and tolerant soybean roots, it was found that the tolerant roots exhibited an increased level of the organic acid citrate, mediating detoxification of aluminium, but aluminium stress-responsive proteins were found to be upregulated in the sensitive genotype (Duressa et al. 2011).

Proteomic studies in leaf tissues of manganese-exposed cowpea plants revealed the potential role of peroxidases in the mediation of hydrogen peroxide detoxification and phenol oxidation in the apoplast of leaves (Fecht-Christoffers et al. 2003). In addition, rise in the level of peroxidases upregulated the secretion of pathogenesis-related enzymes (chitinase, thaumatin, and glucanase) and wound-induced proteins, as a part of manganese-mediated late response. Additionally, manganese stress in cowpea also resulted in the downregulation of proteins regulating carbon assimilation and photosynthesis (Führs et al. 2007). An increased abundance of RuBisCO activase enzyme and other proteins regulating amino acid metabolism was observed in chromate-treated *Pseudokirchneriella subcapitata* (Vannini et al. 2009). In addition, chromium-responsive proteins mediating tolerance against heavy metals were found to be over-accumulated in chromium-stressed *Miscanthus sinensis* roots (Sharmin et al. 2012). Upon chromium treatment, appreciable proteomic changes were observed, marked by the upregulation of proteins involved in defence pathways, photosynthesis, ROS detoxification and chloroplast organisation, indicating the modification of the entire plant metabolism as a part of plant adaptive response against chromium (Wang et al. 2013). Boron deficiency in plants also leads to widespread alteration in the entire plant proteome, leading to downregulation of proteins involved in cell division, energy production and other crucial metabolic processes (Alves et al. 2011). However, in a comparative proteomic analysis involving boron sensitive and tolerant barley genotypes, enzymes mediating phytosiderophore synthesis involved in chelating activity were upregulated in the tolerant barley genotype (Patterson et al. 2007).

6.5   CONCLUSION AND FUTURE PERSPECTIVES

The present chapter highlights the effects of heavy metal toxicity on proteomic constituents in plants. Most of the reports available till date outline the differential expression of plant defence-related proteins, as well as proteins involved in heavy metal detoxification cascades, including metal compartmentalisation and chelation and scavenging of toxic ROS. Moreover, modulation of protein levels regulating plant mitochondrial respiratory pathways and carbon assimilation, along with upregulation of pathogenesis-related proteins, forms a crucial part of heavy metal stress response. The classical 2-DE in conjunction with MS technology seems to be the most widely employed proteomic tool for deciphering facts related to heavy metal tolerance in plants. However, proteomic studies are rather limited to model plants and implementation of this technique is less implemented in plants, whose whole genome sequence is yet to be retrieved and functional annotation of proteins is at a preliminary stage. Some physiological and biological events are part and parcel of certain plant species for which the entire sequence of the genome is not available. This seems to be the greatest drawback of this technology. Moreover, high proteome coverage, encompassing the entire proteome profile, can even lack information related to model plants. Additionally, protein identification from certain biological samples often fails to reveal the actual function of the protein. It is expected that proteomic analysis involving MS and NanoString technique will aid in the identification of novel protein candidates, mediating improved plant tolerance response against heavy metals.

ACKNOWLEDGEMENTS

Financial assistance from the Science and Engineering Research Board, Government of India through the grant [EMR/2016/004799] and the Department of Higher Education, Science and Technology and Biotechnology, Government of West Bengal, through the grant [264(Sanc.)/ST/P/S&T/1G-80/2017] to Prof. Aryadeep Roychoudhury is gratefully acknowledged.

REFERENCES

Ahsan N, Lee DG, Kim KH, Alam I, Lee SH, Lee KW, Lee H, Lee BH (2010) Analysis of arsenic stress-induced differentially expressed proteins in rice leaves by two-dimensional gel electrophoresis coupled with mass spectrometry. Chemosphere 78:224–231.

Ahsan N, Lee DG, Lee SH, Kang KY, Lee JJ, Kim PJ, Yoon HS, Kim JS, Lee BH (2007) Excess copper induced physiological and proteomic changes in germinating rice seeds. Chemosphere 67:1182–1193.

Ahsan N, Nakamura T, Komatsu S (2012) Differential responses of microsomal proteins and metabolites in two contrasting cadmium (Cd)-accumulating soybean cultivars under Cd stress. Amino Acids 42:317–327.

Ahsan N, Renaut J, Komatsu S (2009) Recent developments in the application of proteomics to the analysis of plant responses to heavy metals. Proteomics 9:2602–2621.

Alves M, Moes S, Jenö P, Pinheiro C, Passarinho J, Ricardo CP (2011) The analysis of *Lupinus albus* root proteome revealed cytoskeleton altered features due to long-term boron deficiency. J Proteomics 74:1351–1363.

Anderson L, Seilhamer J (1997) A comparison of selected mRNA and protein abundances in human liver. Electrophoresis 18:533–537.

Aravind P, Prasad MNV (2003) Zinc alleviates cadmium-induced oxidative stress in *Ceratophyllum demersum* L.: A free floating freshwater macrophyte. Plant Physiol Biochem 41:391–397.

Balen B, Tkalec M, Šikić S, Tolić S, Cvjetko P, Pavlica M, Vidaković-Cifrek Ž (2011) Biochemical responses of *Lemna minor* experimentally exposed to cadmium and zinc. Ecotoxicology 20:815–826.

Bohnert HJ, Gong Q, Li P, Ma S (2006) Unravelling abiotic stress tolerance mechanisms–Getting genomics going. Curr Opin Plant Biol 9:180–188.

Bona E, Marsano F, Cavaletto M, Berta G (2007) Proteomic characterization of copper stress response in *Cannabis sativa* roots. Proteomics 7:1121–1130.

Bubb JM, Lester JN (1991) The impact of heavy metals on low-land rivers and the implications for man and the environment. Sci Total Environ 100:207–233.

Cho K, Torres NL, Subramanyam S, Deepak SA, Sardesai N, Han O, Williams CE, Ishii H, Iwahashi H, Rakwal R (2006) Protein extraction/solubilization protocol for monocot and dicot plant gel-based proteomics. J Plant Biol 49:413–420.

Cobon GS, Verrills N, Papakostopoulos P, Eastwood H, Linnane AW (2002) The proteomics of ageing. Biogerontology 3:133–136.

Cuypers A, Vangronsveld J, Clijsters H (2001) The redox status of plant cells (AsA and GSH) is sensitive to zinc imposed oxidative stress in roots and primary leaves of *Phaseolus vulgaris*. Plant Physiol Biochem 39:657–664.

Cvjetko P, Tolić S, Šikić S, Balen B, Tkalec M, Vidaković-Cifrek Ž, Pavlica M (2010) Effect of copper on the toxicity and genotoxicity of cadmium in duckweed (*Lemna minor* L.). Arh Hig Rada Toksikol 61:287–296.

Das K, Roychoudhury A (2014) Reactive oxygen species (ROS) and response of antioxidants as ROS-scavengers during environmental stress in plants. Front Environ Sci 2:53.

de Hoog CL, Mann M (2004) Proteomics. Annu Rev Genomics Hum Genet 5:267–293.

Dietz KJ, Jacob S, Oelze ML, Laxa M, Tognetti V, de Miranda SM, et al. (2006). The function of peroxiredoxins in plant organelle redox metabolism. J Exp Bot 57:1697–1709.

Duquesnoy I, Goupil P, Nadaud I, Branlard G, Piquet-Pissaloux A, Ledoigt G (2009) Identification of *Agrostis tenuis leaf* proteins in response to As(V) and As(III) induced stress using a proteomics approach. Plant Sci 176:206–213.

Duressa D, Soliman K, Taylor R, Senwo Z (2011) Proteomic analysis of soybean roots under aluminum stress international. Int J Plant Genomics 2011:2825–2831.

Durrant WE, Dong X (2004) Systemic acquired resistance. Annu Rev Phytopathol 42:185–209.

Fecht-Christoffers MM, Braun HP, Lemaitre-Guillier C, Van Dorsselaer A, Horst WJ (2003) Effect of manganese toxicity on the proteome of the leaf apoplast in cowpea. Plant Physiol 133:1935–1946.

Führs H, Hartwig M, Molina LE, Heintz D, Van Dorsselaer A, Braun HP, Horst WJ (2008) Early manganese-toxicity response in *Vigna unguiculata* L. – A proteomic and transcriptomic study. Proteomics 8:149–159.

Gratão PL, Monteiro CC, Antunes AM, Peres LEP, Azevedo RA (2008) Acquired tolerance of tomato (*Lycopersicon esculentum* cv. Micro-Tom) plants to cadmium induced stress. Ann Appl Biol 153:321–333.

Hakeem KR, Chandna R, Ahmad P, Iqbal M, Ozturk M (2012) Relevance of proteomic investigations in plant abiotic stress physiology. OMICS 16:621–635.

Hassan MJ, Zhang G, Wu F, Wie K, Chen Z (2005) Zinc alleviates growth inhibition and oxidative stress caused by cadmium in rice. J Plant Nutr Soil Sci 168:255–261.

Hirano H, Islam N, Kawasaki H (2004) Technical aspects of functional proteomics in plants. Phytochemistry 65:1487–1489.

Horvat T, Vidaković-Cifrek Ž, Oreščanin V, Tkalec M, Pevalek-Kozlina B (2007) Toxicity assessment of heavy metal mixtures by *Lemna minor* L. Sci Total Environ 384:229–238.

Hossain Z, Komatsu S (2013a) Contribution of proteomic studies towards understanding plant heavy metal stress response. Front Plant Sci 3:310.

Hossain Z, Komatsu S (2013b) Contribution of proteomic studies towards understanding plant heavy metal stress response. Front Plant Sci 3:1–12.

Hossain MA, Piyatida P, Teixeira da Silva JA, Fujita M (2012) Molecular mechanism of heavy metal toxicity and tolerance in plants: Central role of glutathione in detoxification of reactive oxygen species and methylglyoxal and in heavy metal chelation. J Bot 2012:1–37.

Hradilová J, Rehulka P, Rehulková H, Vrbová M, Griga M, Brzobohatý B (2010) Comparative analysis of proteomic changes in contrasting flax cultivars upon cadmium exposure. Electrophoresis 31:421–431.

Kieffer P, Dommes J, Hoffmann L, Hausman JF, Renaut J (2008) Quantitative changes in protein expression of cadmium-exposed poplar plants. Proteomics 8:2514–2530.

Kieffer P, Planchon S, Oufir M, Ziebel J, Dommes J, Hoffmann L (2009) Combining proteomics and metabolite analyses to unravel cadmium stress-response in poplar leaves. J Proteome Res 8:400–417.

Komatsu S, Ahsan N (2009) Soybean proteomics and its application to functional analysis. J Proteomics 72:325–336.

Lee K, Bae DW, Kim SH, Han HJ, Liu X, Park HC, Lim CO, Lee SY, Chung WS (2010) Comparative proteomic analysis of the short-term responses of rice roots and leaves to cadmium. J Plant Physiol 167:161–168.

Li F, Shi J, Shen C, Chen G, Hu S, Chen Y (2009) Proteomic characterization of copper stress response in *Elsholtzia splendens* roots and leaves. Plant Mol Biol 71:251–263.

Liao XY, Chen TB, Xie H, Liu YR (2005) Soil as contamination and its risk assessment in areas near the industrial districts of Chenzhou City, Southern China. Environ Int 31:791–798.

Mazzucotelli E, Mastrangelo AM, Crosatti C, Guerra D, Stanca AM, Cattivelli L (2008) Abiotic stress response in plants: When post-transcriptional and post-translational regulations control transcription. Plant Sci 174:420–431.

Pandey S, Rai R, Rai LC (2012) Proteomics combines morphological, physiological and biochemical attributes to unravel the survival strategy of Anabaena sp.PCC7120 under arsenic stress. J Proteomics 75:921–937.

Patterson J, Ford K, Cassin A, Natera S, Bacic A (2007) Increased abundance of proteins involved in phytosiderophore production in boron-tolerant barley. Plant Physiol 144:1612–1631.

Perfus-Barbeoch L, Leonhardt N, Vavasseur A, Forestier C (2002) Heavy metal toxicity: Cadmium permeates through calcium channels and disturbs the plant water status. Plant J 32:539–548.

Rose JKC, Bashir S, Giovannoni JJ, Jahn MM, Saravanan RS (2004) Tackling the plant proteome: Practical approaches, hurdles and experimental tools. Plant J 39:715–733.

Roychoudhury A, Basu S, N, Sengupta DN (2012a) Antioxidants and stress-related metabolites in the seedlings of two indica rice varieties exposed to cadmium chloride toxicity. Acta Physiol Plant 34(3):835–847.

Roychoudhury A, Datta K, Datta SK (2011) Abiotic Stress in Plants: From Genomics to Metabolomics. In: Tuteja N, Gill SS, Tuteja R (Eds.),Omics and Plant Abiotic Stress Tolerance, Bentham Science Publishers, UAE, Pp. 91–120.

Roychoudhury A, Krishnamoorthi S, Paul R (2020) Arsenic Toxicity and Molecular Mechanism of Arsenic Tolerance in Different Members of Brassicaceae. In: Wani SH, Thakur AK, Khan YJ (Eds.), *Brassica Improvement*, Springer Nature, Switzerland, Pp. 159–186.

Roychoudhury A, Pradhan S, Chaudhuri B, Das K (2012b) Phytoremediation of Toxic Metals and the Involvement of *Brassica* Species. In: Anjum NA, Pereira ME, Ahmad I, Duarte AC, Umar S, Khan NA (Eds.), Phytotechnologies: Remediation of Environmental Contaminants, CRC press, Taylor and Francis Group, Boca Raton, Pp. 219–251.

Samanta S, Roychoudhury A (2021) Transporters Involved in Arsenic Uptake, Translocation and Efflux in Plants. In: Roychoudhury A, Tripathi DK, Deshmukh R (Eds.), Metal and Nutrient Transporters in Abiotic Stress, Academic Press (Elsevier), UK, Pp. 77–86.

Schneider T, Schellenberg M, Meyer S, Keller F, Gehrig P, Riedel K, et al. (2009) Quantitative detection of changes in the leaf-mesophyll tonoplast proteome in dependency of a cadmium exposure of barley (*Hordeum vulgare* L.) plants. Proteomics 9:2668–2677.

Semane B, Dupae J, Cuypers A, Noben JP, Tuomainen M, Tervahauta A, Kärenlampi S, Van Belleghem F, Smeets K, Vangronsveld J (2010) Leaf proteome responses of *Arabidopsis thaliana* exposed to mild cadmium stress. J Plant Physiol 167:247–254.

Sharma SS, Dietz KJ (2009) The relationship between metal toxicity and cellular redox imbalance. Trends Plant Sci 14:43–50.

Sharmin SA, Alam I, Kim KH, Kim YG, Kim PJ, Bahk JD, Lee BH (2012) Chromium-induced physiological and proteomic alterations in roots of *Miscanthus sinensis*. Plant Sci 187:113–126.

Swinbanks D (1995) Government backs proteome proposal. Nature 378:653.

Timperio AM, Egidi MG, Zolla L (2008) Proteomics applied on plant abiotic stresses: Role of heat shock proteins (HSP). J Proteomics 71:391–411.

Tkalec M, Prebeg T, Roje V, Pevalek-Kozlina B, Ljubešić N (2008) Cadmium induced responses in duckweed *Lemna minor* L. Acta Physiol Plant 30:881–890.

Vannini C, Marsoni M, Domingo G, Antognoni F, Biondi S, Bracale M (2009) Proteomic analysis of chromate-induced modifications in *Pseudokirchneriella subcapitata*. Chemosphere 76:1372–1379.

Verbruggen N, Hermans C, Schat H (2009) Mechanisms to cope with arsenic or cadmium excess in plants. Curr Opin Plant Biol 12:364–372.

Wang R, Gao F, Guo BQ, Huang JC, Wang L, Zhou YJ (2013) Short-term chromium-stress-induced alterations in the maize leaf proteome. Int J Mol Sci 14:11125–11144.

Wang W, Vinocur B, Shoseyov O, Altman A (2004) Role of plant heat-shock proteins and molecular chaperones in the abiotic stress response. Trends Plant Sci 9:244–252.

Xi J, Wang X, Li S, Zhou X, Yue L, Fan J, Hao D (2006) Polyethylene glycol fractionation improved detection of low-abundant proteins by two dimensional electrophoresis analysis of plant proteome. Phytochemistry 67:2341–2348.

Yang Q, Wang Y, Zhang J, Shi W, Qian C, Peng X (2007) Identification of aluminium-responsive proteins in rice roots by a proteomic approach: Cysteine synthase as a key player in Al response. Proteomics 7:737–749.

Zhen Y, Qi JL, Wang SS, Su J, Xu GH, Zhang MS, Miao L, Peng XX, Tian D, Yang YH (2007) Comparative proteome analysis of differentially expressed proteins induced by al toxicity in soybean. Physiol Plant 131:542–554.

Zhou S, Sauvé R, Thannhauser TW (2009) Proteome changes induced by aluminium stress in tomato roots. J Exp Bot 60:1849–1857.

7 Proteomic Responses of Plants to Nutritional Stress

Debabrata Panda, Prafulla K. Behera,
Suraj K. Padhi and Jayanta K. Nayak

7.1 INTRODUCTION

Plants frequently encounter a number of abiotic stresses throughout their life cycle (Ahmad et al. 2016). Nutritional stress in the form of nutrition deficiency or excess is a principal abiotic factor which significantly brings down plant growth, quality and yield to a greater extent (Ahmad et al. 2016; Kerry et al. 2018). A plant requires food for its growth and development, and it uses inorganic minerals as nutrients to complete its life cycle (Kerry et al. 2018). Each plant is distinct, with a minimal requirement level and an optimal nutrient range. The numerous processes driving crop yield formation, like biomass accumulation and partitioning, are directly regulated by nutrient supply. During the mid-nineteenth century, a German scientist, Baron Justus von Liebig, demonstrated that nutrients are the main requirement for plant growth and that if a plant is deprived of any required elements/inorganic minerals, it will lead to plant death (Liebig 1840; Kerry et al. 2018). The literature regarding plant nutrients has described that 23 basic elements are required by different plants to complete their metabolic activities and life cycle, and these elements are further categorised into (a) minerals and (b) non-minerals (Kerry et al. 2018). The elements like carbon (C), hydrogen (H) and oxygen (O) are non-minerals used by plants from water and air, but the remaining 20 elements are considered to be minerals absorbed from soil. The essential minerals are further subcategorised into macronutrients and micronutrients based on plant requirements (Figure 7.1). N, P and K are primary macronutrients, and Mg, S and Ca are secondary or tertiary macronutrients (Kerry et al. 2018). The micronutrients such as Cu, Fe, B, Cl, Zn, Ni, Na, Se, Si, Al, Ni, Co and Mo (known as trace or minor elements) are required by plants in small quantities (Tucker 1999; Kerry et al. 2018; Banerjee and Roychoudhury 2018) (Figure 7.1). The beneficial trace elements play an essential role in plants and have multiple functions; however, the concentration above the threshold level is considered to be detrimental to the growth of plants, with harmful effects on biochemical, physiological and molecular activities in several plants (Kerry et al. 2018; Roychoudhury and Bhowmik 2020). The technological advancements in "omics" approaches are used in various plants for the identification of vital proteins or metabolites. These technologies, which cover metabolomics, proteomics and genomics, deal with plant stress resistance and genes controlling such biomolecules (Srivastava et al. 2013; Ahmad et al. 2013, 2016). Proteomics is concerned with the identification and determination of proteins, as well as their expression, protein-protein interaction and modification of proteins after the post-translation process subjected to stress and control conditions (Nam et al. 2012; Ahmad et al. 2016). Under abiotic stress constraints, a significant alteration and modification in protein expression occurs in different plants. Thus, the proteomics approach is valuable in explaining the function of protein accumulation and its relationship with stress resistance in different plants (Hossain and Komatsu, 2013; Ahmad et al. 2016).

The proteomics technique is used to explore plant responses to stress and also the complexity of biochemical process involved (Ahmad et al. 2016). Proteomics study in plants under different stress offers identification of potential candidate genes for genetic improvement of plants against stress (Barkla et al. 2016). In response to nutritional stress, several signalling pathways get activated, which leads to a complex regulatory network comprising antioxidants, reactive oxygen

DOI: 10.1201/b23255-7

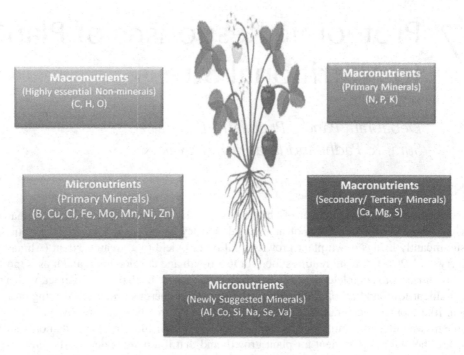

FIGURE 7.1 The beneficial macronutrient and micronutrient essential for growth and various functions in plants.

species (ROS), ion balance, hormones, etc. (Ahmad et al. 2016). The plant cell responses to abiotic stresses like nutritional stress vary between different organs of plants. Proteomics studies based on specific organs, in conjunction with subcellular organelles, can provide more knowledge regarding cellular mechanisms which control nutritional stress responses and signal transduction, which could be helpful to improve nutritional stress tolerance (Ahmad et al. 2016). The wide utilisation of proteomics study in the evaluation of physiology and genetic aberration in response to nutritional stress is outstanding in gaining knowledge regarding plant life. This insight could lead to the development of new concepts about how to protect a plant from element/nutrient toxicity and its tolerance to nutritional stress (Kerry et al. 2018). Henceforth, the current chapter focuses primarily on the proteome responses and tolerance of different plants under nutritional stress. Furthermore, the impact of these studies on crop improvement subjected to nutritional stress is discussed in the chapter.

7.2   CONCEPT OF PROTEOMICS

Proteins are essential molecules that have a direct role in cellular activity. In molecular biology, the central dogma defines that DNA synthesises RNA, from which genetic information flows into protein (Vaz and Tanavde 2018). Although DNA consists of the blueprint for cell assembly, the proteins eventually function as the building blocks (Vaz and Tanavde 2018). In 1994, Marc Wilkins was the first to define the term "proteome", which refers to the investigation of different proteins in a single cell (Vaz and Tanavde 2018). Furthermore, James, in 1997, was the first to define the term "proteomics" to describe the large-scale study and exploration of proteomes related to its structure, function, activity pattern and its composition (Vaz and Tanavde 2018). "Proteomics" is the next significant step following "genomics (genome study)" and "transcriptomics (transcriptome study)" in the investigation of bio-molecular systems. Nowadays, molecular biology has evolved as advanced technology and developed significant capabilities for high-throughput nucleotide sequence analysis, which can be reflected in the protein world. Two modes of proteomics have been existed,

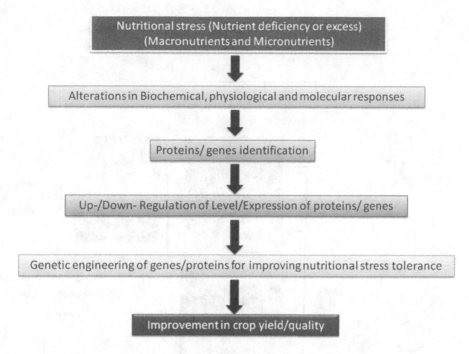

FIGURE 7.2 Improvement of yield in plants under nutritional stress applying proteomics approach.

such as (a) proteomics involving the investigation of only proteins as gene product analysis; (b) proteomics comprising a combination of protein analysis and genomics/transcriptomics (Vaz and Tanavde 2018). Hence, investigating a proteome is similar to getting a snapshot of the protein environment at a specific point in time. Haynes and Yates (2000) showed that proteomics technology must include the following characteristics such as high sensitivity and throughput, capacity to differentiate altered proteins and quantitative display and analysis of all proteins in a required sample. Proteomics analysis is conducted through various approaches; (a) protein separation and identification are accomplished utilising 2-DE or LC-MS/MS technologies; (b) unravelling the role of proteins and their networks of metabolic and signalling pathways in various plants through protein mapping, protein-protein interaction and post-translational protein modification; (c) bioinformatics techniques and database utilisation for model and non-model plant genotypes (Karthikeyan et al. 2022) (Figure 7.2). In recent years, proteomics technologies such as iTRAQ, MudPIT, ICATs and TMTs have been continuously used for investigating plant development and metabolic activities in abiotic stress resistance and adaptation by applying proteomics analysis (Karthikeyan et al. 2022). The flow chart below depicts the enhancement of yield and quality traits in different plants subjected to nutritional stress using proteomics approach (Figure 7.2).

7.3 PROTEOMICS RESPONSES IN DIFFERENT PLANTS SUBJECTED TO NUTRITIONAL STRESS

Proteomics is an effective method of "Omics technology" for investigating and comprehending plant stress responses which facilitates large-scale protein expression in different plants encoded by its genome (Kerry et al. 2018). It is not only a specific tool for characterising entire protein alteration but it also assists in comparing the variance in protein profiling at cellular and organelle levels in the presence of different nutrients (Kerry et al. 2018). In recent years, proteomics has been used extensively under heavy metal toxicity and stress impact on different plants. Many researches rely on gel-based or gel-free proteomic techniques to identify differentially expressed and sensitive proteins

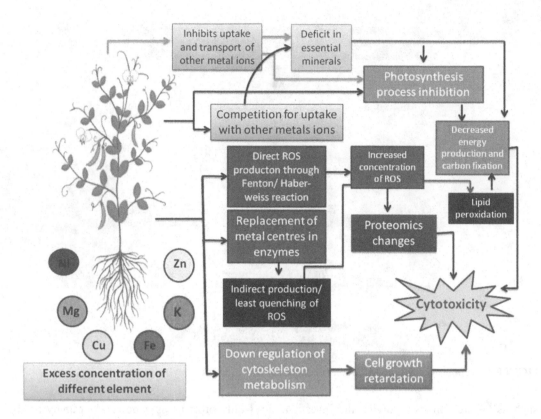

FIGURE 7.3 Biochemical and molecular mechanisms associated with nutritional stress-regulated cytotoxicity in plants.

in various plant species (Komatsu et al. 2017; Kerry et al. 2018). Techniques such as MALDI-TOF, SELDI-TOF and 2D-gel electrophoresis in addition to Edman sequencing and mass spectrometry are gel-based approach, whereas tandem mass tags, isotope-coded affinity tag, multidimensional protein identification technology, etc., are included under gel-free approach techniques (Hu et al. 2015; Kerry et al. 2018). Although the trace elements (macronutrients and micronutrients) and minerals are necessary for plant growth and development, concentrations of trace elements more than the threshold level are hazardous to plants. Studies on enhancing metal levels has revealed that the majority of plant damage is due to oxidative stress caused by the generation of ROS (Kosova et al. 2015; Minic 2015; Kerry et al. 2018). ROS such as H_2O_2, O_2^- and hydroxyl radicals are released during cell metabolism. ROS have a negative impact on nucleic acids, proteins and lipids which ultimately lead to plant cell death (Figure 7.3). Antioxidative defence systems like antioxidant enzymes (superoxide dismutase, glutathione reductase, etc.) protect plants from excessive ROS generation (Meng et al. 2016; Kerry et al. 2018). The most prevalent stress is the stress caused by nutrient deficit or excess in plants. Thus, despite their varied responses, the proteomic techniques explore plant responses and also the complexity of biochemical processes helpful in understanding plant systems at a molecular level (Kerry et al. 2018). The differential protein expressions in various plants under nutritional stress, i.e., nutrient excess or deficit, have been presented in Table 7.1.

Nitrogen (N) is a major limiting element related to agriculture productivity, and the use of N in different plants has a significant impact on the growth and yield of plants and also cost reduction (Kerry et al. 2018). Nitrogen deficiency leads to stiff and erect plant leaves. Several researchers like Chandna and Ahmad (2015) and Nazir et al. (2016) studied proteomic analysis of *Zea mays* and *Oryza sativa* L. plants, etc., subjected to nitrogen stress and described about the overall responses of proteins. The authors reported about the proteomic expression of N-related proteins. Nazir et al. (2016)

TABLE 7.1
Proteomics Responses of Different Plants under Nutritional Stress

Elements	Authors	Plant Species (Used Plant Part)	Proteins Identified/Reported and Their Regulation	Functions in Plants
Nitrogen	Ding et al. (2010)	*Oryza sativa* (root)	Up-regulation of putative subtilisin-like serine protease and down-regulation of adenosine kinase, adenylate kinase and phenylalanine ammonia-lyase protein	Energy and phenylpropanoid metabolism
	Nazir et al. (2016)	*Zea mays* (leaf)	Down-regulation of malate dehydrogenase, ribulose bisphosphate carboxylase, ATP synthase gamma chain protein, whereas up-regulation of 6-phosphogluconate dehydrogenase	TCA cycle, photosynthesis, ATP synthesis and as pentose phosphate shunt
Magnesium	Fukao et al. (2016)	*Arabidopsis thaliana* (root)	Up-regulation of nucleolin-like 1 and DEAD-box RNA helicase protein, but down-regulation of ARP protein, peroxin 1 lc and protein phosphatase 2A-3	Transcription, cytoskeleton, nucleic acid metabolism, etc.
	Peng et al. (2015)	*Citrus sinensis* (root)	Down-regulation of helicase protein, but up-regulation of Ribosomal protein S3 and ascorbate peroxidase 2	Protein metabolism, energy metabolism, nucleic acid metabolism and stress response
Zinc	Lucini and Bernardo (2015)	*Lactuca sativa* (leaf)	Proteins like putative cellulose synthase and phenylalanine ammonia-lyase down-regulated, but cytosolic fructose-1,6-bisphosphate and MPBQ/MSBQ transferase get up-regulated	Photosynthesis, cell wall and energy metabolism, etc.
Nickel	Wang et al. (2012)	*Brassica juncea* (leaf)	Down-regulation of ascorbate peroxidase and S-adenosylmethionine synthetase and up-regulation of putative actin 1 and Aconitate hydrates	Sulphur metabolism, carbohydrate metabolism and cell wall synthesis
Aluminium	Dai et al. (2013)	*Hordeum vulgare* (whole plants)	Up-regulation of vacuolar proton-ATPase D and down-regulation the proteins such as methionine synthase and phosphoglycerate mutase	Photosynthesis, metabolism, signal transduction and transporters
	Jiang et al. (2015)	*C. sinensis* (root)	Up-regulation of catalase, ATP sulphurylase 1 and down-regulation of blue copper protein	Defence and sulphur metabolism
Potassium	Zeng et al. (2015)	*Hordeum spontaneum, H. vulgare* (whole plant)	Phenylalanine ammonia-lyase and pectin esterase down-regulated, but 40S ribosomal protein S4, NADPH-cytochrome P450 reductase proteins up-regulated	Protein synthesis, cell structure, defence, metabolism and transporter
	Ren et al. (2016)	*Nicotiana tabacum* (root)	The proteins tubulin, plant peroxidase and glutathione S-transferase were down-regulated but lipase protein up-regulated	Transport/defence, metabolism and cytoskeleton
Calcium	Yina et al. (2015)	*Glycine max* (whole plant)	Up-regulation of maturation polypeptide, lactoylglutathione lyase and down-regulation of cysteine proteinase RD21a and calreticulin-1 precursor	Cell growth/division, proteolysis, transportation and disease/defence
	Wang et al. (2015)	*A. thaliana* (whole plant)	Down-regulation of nitrilase 1 and proteins like ACC oxidase 2, annexin 1 and gamma-glutamyl peptidase	Cell signalling, metabolism and hormone biosynthesis

(Continued)

TABLE 7.1 *(Continued)*
Proteomics Responses of Different Plants under Nutritional Stress

Elements	Authors	Plant Species (Used Plant Part)	Proteins Identified/Reported and Their Regulation	Functions in Plants
Cadmium	Sun et al. (2016)	*Cucumis sativus* (leaf)	Glutathione S-transferase protein was up-regulated and glutelin type-B-2-like and PsbP domain-containing protein 6 were down-regulated	Photosynthesis, primary metabolism and storage protein
Copper	Qin et al. (2015)	*Allium cepa* (root)	Up-regulation of glutaredox in and proliferation-associated 2 g4 protein, whereas Ran-binding protein 1 and cinnamoyl-CoA reductase 1 down-regulated	Protein and cell wall synthesis and cell cycle and DNA replication
	Roy et al. (2016)	*Sorghum bicolor* (leaf)	Thymidine kinase and maturase K were down-regulated but thaumatin-like protein and alcohol dehydrogenase were up-regulated	Protein translation and synthesis, growth and development
Boron	Chen et al. (2014)	*A. thaliana* (leaf)	Down-regulation of actin 7 and glycolate oxidase, but up-regulation of proteins such as RuBisCo activase and fructose bisphosphate aldolase	Photosynthesis and energy metabolism
	Yang et al. (2016)	*Citrus grandis* (Root)	Up-regulation of clathrin light chain proteins, but down-regulation protein such as phospholipase C2, alcohol dehydrogenase, peroxiredoxin IIF was observed	Energy metabolism, signal transduction, nucleic acid metabolic process, cellular responses to stress
Silicon	Muneer and Jeong (2015a, b)	*Lycopersicum esculentum* (root)	UDP-glucosyltransferase and cinnamoyl CoA reductase-like 1 were up- and down-regulated, respectively	Stress response, cellular metabolic process and transferase activity
Iron	Li and Lan (2017)	*A. thaliana* (root)	Up-regulation of oxidoreductase, WAK-like receptor-like kinase, but FRO3, IRT3, stomatal ascorbate peroxidase were down-regulated	Signalling, hormone metabolism, transport and metal handling
	Donnini et al. (2010)	*C. sativus* (root)	Proteins like alanine aminotransferase, phosphoglycerate kinase and malate dehydrogenase showed Up-regulation but xylan 1,4-beta-xylosidase was down-regulated	Glycolysis, carbohydrate, nitrogen and sucrose metabolism
Manganese	You et al. (2014)	*C. grandis* (root)	Maturase K and alcohol dehydrogenase up-regulated, iron-superoxide dismutase and valosin-containing protein down-regulated	Cell transport, protein, nucleic acid and carbohydrate metabolism
Phosphorous	Zhang et al. (2014)	*Z. mays* (leaf)	Ferredoxin and ascorbate peroxidise were up-regulated, but RNA-binding protein and Ribulose-1,5-bisphosphate carboxylase down-regulated	Defence, signal transduction and energy metabolism
	Zhang et al. (2016b)	*Populus cathayana* (whole plant)	Apocytochrome f and alanine aminotransferase down-regulated, but alpha-galactosidase and Elongation factor G was Up-regulated	Photosynthesis, protein degradation, translation, transport of amino acid

(Continued)

TABLE 7.1   *(Continued)*
Proteomics Responses of Different Plants under Nutritional Stress

Elements	Authors	Plant Species (Used Plant Part)	Proteins Identified/Reported and Their Regulation	Functions in Plants
Sulphur	Liu et al. (2014)	*A. thaliana* (whole plant)	Lipoamide dehydrogenase, 6-phosphogluconate dehydrogenase and methionine synthase were up-regulated but phosphoglycerate dehydrogenase was down-regulated	Metabolism of glutamate, sulphur, methionine, glycine and glutathione
	Yildiz and Terzi (2016)	*Brassica napus* (leaf)	Up-regulation of fructose-bisphosphate aldolase, 20-kDachaperonin, whereas down-regulation of proteins like RuBisCo activase 2 and glutathione S-transferase 103-1A	Photosynthesis, sulphur and energy metabolism and protein folding and stabilisation

reported up-regulation of ribulose-1,5-bisphosphate carboxylase and 5-deoxyuridine-5′-triphosphate-nucleotide hydrolase protein whereas down-regulation of 6-phosphogluconate dehydrogenase, malate dehydrogenase and ATP synthase gamma chain proteins in *Z. mays*. In the case of wheat, Chandna and Ahmad (2015) reported down-regulation of proteins like glutathione S-transferase and glutamate synthase but up-regulation of cytochrome P450, ATPase alpha subunit and putative glyoxalase proteins under N deficiency (Table 7.1). Nitrogen has the ability to improve photosynthetic efficiency by enhancing chlorophyll synthesis and improving antioxidant defence enzymes present in nitrate (NO_3^-) and ammonium (NH_4^+) ions (Kerry et al. 2018). Marino et al. (2016) described that under N stress, the addition of NO_3^- and NH_4^+ has activated TGG1 and TGG2 myrosinases and controlled differential regulation of glucosinolate metabolism in *Arabidopsis thaliana*.

In the case of plants, magnesium (Mg) is considered as the major cation in living plant cells, where this Mg^{2+} ion participates in a variety of processes such as photosynthesis, nucleic acid synthesis and catalytic activity of enzymes (Kerry et al. 2018). It is also a major component of chlorophyll production and has a critical function in heavy metal toxicity in *A. thaliana* and *Citrus sinensis* (Peng et al. 2015; Fukao et al. 2016; Kerry et al. 2018). To understand the molecular regulation of Mg resistance in plants, Peng et al. (2015) carried out a proteomics investigation of Mg stress changes in *C. sinensis* using 2-DE. The authors showed that identified proteins were involved in cell wall, carbohydrate and lipid metabolism, etc., during the Mg deficit condition (Table 7.1). Furthermore, proteomics analysis in roots of *A. thaliana* showed down-regulation of peroxin 11c, ARP protein and protein phosphatise 2A-3 and up-regulation of nucleolin-like 1 protein showing nucleic acid metabolism, cytoskeleton, etc., by applying iTRAQ-based techniques (Fukao et al. 2016). Furthermore, Guo et al. (2015) reported that long-term Mg deficit impacted the proteomes more in leaves than in roots.

Zinc (Zn) is one of the micronutrients required by plants for enzyme activation related to auxin and nucleic acid synthesis, chlorophyll formation, seed formation, etc. (Kerry et al. 2018). Hossain et al. (2016) studied the nano-particle-based experiment on *Glycine max*, and the authors reported that Zn absorption boosts oxidative burst, where differentially expressed proteins are found to be linked to cell organisation, secondary metabolism, etc. The study on leaves of *Lactuca sativa* via LC-MS/MS and MALDI-TOF techniques identified many proteins such as putative cellulose synthase protein, regulating cell wall metabolism, MPBQ/MSBQ transferase protein in photosynthesis, along with down-regulation of phenylalanine ammonia-lyase in synthesis of phenylpropanoids (Lucini and Bernardo 2015) (Table 7.1).

Nickel (Ni) is an essential element in plants; however, the least availability of Ni has negative impacts on the growth and metabolic activities of plants (Kerry et al. 2018). Brown et al. (1987)

showed that the average concentration of Ni is about 0.05–5.0 ppm in different plants. The vital function regulated by Ni in plants includes nitrogen fixation and nodulation, and germination of seeds, it also acts as a cofactor of urease (Kerry et al. 2018). Proteomic study of *Brassica juncea* using 2-DE revealed about 61 differentially expressed proteins, 37 of which were ambiguously identified by MALDI-TOF MS technique. The identified proteins were involved in carbohydrate and S metabolism and defence against oxidative stress (Wang et al. 2012; Kerry et al. 2018).

Aluminium (Al) is the third most prevalent element, comprising 7% of the Earth's crust. As a result, roots are exposed to Al in Al oxide or aluminosilicate form in the soil, but these Al forms are inaccessible to plants for metabolic processes. At lower pH, Al is converted into Al^{3+} cation, which plants may absorb (Matsumoto and Motoda 2012; Tomioka et al. 2012). Al toxicity is responsible for stunted growth in plants. To understand the tolerance capacity of rice to higher concentrations of Al stress, Yang et al. (2007) performed a proteomic experiment in Xiangnuo 1 (Al tolerant rice) cultivar using 2-DE coupled with MS techniques, and the authors reported about a set of 17 Al-responsive proteins functionally associated with signal transduction and antioxidant defence (Zheng et al. 2014). Liu et al. (2017) studied the iTRAQ-based proteomic approach under Al stress and Al with IAA stress in the wheat (Al-resistance) line; the results reported that MAPK 1a and MAPK 2c proteins were expressed under both the stresses; however, under only single Al stress, proteins like ARF were up-regulated and were Aux down-regulated.

Potassium (K) is one of the important macronutrients that promote leaf growth by regulating anion-cation equilibrium in plant cells and accounts for 2–10% of dry weight in plants (Li et al. 2019). It is involved in various physiological functions like transport processes, photosynthesis, signal transduction and enzyme activation (Li et al. 2019). Although K^+ represents the fourth most abundant element of Earth's lithosphere, only 1–4% of soil particles are accessible to plants (Li et al. 2019). The deficiency of K may have an impact on plant productivity and quality, which may aggravate oxidative stress in different plants (Torregrosa et al. 2016). The proteomic analysis in roots of *Nicotiana tabacum*, and whole the plants of *Hordeum vulgare* and *Hordeum spontaneum* reported down-regulation of tubulin, glutathione S-transferase, 40S ribosomal protein S4 and pectin esterase proteins, whereas up-regulation of lipase and phenylalanine ammonia-lyase in K-dependent manner (Zeng et al. 2015; Ren et al. 2016). Many studies also reported negative effects of K deficit in plants such as growth decline, pathogen resistance decrement and nutritional imbalance (Kerry et al. 2018). Furthermore, Zhang et al. (2016c) identified 258 peptides by applying proteomics approaches in the xylem sap of the cotton plant that brought about the decreased expression of peptides under K deficiency (Table 7.1).

The plants absorb calcium (Ca) from the soil in Ca^{2+} ion form which is responsible for specific enzyme activation that controls water transport and salt balance (Kerry et al. 2018). The proteomic studies on the effect of calcium on different plants such as *G. max*, *H. vulgare* and *A. thaliana* were carried out by Oh et al. (2014), Ciesla et al. (2016) and Wang et al. (2015), respectively (Table 7.1). The experiment on *Cucumis sativus* reported overexpression of enzymes linked with TCA cycle and glycolysis, and nitrogen metabolism, when Ca was applied exogenously (He et al. 2012). The Ca-dependent proteins like annexin 1, protein kinase and ACC oxidase 2 were reported applying the proteomics approach where they regulate many useful functions such as energy pathways and signal transduction (Kerry et al. 2018).

Excess presence of copper (Cu) in plants is found to be detrimental, although it is a key micronutrient for plant growth (Rout et al. 2013; Kerry et al. 2018). The average concentration of Cu is about 2–50 ppm in plants and involved in various functions in plants like protein synthesis, water transport regulation, chlorophyll synthesis, reproduction, seed production and enhancement of flavour in fruits and vegetables (Kerry et al. 2018). The proteomics analysis in different plants such as *Allium cepa*, *Sorghum bicolor* and *O. sativa* was carried out using 2-DE and MALDI-TOF MS by Qin et al. (2015), Roy et al. (2016) and Chen et al. (2015), respectively (Table 7.1). Under higher concentrations of Cu, the overexpression of Cu-binding proteins was reported responsible for protein distortion and antioxidative defence system (Zhang et al. 2016a). For instance, Qin et al. (2015) identified various proteins such as glutaredoxin, and Ran-binding protein-1-regulating protein synthesis and DNA replication were reported in *A. cepa*.

Boron (B) is a useful microelement for plant growth and yield; however, an excessive amount of B is toxic to various plant species (Guo et al. 2016). Yang et al. (2016) conducted the proteomics analysis of the roots of *Citrus grandis* plant by applying iTRAQ method which reported down-regulation of alcohol dehydrogenase 1, phospholipase C2 proteins and up-regulation of clathrin light-chain protein. Similarly, the effect of B toxicity on *citrus* species analysed through MS techniques by Sang et al. (2015) showed B-toxicity-responsive proteins associated with various plant cell functions such as coenzyme synthesis, cell transport and cell cycle, amino acid metabolism and nucleotide metabolism.

Silicon (Si) is the most abundant metal and uniformly distributed in soils, and it is involved in various functions of plant cells like stress and disease tolerance and plant growth (Kerry et al. 2018). Okuda and Takahashi were the first to discover Si in the year 1965 and the average concentration is about 0.1–10% on dry weight basis in plants. The experiment on *Dianthus caryophyllus* by Soundararajan et al. (2017) depicted that the exogenous application of Si helped in the molecular regulation of the hyperhydricity, and the identified proteins were involved in ribosomal binding, hormone/cell signalling and photosynthesis for recovery of shoots from hyperhydricity in *D. caryophyllus*.

Iron (Fe) is one of the micronutrients necessary for plant growth; however, in aerobic and neutral pH conditions, Fe is less soluble in nature, and hence, Fe accessibility in soil is limited for plant growth (Liang et al. 2013). DNA synthesis, respiration and photosynthesis are some of the essential functions carried out by Fe in plants (Kerry et al. 2018). The long-term effect of Fe deficiency in *Pisum sativum* was conducted by Meisrimler et al. (2016), and the result of proteome investigation revealed that neutralising proteins were produced for the electron transport component of mitochondria, regulating cell energy proliferation, maintaining low ROS generation and delaying cell death of mitochondria (Meisrimler et al. 2016). The regulations of novel proteins are Fe-dependent, which is evident from the Fe-deficiency consequences experiment conducted on the roots of *C. sativus* (Donnini et al. 2010) and *A. thaliana* (Li and Lan 2017).

Manganese (Mn) plays an important role in nitrogen assimilation, chlorophyll generation, photosynthesis, organic acid metabolism, plant growth, etc. Mn was discovered by Mchargue in (1922), and the average concentration in a plant is 10–600 ppm on a dry weight basis (Kerry et al. 2018). A study on *Camellia sinensis* regarding Mn toxicity using proteomics techniques reported the tolerance capacity of *C. sinensis* to Mn toxicity due to CsMTP8 protein located in plasma membrane (Kerry et al. 2018). Celma et al. (2016) investigated an Mn deficiency experiment in *A. thaliana* that reported about 11,606 proteins in the roots by applying LC MS/MS-based iTRAQ proteomic technology which signifies the importance of manganese in a plant cell (Table 7.1).

Selenium (Se) is a vital micronutrient for plants and protects from different abiotic stress (Kerry et al. 2018). Se can be taken by plants in selenate, selenite and organic Se form. The selenite is prevalent in acid and neutral soil, but selenate is abundant in alkaline soil (Feng and Ma 2021). Smolen et al. (2016) revealed that higher concentration of Se causes alteration of protein structure and function in plants. The research by Sun et al. (2016) revealed that cadmium combined with Se stimulates several proteins relevant to photosynthesis and metabolism, which are detected in *C. sativus* plant applying 2-DE coupled with MS, but no direct link between Se stress and proteome function alteration was observed (Kerry et al. 2018).

Sulphur (S) acts as the key component of various coenzymes, vitamins and mineral nutrients required for plant growth, and its deficiency or assimilation inhibits chlorophyll and protein generation (Kerry et al. 2018). Some reports showed regulation of ethylene signalling by sulphur during heavy metal toxicity (Zierer et al. 2016). The proteins associated with photosynthesis, protein folding, ATP synthesis, stress defence and enzyme metabolism in response to S stress were found in *A. thaliana* and *Brassicanapus* plants (Yildiz and Terzi 2016; Kerry et al. 2018).

Phosphorous (P) is another vital mineral nutrient necessary for plant growth, accounting for 0.5% of plant dry weight basis; however, it is one of the least accessible mineral nutrients in soils due to its strong binding by soil particles (Liang et al. 2013; Zhou et al. 2022). P is an important requirement and structural component for nucleic acid, photosynthesis process and the cell membranes of plants (Kerry et al. 2018; Zhou et al. 2022). Various authors such as Zhang et al. (2014) and Muneer and Jeong (2015b) reported about the effects of P stress on the photosynthesis process, including influencing energy transfer across the thylakoid membrane, decreasing electron transfer system efficiency and inactivation of enzymes associated with Calvin cycle. Plants have evolved several ways to adapt to P deprivation, including modification in morphology and architecture of root, enhanced phosphatases and organic acid generation and recycling of internal P through bypass channel (Liang et al. 2013). The experiment on *A. thaliana* under low phosphorous stress was conducted applying 2-D DIGE techniques and about 59 differentially expressed proteins were identified such as ATP synthase and carbonic anhydrases 2 under lower P deficiency (Kerry et al. 2018) (Table 7.1).

7.4 CONCLUSION AND FUTURE PROSPECTS

Plant metabolism is well regulated by nutrients in the form of macronutrients and micronutrients. Deficiency of nutrients and excess concentration of any beneficial elements is detrimental to plant growth, which activates a number of pathways in plants. The most promising pathways include protein homeostasis, glucose metabolism, signal transduction and antioxidant defence and substantially reduce the quality of agriculture output. Hence, it appears possible to find proteins which are expressed in nutrient-deficit conditions and use them as a possible signal of mineral insufficiency in plants. Moreover, substantial research work has been conducted to determine the concentration, symptoms and mode of actions of many nutrients, but relatively few publications are available regarding the introgression of nutrient-associated genes to overcome nutrient deficits within plant cells/tissues. As a result, research regarding the development of crop plant tolerance to the nutrient deficit would be more beneficial. Proteomics has been shown to be quite useful in helping plant physiologists comprehend what is going on in a cell as a result of an external stimulus. Furthermore, more emphasis should be paid to the use of novel methodological approaches with high-throughput technologies for protein profiling, identification and characterisation of important proteins and genes related to plant tolerance to nutritional stress.

REFERENCES

Ahmad, P.; Abdel Latef, A. A. H.; Rasool, S.; Akram, N. A.; Ashraf, M.; Gucel, S. Role of Proteomics in Crop Stress Tolerance. *Front. Plant Sci.* 2016, *7*, 1336.

Ahmad, F.; Rai, M.; Ismail, M. R.; Juraimi, A. S.; Rahim, H. A.; Asfaliza, R.; Latif, M. A. Waterlogging Tolerance of Crops: Breeding, Mechanism of Tolerance, Molecular Approaches, and Future Prospects. *Biomed Res. Int.* 2013, 1–10.

Banerjee, A.; Roychoudhury, A. Role of Beneficial Trace Elements in Salt Stress Tolerance of Plants. In: Hasanuzzaman, M.; Fujita, M.; Oku, H.; Nahar, K.; Hawrylak-Nowak, B. (Eds.) Plant Nutrients and Abiotic Stress Tolerance, Springer Nature, Singapore, 2018, pp. 377–390.

Barkla, B. J.; Vera-Estrella, R.; Raymond, C. Single-Cell-Type Quantitative Proteomic and Ionomic Analysis of Epidermal Bladder Cells from the Halophyte Model Plant *Mesembryanthemum crystallinum* to Identify Salt-Responsive Proteins. *BMC Plant Biol.* 2016, *16*, 110.

Brown, P. H.; Welch, R. M.; Cary, E. E.; Checkai, R. T. Beneficial Effect of Nickel on Plant Growth. *J. Plant Nutr.* 1987, *10*, 2125–2135.

Celma, J. R.; Tsai, Y. H.; Wen, T. N.; Wu, Y. C.; Curie, C.; Schmidt, W. Systems-Wide Analysis of Manganese Deficiency Induced Changes in Gene Activity of *Arabidopsis* Roots. *Sci. Rep.* 2016, *6*, 35846.

Chandna, R.; Ahmad, A. Nitrogen Stress-Induced Alterations in the Leaf Proteome of Two Wheat Varieties Grown at Different Nitrogen Levels. *Physiol. Mol. Boil. Plants* 2015, *21*, 19–33.

Chen, M.; Mishra, S.; Heckathorna, S.; Frantzb, J. M.; Krause, C. Proteomic Analysis of *Arabidopsis thaliana* Leaves in Response to Acute Boron Deficiency and Toxicity Reveals Effects on Photosynthesis, Carbohydrate Metabolism, and Protein Synthesis. *J. Plant Physiol.* 2014, *171*, 235–242.

Chen, C.; Song, Y.; Zhuang, K.; Li, L.; Xia, Y.; Shen, Z. Proteomic Analysis of Copper-Binding Proteins in Excess Copper-Stressed Roots of Two Rice (*Oryza sativa* L.) Varieties with Different Cu Tolerances. *PLoS ONE* 2015, *10*(4), e0125367.

Ciesla, A.; Mitula, F.; Misztal, L.; Stronska, O. F.; Janicka, S.; Zielinska, M. T.; Marczak, M.; Janicki, M.; Ludwikow, A.; Sadowski, J. A. Role for Barley Calcium-Dependent Protein Kinase CPK2a in the Response to Drought. *Front. Plant Sci.* 2016, *7*, 1–15.

Dai, H.; Cao, F.; Chen, X.; Zhang, M.; Ahmed, I. M.; Chen, Z. H.; Li, C.; Zhang, G.; Wu, F. Comparative Proteomic Analysis of Aluminum Tolerance in Tibetan Wild and Cultivated Barleys. *PLoS ONE* 2013, *8*(5), e63428.

Ding, C.; You, J.; Liu, Z.; Rehmani, M. I. A.; Wang, S.; Li, G.; Wang, Q.; Ding, Y. Proteomic Analysis of Low Nitrogen Stress Responsive Proteins in Roots of Rice. *Plant Mol. Biol. Rep.* 2010, *29*, 618–625.

Donnini, S.; Prinsi, B.; Negri, A. S.; Vigani, G.; Espen, L.; Zocchi, G. Proteomic Characterization of Iron Deficiency Responses in *Cucumis sativus* L. Roots. *BMC Plant Biol.* 2010, *10*, 1–15.

Feng, X.; Ma, Q. Transcriptome and Proteome Profiling Revealed Molecular Mechanism of Selenium Responses in Bread Wheat (*Triticum aestivum* L.). *BMC Plant Biol.* 2021, *21*, 584.

Fukao, Y.; Kobayashi, M.; Zargar, S. M.; Kurata, R.; Fukui, R.; Mori, I. C.; Ogata, Y. Quantitative Proteomic Analysis of the Response to Zinc, Magnesium, and Calcium Deficiency in Specific Cell Types of Arabidopsis Roots. *Proteomes* 2016, *4*, 1–13.

Guo, W.; Chen, S.; Hussain, N.; Cong, Y.; Liang, Z.; Chen, K. Magnesium Stress Signaling in Plant: Just a Beginning. *Plant Signal. Behav.* 2015, *10*, e992287.

Guo, P.; Qi, Y. P.; Yang, L. T.; Ye, X.; Huang, J. H.; Chen, L. S. Long-Term Boron Excess Induced Alterations of Gene Profiles in Roots of Two Citrus Species Differing in Boron Tolerance Revealed by cDNA-AFLP. *Front. Plant Sci.* 2016, *7*, 898.

Haynes, P. A.; Yates, J. R. Proteome Profiling-Pitfalls and Progress. *Yeast* 2000, *17*(2), 81–87.

He, L.; Lu, X.; Tian, J.; Yang, Y.; Li, B.; Li, J.; Guo, S. Proteomic Analysis of the Effects of Exogenous Calcium on Hypoxia-Responsive Proteins in Cucumber Roots. *Proteome Sci.* 2012, *10*, 1–15.

Hossain, Z.; Komatsu, S. Contribution of Proteomic Studies Towards Understanding Plant Heavy Metal Stress Response. *Front. Plant Sci.* 2013, *3*, 310.

Hossain, Z.; Mustafaa, G.; Sakatac, K.; Komatsua, S. Insights into the Proteomic Response of Soybean Towards Al_2O_3, ZnO and Ag Nanoparticles Stress. *J. Hazard. Mater.* 2016, *304*, 291–305.

Hu, J.; Rampitsch, C.; Bykova, N. V. Advances in Plant Proteomics Toward Improvement of Crop Productivity and Stress Resistance. *Front. Plant Sci* 2015, *6*, 209.

Jiang, H. X.; Yang, L. T.; Qi, Y. P.; Lu, Y. B.; Huang, Z. R.; Chen, L. S. Root iTRAQ Protein Profile Analysis of Two Citrus Species Differing in Aluminum Tolerance in Response to Long-Term Aluminum-Toxicity. *BMC Genomics* 2015, *16*, 949.

Karthikeyan, A.; Renganathan, V. G.; Senthil, N. Role of Proteomics in Understanding the Abiotic Stress Tolerance in Minor Millets. In: Pudake, R. N. et al. (Eds.) Omics of Climate Resilient Small Millets, Springer Nature Singapore Pte, 2022, pp 125–139.

Kerry, R. G.; Mahapatra, G. P.; Patra, S.; Sahoo, S. L.; Pradhan, C.; Padhi, B. K.; Rout, J. R. Proteomic and Genomic Responses of Plants to Nutritional Stress. *BioMetals* 2018, *31*, 161–187.

Komatsu, S.; Wang, X.; Yin, X.; Nanjo, Y.; Ohyanagi, H.; Sakata, K. Integration of Gel-Based and Gel-Free Proteomic Data for Functional Analysis of Proteins through Soybean Proteome Database. *J. Proteomics* 2017, *163*, 52–66.

Kosova, K.; Vitamvas, P.; Urban, M. O.; Klima, M.; Roy, A.; Prasil, I. T. Biological Networks Underlying Abiotic Stress Tolerance in Temperate Crops—A Proteomic Perspective. *Int. J. Mol. Sci.* 2015, *16*, 20913–20942.

Li, W.; Lan, P. The Understanding of the Plant Iron Deficiency Responses in Strategy I Plants and the Role of Ethylene in this Process by Omic Approaches. *Front. Plant Sci.* 2017, *8*, 40.

Li, L.; Lyu, C.; Huang, L.; Chen, Q.; Zhuo, W.; Wang, X.; Lu, Y.; Zeng, F.; Lu, L. Physiology and Proteomic Analysis Reveals Root, Stem and Leaf Responses to Potassium Deficiency Stress in Alligator Weed. *Sci. Rep.* 2019, *9*, 17366.

Liang, C.; Tian, J.; Liao, H. Proteomics Dissection of Plant Responses to Mineral Nutrient Deficiency. *Proteomics* 2013, *13*, 624–636.

Liebig, B. J. Chemistry in Its Application to Agriculture and Physiology, Taylor and Walton, London, 1840, pp. 92–165.

Liu, C.; Cheng, Y. J.; Wang, J. W.; Weigel, D. Prominent Topologically Associated Domains Differentiate Global Chromatin Packing in Rice from Arabidopsis. *Nat. Plants* 2017, *3*, 742–748.

Liu, T.; Chen, J. A.; Wang, W.; Simon, M.; Wu, F.; Hu, W.; Chen, J. B.; Zheng, H. A Combined Proteomic and Transcriptomic Analysis on Sulfur Metabolism Pathways of *Arabidopsis thaliana* Under Simulated Acid Rain. *PLoS ONE* 2014, *9*(3), e90120.

Lucini, L.; Bernardo, L. Comparison of Proteome Response to Saline and Zinc Stress in Lettuce. *Front. Plant Sci.* 2015, *6*, 240.

Marino, D.; Ariz, I.; Lasa, B.; Santamaria, E.; Fernandez-Irigoyen, J.; Gonzalez-Murua, C.; Tejo, P. M. A. Quantitative Proteomics Reveals the Importance of Nitrogen Source to Control Glucosinolate Metabolism in *Arabidopsis thaliana* and *Brassica oleracea*. *J. Exp. Bot.* 2016, *67*, 3313–3323.

Matsumoto, H.; Motoda, H. Aluminum Toxicity Recovery Processes in Root Apices. Possible Association with Oxidative Stress. *Plant Sci.* 2012, *185*, 1–8.

Meisrimler, C. N.; Wienkoop, S.; Lyone, D.; Geilfus, C. M.; Luthje, S. Long-Term Iron Deficiency: Tracing Changes in the Proteome of Different Pea (*Pisum sativum* L.) cultivars. *J. Proteomics* 2016, *140*, 13–23.

Meng, F.; Luo, Q.; Wang, Q.; Zhang, X.; Qi, Z.; Xu, F.; Lei, X.; Cao, Y.; Chow, W. S.; Sun, G. Physiological and Proteomic Responses to Salt Stress in Chloroplasts of Diploid and Tetraploid Black Locust (*Robinia pseudoacacia* L.). *Sci. Rep.* 2016, *6*, 23098.

Minic, Z. Proteomic Studies of the Effects of Different Stress Conditions on Central Carbon Metabolism in Microorganisms. *J. Proteomics Bioinf.* 2015, *8*, 80–90.

Muneer, S.; Jeong, B. R. Proteomic Analysis of Salt-Stress Responsive Proteins in Roots of Tomato (*Lycopersicon esculentum* L.) Plants Towards Silicon Efficiency. *Plant Growth Regul.* 2015a, *77*, 133–146.

Muneer, S.; Jeong, B. R. Proteomic Analysis Provides New Insights in Phosphorus Homeostasis Subjected to Pi (Inorganic Phosphate) Starvation in Tomato Plants (*Solanum lycopersicum* L.). *PLoS ONE* 2015b, *10*(7), e0134103.

Nam, M. H.; Huh, S. M.; Kim, K. M.; Park, W. J.; Seo, J. B.; Cho, K.; Kim, D. Y.; Kim, B. G.; Yoon, I. S. Comparative Proteomic Analysis of Early Salt Stress-Responsive Proteins in Roots of SnRK2 Transgenic Rice. *Proteome Sci.* 2012, *10*, 25.

Nazir, M.; Pandey, R.; Siddiqi, T. O.; Ibrahim, M. M.; Qureshi, M. I.; Abraham, G.; Vengavasi, K.; Ahmad, A. Nitrogen-Deficiency Stress Induces Protein Expression Differentially in Low-N Tolerant and Low-N Sensitive Maize Genotypes. *Front. Plant Sci.* 2016, *7*, 298.

Oh, M. W.; Nanjo, Y.; Komatsu, S. Gel Free Proteomic Analysis of Soybean Root Proteins Affected by Calcium Under Flooding Stress. *Front. Plant Sci.* 2014, *5*, 559.

Peng, H. Y.; Qi, Y.; Lee, J.; Yang, L. T.; Guo, P.; Jiang, H. X.; Chen, L. S. Proteomic Analysis of *Citrus sinensis* Roots and Leaves in Response to Long-Term Magnesium-Deficiency. *BMC Genomics* 2015, *16*, 253.

Qin, R.; Ning, C.; Bjorn, L. O.; Li, S. Proteomic Analysis of *Allium cepa* var. agrogarum L. Roots Under Copper Stress. *Plant Soil* 2015, *401*, 197–212.

Ren, X. L.; Li, L. Q.; Xu, L.; Guo, Y. S.; Lu, L. M. Identification of Low Potassium Stress-Responsive Proteins in Tobacco (*Nicotiana tabacum*) Seedling Roots Using an iTRAQ-Based Analysis. *Genet. Mol. Res.* 2016, *15*, 1–13.

Rout, J. R.; Ram, S. S.; Das, R.; Chakraborty, A.; Sudarshan, M.; Sahoo, S. L. Copper-Stress Induced Alterations in Protein Profile and Antioxidant Enzymes Activities in the *in vitro* Grown *Withania somnifera* L. *Physiol. Mol. Biol. Plants* 2013, *19*, 353–361.

Roy, S. K.; Kwon, S. J.; Cho, S. W.; Kamal, A. H. M.; Kim, S. W.; Sarker, K.; Oh, M. W.; Lee, M. S.; Chung, K. Y.; Xin, Z.; Woo, S. H. Leaf Proteome Characterization in the Context of Physiological and Morphological Changes in Response to Copper Stress in Sorghum. *Biometals* 2016, *29*, 495–513.

Roychoudhury, A.; Bhowmik, R. Genetic Engineering of Rice to Fortify Micronutrients. In: Roychoudhury, A. (Ed.) Rice Research for Quality Improvement: Genomics and Genetic Engineering (Volume 2), Springer Nature Singapore Pte Ltd, 2020, pp. 563–579.

Sang, W.; Huang, Z. R.; Qi, Y. P.; Yang, L. T.; Guo, P.; Chen, L. S. Two-Dimensional Gel Electrophoresis Data in Support of Leaf Comparative Proteomics of Two Citrus Species Differing in Boron-Tolerance. *Data Brief* 2015, *4*, 44–46.

Smolen, S.; Kowalska, I.; Czernicka, M.; Halka, M.; Keska, K.; Sady, W. Iodine and Selenium Biofertification with Additional Application of Salicylic Acid Affects Yield, Selected Molecular Parameters and Chemical Composition of Lettuce Plants (*Lactuca sativa* L. var. capitata). *Front. Plant Sci.* 2016, *7*, 1553.

Soundararajan, P.; Manivannan, A.; Cho, Y. S.; Jeong, B. R. Exogenous Supplementation of Silicon Improved the Recovery of Hyperhydric Shoots in *Dianthus caryophyllus* L. by Stabilizing the Physiology and Protein Expression. *Front. Plant Sci.* 2017, *8*, 738.

Srivastava, V.; Obudulu, O.; Bygdell, J.; Löfstedt, T.; Rydén, P.; Nilsson, R. et al. On PLS Integration of Transcriptomic, Proteomic and Metabolomic Data Shows Multi-Level Oxidative Stress Responses in the Cambium of Transgenic hipI-Superoxide Dismutase Populus Plants. *BMC Genomics* 2013, *14*, 893.

Sun, H.; Dai, H.; Wang, X.; Wang, G. Physiological and Proteomic Analysis of Selenium-Mediated Tolerance to Cd Stress in Cucumber (*Cucumis sativus* L.). *Ecotoxicol. Environ. Saf.* 2016, *133*, 114–126.

Tomioka, R.; Takenaka, C.; Maeshima, M.; Tezuka, T.; Kojima, M.; Sakakibara, H. Stimulation of Root Growth Induced by Aluminum in *Quercus serrata* Thunb. Is Related to Activity of Nitrate Reductase and Maintenance of IAA Concentration in Roots. *Am. J. Plant Sci.* 2012, *3*, 1619.

Torregrosa, V. L.; Husekova, B.; Sychrova, H. Potassium Uptake Mediated by Trk1 Is Crucial for *Candida glabrate* Growth and Fitness. *PLoS ONE* 2016, *11*(4), e0153374.

Tucker, M. R. Essential Plant Nutrients: Their Presence in North Carolina Soils and Role in Plant Nutrition, Agronomic Division, NCDA & CS bulletin-Agronomic Division, USA, 1999, pp. 1–9.

Vaz, C.; Tanavde, V. Proteomics. In: Arivaradarajan, P.; Misra, G. (Eds.) Omics Approaches, Technologies and Applications, Springer, Singapore, 2018, pp. 57–73.

Wang, Y.; Hu, H.; Zhu, L. Y.; Li, X. X. Response to Nickel in the Proteome of the Metal Accumulator Plant *Brassica juncea. J. Plant Interact.* 2012, *7*, 230–237.

Wang, X.; Ma, X.; Wang, H.; Li, B.; Clark, G.; Guo, Y.; Roux, S.; Sun, D.; Tang, W. Proteomic Study of Microsomal Proteins Reveals a Key Role for Arabidopsis annexin 1 in Mediating Heat Stress Induced Increase in Intracellular Calcium Levels. *Mol. Cell Proteomics* 2015, *14*, 686–694.

Yang, L. T.; Lu, Y. B.; Zhang, Y.; Guo, P.; Chen, L. S. Proteomic Profile of *Citrus grandis* Roots Under Long-Term Boron-Deficiency Revealed by iTRAQ. *Trees* 2016, *30*, 1057–1071.

Yang, Q. S.; Wang, Y. Q.; Zhang, J. J.; Shi, W. P.; Qian, C.; Peng, X. Identification of Aluminum Responsive Proteins in Rice Roots by a Proteomic Approach: Cysteine Synthase as a Key Player in Al Response. *Proteomics* 2007, *7*, 737–749.

Yildiz, M.; Terzi, H. Proteomic Analysis of Chromium Stress and Sulfur Deficiency Responses in Leaves of Two Canola (*Brassica napus* L.) Cultivars Differing in Cr (VI) Tolerance. *Ecotoxicol. Environ. Saf.* 2016, *124*, 255–266.

Yina, Y.; Yanga, R.; Hana, Y.; Gua, Z. Comparative Proteomic and Physiological Analyses Reveal the Protective Effect of Exogenous Calcium on the Germinating Soybean Response to Salt Stress. *J. Proteomics* 2015, *113*, 110–126.

You, X.; Yang, L. T.; Lu, Y. B.; Li, H.; Zhang, S. Q.; Chen, L. A. Proteomic Changes of Citrus Roots in Response to Long-Term Manganese Toxicity. *Trees* 2014, *28*, 1383–1399.

Zeng, J.; He, X.; Quan, X.; Cai, S.; Han, Y.; Nadira, U. A.; Zhang, G. Identification of the Proteins Associated with Low Potassium Tolerance in Cultivated and Tibetan Wild Barley. *J. Proteomics* 2015, *126*, 1–11.

Zhang, Z.; Chao, M.; Wang, S.; Bu, J.; Tang, J.; Li, F.; Wang, Q.; Zhang, B. Proteome Quantification of Cotton Xylem Sap Suggests the Mechanisms of Potassium Deficiency-Induced Changes in Plant Resistance to Environmental Stresses. *Sci. Rep.* 2016c, *6*, 21060.

Zhang, K.; Liu, H.; Tao, P.; Chen, H. Comparative Proteomic Analyses Provide New Insights into Low Phosphorus Stress Responses in Maize Leaves. *PLoS ONE* 2014, *9*(5), e98215.

Zhang, H.; Xia, Y.; Chen, C.; Zhuang, K.; Song, Y.; Shen, Z. Analysis of Copper Binding Proteins in Rice Radicles Exposed to Excess Copper and Hydrogen Peroxide Stress. *Front. Plant Sci.* 2016a, *7*, 1216.

Zhang, S.; Zhou, R.; Zhao, H.; Korpelainen, H.; Li, C. iTRAQ Based Quantitative Proteomic Analysis Gives Insight into Sexually Different Metabolic Processes of Poplars Under Nitrogen and Phosphorus Deficiencies. *Proteomics* 2016b, *16*, 614–628.

Zheng, L.; Lan, P.; Shen, R. F.; Li, W. F. Proteomics of Aluminum Tolerance in Plants. *Proteomics* 2014, *14*, 566–578.

Zhou, M.; Zhu, S.; Mo, X.; Guo, Q.; Li, Y.; Tian, J.; Liang, C. Proteomic Analysis Dissects Molecular Mechanisms Underlying Plant Responses to Phosphorus Deficiency. *Cell* 2022, *11*, 651.

Zierer, W.; Hajirezaei, M. R.; Eggert, K.; Sauer, N.; Wiren, N.; Pommerrenig, B. Phloem-Specific Methionine Recycling Fuels Polyamine Biosynthesis in a Sulfur-Dependent Manner and Promotes Flower and Seed Development. *Plant Physiol.* 2016, *170*, 790–806.

8 Phosphoproteomic Study of Plants under Different Abiotic Stress

Aditya Banerjee and Aryadeep Roychoudhury

8.1 INTRODUCTION

Abiotic stresses like salinity, drought, high or low temperature, high or low light intensity, irradiation and heavy metal toxicity negatively affect plant growth and crop productivity and quality since these sub-optimal conditions inhibit normal systemic development (Banerjee and Roychoudhury 2021a,b, 2022a,b). In order to adapt to such environmentally challenging conditions, plants usually reprogram the entire metabolome and proteome to ensure effective maintenance of anti-stress metabolites and proteins within cells and tissues (Levitt 1980; Zhu 2016; Banerjee and Roychoudhury 2018; Zhang et al. 2022).

The collection of proteins encoded by the genome was coined as 'proteome' by Marc Wilkins such that the hypothetical static structure consisting of diverse proteins in the system can be referred to altogether at once (Kosova et al. 2011). Abiotic stresses are widely known to alter the proteomes variably in different plant species (Banerjee and Roychoudhury 2022c). During such alterations, the proteome undergoes multiple post-translational modifications (PTMs) like phosphorylation, acetylation, glycosylation, methylation and sumoylation, which altogether alters and modulates protein activity at the molecular level (Friso and Wijk 2015; Grabsztunowicz et al. 2017; Arsova et al. 2018; Trinidad et al. 2020). Vega et al. (2018) reported that temporary or trans-generational priming of the plant immune system also promotes PTMs like phosphorylation, ubiquitination and sumoylation. Recently, Xue et al. (2022) presented an integrative platform based on quantitative PTMs across 43 plant species as 'qPTM plants'. At the time of formation, this database hosted 12,42,365 experimentally validated events of PTMs, which may be accessed through a user-friendly platform. Presently, study of the dynamism of the proteome is a gradually advancing branch of plant proteomics. There have been manuscripts investigating phosphoproteomics, redox proteomics and proteomics at the sub-cellular level in several major plant species and important crops like rice, wheat, maize and barley (Rampitsch and Bykova 2012; Hossain et al. 2012; Romero-Puertas et al. 2013; Mock and Dietz 2016; Tan et al. 2017; Kosova et al. 2018). Phosphorylation is a widely occurring PTM in the Ser, Thr and Tyr residues of proteins in plants (Kumar et al. 2018). This PTM crucially regulates the activity of proteins during systemic response to environmentally challenging conditions (Li et al. 2022). Cheng et al. (2014) reported the formation of a comprehensive database containing phosphorylation events in proteins in plants. The present chapter concisely discusses the importance and dynamics of phosphoproteomics in plant adaptation to abiotic stresses.

8.2 IMPORTANCE OF PHOSPHOPROTEOMICS

Phosphorylation is one of the most pre-dominant types of PTMs in proteins in response to abiotic stresses in plants (Banerjee and Roychoudhury 2020a). Phosphorylation of proteins aids in the relay of signals after perception of cell surface-localized receptors to downstream transcription factors (TFs) and even proteins required for translation (Kersten et al. 2009). Multiple signaling pathways, dependent upon phosphorylation and regulated by several protein kinases, have been reported in

DOI: 10.1201/b23255-8

plants like rice and *Arabidopsis* (Asai et al. 2002; Bentem and Hirt 2007; Pitzschke et al. 2009; Chen and Ronald 2011; Roychoudhury and Banerjee 2017). Thus, the activity of these kinases on various target proteins has profound downstream effects on the entire signaling pathway and hence requires thorough introspection from the proteomic perspectives (Schulze et al. 2010). Crosstalks between phosphorylated and/or de-phosphorylated proteins with phytohormonal and stress-responsive pathways are also extensively reported in crops as well as model plant species (Roychoudhury et al. 2013). Overall, the interaction of phosphoproteomics with other stress-adaptive pathways is complex and requires integrated investigations for ultimate elucidation and crop development under sub-optimal conditions (Nirmala et al. 2010).

Phosphorylation of crucial proteins involved in stress signaling is usually regulated by different kinases like calcium-dependent protein kinases (CDPKs), calcineurin B-like protein (CBL)-interacting protein kinases (CIPKs) and mitogen-activated protein kinases (MAPKs) (Wan et al. 2007; Lee et al. 2008; Batistic and Kudla 2009; Luan 2009; Popescu et al. 2009). Likewise to ensure de-phosphorylation of proteins, 112 and 132 protein phosphatases have been reported in model plants like *Arabidopsis* and rice, respectively (Kerk et al. 2002; Singh et al. 2010). Analysis of non-synonymous single nucleotide polymorphisms (nsSNPs) within potential sites of phosphorylation revealed higher localization of phosphorylation hotspots outside domains having conserved sequences (Riano-Pachon et al. 2010). It was also found that the nsSNPs which affected phosphorylation sites affected the activity of receptors, stress-adaptive proteins and adaptor proteins, although the proteins participating in metabolism were comparatively affected at a lower extent (Rampitsch and Bykova 2012). Thus, this chapter aims to understand the modulatory effects of phosphorylation and de-phosphorylation in signal relay during abiotic stresses from a proteomic approach.

8.3 IMPACT OF PHOSPHOPROTEOMICS ON SALINITY STRESS IN PLANTS

Unregulated intrusion of Na^+ into cells during salt stress triggers widespread tissue necrosis and cell death in plants (Banu et al. 2009). In order to adapt to such harsh conditions, plants modulate the overall signaling network for activating stress-responsive genes and encoded proteins (Yang and Guo 2018; Isayenkov and Maathuis 2019). The phosphorylation cascade triggered by plants in response to salt stress activates the salt overly sensitive (SOS) signaling pathway, which enables the extrusion of excess Na^+ (Ji et al. 2013). It was found that phosphorylation of MAPKs like MEKK1, MEK1 and MPK4/MPK6 ensured the relay of salt-adaptive signaling in *Arabidopsis*. Furthermore, overexpression of *MKK2* and *MPK6* promoted the ability of transgenic *Arabidopsis* seedlings to endure salinity and freezing conditions, whereas salt hypersensitive phenotype was developed in the plants wherein *MKK2* was downregulated (Teige et al. 2004). Similarly, overexpression of *MAPK5* promoted salt tolerance in transgenic rice seedlings due to the activation of salt-responsive signaling cascade, thus illustrating the essentiality of protein phosphorylation in generating abiotic stress tolerance (Xiong and Yang 2003; Zhu 2003). Roychoudhury et al. (2008) inferred that activation of the master-regulatory trans-acting factor OSBZ8 required for upregulation of crucial *osmotic stress responsive* (OR) genes was dependent upon phosphorylation probably by casein kinase II (CKII)-like proteins. Abscisic acid (ABA)-dependent activation of the ABA-responsive element binding factor 1 (AREB1) occurred via phosphorylation by 42 kDa kinase (Uno et al. 2000). Multiple SnRK2-type kinases have been found to modulate the activation of target ABA binding factors (ABFs) via phosphorylation (Roychoudhury and Paul 2012).

In-gel staining of proteins isolated from rice roots exposed to salt stress using the Pro-Q Diamond phosphoprotein stain characterized the upregulation of stress-responsive proteins like salt-induced protein, mannose-binding rice lectin, glutathione-S-transferase (GST), dimethylaniline monooxygenase and DnaK-type heat shock protein 70 (HSP70) (Schulenberg et al. 2003; Chitteti and Peng 2007). In another study, Prak et al. (2008) investigated the regulation of phosphorylation in the aquaporin-like plasma membrane intrinsic proteins (PIPs) in response to salt stress in the roots of *Arabidopsis*. Analysis using stable isotope labeling showed that phosphorylation was reduced in PIP2;1 during stress in *Arabidopsis*. It was also reported that the majority of phosphorylation is localized toward the C-terminus of aquaporins (Maurel et al. 2008). Prak et al. (2008) found that phosphorylation at one of the sites was essential for targeting PIP2;1 to the cell membrane in *Arabidopsis* seedlings. Application of short-term salt stress phosphorylated ten proteins and de-phosphorylated six proteins in the roots of maize when the proteome was analyzed using two-dimensional gel electrophoresis (2-DE). It was further ascertained that phosphorylation of enzymes like fructokinase 2 and sucrose synthase could be considered markers for determining early response to salt stress (Zorb et al. 2010). Salinity also triggers phosphorylation of 14-3-3 proteins, which in turn phosphorylates xyloglucan endotransglucosylase (XET) hydrolases, which control and regulate the growth and adaptation of the cell wall (Rampitsch and Bykova 2012).

In another study, Hsu et al. (2009) showed the majority of phosphorylation in membrane proteins in salt-stressed *Arabidopsis*, among which phosphorylation sites for PIP2;4 and MKK2 were identified. Brader et al. (2007) reported that MKK2 was phosphorylated by the upstream kinase MEKK1 during salt stress in *Arabidopsis*. Reduction in hydraulic conductivity in the roots of salt-stressed barley seedlings has been reported to promote the staurosporine-sensitive kinase-dependent phosphorylation of PIP2 aquaporin (Horie et al. 2011). Nakagami et al. (2010) also showed the conservation of phosphorylation at specific sites of physiological and stress-responsive proteins in salt-stressed rice plants. Interestingly, Chang et al. (2012) showed the negative correlation of rice PIPs with water flux during salinity.

8.4 IMPACT OF PHOSPHOPROTEOMICS ON DESICCATION AND FLOODING STRESS IN PLANTS

The absence of water leading to drought and desiccation and the excess of water leading to flooding and anoxia or hypoxia are two predominant types of environmental challenges that are experienced by crops as a result of global climate change (Eichelmann et al. 2022; Muller and Bahn 2022). These stresses have been reported as one of the major constraints for sustainable crop production across Asia, and some significant studies at the proteomic level have been conducted to identify the underlying basis of stress tolerance at the protein level (Bin Rahman and Zhang 2022). Phosphate-specific western blotting and mass spectrometric analysis following 2-DE identified differential phosphorylation of ten proteins during drought stress in rice (Ke et al. 2009). These included ABA-dependent stress-inducible proteins, ethylene-responsive proteins and drought-responsive S-like ribonuclease. Three proteins associated with drought signaling were also found to be de-phosphorylated in stressed rice plants (Ke et al. 2009).

He and Li (2008) identified six proteins that showed differential phosphorylation when treated with short exposures of ABA in rice seedlings. These proteins were detected as ascorbate peroxidase, superoxide dismutase, triosephosphate isomerase and other unidentified putative proteins. Furthermore, the malate dehydrogenase, found in the glyoxysome, was phosphorylated in response to ABA treatment in rice (He and Li 2008). In another study, Kline et al. (2010) reported increased phosphorylation of SnRK2 kinases (involved in ABA signal transduction) in the *Arabidopsis thaliana* phosphoproteome when the seedlings were exposed to drought stress.

Excessive water-logging or flooding leads to oxygen deprivation or anoxia in plant roots and parts of the shoots along with the generation of phytotoxic chemicals in flooded soil (Setter and Waters 2003; Nanjo et al. 2010). Nanjo et al. (2012) found that de-phosphorylation of proteins regulating protein synthesis and folding was an early response to flooding in soybean seedlings. Differential phosphorylation of proteins, associated with energy production, protein synthesis and maintenance of cellular scaffold, was reported in flooding-stressed soybean roots. Furthermore, in these stressed roots, decreased phosphorylation and expression were observed in proteins, which regulated elongation of roots in soybean plants (Nanjo et al. 2012).

8.5 IMPACT OF PHOSPHOPROTEOMICS ON TEMPERATURE STRESS IN PLANTS

Extremes of temperature like heat or chilling stress crucially affect crop productivity and yield due to compromised reproductive development (Hatfield and Prueger 2015; Zhao et al. 2017). Increased production of greenhouse gases has steeply increased the global temperature in the tropical and equatorial zones (Wang and Zhang 2019; Thornton et al. 2022). Thus, heat stress is a major hurdle for sustainable crop cultivation in these areas (Banerjee and Roychoudhury 2020b). Exposure to heat has been reported to reprogram the phosphoproteome of susceptible rice cultivars (Lee et al. 2007; Scafaro et al. 2010; Chen et al. 2011). High temperature induced the translation of ten phosphoproteins, which were found to regulate the Calvin cycle, reactive oxygen species (ROS) metabolism, ATP production and microtubular locomotion (Chen et al. 2011). De-phosphorylation of ribulose bisphosphate carboxylase (RuBisCo) and phosphorylation of the β-subunit of ATP synthase were also observed in the rice seedlings exposed to heat stress (Chen et al. 2011).

Similar to heat stress, low temperature induced the phosphorylation in the tonoplast monosaccharide transporters (TMTs) in *Arabidopsis* (Schulze et al. 2012). Furthermore, it was observed that exposure to 4°C stimulated a mitogen-activated triple kinase-like kinase-mediated phosphorylation at a specific Ser residue located within the central hydrophilic loop of TMT1 and TMT2 in *Arabidopsis*, and sequential conservation in this loop has been found to identify TMTs among other sugar transporters (Wingenter et al. 2011; Schulze et al. 2012). Wormit et al. (2006) inferred that these TMTs mediated the crucial import of carbohydrates like glucose and fructose within the mesophyll cell vacuoles during cold stress. Differential phosphorylation of TMTs in response to low-temperature stress has also been detected in rice and barley (Whiteman et al. 2008; Endler et al. 2009).

8.6 CONCLUSION

Phosphorylation of proteins is an important PTM that is responsible for the activation and/or de-activation of a large number of vital stress signaling and adaptive pathways. Modulation of these pathways results in the development of abiotic stress tolerance in plants and crops. The majority of these pathways are associated with the ABA-dependent signaling. Phosphorylation is also triggered in downstream proteins and transporters, which lead to better active or passive transport of electrolytes and metabolites during harsh environmental conditions. Thus, detailed elucidation of the phosphorylation status of the proteome is essential to identify molecular candidates that may be targeted and biotechnologically switched to constitutively activated or de-activated mode in order to develop abiotic stress tolerance.

8.7 FUTURE PERSPECTIVES

Phosphoproteomics is a rapidly developing yet naive field that requires much more extensive investigative approaches for complete understanding of the underlying mechanism. Future perspectives would involve the use of epigenomic approaches to map the regulation of genes encoding the regulatory kinases, which control the phosphorylation status of the proteome. Furthermore, high-throughput quantitative proteomics supported with intricate mass spectrometric technology should be adopted to visualize the alterations in the phosphoproteome in a more holistic manner. More studies involving crops, exposed to abiotic stresses at the field level should be conducted to analyze the effects of environmental challenges on variations in the phosphoproteome and to record advancement in the newer field of plant environmental-phosphoproteomics.

ACKNOWLEDGMENTS

Financial assistance from the Science and Engineering Research Board, Government of India through the grant [EMR/2016/004799] and the Department of Higher Education, Science and Technology and Biotechnology, Government of West Bengal, through the grant [264(Sanc.)/ ST/P/S&T/1G-80/2017] to Prof. Aryadeep Roychoudhury is gratefully acknowledged. Mr. Aditya Banerjee is thankful to the University Grants Commission, Government of India, for providing Senior Research Fellowship during the course of this work.

REFERENCES

Arsova B, Watt M, Usadel B (2018) Monitoring of plant protein post-translational modifications using targeted proteomics. Front Plant Sci 9:1168.

Asai T, Tena G, Plotnikova J, Willmann MR, Gomez-Gomez L, Boller T, Ausubel FM, Sheen J (2002) MAP kinase signalling cascade in *Arabidopsis* innate immunity. Nature 415:977–983.

Banerjee A, Roychoudhury A (2018) Abiotic stress, generation of reactive oxygen species, and their consequences: An overview. In: Singh VP, Singh S, Tripathi D, Mohan Prasad S, Chauhan DK (Eds.) Revisiting the role of reactive oxygen species (ROS) in plants: ROS Boon or bane for plants? John Wiley & Sons, Inc., USA, pp. 23–50.

Banerjee A, Roychoudhury A (2020a) Deciphering the roles of protein phosphatases in the regulation of salt-induced signaling responses in plants. In: Pandey GK (Ed.) Protein phosphatases and stress management in plants: Functional and genomic perspective. Springer Nature, Switzerland AG, pp. 149–162.

Banerjee A, Roychoudhury A (2020b) Seed priming as a method to generate heat-stress tolerance in plants: A minireview. In: Wani SH, Kumar V (Eds.) Heat stress tolerance in plants: Physiological, molecular and genetic perspectives. John Wiley & Sons, United Kingdom, pp. 23–28.

Banerjee A, Roychoudhury A (2021a) Differential lead-fluoride and nickel-fluoride uptake in co-polluted soil variably affects the overall physiome in an aromatic rice cultivar. Environ Pollut 268: 115504.

Banerjee A, Roychoudhury A (2021b) Maghemite nano-fertilization promotes fluoride tolerance in rice by restoring grain yield and modulating the ionome and physiome. Ecotoxicol Environ Saf 215: 112055.

Banerjee A, Roychoudhury A (2022a) Rhizofiltration of combined arsenic-fluoride or lead-fluoride polluted water using common aquatic plants and use of the 'clean' water for alleviating combined xenobiotic toxicity in a sensitive rice variety. Environ Pollut 304:119128.

Banerjee A, Roychoudhury A (2022b) Assessing the rhizofiltration potential of three aquatic plants exposed to fluoride and multiple heavy metal polluted water. Vegetos 35:1158–1164.

Banerjee A, Roychoudhury A (2022c) Dissecting the phytohormonal, genomic and proteomic regulation of micronutrient deficiency during abiotic stresses in plants. Biologia 77:3037–3058.

Banu MNA, Hoque MA, Watanabe-Sugimoto M, Matsuoka K, Nakamura Y, Shimoishi Y, Murata Y (2009) Proline and glycinebetaine induce antioxidant defense gene expression and suppress cell death in cultured tobacco cells under salt stress. J Plant Physiol 166:146–156.

Batistic O, Kudla J (2009) Plant calcineurin B-like proteins and their interacting protein kinases. Biochim Biophys Acta 1793:985–992.

Bentem S, Hirt H (2007) Using phosphoproteomics to reveal signalling dynamics in plants. Trends Plant Sci 12:404–411.

Bin Rahman ANMR, Zhang J (2022) The coexistence of flood and drought tolerance: An opinion on the development of climate-smart rice. Front Plant Sci 13:860802.

Brader G, Djamei A, Teige M, Palva ET, Hirt H (2007) The MAP kinase MKK2 affects disease resistance in *Arabidopsis*. Mol Plant Microbe Interact 20:589–596.

Chang IF, Hsu JL, Hsu PH, Sheng WA, Lai SJ, Lee C, Chen CW, Hsu JC, Wang SY, Wang LY, Chen CC (2012) Comparative phosphoproteomic analysis of microsomal fractions of *Arabidopsis thaliana* and *Oryza sativa* subjected to high salinity. Plant Sci 185–186:131–142.

Cheng H, Deng W, Wang Y, Ren J, Liu Z, Xue Y (2014) dbPPT: A comprehensive database of protein phosphorylation in plants. Database (Oxford) 2014:bau121.

Chen X, Ronald PC (2011) Innate immunity in rice. Trends PlantSci 16:451–459.

Chen X, Zhang W, Zhang B, Zhou J, Wang Y, Yang Q, Ke Y, He H (2011) Phosphoproteins regulated by heat stress in rice leaves. Proteome Sci 9:37.

Chitteti BR, Peng Z (2007) Proteome and phosphoproteome differential expression under salinity stress in rice (*Oryza sativa*) roots. J Proteome Res 6:1718–1727.

Eichelmann E, Mantoani MC, Chamberlain SD, Hemes KS, Oikawa PY, Szutu D, Valach A, Verfaillie J, Baldocchi DD (2022) A novel approach to partitioning evapotranspiration into evaporation and transpiration in flooded ecosystems. Global Change Biol 28: 990–1007.

Endler A, Reiland S, Gerrits B, Schmidt UG, Baginsky S, Martinoia E (2009) *In vivo* phosphorylation sites of barley tonoplast proteins identified by a phosphoproteomics approach. Proteomics 9:310–321.

Friso G, Wijk KJ (2015) Posttranslational protein modifications in plant metabolism. Plant Physiol 169:1469–1487.

Grabsztunowicz M, Koskela MM, Mulo P (2017) Post-translational modifications in regulation of chloroplast function: recent advances. Front Plant Sci 8:240.

Hatfield JL, Prueger JH (2015) Temperature extremes: effect on plant growth and development. Weather Climate Extremes 10:4–10.

He H, Li J (2008) Proteomic analysis of phosphoproteins regulate by abscisic acid in rice leaves. Biochem Biophys Res Commun 371:883–888.

Horie T, Kaneko T, Sugimoto G, Sasano S, Panda SK, Shibasaka M, Katsuhara M (2011) Mechanisms of water transport mediated by PIP aquaporins and the irregulation via phosphorylation events under salinity stress in barley roots. Plant Cell 52:663–675.

Hossain Z, Nouri MZ, Komatsu S (2012) Plant cell organelle proteomics in response to abiotic stress. J Proteome Res 11:37–48.

Hsu JL, Wang LY, Wang SY, Lin CH, Ho KC, Shi FK, Chang IF (2009) Functional phosphoproteomic profiling of phosphorylation sites in membrane fractions of salt-stressed *Arabidopsis thaliana*. Proteome Sci 7:42.

Isayenkov SV, Maathuis FJM (2019) Plant salinity stress: many unanswered questions remain. Front Plant Sci 10:80.

Ji H, Pardo JM, Batelli G, Oosten MJV, Bressan RA, Li X (2013) The salt overly sensitive (SOS) pathway: established and emerging roles. Mol Plant 6:275–286.

Ke Y, Han G, He H, Li J (2009) Differential regulation of proteins and phosphoproteins in rice under drought stress. Biochem Biophys Res Commun 379:133–138.

Kerk D, Bulgrien J, Smith DW, Barsam B, Veretnik S, Gribskov M (2002) The complement of protein phosphatase catalytic sub-units encoded in the genome of *Arabidopsis*. Plant Physiol 129:908–925.

Kersten B, Agrawal G, Durek P, Neigenfind J, Schulze W, Walther D, Rakwal R (2009) Plant phosphoproteomics: an update. Proteomics 9:964–988.

Kline KG, Barrett-Wilt GA, Sussman MR (2010) In planta changes in protein phosphorylation induced by the plant hormone abscisic acid. Proc Natl Acad Sci USA 107:15986–15991.

Kosova K, Vitamvas P, Prasil IT, Renaut J (2011) Plant proteome changes under abiotic stress – Contribution of proteomics studies to understanding plant stress response. J Proteomics 74:1301–1322.

Kosova K, Vítámvás P, Urban MO, Prášil IT, Renaut J (2018) Plant abiotic stress proteomics: The major factors determining alterations in cellular proteome. Front Plant Sci 9:122.

Kumar V, Khare T, Sharma M, Wani SH (2018) Engineering crops for the future: A phosphoproteomics approach. Curr Protein Peptide Sci 19:413–426.

Lee D, Ahsan NB, Lee S, Kang K, Bahk J, Lee I, Lee B (2007) A proteomic approach in analyzing heat-responsive proteins in rice leaves. Proteomics 7:3369–3383.

Lee MO, Cho K, Kim SH, Jeong SH, Kim JA, Jung YH, Shim J, Shibato J, Rakwal R, Tamogami S (2008) Novel rice OsSIPK is a multiple stress responsive MAPK family member showing rhythmic expression at mRNA level. Planta 227:981–990.

Levitt J (1980) Responses of plants to environmental stresses: Chilling, freezing and high temperature stresses. Academic Publisher, New York, NY.

Li W, Han X, Lan P (2022) Emerging roles of protein phosphorylation in plant iron homeostasis. Trends Plant Sci 27:908–921.

Luan S (2009) The CBL-CIPK network in plant calcium signalling. Trends Plant Sci 14:37–42.

Maurel C, Verdoucq L, Luu D, Santoni V (2008) Plant aquaporins: membrane channels with multiple integrated functions. Annu Rev Plant Biol 59:595–624.

Mock HP, Dietz KJ (2016) Redox proteomics for the assessment of redox related posttranslational regulation in plants. Biochim Biophys Acta Proteins Proteomics 1864:967–973.

Muller LM, Bahn M (2022) Drought legacies and ecosystem responses to subsequent drought. Global Change Biol 28: 5086–5103.

Nakagami H, Sugiyama N, Mochida K, Daudi A, Yoshida Y, Toyoda T, Tomita M, Ishihama Y, Shirasu K (2010) Large-scale comparative phosphoproteomics identifies conserved phosphorylation sites in plants. Plant Physiol 153:1161–1174.

Nanjo Y, Skultety L, Ashraf Y, Komatsu S (2010) Comparative proteome analysis of early-stage soybean seedlings responses to flooding by using gel and gel-free techniques. J Proteome Res 9:3989–4002.

Nanjo Y, Skultety L, Uváčková L, Klubicová K, Hajduch M, Komatsu S (2012) Mass spectrometry-based analysis of proteomic changes in the root tips of flooded soybean seedlings. J Proteome Res 11:372–385.

Nirmala J, Drader T, Chen X, Steffenson B, Kleinhofs A (2010) Stem rust spores elicit rapid RPG1 phosphorylation. Mol Plant Microbe Interact 23:1635–1642.

Pitzschke A, Schikora A, Hirt H (2009) MAPK cascade signalling networks in plant defence. Curr Opin Plant Biol 12:421–426.

Popescu SC, Popescu GV, Bachan S, Zhang Z, Gerstein M, Snyder M, Dinesh Kumar SP (2009) MAPK target networks in *Arabidopsis thaliana* revealed using functional protein microarrays. Genes Dev 23:80–92.

Prak S, Hem S, Boudet J, Viennois G, Sommerer N, Rossignol M, Maurel C, Santoni V (2008) Multiple phosphorylations in the C-terminal tail of plant plasma membrane aquaporins: role in subcellular trafficking of AtPIP2;1 in response to salt stress. Mol Cell Proteomics 7:1019–1030.

Rampitsch C, Bykova NV (2012) The beginnings of crop phosphoproteomics: exploring early warning systems of stress. Front Plant Sci 3:144.

Riano-Pachon DM, Kleessen S, Neigenfind J, Durek P, Weber E, Engelsberger WR, Walther D, Selbig J, Schulze WX, Kersten B (2010) Proteome-wide survey of phosphorylation patterns affected by nuclear DNA polymorphisms in *Arabidopsis thaliana*. BMC Genomics 11:411.

Romero-Puertas MC, Rodríguez-Serrano M, Sandalio LM (2013) Protein S-nitrosylation in plants under abiotic stress: an overview. Front Plant Sci 4:373.

Roychoudhury A, Banerjee A (2017) Abscisic acid signaling and involvement of mitogen activated protein kinases and calcium-dependent protein kinases during plant abiotic stress. In: Pandey G (Ed.) Mechanism of Plant Hormone Signaling under Stress. John Wiley & Sons, Inc., USA, Vol. 1, pp. 197–241.

Roychoudhury A, Gupta B, Sengupta DN (2008) Trans-acting factor designated OSBZ8 interacts with both typical abscisic acid responsive elements as well as abscisic acid responsive element-like sequences in the vegetative tissues of indica rice cultivars. Plant Cell Rep 27:779–794.

Roychoudhury A, Paul A (2012) Abscisic Acid-Inducible Genes during Salinity and Drought Stress. In: Berhardt LV. (Ed.), Advances in Medicine and Biology, Volume 51, Nova Science Publishers, New York, pp. 1–78.

Roychoudhury A, Paul S, Basu S (2013) Cross-talk between abscisic acid-dependent and abscisic acid-independent pathways during abiotic stress. Plant Cell Rep 32:985–1006.

Scafaro AP, Haynes PA, Atwell BJ (2010) Physiological and molecular changes in *Oryza meridionalis* Ng., a heat-tolerant species of wild rice. J Exp Bot 61:191–202.

Schulenberg B, Aggeler B, Beechem J, Capaldi R, Patton W (2003) Analysis of steady-state protein phosphorylation in mitochondria using a novel fluorescent phosphosensor dye. J Biol Chem 278:27251–27255.

Schulze B, Menzel T, Jehle AK, Müller K, Beeler S, Boller T, Felix G, Chinchilla D (2010) Rapid heteromerization and phosphorylation of ligand-activated plant trans-membrane receptors and their associated kinase BAK1. J Biol Chem 285:9444–9451.

Schulze WX, Schneider T, Starck S, Martinoia E, Trentman O (2012) Cold acclimation induces changes in *Arabidopsis* tonoplast protein abundance and activity and alters phosphorylation of tonoplast monosaccharide transporters. Plant J 69:529–541.

Setter TL, Waters I (2003) Review of prospect for germplasm improvement for waterlogging tolerance in wheat, barley and oats. Plant Soil 253: 1–34.

Singh A, Giri J, Kapoor S, Tyagi A, Pandey G (2010) Protein phosphatase complement in rice: genome-wide identification and transcriptional analysis under abiotic stress conditions and reproductive development. BMC Genomics 11:435.

Tan BC, Lim YS, Lau SE (2017) Proteomics in commercial crops: an overview. J Proteomics 169:176–188.

Teige M, Scheikl E, Eulgem T, Dóczi R, Ichimura K, Shinozaki K, Dangl JL, Hirt H (2004) The MKK2 pathway mediates cold and salt stress signalling in *Arabidopsis*. Mol Cell 15:141–152.

Thornton P, Nelson G, Mayberry D, Herrero M (2022) Impacts of heat stress on global cattle production during the 21st century: a modelling study. Lancet Planet Health 6:E192–E201.

Trinidad JL, Pabuyon ICM, Kohli A (2020) Harnessing protein posttranslational modifications for plant improvement. In: Tuteja N, Tuteja R, Passricha N, Saifi SK (Eds) Advancement in Crop Improvement Techniques. Woodhead Publishing, Elsevier, USA, pp. 385–401.

Uno Y, Furihata T, Abe H, Yoshida R, Shinozaki K, Yamaguchi-Shinozaki K (2000) *Arabidopsis* basic leucine zipper transcription factors involved in an abscisic acid dependent signal transduction pathway under drought and high-salinity conditions. Proc Natl Acad SciUSA 97:11632–11637.

Vega D, Newton AC, Sadanandom A (2018) Post-translational modifications in priming the plant immune system: ripe for exploitation? FEBS Lett 592:1929–1936.

Wang F, Zhang J (2019) Heat stress response to national-committed emission reductions under the Paris Agreement. Int J Environ Res Public Health 16:2202.

Wan B, Lin Y, Mou T (2007) Expression of rice Ca^{2+}-dependent protein kinases (CDPKs) genes under different environmental stresses. FEBS Lett 581:1179–1189.

Whiteman SA, Nuhse TS, Ashford DA, Sanders D, Maathuis FJ (2008) A proteomic and phosphoproteomics analysis of *Oryza sativa* plasma membrane and vacuolar membrane. Plant J 56:146–156.

Wingenter K, Trentmann O, Winschuh I, Hoermiller II, Heyer AG, Reinders J, Schulz A, Geiger D, Hedrich R, Neuhaus HE (2011) A member of the mitogen-activated protein 3-kinase family is involved in the regulation of plant vacuolar glucose uptake. Plant J 68:890–900.

Wormit A, Trentmann O, Feifer I, Lohr C, Tjaden J, Meyer S, Schmidt U, Martinoia E, Neuhaus HE (2006) Molecular identification and physiological characterization of a novel monosaccharide transporter from *Arabidopsis* involved in vacuolar sugar transport. Plant Cell 18:3476–3490.

Xiong L, Yang Y (2003) Disease resistance and abiotic stress tolerance in rice are inversely modulated by an abscisic acid-inducible mitogen-activated protein kinase. Plant Cell 15:745–759.

Xue H, Zhang Q, Wang P, Cao B, Jia C, Cheng B, Shi Y, Guo W-F, Wang Z, Liu Z-X, Cheng H (2022) qPTM-plants: an integrative database of quantitative post-translational modifications in plants. Nucleic Acids Res 50:D1491–D1499.

Yang Y, Guo Y (2018) Unraveling salt stress signaling in plants. J Integr Plant Biol 60:796–804.

Zhang H, Zhu J, Gong Z, Zhu J-K (2022) Abiotic stress responses in plants. Nat Rev Genet 23:104–119.

Zhao C, Liu B, Piao S, Wang X, Lobell DB, Huang Y et al. (2017) Temperature increase reduces global yields of major crops in four independent estimates. Proc Natl Acad Sci USA 114:9326–9331.

Zhu JK (2003) Regulation of ion homeostasis under salt stress. Curr Opin Plant Biol 6:441–445.

Zhu J-K (2016) Abiotic stress signaling and responses in plants. Cell 167:313–324.

Zorb C, Schmitt S, Mühling KH (2010) Proteomic changes in maize roots after short-term adjustment to saline growth conditions. Proteomics 10:4441–4449.

9 Advancement in Proteomics Research in Response to Biotic Stress in Plants

Soumita Mitra and Nilanjan Chakraborty

9.1 INTRODUCTION

The biological entities from all kingdoms of life have always influenced the productivity and well-being of plants. Sometime the interaction between the biological organism and the plant results in negative effects and causes certain level of metabolic disturbance which can be called biotic stress (Sergeant and Renaut 2010). The massive devastation to agricultural crops caused by varieties of organisms like viruses, fungi, bacteria, nematodes and herbivorous insects is well documented, which in turn lower the economic growth of the human population, depending on those crops. In the middle of the 19th century, the massive attack of potato fields by *Phytophthora infestans* killing or displacing 25% of the Irish population resulted in the migration of Irish farmers to North America (Fraser 2003). Pesticides and fungicides are widely used to control the yield reductions caused by biotic stress, but their harmful effects on environment and human health are now a great concern and largely debated. To overcome this, it is important to understand the mechanisms involved in disease resistance so as to control the interactions between plants and pathogens (Chakraborty et al. 2022).

The proteins are the key players in plant biotic stress response. In fact, it is believed that the plant immune system responds to biotic stress as a complex interaction and crosstalk between multiple signals and a diverse set of stress-tolerant-related proteins (Liu et al. 2019). No individual proteins reflect the complex network of signals; rather combined proteins are likely to function together and play multiple roles in stress response. Protein function depends not only on the molecular structure of the protein but also on its subcellular localization and post-translational modifications (PTMs). In this connection, proteomic analysis is one of the best procedures in the identification, quantification, characterization, interactions and functions of a large set of proteins in a living cell (Wang et al. 2019). The proteomics studies have progressed from simple identification of individual proteins to comparative and quantitative protein profiling, peptide mass fingerprinting, PTM analysis of proteins, subcellular localization, quantification of protein complexes, analysis of protein signalling pathways and protein–protein interactions (Sergeant and Renaut 2010). The mass spectrometry (MS) technique used in the complex proteome analysis can result in up to several thousands of peptide spectra which cannot be easily analysed using any standard biochemical technique. The present decade has recorded great achievements in the field of proteomics in response to biotic stress towards crop improvement, especially in phase chromatography or two-dimensional gel electrophoresis (2DGE), tandem and non-tandem MS, isobaric tags for relative and absolute quantification labelling technique (iTRAQ), X-ray crystallography and progression from gel-based proteomics to shot-gun proteomics. The study of these technologies as response of plants to biotic stress remains a research field with great potential for growth.

9.2 HOW DO PLANTS PROTECT THEMSELVES FROM STRESS?

Based on the mode of nutrition acquisition, the plant pathogens may be of various types necrotrophs (organisms derive their energy from dead or dying cells), biotrophs (organisms that derive energy and multiply on another living organism) and hemi-biotrophs (organism that is parasitic in

DOI: 10.1201/b23255-9

living tissue for some time and then continues to live in dead tissue) (Sergeant and Renaut 2010). Despite lacking the adaptive immune system, plants can counteract biotic stresses by adapting against diverse sets of physical and biochemical strategies that may be constitutive or inducible. The presence of waxy cuticle, formation of cork layer, abscission layer and of tyloses, deposition of gums and cellulose and barks provides constitutive defence mechanisms, which, along with structural firmness, act as a first line of defence (Dangl and Jones 2001). Pathogen-associated molecular pattern (PAMP)-triggered immunity (PTI) and effector-triggered immunity (ETI) are inducible defence mechanisms in plants as regulated by chitin originating from fungal cell wall or flagellin from bacteria, a recognition that initiates protective measures (Tang et al. 2017). Biotic stressor and plant interaction can result in transcription of specific resistance (*R*) gene and this gene specifically enhances pathogenicity to specific host plants, which is counteracted by a plant factor often inducing localized cell death through a hypersensitive response. The interaction between plant and pathogens are driven out by molecular interactions between pathogen *avr* (*avirulence*) gene loci and alleles of the corresponding plant disease *resistance* (*R*) gene locus. When the interaction develops into a disease-resistance phenomenon, it means that corresponding *R* and *avr* genes are present in respective hosts and pathogens (Stukenbrock and McDonald 2009). Alternately, when *R* and *avr* genes are not compatible or absent, it leads to a diseased condition. Following pathogen attack, the plants induce hypersensitive response which is characterized by several biochemical changes such as rapid changes in ion fluxes, lipid peroxidation, apoplast acidification, protein phosphorylation, nitric oxide generation and a rapid induction of reactive oxygen species (ROS) and antimicrobial compounds, such as phytoalexins. The plant-pathogen interaction activates the genes related to the production of lignin, flavonoids, benzoic acid, abscisic acid, jasmonic acid (JA), ethylene and salicylic acid (SA). PR (pathogenesis related) proteins play a critical role in modulating cellular mechanisms and activating plant immunity (Dangl and Jones 2001). Another type of defence mechanism is the systemic acquired resistance (SAR) induced by prior treatment with biological agent and subsequent pathogen attacks that can last for days. The biotic elicitor through SAR leads to the accumulation of PR proteins in plants (Sergeant and Renaut 2010). Basic overview of host reactions after pathogen invasion or interaction has been depicted in Figure 9.1.

9.3 IMPORTANCE OF PROTEOMICS STUDY AGAINST BIOTIC STRESS

As proteins are directly involved in plant biotic stress response, it is essential to study the proteome changes under differential stress conditions. Several biotic stress factors result in profound alterations in protein network covering signalling, energy metabolism (glycolysis, Krebs cycle, ATP biosynthesis, photosynthesis), storage proteins, protein metabolism, several other biosynthetic pathways (e.g., S-adenosyl methionine metabolism, lignin metabolism), transporters, proteins involved in protein folding and chaperone activities, other protective proteins such as late embryogenesis abundant (LEA), PR proteins, ROS scavenging enzymes as well as proteins affecting regulation of plant growth and development (Roychoudhury et al. 2007, 2013; Kosová et al. 2018). Accordingly, studying protein profile is critical to understand the sophisticated plant responses to different biotic stressors under the view of proteome because (1) proteins participate directly in the formation of new plant phenotypes by regulating physiological characteristics to adapt to changes in the environment; and (2) proteins are the critical executors of cellular mechanisms and key players in the maintenance of cellular homeostasis (Liu et al. 2019). The chemical and physical diversity of proteins, from large to small and from acidic to basic, are much larger than the diversity of nucleotide-based macromolecules. Proteomes can identify wide array of proteins including observation of any change in protein level during specific developmental stage of plants or plants under stresses. Moreover, proteomics can reflect the metabolic processes and their possibilities to interact with important regulatory pathways (Tang et al. 2017; Liu et al. 2019). The results of proteomic studies aimed at comparison of stress response in plant genotypes differing in stress adaptability in the stress-tolerant genotypes with respect to the susceptible ones.

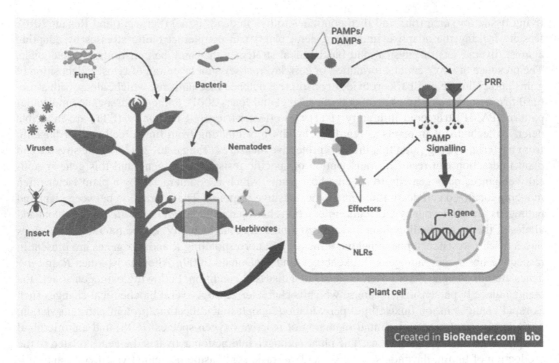

FIGURE 9.1 A molecular road map to plant immune response against various kinds of biotic stressors. Plants are the target of a variety of pathogens and cause disease. The invaders release PAMP or DAMP pattern recognition receptor (PRR) on the plant cell surface and initiate cell surface immunity in plants. PAMP/DAMP-triggered pathways can lead to changes in gene expression. Pathogens can deliver effectors to the inside of cells and can interfere with PAMP-triggered immunity by the suppression of PAMP signalling pathways. The intracellular immune receptors (NLRs) bind to those effectors and set off intracellular immunity that boosts the PAMP-triggered responses. *Abbreviations*: DAMP: damage-associated molecular pattern; NLR: nucleotide-binding domain leucine-rich repeat receptors; PAMP: pathogen-associated molecular pattern; PRR: pathogen recognition receptor; R: resistance.

9.4 BASIC TECHNIQUES OF PROTEOME ANALYSIS IN PLANTS

A general overview of a standard proteomics experiment in plant research is given in Figure 9.2. In order to analyse proteome of an organism or tissue after the sequential step of protein extraction and purification, it is too complex to directly identify the proteins. Therefore, developments along two routes, gel-based and gel-free, were essential. Recent researches on the role of proteomics against biotic stress have shown that the gel-based method was primarily used (Renaut et al. 2008).

Currently, 2D gel electrophoresis followed by MS technique constitutes the state of the art in the gel-based approach to proteome studies (Kjellsen et al. 2010); data analysis is done using one of the available software packages DeCyder v7.0 (GE-Healthcare), Progenesis Same Spots v3.0 (Nonlinear Dynamics), Dymension 3 (Syngene), Melanie v7.0 DIGE (SIB) or Delta 2D (Decodon) (Wilhelm et al. 2014). Several studies have successfully used this recent proteomics technique as a discovery tool to unravel the adaptive mechanisms used by plants in response to biotic stress.

9.4.1 PROTEOME ANALYSIS USING 2D GEL ELECTROPHORESIS

Proteomics analysis based on 2 DGE is the sensitive and commonly used method to analyse proteomes and identify differentially accumulated proteins among various mutant populations of crop plants (Reuben-Kalu and Eke-Okoro 2020). Two-DGE is useful to identify individual protein in a given sample, allowing detection of low and high abundant proteins without reaching saturation;

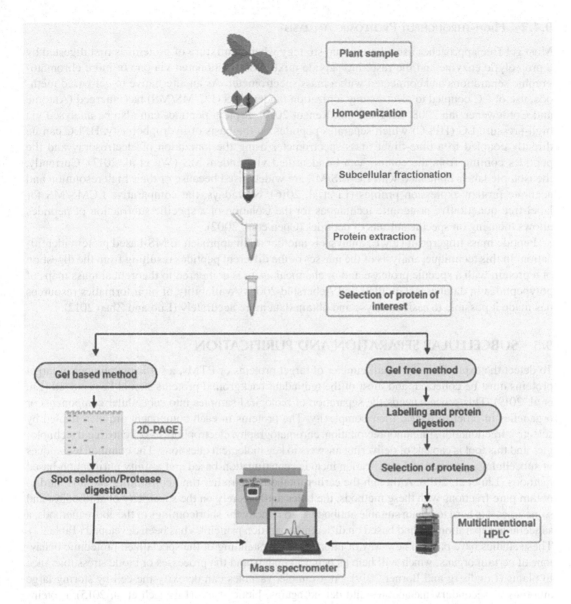

FIGURE 9.2 A short overview of plant proteome analysis through mass-spectrometry. Plant sample is collected and proteins of interest are selected step by step. Then the protein is analysed in mass spectrometer using either gel-based method or gel-free method. (Photography is created with BioRender.com). *Abbreviations*: 2D: two-dimensional; HPLC: high-pressure liquid chromatography; PAGE: polyacrylamide gel electrophoresis.

running two samples on the same gel limits the number of gels and increases reproducibility. Moreover, there is no requirement of post-staining steps; the use of an internal standard simplifies the 2-DGE analysis and allows normalization of the spot intensities over different gels, improving statistical analysis (Sergeant and Renaut 2010).

Two-DGE has some technical limitations which include non-amenability to high throughput processing of multiple samples at the same time, detection of only the most abundant proteins in the sample, and requirement of expertise in the generation of consistent results, which therefore makes it difficult to identify some lesion mimic mutants involving various proteins at the whole proteome level (Sergeant and Renaut 2010).

9.4.2 High-throughput Proteomic Analysis

Most gel-free approaches use a bottom–up strategy where a mixture of proteins is first digested by a proteolytic enzyme and the resulting peptide mixture is fractionated via one or more chromatographic separations and connected with a mass spectrometer. As an alternative to gel-based methods, use of LC coupled to electrospray ionization tandem MS (LC/MS/MS) has emerged (Antoine and Cordewener Jan 2008; Sergeant and Renaut 2010). Protein peptides can also be analysed via high-pressure LC (HPLC) which separates peptides on the basis of hydrophobicity. HPLC can be directly coupled to a time-flight mass spectrometer using the ionization of electrospray and the peptides coming from the column can be identified via tandem MS (Wu et al. 2017). Currently, the isotopic labels combined with LC-MS/MS are widely used because of their high resolution and accurate protein expression profiles (Li et al. 2016). Nowadays, the comparative LCMS/MS for label-free quantitative proteomic techniques for the isolation of a specific subfraction of peptides allows focusing on specific subsets of peptides (Chen et al. 2002).

Peptide mass fingerprint (PMF) analysis is another main approach to MS-based protein identification. In this technique, analysis of the masses of the different peptides resulting from the digestion of a protein with a specific protease and/or chemical agent is compared to theoretical mass maps of polypeptides in databases (Goodlett and Aebersold 2001). Availability of bioinformatics resources has made it possible to easily analyse and obtain data more accurately (Luo and Zhao 2012).

9.5 SUBCELLULAR SEPARATION AND PURIFICATION

To detect the expression of a small number of target proteins or PTMs, a sufficient number of target proteins must be collected, and most of the redundant background proteins should be removed (Liu et al. 2019). This process needs the separation of biological samples into subcellular components or organelles in order to reduce their complexity. The proteins in each component are then treated by selective fractionation, immunoprecipitation, chromatography, electrophoresis or centrifugal technologies and this tool is capable of delivering answers to key biological questions. The common techniques of subcellular separation and purification include centrifugation-based and affinity purification-based methods (Liu et al. 2019). Although the earlier methods are rather time-consuming and it is hard to obtain pure fractions with these methods, the latter methods rely on the specificity of antibodies, and sometimes it is hard to obtain suitable antibodies. To remedy the shortcomings of the above methods, a subcellular separation method based on different separation principles has been developed (Table 9.1). These studies have paved a new way of improved understanding of the specialized proteomic behaviour of certain organs, which will help us better to understand the processes of biotic stress tolerance in plants (Lundberg and Borner 2019). For example, vacuoles can detoxify the cell by storing large amounts of secondary metabolites and defend against biotic stress (Eisenach et al. 2015). Proteins exhibit spatiotemporal changes and enable the cell to adapt to biotic perturbations and form the proteome network, thereby defining the cell function and phenotype. The nuclei can sense signals from pathogens and translate these signals into molecular responses that act as warnings to allow the early detection of pathogen attack, notably through rapid modulation of the proteome (Howden et al. 2017). Chloroplasts are involved in the biosynthesis of defence phytohormones such as SA and JA, upon pathogen attack, and play a role in the plant stress response. As plasma membrane is a communication interface with the extracellular environment, the proteomic studies on plasma membrane proteins are essential to explore the pathways of signal transduction and elucidate the mechanism of the plant defence system against biotic stress (Jones and Hammond-Kosack 2000).

9.6 POST-TRANSLATIONAL MODIFICATION AND PROTEOMES

After translation, proteins can be further modified by covalent addition of chemical units or by changing the structures of the amino acids themselves within a protein (Liu et al. 2019). PTMs are

TABLE 9.1

Subcellular Isolation Techniques Used in Plant Proteomic Studies

Subcellular Organelles/ Proteome	Techniques	Plants	Organs	Reference
Plasma membrane	Two-phase partitioning combined with free-flow electrophoresis	*Arabidopsis thaliana*	Seedlings	Michele et al. (2016)
Nucleus	Flow cytometry with a mild formaldehyde-based fixation	*Hordeum vulgare*	Root	Petrovska et al. (2014)
Apoplastic fluid	Vacuum infiltration-centrifugation	*Gossypium barbadense*	Root	Wang et al. (2019)
Mitochondria	Centrifugation combined with 1D blue native polyacrylamide gel electrophoresis	*Arabidopsis thaliana*	Leaf	Senkler et al. (2017)
Chloroplasts	Homogenization in sorbitol-based isolation medium with blender and centrifugation	*Malus domestica*	In vitro material	Morkunaite-Haimi et al. (2018)
Ribosome	Asymmetric flow field-flow fractionation	*Nicotiana benthamiana*	Transgenic lines	Pitkanen et al. (2014)

regulatory processes that can influence protein structure, function, subcellular localization, activity and stability, which further affect the role of protein in plant defence. It can lead to tremendous changes in the regulation of biological processes even when there are no changes at the total protein or transcript levels (Bulko et al. 2015; Banerjee and Roychoudhury 2018). Proteome definition is incomplete without the map of PTMs. Although PTMs are extremely important cellular processes, their analysis remains challenging because of their complex structures, dynamic nature and low abundance (Liu et al. 2019). Fortunately, substantial progress has been made in enrichment and fractionation strategies for global studies of PTMs (Table 9.2). For example, after pathogen invasion, the elicitor initiates signalling events such as the production of ROS, changes in transcript levels and activation of mitogen-activated protein kinases (MAPKs) and calcium-dependent

TABLE 9.2

Separation Techniques for Study of PTMs of Proteins

Sl No	Nature of the Post-Translational Modification of Protein	Separation Method and Technology	Reference
1	Phosphorylation	Polymer-supported metal-ion-affinity capture	Iliuk et al. (2015)
2	Ubiquitination	Combined fractional diagonal chromatography	Walton et al. (2016)
3	Sumoylation	Replaced SUMO1 and SUMO2 isoforms with a variant and purification	Rytz et al. (2018)
4	Acetylation	Peptide prefractionation, immunoaffinity enrichment	Jiang et al. (2018)
5	Glycosylation	Dendrimer-conjugated benzoboroxole	Xiao et al. (2018)
6	Methylation	Combining SCX, IMAC and H-pH-RPLC	Liu et al. (2019)

Note: SCX strong cation exchange, IMAC immobilized metal ion affinity chromatography, H-pH-RPLC high-pH reversed-phase liquid chromatography, SUMO1/2 small ubiquitin-like modifier isoform 1/2

protein kinases (CDPKs). MAPKs and some CDPKs are regulated by phosphorylations upon biotic stress (Peck et al. 2003; Roychoudhury and Banerjee, 2017). The phosphoproteomics study allows the detection of proteins whose phosphorylation status is altered and provides knowledge on the regulation of other phosphoproteins directly involved in the biotic stress response (Kersten et al. 2009). Apart from phosphorylation, ubiquitination, acetylation, methylation and protein carbonylation cause direct oxidation of proline, lysine, arginine and threonine residues (Rao and Moller 2011; Lounifet al. 2013). Protein carbonylation leads to the loss of protein function and ultimately to the degradation of oxidized proteins (Polge et al. 2009; Roychoudhury et al. 2011). When a pathogen attacks, there is a significant increase in the carbonylation of proteins in plants as an early response (Xu et al. 2008) and the identification of oxidized amino acid in proteins may be important source of information for the oxidation mechanism and metabolic pathway of cell vitality loss under stress (Fedorova et al. 2014). A thorough understanding of mechanisms by which PTMs promote an effective immune response in plants will provide a dynamic and comprehensive perspective on plant immunity and provide new strategies for improving the performance of plant responses to biotic stress.

9.7 APPLICATION OF PROTEOMICS TO IMPROVE RESISTANCE IN PLANTS AGAINST BIOTIC STRESS

The vast knowledge generated through proteomics study has multiple applications. The identification of candidate resistance metabolites, proteins and genes can be used as potential biomarkers in breeding (Machado et al. 2012; Quanbeck et al. 2012). Resistance-related (RR) metabolites identified through flow injection nano-spray ionization MS and HPLC can be directly used in screening of genotypes (Parker et al. 2009; Aldini et al. 2011). The identified candidate genes, RR genes, and associated trans- and/or cis-acting elements can be used in marker-assisted selection (MAS) breeding programme for crop improvement, but it is very challenging when several genes are involved (Zhang et al. 2010). Alternately, several RR genes from crop plants can be carefully stacked to a superior cultivar based on cis genics or genome analysis. Alternately, several RR genes identified in economically important plants can be carefully stacked to an elite cultivar based on cis genics or genome modification, if the recipient species lacks the required genes (Jacobsen and Schouten 2009). The performance of recipient cultivars should be verified in the field after gene transfer into elite cultivars and significant resistance effect can be identified and utilized in breeding programs to improve resistance against biotic stress (Kushalappa and Gunnaiah 2013).

9.8 OVERVIEW OF THE RECENT ADVANCEMENT OF PROTEOMICS IN HOST-PATHOGEN INTERACTION

An overview of recent articles dealing with proteomic responses of plants to pathogen is summarized in Table 9.3. The relationship between plant and pathogen at the proteome level is detected using different proteomic techniques such as 1DE, 2DE, LC, time of flight mass spectrometry (MS-TOF), focussing on subcellular compartmentation, etc. Resistant and tolerant plants can also be compared in the scope of a study. Table 9.3 gives a clear picture of fluctuation of proteins involved in response to attack by a particular type of pathogen. The comparative study between resistant cultivar and susceptible variant using model plant enables scientists to screen the proteomic response to different pathogens and evaluate a common response and describe specificity to each pathogen (Sergeant and Renaut 2010). Structures of few important proteins which are activated during host-pathogen interactions are elaborated in Figure 9.3.

TABLE 9.3
Overview of Recent Publications on Proteomics and Plant–Pathogen Relationships

Plant Species	Pathogen	Proteomics Techniques	Proteins Present	Proteins Absent	Reference
			Insects		
Arabidopsis thaliana	*Spodoptera exigua*	2DE-PAGE, MALDI TOF	Transketolase, PGK, SAM synthase 3, 2,3-biphosphoglycerate-independent phosphoglycerate mutase, SBPase, ureidopropionase, GDP-D-mannose 3, 5 – epimerase, fatty acid synthase, Uridylyl transferase, EF1B-gamma, DNA-damage resistance proteins	Dihydrolipoamide dehydrogenase 2, aspartate aminotransferase 5, NADP-dependent oxidoreductase, resistance protein RPP13, thylakoid membrane phosphoprotein of 14 kDa, 29 kDa ribonucleoprotein, hypothetical protein	Zhang et al. (2010)
Nicotiana attenuata	*Manduca sexta*	2DE-PAGE, MALDI (PMF) if not identified LCMS/MS	OEE, RuBisCo activase, ACS7, OSJNBa0039C07.4, Vacuolar H1-ATPase A2, SAM synthetase, G3PD, O acetylserine thiol lyase, ChlGln synthetase, 3-dehydroquinate dehydratase, RNA binding-like protein.	G3PD, QR-like protein.	Giri et al. (2006)
Oryzasativa	*Nilaparvata lugens*	iTRAQ, LC MS/MS	Peroxidase 12 prec, Cyt P450, allene oxide cyclase 4, peroxidase 2 precursor, OsAPx2, –DOX2, Catalase isozyme A, Stromal 70 kDa HSP, OsFtsH2-*Oryza sativa* FtsH protease, a typical receptor-like kinase MARK, 60S ribosomal protein L6, CRK6, 40S ribosomal protein S14, CRK5, clathrin heavy chain, NADP-dependent malic enzyme, glucan endo-1, 3- glucosidase, Lichenase-2 precursor, FBA, Glycine cleavage system H-protein (difference between resistant and susceptible).	Peroxidase 2 precursor, DREPP2 protein, phenylalanine ammonia-lyase, TPI, G3PD, -L fucosidase precursor, Peptidyl-prolyl *cis–trans* isomerase, Mit-processing peptidase beta subunit, Proteasome subunit beta type 7-A prec, 50S ribosomal protein L12-1, glucan endo-1,3-glucosidase GII prec, phosphoglucomutase, sucrose synthase 1, quercetin 3-*O* methyltransferase, glycine-rich RNA-binding protein 2, GST, GST F2, Lactoyl glutathione lyase, SAM synthetase 1, anthocyanidin 3-*O* glucosyltransferase, aquaporins.	Wei et al. (2009)
1. *Vicia faba,* *Brassica* 2. *napus* and *Solanum* 3. *tuberosum*	*Myzus persicae*: Transfer aphids from bean to bean (control), to rape or to potato	2DE-PAGE, nLC + ESI MS/MS	Tubulin α, ENSANGP00000009311 (from aphid symbionts), chlorophyll a/b binding protein, chaperonin.	PGK, mitochondrial aconitase pyruvate carboxylase, transketolase, succinyl CoA ligase, ATP citrate lyase, citrate synthase, EF 2, eIF-4A, actin.	Francis et al. (2006)

(Continued)

TABLE 9.3 *(Continued)*
Overview of Recent Publications on Proteomics and Plant–Pathogen Relationships

Plant Species	Pathogen	Proteomics Techniques	Proteins Present	Proteins Absent	Reference
			Virus		
Oryza sativa indica and *O. sativa japonica*	Rice yellow mottle virus	2DE, MALDI, LC MS/MS	*O. sativa japonica*: SOD, chaperonin CPN60-2 mitprec, Low-molecular weight HSP 17.9, G3PD, Nuclear RNA binding Pro A, -Amylase isozyme 3D, –Amylase isozyme 3E, Translation Initiation factor 5A; *O. sativa indica*: SOD, HSP 70 kDA, dehydrin RAB25, salt-stress induced protein, PR protein PR-10A, chaperonin 10 kDa, Aldolase C1, P-glycerate dehydrogenase, ubiquitin-like protein, -amylase isozyme 3D, Ribosomal 40S.	*O. sativa japonica*: Ubiquitin-like protein, EF1b' *O. sativa* indica: Ethylene inducible protein, HSP 70 kDa, dehydrin RAB25, 2,3-Biphosphoglycerate independent phosphoglycerate mutase, RNA binding protein	Ventelon-Debout et al. (2004)
Thylakoid membranes of *Nicotiana benthamiana*	Tobamo virus (PMMoV-S)	2DE-PAGE, Edman sequencing	-	Isoforms of PsbP protein are more reduced than PsbO subunit.	Perez-Bueno et al. (2004)
			Fungi		
Arabidopsis thaliana	*Alternaria brassicicola*	2DE-PAGE, LC-MS/MS	GST (class-phi), CA1-carbonate dehydratase, DSBA oxidoreductase family protein 0, osmotin 34, PR4-PR protein, alcohol oxidase, phosphoenolpyruvate carboxykinase, glycosyl hydrolase family 17, ATP synthase CF0 B chain, Lectin.	3-Hydroxyacyl-[acyl-carrier protein], PBF1-20S proteasome subunit F1, FNR 2-NADPH dehydrogenase, myrosinase TGG2, glucose-6-phosphate isomerase cytosolic, NDH complex subunit, ribosome recycling factor RRF chl.	Mukherjee et al. (2010)
Oryza sativa (resistant and susceptible lines)	*Rhizoctonia solani*	2DE-PAGE, LC-MS/MS	Increased in both lines: β-1, 3-glucanase, 20S proteasome beta subunit, Put 26S proteasome non-ATPase regulatory subunit 14, RuBisCo LSU. Only in resistant line: Stromal APX, 3-beta hydroxysteroiddehydrogenase/isomerase protein, Put chitinase, G3PD A subunit, Put 1 subunit of 20S proteasome, RuBisCo activase.	Only in resistant line: Put chaperonin 60 precursor, 14-3-3-like protein, Endosperm lumenal binding protein.	Lee et al. (2006)
Brassica napus	*Sclerotinia sclerotiorum*	2DE-PAGE, ESI-q-TOF MS/MS	RuBisCo, RuBisCo activase, PRK, ADP-glucose pyrophosporylase, GS, FNR, jasmonic acid-responsive protein, uroporphyrinogen decarboxylase, ATPase, glutamate ammonia ligase, ribosomal protein L12, peroxidase, glyoxalase, endopeptidase, plastid-dividing ring protein.	G3PD, FBA, TPI, malate dehydrogenase, protein disulfide isomerase, chaperonin, mRNA binding, methionine adenosyl transferase, eukaryotic translation initiation factor-5A, AT4g38970.	Liang et al. (2008)

(Continued)

TABLE 9.3 *(Continued)*

Overview of Recent Publications on Proteomics and Plant–Pathogen Relationships

Plant Species	Pathogen	Proteomics Techniques	Proteins Present	Proteins Absent	Reference
Zea mays germinating embryo	*Fusarium verticillioides*	2DE-PAGE MALDI & ESI	FBA, GST, glucan endo-1, 3--D-glucosidase, SOD [Cu–Zn] 4AP, HSP 17.2, Initiation factor 5A, peptidyl-prolyl cis–trans isomerase, Catalase 2.	Adenosine kinase, G3PD, Late embryogenesis abundant protein group 3, Globulin-2 precursor.	Campo et al. (2004)
Medicago truncatula	*Aphanomyces euteiches*	2DE-PAGE, MALDI and PMF	Disease-resistance-response protein pi 49, ABA-responsive protein ABR17, PR protein class 10, 18.2 kDa class I HSP, Proline-rich protein, glycine-rich cell wall structural protein, Isoliquiritigenin 2-O-methyltransferase.	Cold acclimation-specific protein CAS18/ dehydrin-like protein.	Colditz et al. (2004)
Bacteria					
Olea europaea subsp. *Europaea*	*Pseudomonas savastanoi* pv. *savastanoi*	2DE-PAGE, PMF MALDI-TOF	Pathogen: OprF, aconitase hydratase 2, hypothetical protein ytlR, TerEPlant: enolase.	Plant: CDPK.	Campos et al. (2009)
Arabidopsis thaliana	*Pseudomonas syringae* pv. *tomato* DC3000	nitration of proteins, 2DE-PAGE, LC-MS/MS	PsbO2, PsbO1, ATP synthase CF1 subunit, RuBisCo LSU, RuBisCo activase, FBA 1 & 2, GS 2.	–	Cecconi et al. (2009)
Solanum lycopersicum	*Ralstonia solanacearum*	2DE-PAGE, n and cLC, de novo	*Plant*: NAD-malate dehydrogenase, PR protein STH-21, Enoyl-ACP reductase prec, fructokinase, cyclophilin-like protein, BTF-3-like transcription factor, Put lactoylglutathione lyase, Put 3-keto-acylACP dehydratase, peptidyl-prolyl cis-trans isomerase, ATP synthase subunit; *Bacteria*: Put hemolysin-type protein, molecular chaperone, Peroxiredoxin, Alkyl hydroperoxide reductase C, ABC transporter substrate-binding protein, Probable signal peptide protein, GroEL.	Plant: TSI-1 protein (PR protein), HSP 20.0 protein.	Dahal et al. (2009)
Parasitic Plant					
Medicago truncatula	*Orobranche crenata*	2DE-PAGE, PMF + MS/MS MALDI-TOF	Late-resistant *Medicago*: proteinase inhibitor, trypsin inhibitors. Early-resistant *Medicago*: chitinase, GST, glycine-rich RNA-binding prot, trypsin inhibitor, FBA, guanine nucleotide-binding protein.	Late-resistant *Medicago*: glycine-rich RNA binding prot, put transaldolase, reverse transcriptase-like protein. Early-resistant *Medicago*: chitinase, glycoside hydrolase, proteasome subunit alpha type 7, TH65-like prot, kinesin motor protein.	Simpson et al. (2000)

(Continued)

TABLE 9.3   (Continued)
Overview of Recent Publications on Proteomics and Plant–Pathogen Relationships

Plant Species	Pathogen	Proteomics Techniques	Proteins Present	Proteins Absent	Reference
Pisum sativum	*Orobranche crenata*	2DE-PAGE, PMF MALDI-TOF	GS root isozyme B, GS root isozyme A, Isovaleryl-CoA-dehydrogenase, Peroxidase 43 precursor, Glucan endo- 1, 3- glucosidase prec, Profucosidase.	Fructokinase, alternative oxidase 2 mitprec, FBA cytoplasmic isozyme, FBA cytoplasmic isozyme 2, FNR root isozyme, putative enoyl-CoA hydratase.	Castillejo et al. (2004)

Note: APX: Ascorbate Peroxidase; CF1: ATP Synthase Catalytic Portion; chl: Chloroplastic; CP: Viral-Coat Protein; Cyt: Cytochrome; EF: Translation Elongation Factor; FBA: Fructose-Bisphosphate Aldolase; FNR: Ferredoxin-NADP⁺ Reductase; G3PD: Glyceraldehyde-3-Phosphate Dehydrogenase; GS Glutamine Synthetase; GST: Glutathione S-Transferase; HSP: Heat Shock Protein; Lhca: PSI Chlorophyll a/b Binding Proteins; Lhcb: PSII Chlorophyll a/b Binding Proteins; LSU: Large Subunit; mit: Mitochondrial; OEE: Oxygen-Evolving Enhancer Protein; Orth: Orthologue; PDI: Protein Disulphide Isomerase; PEPc: Phosphoenolpyruvate Carboxylase; PGK: Phosphoglycerate Kinase; PR: Pathogenesis-Related: Prec: Precursor; PRK: Phosphoribulo-Kinase; Psa Proteins: PSI Proteins Codified by Psa Genes; Psb Proteins: PSII Proteins Codified by Psb Genes; Put: Putative; SAM Synthase: S-Adenosylmethionine Synthetase; SBPase: Sedoheptulose-1,7-Bisphosphatase; SFBP: Sedoheptulose/Fructose-Bisphosphate Aldolase; SOD: Superoxide Dismutase; SSU: Small Subunit; Thr: Threonine; TPI: Triosephosphate Isomerase; VSP: Vegetative Storage Protein

9.9   CASE STUDIES

9.9.1   PROTEOMIC ANALYSIS OF *Nicotiana* IN RESPONSE TO *Manduca sexta* ATTACK ON ITS HOST

When *Manduca sexta,* a moth of the family Sphingidae, attacks *Nicotiana attenuata*, it introduces fatty acid-amino acid conjugates (FACs) through oral secretions into the feeding wound. The FACs trigger a huge transcriptional fluctuation in its host plant. Using 2 DGE, matrix-assisted laser desorption ionization-time of flight (MALDI-TOF), and LC-tandem MS, Giri et al. (2006) have shown decreases in photosynthetic-related processes and increases in defence-related proteins. Upon pathogen attack, there is a continuous degradation of photosynthesis-related proteins such as RuPBCase activase leading to reduced photosynthetic rates and RuBPCase activity, hence providing less biomass. From the protein spot analyses over gels, it was noted that several amino acid metabolism proteins were upregulated such as threonine deaminase, S-adenosylmethionine synthetase, O-acetylserine thiol lyase, spermidine synthase, Glnsynthetase and 3-dehydroquinate dehydratase. Threonine deaminase acts in antinutritional defences by limiting amino acid supply to the invading insect. This study identified several well-characterized proteins having direct and indirect roles in insect-elicited responses in *Nicotiana attenuata*, as well as several proteins of unknown function (Giri et al. 2006).

9.9.2   IDENTIFICATION OF PROTEINS LINKED TO *Alternaria brassicicola* INFECTION IN *Arabidopsis* VIA PROTEOMICS ANALYSIS

Mukherjee et al. (2010) have identified different proteins through proteomic analysis via 2DE and LCMS, during the invasion of a necrotrophic fungal pathogen *Alternaria brassicicola*, in the model plant *Arabidopsis thaliana*. The signalling network between host and pathogen pair induces defence response in plants at the transcriptional level. The pathogenesis-related protein PR4, a glycosyl hydrolase and the antifungal protein osmotin are strongly up regulated

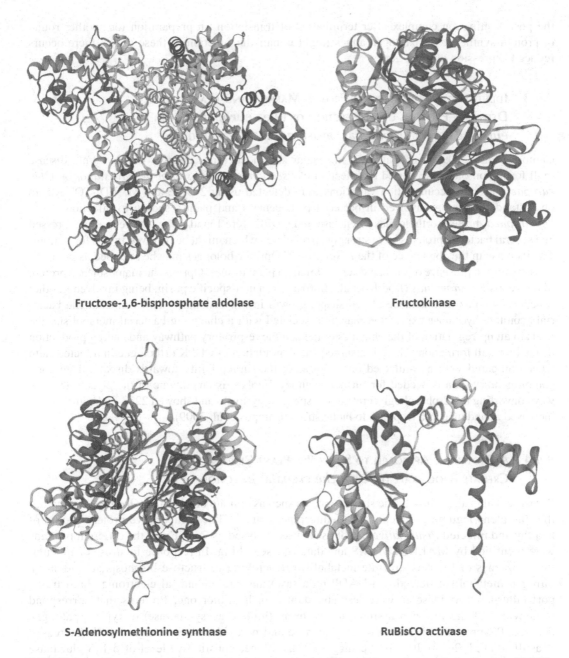

Fructose-1,6-bisphosphate aldolase

Fructokinase

S-Adenosylmethionine synthase

RuBisCO activase

FIGURE 9.3 These are probable 3-D tertiary structures of four major enzymes derived using SWISS-MODEL software. These enzymes are highly expressed in plants upon attack by different kinds of biotic stressors. The above enzymes are (1) Fructose-1, 6-bisphosphate aldolase (FBA) in *Zea mays*, (2) Fructokinase in *Solanum lycopersicum*, (3) S-adenosyl methionine synthase in *Arabidopsis thaliana*, (4) Ribulose-1, 5-bisphosphate carboxylase/oxygenase (RuBisCo) activase in *Oryza sativa* subsp. *japonica*.

(Mukherjee et al. 2010). Glucanase is an enzyme involved in cell-wall degradation providing defence against the fungal pathogen. Osmotin acts as an antifungal peptide (Xu et al. 2008). In response to *A. brassicicola*, Mukherjee et al. (2010) found evidence from peptide number and coverage for a decrease in abundance of ferredoxin NADP oxidoreductase (FNR), beta-hydroxyacyl-acyl carrier protein dehydratase, myrosinase and ribosome recycling factor (RRF). RRF dissociates the post-termination complex after termination of translation, in preparation for another round of protein synthesis. As pathogen attack might impair the function of these genes, there occurs reduced expression of these genes.

9.9.3 IDENTIFICATION OF BACTERIAL PROTEIN MARKERS AND PLANT DEFENSIVE PROTEINS UPON INFECTION OF *Olea europaea* SUBSP. *europaea* BY *Pseudomonas savastanoi* PV. *savastanoi*

Campos et al. (2009) investigated the molecular and biochemical changes of Olive knot disease (gall formation on stem) caused by infection of *Pseudomonas savastanoi* pv. *savastanoi* on *Olea europaea*. Protein accumulation variations were detected using 2 DGE and MALDI-TOF MS to understand the factors involved in bacterial pathogenesis and plant response to infection. From *O. europaea*, protein orthologues of enolase and CDPK were found to be differentially expressed and several bacterial proteins were strongly up regulated which might be involved in gall formations. The increase in the abundance of the *P. savastanoi* OprF orthologues may be regarded as potential markers for the presence of the bacterium in infected plant tissues. OprF is the major surface protein of the genus *Pseudomonas* (Bodilis et al. 2006) and is a non-specific porin, being involved in other functions such as maintenance of cell shape, growth in low-osmolarity environment. The bacterial aconitate hydratase expression may be associated with a change in bacterial metabolism that leads to an up-regulation of the enzymes of the aerobic respiratory pathway and energy production needed for gall formation. There is a subsequent degradation of CDPK orthologue in infected stem when compared with non-infected one suggesting the susceptibility towards disease. Therefore, complimentary data is needed for further accuracy. Enolase is an enzyme of the glycolytic pathway converting the 2-phosphoglycerate to phosphoenolpyruvate and showed 2277-fold significantly increased abundance as a response to biotic stress (Campos et al. 2009).

9.9.4 A PROTEOMIC APPROACH TO STUDY PEA PLANT RESPONSE TO CRENATE BROOMRAPE (*Orobanche crenata*) INFECTION

Parasitic plants also cause biotic stress to plants such as crenate broomrape (*Orobanche crenata*) that threatens legume production in Mediterranean areas. The comparative protein profile of healthy and infected *Pisumsativum* root tissue was analysed by 2 DGE and the different proteins were identified by MALDI-TOF MS and database searching. In response to stress or inoculation, enzymes of the carbohydrate metabolism (fructokinase, fructose-bisphosphate aldolase), nitrogen metabolism (ferredoxin-NADP reductase) and mitochondrial electronic chain transport (alternative oxidase 2) were less abundant than healthier one. Proteins that correspond to enzymes of the nitrogen assimilation pathway (glutamine synthetase) or typical pathogen defence, PR proteins, including β-1,3-glucanase and peroxidases, were significantly increased (Castillejo et al. 2004). It could be suggested that higher constitutive level of β-1, 3-glucanase and chitinase in the resistant genotype could help the plant in priming defence reactions against pathogens more rapidly. The peroxidase enzyme activity is extremely high. They are concerned with a variety of functional roles, including lignification and cell wall phenol deposition, suberization, hormone catabolism, developmental-related processes, defence against pathogens and response to other stresses (Penel et al. 1992).

9.10 LIMITATIONS IN PROTEOMICS ANALYSIS

Despite the significant advances in proteomics over the decades, some limitations still persist including the number of genes that code for proteins, total number of proteins observed in a complex protein sample and the complex analytical techniques leading to variation in data collection which creates problems in obtaining accurate results (Min et al. 2019). The proteomic data analysis may have some limitations at each phase of the analysis which may include: (1) preparation of protein sample and extraction of the proteins, (2) peptide separation, (3) MS-based analysis, and (4) bioinformatics tools employed for the interpretation of data obtained (Liu et al. 2019). Furthermore, proteomic analysis shows inability to directly analyse proteins because of their large size and unavailability of precise libraries (Sergeant and Renaut 2010).

9.11 CONCLUSION AND FUTURE PERSPECTIVES

The yield reduction in agricultural crops due to biotic stress is an extensive and increasing concern for scientists. It is important to elucidate how plants respond to biotic stress. Advances in proteomics at the cellular and subcellular levels will provide precise regulation of targets for plant immunity. Today, proteomic research also has a role to unravel the underlying mechanisms at the frontier between resistance and susceptibility towards plant-pathogenic attack. The readily available and detailed information on protein databases and other bioinformatic tools have significantly taken the proteome coverage of many plants and crop species to another level towards crop improvement. MS-based proteomic strategies coupled with subcellular fractionation have a role in measurement of proteins with critical functions and contributions to biotic stress mechanisms.

There are multiple future challenges towards proteomics application in plant stress resistance improvement. Research should focus on a major challenge that involves proper matching characteristics of peptides and the spectra and credibility of the peptide matching results. The transfer of a set of significant effect RR genes, along with regulatory genes from one genotype to another within sexually compatible plant species, through cis genics opposed to transgenic approaches, should be more acceptable to the public, as no foreign DNA is incorporated. The spatial proteomics and PTM proteomics with accurate design and appropriate bioinformatic assistance are invaluable tools to uncover plant resistance mechanisms against biotic stress. Therefore, proteomics studies need adequate dry laboratory, time and carefulness in data mining so that even missing proteins can be properly identified in any proteome analysis. We are optimistic that an increasing number of adequate laboratories will adopt, practice and advance MS-based proteomics to answer important biological questions.

REFERENCES

Aldini G, Regazzoni L, Pedretti A, Carini M, Cho SM, Park KM, Yeum KJ (2011) An integrated high resolution mass spectrometric and informatics approach for the rapid identification of phenolics in plant extract. J Chromatogr A 1218:2856–2864

America Antoine HP, Cordewener Jan HG (2008) Comparative LC-MS: A landscape of peaks and valleys. Proteomics 8(4):731–749

Banerjee A, Roychoudhury A (2018) The gymnastics of epigenomics in rice. Plant Cell Rep 37:25–49

Bodilis J, Hedde M, Orange N, Barray S (2006) OprF polymorphism as a marker of ecological niche in *Pseudomonas*. Environ Microbiol 8(9):1544–1551

Bulko TV, Shumyantseva VV, Suprun EV, Archakov AI (2015) Electrochemical methods for detection of post-translational modifications of proteins. Biosens Bioelectron 61:131–139

Campo S, Carrasca lM, Coca M, Abian J, San SB (2004) The defense response of germinating maize embryos against fungal infection: A proteomics approach. Proteomics 4(2):383–396

Campos A, Da Costa G, Coelho AV, Fevereiro P (2009) Identification of bacterial protein markers and enolase as a plant response protein in the infection of *Olea europaea* subsp. europaea by *Pseudomonas savastanoi* pv *savastanoi*. Eur J Plant Pathol 125(4):603–616

Castillejo MA, Amiour N, Dumas-Gaudot E, Rubiales D, Jorrin-Novo JV (2004) A proteomic approach to studying plant response to crenate broomrape (*Orobanche crenata*) in pea (*Pisum sativum*). Phytochem 65(12):1817–1828

Cecconi D, Orzetti S, Vandelle E, Rinalducci S, Zolla L, Delledonne M (2009) Protein nitration during defense response in *Arabidopsis thaliana*. Electrophoresis 30(14):2460–2468

Chakraborty N, Sarkar A, Dasgupta A, Paul A, Mukherjee K, Acharya K (2022) In planta validation of nitric oxide mediated defense responses in common bean against *Colletotrichum gloeosporioides* infection. Indian Phytopathol 75:15–24. https://doi.org/10.1007/s42360-021-00425-0

Chen G, Gharib TG, Huang CC, Thomas DG, Shedden KA, Taylor JMG, Kardia SLR, Misek DE, Giordano TJ, Iannettoni MD, Orringer MB, Hanash SM, Beer DG (2002) Proteomic analysis of lung adenocarcinoma: Identification of a highly expressed set of proteins in tumors. Clin Cancer Res 8(7): 2298–2305

Colditz F, Nyamsuren O, Niehaus K, Eubel H, Braun HP, Krajinski F (2004) Proteomic approach: Identification of *Medicago truncatula* proteins induced in roots after infection with the pathogenic oomycete *Aphanomyces euteiches*. Plant Mol Biol 55(1):109–120

Dahal D, Heintz D, Van Dorsselaer A, Braun HP, Wydra K (2009) Pathogenesis and stress related, as well as metabolic proteins are regulated in tomato stems infected with *Ralstonia solanacearum*. Plant Physiol Biochem 47(9):838–846

Dangl L, Jones D (2001) Plant pathogens and integrated responses to infection. Nature 411(6839):826–833

Eisenach C, Francisco R, Martinoia E (2015) Plant vacuoles. Curr Biol 25(4):R136–R137

Fedorova M, Bollineni RC, Hoffmann R (2014) Protein carbonylation as a major hallmark of oxidative damage: Update of analytical strategies. Mass Spectrom Rev 33(2):79–97

Francis F, Gerkens P, Harmel N, Mazzucchelli G, De Pauw E, Haubruge E (2006) Proteomics in *Myzus persicae*: Effect of aphid host plant switch. Insect Biochem Mol Biol 36(3):219–227

Fraser EDG (2003) Social vulnerability and ecological fragility: Building bridges between social and natural sciences using the Irish Potato Famine as a case study. Conserv Ecol 7(2)

Giri AP, Wunsche H, Mitra S, Zavala JA, Muck A, Svatos A, Baldwin IT (2006) Molecular interactions between the specialist herbivore *Manduca sexta* (Lepidoptera, Sphingidae) and its natural host *Nicotiana attenuata*. VII. Changes in the plant's proteome. Plant Physiol 142(4):1621–1641

Goodlett DR, Aebersold R (2001) Mass spectrometry in proteomics. Chem Rev 101(2):269–295

Howden AJM, Stam R, Martinez Heredia V, Motion GB, Ten Have S, Hodge K, Marques Monteiro Amaro TM, Huitema E (2017) Quantitative analysis of the tomato nuclear proteome during *Phytophthora capsici* infection unveils regulators of immunity. New Phytol 215(1):309–322

Iliuk A, Jayasundera K, Wang WH, Schluttenhofer R, Geahlen RL, Tao WA (2015) In-depth analyses of B cell signaling through tandem mass spectrometry of phosphopeptides enriched by poly MAC. Int J Mass Spectrometry 377:744–753

Jacobsen E, Schouten H (2009) Cisgenesis: An important sub-invention for traditional plant breeding companies. Euphytica 170:235–247

Jiang J, Gai Z, Wang Y, Fan K, Sun L, Wang H, Ding Z (2018) Comprehensive proteome analyses of lysine acetylation in tea leaves by sensing nitrogen nutrition. BMC Genomics 19(1):840

Jones J, Hammond-Kosack K (2000) Responses to plant pathogens. In: Buchanan B, Gruissen W, Jones R (Eds.) Biochemistry and Molecular Biology of Plants, ASPP, Rockville, MD, pp. 1102–1157

Kersten B, Agrawal K, Durek P, Neigenfind J, Schulze W, Walther D, Rakwal R (2009) Plant phosphoproteomics: An update. Proteomics 9(4):964–988

Kjellsen TD, Shiryaeva L, Schroder PW, Strimbeck GR (2010) Proteomics of extreme freezing tolerance in Siberian spruce (Piceaobovata). J Proteomics 73(5):965–975

Kosová K, Vítámvás P, Urban M, Prášil I, Renaut J (2018) Plant abiotic stress proteomics: The major factors determining alterations in cellular proteome. Front Plant Sci 9:122

Kushalappa AC, Gunnaiah R (2013) Metabolo-proteomics to discover plant biotic stress resistance genes. Trends Plant Sci 18(9):522–531

Lee J, Bricker TM, Lefevre M, Pinson SR, Oard JH (2006) Proteomic and genetic approaches to identifying defence-related proteins in rice challenged with the fungal pathogen *Rhizoctonia solani*. Mol Plant Pathol 7(5):405–416

Liang Y, Srivastava S, Rahman MH, Strelkov SE, Kav NN (2008) Proteome changes in leaves of *Brassica napus* L. as a result of *Sclerotinia sclerotiorum* challenge. J Agric Food Chem 56(6):1963–1976

Li X, Jackson A, Xie M, Wu D, Tsai WC, Zhang S (2016) Proteomic insights into floral biology. BBA Proteins Proteomics 1864:1050–1060

Liu Y, Song L, Liu K, Sheng W, Luqi H, Lanping G (2019) Proteomics: A powerful tool to study plant responses to biotic stress. Plant Methods 15:135

Lounif I, Arc E, Molassiotis A, Job D, Rajjou L, Tanou G (2013) Interplay between protein carbonylation and nitrosylation in plants. Proteomics 13(3–4):568–578

Lundberg E, Borner GHH (2019) Spatial proteomics: A powerful discovery tool for cell biology. Nat Rev Mol Cell Biol 20(5):285–330

Luo R, Zhao H (2012) Protein quantitation using iTRAQ: Review on the sources of variations and analysis of nonrandom missingness. Stat Interface 5(1):99–107

Machado ART, Campos VPVAC, da Silva WJR, Zeri ACdM, Oliveira DF (2012) Metabolic profiling in the roots of coffee plants exposed to the coffee root-knot nematode, *meloidogyne exigua*. Eur J Plant Pathol 134:431–441

Michele R, McFarlane HE, Parsons HT, Meents MJ, Lao JM, Fernandez Nino SMG, Petzold CJ, Frommer WB, Samuels AL, Heazlewood JL (2016) Free-flow electrophoresis of plasma membrane vesicles enriched by two-phase partitioning enhances the quality of the proteome from *Arabidopsis* seedlings. J Proteome Res 15(3):900–913

Min C, Gupta W, Agrawal R, Rakwal GK, Kim T (2019) Concepts and strategies of soybean seed proteomics using the shotgun proteomics approach. Expert Rev Proteomics 16(9):795–804

Morkunaite-Haimi S, Vinskiene J, Staniene G, Haimi P (2018) Efficient isolation of chloroplasts from in vitro shoots of *Malus* and *Prunus*. Zemdirbyste 105(2):171–176

Mukherjee AK, Carp MJ, Zuchman R, Ziv T, Horwitz BA, Gepstein S (2010) Proteomics of the response of *Arabidopsis thaliana* to infection with *Alternaria brassicicola*. J Proteomics 73(4):709–720

Parker D, Beckmann M, Zubair H, Enot DP, Caracuel-Rios Z, Overy DP, Snowdon S, Talbot NJ, Draper J (2009) Metabolomic analysis reveals a common pattern of metabolic re-programming during invasion of three host plant species by *Magnaporthe grisea*. Plant J 59:723–737

Peck C (2003) Early phosphorylation events in biotic stress. Curr Opin Plant Biol 6(4):334–338

Penel C, Gaspar T, Greppin H (1992) Plant Peroxidases: 1980–1990: Topics and Detailed Literature on Molecular, Biochemical, and Physiological Aspects. University of Geneva, Geneva, Switzerland

Perez-Bueno ML, Rahoutei J, Sajnani C, Garcia-Luque I, Baron M (2004) Proteomic analysis of the oxygen-evolving complex of photosystem II under biotec stress: Studies on *Nicotiana benthamiana* infected with tobamoviruses. Proteomics 4(2):418–425

Petrovska B, Jerabkova H, Chamrad I, Vrana J, Lenobel R, Urinovska J, Sebela M, Dolezel J (2014) Proteomic analysis of barley cell nuclei purified by flow sorting. Cytogenet Genome Res 143(1–3):78–86

Pitkanen L, Tuomainen P, Eskelin K (2014) Analysis of plant ribosomes with asymmetric flow field-flow fractionation. Anal Bioanal Chem 406(6):1629–1637

Polge C, Jaquinod M, Holzer F, Bourguignon J, Walling L, Brouquisse R (2009) Evidence for the existence in *Arabidopsis thaliana* of the proteasome proteolytic pathway activation in response to cadmium. J Biol Chem 284(51):35412–35424

Quanbeck SM, Brachova L, Campbell AA, Guan X, Perera A, He K, Rhee SY, Bais P, Dickerson JA, Dixon P, Wohlgemuth G, Fiehn O, Barkan L, Lange I, Lange BM, Lee I, Cortes D, Salazar C, Shuman J, Shulaev V, et al. (2012) Metabolomics as a hypothesis-generating functional genomics tool for the annotation of *Arabidopsis thaliana* genes of "unknown function". Front Plant Sci 3: 1–12.

Rao RSP, Moller IM (2011) Pattern of occurrence and occupancy of carbonylation sites in proteins. Proteomics 11(21):4166–4173

Renaut J, Sergeant K, Carpentier SC, Panis B, Vertommen A, Swennen R, Laukens K, Witters R, Samyn B, Devreese B (2008) Proteome analysis of non-model plants: A challenging but powerful approach. Mass Spectrom Rev 27(4):354–377

Reuben-Kalu JI, Eke-Okoro ON (2020) Recent advances in plant proteomics towards stress adaptation and crop improvement. Int J Agric Sci Res 10(4):71–80

Roychoudhury A, Banerjee A (2017) Abscisic acid signaling and involvement of mitogen activated protein kinases and calcium-dependent protein kinases during plant abiotic stress. In: Pandey GK (Ed.) Mechanism of Plant Hormone Signaling Under Stress (Vol. 1), John Wiley & Sons, Inc, Hoboken, NJ, Pp. 197–241

Roychoudhury A, Basu S, Sengupta DN (2011) Amelioration of salinity stress by exogenously applied spermidine or spermine in three varieties of *indica* rice differing in their level of salt tolerance. J Plant Physiol 168:317–328

Roychoudhury A, Paul S, Basu S (2013) Cross-talk between abscisic acid-dependent and abscisic acid-independent pathways during abiotic stress. Plant Cell Rep 32(7):985–1006

Roychoudhury A, Roy C, Sengupta DN (2007) Transgenic tobacco plants overexpressing the heterologous *lea* gene *Rab16A* from rice during high salt and water deficit display enhanced tolerance to salinity stress. Plant Cell Rep 26(10):1839–1859

Rytz TC, Miller MJ, McLoughlin F, Augustine RC, Marshall RS, Juan YT, Charng YY, Scalf M, Smith LM, Vierstra RD (2018) SUM Oylome profiling reveals a diverse array of nuclear targets modified by the SUMO ligase SIZ1 during heat stress. Plant Cell 30(5):1077–1099

Senkler J, Senkler M, Eubel H, Hildebrandt T, Lengwenus C, Schertl P, Schwarzlander M, Wagner S, Wittig I, Braun HP (2017) The mitochondrial complexome of *Arabidopsis thaliana*. Plant J 89(6):1079–1092

Sergeant K, Renaut J (2010) Plant biotic stress and proteomics. Curr Proteom 7(4):275–297

Simpson AJ, Reinach FC, Arruda P, Abreu FA, Acencio M, Alvarenga R, Alves LM, Araya JE, Baia GS, Baptista CS, Barros MH, Bonaccorsi ED, Bordin S, Bové JM, Briones MR, Bueno MR, Camargo AA, Camargo LE, Carraro DM, Carrer H, et al. (2000) The genome sequence of the plant pathogen *Xylella fastidiosa*. The *Xylella fastidiosa* consortium of the organization for nucleotide sequencing and analysis. Nature 406(6792):151–159

Stukenbrock H, McDonald A (2009) Population genetics of fungal and oomycete effectors involved in gene-for-gene interactions. Mol. Plant Microbe Interact 22:371–380

Tang D, Wang G, Zhou JM (2017) Receptor kinases in plant–pathogen interactions: More than pattern recognition. Plant Cell 29(4):618–637

Ventelon-Debout M, Delalande F, Brizard JP, Diemer H, Van Dorsselaer A, Brugidou C (2004) Proteome analysis of cultivar specific deregulations of *Oryza sativa indica* and *Oryza sativa japonica* cellular suspensions undergoing rice yellow mottle virus infection. Proteomics 4(1):216–225

Walton A, Stes E, Cybulski N, Van Bel M, Inigo S, Durand AN, Timmerman E, Heyman J, Pauwels L, De Veylder L, Goossens A, De Smet I, Coppens F, Goormachtig S, Gevaer K (2016) It's time for some "Site"-seeing: Novel tools to monitor the ubiquitin landscape in *Arabidopsis thaliana*. Plant Cell 28(1):6–16

Wang Q, Liu Z, Wang K, Wang Y, Ye M (2019) A new chromatographic approach to analyze methyl proteome with enhanced lysine methylation identification performance. Anal Chim Acta 1068:111–119

Wei Z, Hu W, Lin Q, Cheng X, Tong M, Zhu L, Chen R, He G (2009) Understanding rice plant resistance to the Brown Planthopper (*Nilaparvata lugens*): A proteomic approach. Proteomics 9(10):2798–2808

Wilhelm M, Schlegl J, Hahne H (2014) Mass-spectrometry-based draft of the human proteome. Nature 509:582–587

Wu X, Yu Y, Xu J, Dong F, Liu X, Du P, Wei D, Zheng Y (2017) Residue analysis and persistence evaluation of fipronil and its metabolites in cotton using high-performance liquid chromatography-tandem mass spectrometry. PLoS ONE 12(3):e0173690

Xiao H, Chen W, Smeekens JM, Wu R (2018) An enrichment method based on synergistic and reversible covalent interactions for large-scale analysis of glycoproteins. Nat Commun 9(1):1692

Xu XB, Qin GZ, Tian SP (2008) Effect of microbial biocontrol agents on alleviating oxidative damage o peach fruit subjected to fungal pathogen. Int J Food Microbiology 126(1–2):153–158

Zhang JH, Sun LW, Liu LL, An SL, Wang X, Zhang J, Jin JL, Li SY, Xi JH (2010) Proteomic analysis of interactions between the generalist herbivore *Spodoptera exigua* (Lepidoptera: Noctuidae) and *Arabidopsis thaliana*. Plant Mol Biol Rep 28:324–333

10 Proteomic-driven Approaches for Unravelling Plant–Microbe Interactions

Raj Kumar Gothwal, Sampat Nehra, Parul Sinha,
Aruna Shekhar N. C., and Purnendu Ghosh

10.1 INTRODUCTION

Growth, reproduction and yield of plants are influenced by biotic variables. In their natural environment, plants are surrounded by a variety of microorganisms. Some microorganisms have positive direct interactions with plants, whereas others colonise the plant only for their personal gain. Microbes can also indirectly impact plants by significantly changing their surroundings (Schenk et al. 2012). To better understand the mechanism of pathogen infection or symbiosis, work is being emphasised on identifying the molecular elements involved in plant–microbe interaction. Interactions between plants and microbes take place within the complex cellular structure of plants. It is simple to comprehend the complex yet integrated cellular processes thanks to several physiological and molecular investigations (Singh & Roychoudhury 2021).

The interaction between plants and pathogens has many facets. Plants create two paths at the start of the interaction to detect and fend against pathogen attacks. After being recognised by certain pathogen-effector molecules, one pathway leads to the production of danger-associated molecular patterns (DAMPs) and pathogen-associated molecular patterns (PAMPs), while the other develops effector-triggered immunity (ETI) and PAMPs-triggered immunity (PTI). Upstream signalling cascades are consequently engaged, resulting in the production of antimicrobial chemicals that eliminate the pathogen and thereby preserve homeostasis. This carefully regulated complex process involves numerous proteins and signalling pathways. Because of the complexity of plant–pathogen interactions, it is very difficult to pinpoint the anatomical characteristics, metabolites and signalling pathways that are activated; for this reason, traditional biochemical and genetic experimental methods are inadequate. In order to better understand the molecular interactions that take place between the hosts, the pathogen, and/or helpful microbes, proteomic methods are being applied. Using proteomic approaches, it has been possible to identify antioxidant, stress-related and pathogenic proteins that are expressed during interactions between plants and microbes. For the successful pathogen recognition, generation of resistance and maintenance of host integrity, it is believed that fine regulation of protein expression occurs. However, our understanding of the molecular interactions between plants and microbes is still limited.

Plants are constantly interacting with bacteria, viruses and fungi. Defence pathways are frequently developed in response to pathogens, although they are strictly controlled to permit advantageous associations with symbionts. In order to accommodate bacteria intracellularly (endo-symbioses), a number of symbioses also need specialised signalling pathways, which can sometimes result in the development of root nodules. Understanding these reactions, how they are regulated, and how they are related is the goal of research in plant–microbe interactions. These studies may provide knowledge that can be used to develop plants with enhanced disease resistance and innovative symbiotic relationships. The use of proteomics for plant protein analysis has been discussed in this chapter, with a focus on applications to study plant–microbe interactions.

10.2 PROTEIN SEPARATION TECHNIQUES

Comparatively to the analysis of DNA or RNA, the analysis of a set of proteins under clear-cut plant–microbe interaction specifies a more precise perspective of physiological and cellular mechanisms. When studying at the proteome level, it is possible to simultaneously examine the entire proteome that an organism possesses, together with its indexing, qualitative presence, quantitative abundance, localisation and population-level variation. Understanding the function of various proteins in development and growth, changes to the proteome caused by biotic and abiotic stimuli, post-translational alterations and connections with other proteins and molecules may all be aided by the application of proteomics. Gel-based methods like two-dimensional polyacrylamide gel electrophoresis (2D-PAGE) and fluorescent two-dimensional "difference gel electrophoresis" (2D-DIGE), as well as gel-free methods like isotope-coded affinity tags (ICAT), isobaric tags for relative and absolute quantitation (iTRAQ), multidimensional protein identification technology (MudPIT) and the widely used primary tool mass spectrophotometry, are all being used to advance our understanding of proteins (Jain et al. 2021). Straightforward gel-based resolution is used in many proteomic studies because separated proteins are simple to identify and describe using mass spectrometry (MS). The most widely used of these is 2D gel electrophoresis, which separates proteins according to their molecular weight and isoelectric point (pI). Fluorescent 2D gel electrophoresis is a protein quantification method that combines gel electrophoresis with illuminating protein spectra obtained by fluorescently labelling materials to resolve protein in complicated mixtures. This method regulates gel-to-gel alteration as differently labelled proteins that can be resolved on the same gel. By resolving peptides after sequence-specific digestion, liquid chromatography (LC) technologies can considerably increase the inclusion of proteins using MS-based sequencing, overcoming the problems of spot overlap. *MudPIT*, or multidimensional protein identification technology, has emerged as a powerful strategy for extensive proteomic research, in recent times, where the peptide combination must be put into a biphasic micro-capillary column filled with reversed-phase and strong-cation-exchange materials.

10.2.1 Two-Dimensional Polyacrylamide Gel Electrophoresis (2D-PAGE)

To distinguish a specific protein from a mixture of proteins, 2D-PAGE uses two distinct characteristics of proteins: their molecular size and pI. In isoelectric focusing, proteins are separated based on the charge that they carry. In SDS-PAGE, proteins are separated based on their size. 2D-PAGE provides a useful platform for analysing the protein mixture. On a large format 2D-gel, up to 5000 protein spots can be found under the right conditions. By using several staining methods on the gel, such as Coomassie Brilliant Blue staining, silver staining or SYPRO staining, the separated protein spots can be seen (Dubey & Grover 2001). SYPRO is a set of molecular probes for the identification of proteins. It discolours luminous protein separation analysis by PAGE detection. Sensitive scanners capture 2D-PAGE digital images, which are then processed by software like Melanie, PDQuest, Phoretix, Progenesis, Z3 or Z4000 (Righetti et al. 2004). The target areas are manually or robotically removed from the gel. MS is used to determine the identity of proteins after digestion with sequence-specific proteases (exopeptidase or endopeptidase) (Rose et al. 2004).

Multiple bits of information, including molecular weight, pI, quantity and post-translational changes, are simultaneously provided by analysis of 2D-PAGE gels (PTMs). By using 2D-PAGE, several PTMs can be profiled. A few examples of specific applications are (a) phosphor-protein expression profiling, which is based on the idea that phosphorylated proteins shift their pIs to a more acidic region on the gel; (b) acetylated-protein expression profiling, which is based on the idea that acetylating proteins at their amino termini or on lysine side-chains will cause them to shift to a more acidic region on the gel; and (c) glycosylated-protein expression profiling that is based on the fact that the association of comparatively simple glycans (O-glycosylation) or more complex glycans

(N-glycosylation) can have varied impacts on protein molecular weight, pI and protein activities (Wittmann-Liebold et al. 2007).

10.2.2 FLUORESCENT TWO-DIMENSIONAL DIFFERENCE GEL ELECTROPHORESIS (2D-DIGE)

As a more quantitative variation of 2DE, fluorescence difference gel electrophoresis (DIGE) was designed. To distinguish separate proteins resolved on the same gel, samples are here selectively covalently tagged with fluorophores (Ünlü et al. 1997). In this method, one of two fluorescent dyes that are distinct but structurally similar is applied to each of the samples to be compared (cy2, cy3, cy5, etc.). Each dye reacts with amino groups, and as a result, the dye binds to lysine residues and the N-terminal amino group, fluorescently labelling each protein. Then, the two protein mixtures for comparison are combined, and a single 2D gel is performed. As a result, each protein in a sample overlays with its identical counterpart, differently labelled in a distinct sample. To determine if a specific location is connected to one dye molecule instead of two, the gel is scanned at two different wavelengths that excite the two dye molecules (Ünlü et al. 1997). By directly comparing hundreds of protein samples run under similar electrophoresis conditions, DIGE demonstrates better sensitivity, linearity, accuracy of measurement and enhanced reproducibility. In a single experiment, the DIGE approach allows for the quantitative and repeatable comparison of hundreds of proteins. Since the DIGE approach permits running an internal strand as part of each separation, more minor abundance variations can be statistically shown with the pooled internal standard methodology and finding real differences in protein expressions.

10.2.2.1 Gel-free Proteomics

For some proteomic applications, gel-based separation is still routinely used, but for large-scale shotgun proteomics, gel-free approaches for separating peptides after sequence-specific digestion have taken over (Roe & Griffin 2006). Since considerable pre-fractionation of peptide mixtures significantly boosts proteome coverage with MS-based peptide sequencing, the majority of gel-free approaches use two (2D-LC) or more complementary dimensions of LC (Mallick & Kuster 2010). The effectiveness of gel-free proteomics for detecting protein alterations in plant–microbe interaction systems has been shown in a number of researches. Three hundred seventy-seven distinct plant proteins were discovered using 2D-LC-MS/MS in the nodules of *Medicago truncatula* plants that had received an inoculation of *Sinorhizobium meliloti* (Larrainzar et al. 2007). Crops like rice and pea have also been treated in general with this method. To determine which cytoplasmic and endoplasmic reticulum (ER)-associated molecular chaperones are differentially regulated in *Nicotiana glutinosa* upon infection with tobacco mosaic virus (TMV), complementary separation techniques 2DE and 2D-LC are frequently utilised in conjunction with 2D-DIGE and MudPIT (Jayaraman et al. 2012).

10.2.3 MULTIDIMENSIONAL PROTEIN IDENTIFICATION TECHNOLOGY (MUDPIT)

Primarily for the study of intricate protein combinations, the MudPIT technique is applied. Trypsin and endo-proteinase lysis are used in this method to enzymatically degrade protein samples in a sequence-specific manner. The resulting peptide mixtures are then separated using strong cation exchange (SCX) and reverse phase (RP) high-performance LC (HPLC) (Issaq et al. 2005). The mass spectrometer receives peptides from the RP column, and the MS data is utilised to search protein databases. The advantage of this technique is that it produces an entire list of all the proteins contained in a specific protein sample while being quick, sensitive and reproducible. MudPIT, which combines SCX and RP-HPLC separation, is successfully used in a wide variety of proteomics experiments, including large-scale protein cataloguing in cells and organisms, membrane and organelle protein profiling, identification of protein complexes, determination of PTMs and quantitative analysis of protein expression (Washburn et al. 2001).

10.2.4 Isotope-coded Affinity Tags (ICAT)

ICAT is a quantitative proteomics technique that uses chemical labelling reagents, instead of gels, to locate and count the majority of the proteins in two cell types. ICAT makes use of the isotope ^{13}C. To produce peptide fragments, few of which are tagged, two populations are labelled with two separate ICAT reagents (one with light and another with heavy), mixed and enzymatically cleaved. The tagged peptides are isolated using avidin affinity chromatography, and their characterisation is then completed using a tandem mass spectrometric analysis using micro-capillary LC and electron spray (Gygi et al. 1999). In a single automated process, the relative amounts of the components of protein mixtures are ascertained by combining the results generated by MS and ICAT-labelled peptide analysis. This method provides a quantitative and qualitative assessment of changes in protein levels brought on by various stress situations (Kav et al. 2007).

10.2.5 Isobaric Tags for Relative and Absolute Quantitation (iTRAQ)

iTRAQ is a non-gel-based technique used to quantify protein from different sources in a single experiment (Rose et al. 2004). This method makes use of isotope-coded covalent tags, just like ICAT. The tagging of proteins makes a difference. This method is based on covalently attaching an isotopic tag with a different mass to the N-terminus and side chain amine of peptides produced during protein digestion. Nano chromatography is used to extract the peptides from the mixture (Zieske 2006). Tandem MS (MS/MS) analysis of the isolated peptides is then performed. The tagged peptides and related proteins are then found via a database search based on the fragmentation data. When a tag is attached, it breaks up, creating a low molecular mass reporter ion that may be used by software to roughly quantify the peptides and proteins that make up those peptides (e.g., i-tracker) (Aggarwal et al. 2006). The benefits of iTRAQ include multiplexing numerous samples, quantification, streamlined analysis, better analytical precision and accuracy and all of these features. In a single mass-spectrometry-based experiment, it has the capacity to compare up to four different samples (Lodha et al. 2013).

10.3 PROTEIN IDENTIFICATION

The most popular method for unbiased protein identification is MS, which has been extensively used in plant–microbe proteomics (Mathesius et al. 2003). Carl-Ove Anderson published the first study on the use of MS to evaluate amino acids and peptides in 1958 (Andersson 1958). The three key steps in an MS analysis are ionisation of the protein or peptide, ion separation and detection. Matrix-assisted laser desorption or ionisation (MALDI) and electrospray ionisation (ESI) are the two main methods for ionising proteins or peptides, whereas mass analysers include time-of-flight (TOF),quadrupole, ion trap, orbitrap and Fourier transform ion cyclotron resonance (FT-ICR) (Chait 2011). Test samples are transformed into gaseous ions in the mass spectrometer during MS, which are then separated and identified based on their mass-to-charge ratio (m/z). Trypsin is used to fragment proteins or peptides in the case of protein samples. LC is then used to separate the broken proteins. The samples are then made into the gaseous phase by being ionised. MS underwent a revolution with the development of TOF-MS and comparatively non-destructive techniques to turn proteins into volatile ions. In the mass analyser, gaseous ions are separated. By employing collision-induced dissociation to fragment peptides with a certain mass, they are then subjected to a second MS, which yields a series of fragment peaks from which the amino acid sequence of the peptide may be deduced. Algorithms for protein identification are used to compare the findings to established benchmarks. These algorithms can be divided into two groups de novo search algorithms and database search algorithms. Plant–microbe proteomics has recently experienced a boom due to the advancement of MS technology. Using MALDI-TOF/TOF, the responses of the ethylene-insensitive mutant sickle and wild-type *M. truncatula* were compared just over a few years ago (Prayitno et al. 2006). The study revealed that ethylene-inducible proteins varied between genotypes. While MALDI has special uses like tissue imaging, ESI is the preferred technique for incorporating large-scale shotgun proteomics into a systems biology approach due to its compatibility with online LC separation.

10.4 PLANT–PATHOGEN INTERACTIONS

Numerous microorganisms interact with plants; some interactions have positive effects, while others have negative ones. Worldwide, plant diseases result in significant agricultural losses each year. Conversely, advantageous interactions between plants and microbes lead to a decrease in disease as well as increased plant growth. Numerous studies have already been conducted from both the plant and the pathogen viewpoints to acquire a deeper knowledge of plant–pathogen interactions. These investigations have shown that plant–pathogen interactions are the consequence of precise communication between the plant and the invasive pathogen (Hammond-Kosack 2000). Numerous pathogens, including bacteria, fungi, nematodes, oomycetes, viruses and insects, can be found within plants. Plants are unable to mount effective anti-infectious defensive responses in encounters that are compatible, allowing viruses to finish their life cycle. Plants launch a series of intricate defence reactions against pathogenic interactions during incompatible interactions to thwart the growth of pathogens (Schenk et al. 2000).

Thus, studies at the proteome level can be utilised to clarify numerous cellular processes involved in plant perception of pathogens in both compatible and incompatible interactions, as well as those events brought by interacting microbes that are advantageous. Studies in this area can assist us in not only identifying proteins but also in learning about the associated genes that may be used to create genetically modified plants for the control of biotic stress and/or enhancing plant growth.

10.4.1 PLANT–BACTERIA INTERACTION

A lot of different signalling molecules are involved in plant–bacteria relationships, and these molecules affect how the plant genes are expressed, which ultimately determines whether the contact is beneficial or harmful. Bacteria now possess a wide variety of specially created macromolecular nanomachines that release different virulence factors (Costa ct al. 2015). The secretion structures (type I–VI) of Gram-negative bacteria, mycobacterial type VII, the chaperone-usher cascade and the curli secretion system are well-defined in bacteria. These structures are distinct due to the structural proteins that build them up (Costa et al. 2015). Toxins, hydrolysing proteins and pathogenic components are released by type-II machinery. During an infection, a pathogenic bacterium mostly exploits the type-III secretion system, which allows it to directly transfer effector proteins into the target cell and disrupt biological functions (Mehta et al. 2008).

A great deal of information has been revealed about the symbiotic relationships between N_2-fixing bacteria and legumes in the proteome-based investigation of plant–bacterial association. It is discovered that *S. meliloti* forms a symbiotic connection with the significant model legume *M. truncatula* for nodulation. Utilising a proteomic method on *M. truncatula* nodules, 377 different plant-related proteins were discovered (Larrainzar et al. 2007). The majority of the proteins found belonged to functional categories involved in protein synthesis, degradation and the production of amino acids. Additionally, proteins implicated in the process of synthesis of sulphur-containing amino acids as well as those with a role in N_2 assimilation in nodules were detected. When *S. meliloti* was used to treat the root, leaf sheath and leaf of rice, comparative proteomic analysis revealed distinct proteins with functions in nine distinct categories (Chi et al. 2010). The interaction resulted in an upregulation of defence-related proteins in the root and leaf sheath and upregulation of photosynthesis-related proteins in the leaf, leading to an overall increase in rice plant growth. Comparing the root phosphor-proteome of the legume *Lotus japonicus* with a conserved pathogen-associated peptide motif (flagellin flg 22) and a symbiotic signalling molecule (nodulation factor).

a few minutes after perception revealed not only signal specific but also significant overlap patterns (Serna-Sanz et al. 2011).

The bacterial blight, the second most serious disease to affect rice, is caused by *Xanthomonas oryzae* pv. *oryzae* (*Xoo*). Twenty distinct protein spots implicated in energy, metabolism, defence and protein synthesis were identified using proteomic analysis of proteins from rice leaves that had been inoculated with compatible or incompatible *Xoo* strains. Ten of these proteins, including peroxiredoxin, the major subunit of RuBisCO and ATP synthase, were identified as distinctive to the compatible association (Mahmood et al. 2006). A different study examined the proteome and transcriptome profiles of *Arabidopsis thaliana* leaves during early responses (1–6 h post-inoculation) to the challenge posed by *Pseudomonas syringae* pv. *tomato*. The proteome alterations in *A. thaliana* in response to the extremely virulent strain DC3000 of *P. syringae* pv. *tomato* were compared. It produced effective parasitism, a DC3000 *hrp* mutant that brought about basal resistance and a transconjugant of DC3000 that expressed *avrRpm1*, triggering gene-for-gene resistance. Glutathione S-transferases (GSTs) and peroxiredoxins (Prxs) were two sub-groups of proteins that consistently displayed distinct changes in abundance following varied stressors and time intervals. The modulation of redox conditions in infected tissue was assumed to be regulated by both of these classes of antioxidant enzymes. The Affymetrix Gene Chip analysis of these results revealed alterations in the expression profiles for the associated *GST* and *Prx* genes. For the Prx family, in contrast to the GST family, a strong connection between alterations found at the transcript and protein levels was generally observed.

By correlating two partially resistant lines with a susceptible control line over a time course (72 and 144 h post-inoculation), it was possible to identify proteins from the wild tomato species *Lycopersicon hirsutum* that are regulated in response to the bacterial canker causal agent (*Clavibacter michiganensis* ssp. *michiganensis*). A total of 26 tomato proteins with differential regulation were found using 2DE and ESI-MS/MS, 12 of which had a direct connection to stress and defence mechanisms.

Phosphoproteomics (large-scale analysis of phosphoproteins) is an effective method for better understanding the events that take place quickly in the host after bacterial sensing because phosphorylation is a key factor in the beginning of the plant response to bacterial signals (Xing et al. 2002). Although it has been demonstrated that phosphorylation pathways alter quickly after receiving signals, only a small number of these phosphoproteins have been found in plant species (Peck et al., 2001). Early alterations in proteins that may have been phosphorylated during the bacterial defence response, such as dehydrin, chaperone, heat shock protein and glucanase, have been characterised utilising a phosphor-proteome approach. These proteins are most likely phosphorylated as part of the initial basal plant defence response (Jones et al., 2006).

10.4.2 PLANT–FUNGUS INTERACTION

Although a huge number of bacteria cause plant disease, fungi are mostly responsible for crop loss. The relationship between plants and fungi belonging to the biotrophic and necrotrophic groups has been defined using the proteomic technique. In the past ten years, significant progress has been made in pinpointing the variables that influence plant–fungus interactions. At the moment, more than 25 fungal genomes have been deciphered, including diseases for both plants and people, such as *Magnaporthe grisea* and *Aspergillus fumigates* (Murad et al. 2006). By employing a proteome approach to clarify the mechanisms underpinning resistance, we can gain a better qualitative and quantitative understanding of the relationship between plants and fungi, which will help us, apply this information to the development of resistant cultivars.

Intracellular and secreted proteins are either up- or down-regulated when pathogenic fungus begins the infection process, thereby enhancing the fungal capacity for predation (Murad et al. 2006). An understanding of the dimorphic change from budding to filamentous growth (Böhmer et al. 2007) as well as appressorium building was the goal of the ground-breaking proteomic investigations. The creation of a successful relationship between the pathogen *Phytophtora infestans* and its host plant potato is thought to depend on the formation of an appressorium (Grenville-Briggs et al. 2005). In another study, the interaction of tomato plants affected by a fungus that causes wilt sap from plant xylem was used to study *Fusarium oxysporum*. While the remaining PR proteins were discovered to be only regulated in compatible interplay, the PR-5 protein was revealed to be stimulated in both compatible and incompatible relationships (Rep et al. 2002). In a different investigation into the relationship between *A. thaliana* and *Plasmodiophora brassicae*, protein from plant roots and stems was examined. Compared to unchallenged controls, the expression of proteins with roles in cellular functions like defence, metabolism and differentiation was altered in challenged plants (Devos et al. 2006).

An additional method that has been commonly used to analyse fungal proteins is the exoproteome, also referred to as the secretome. In this case, hop (*Humulus lupulus*) cell walls were used as a growth medium for *Fusarium graminearum*, a destructive pathogen of wheat, maize and other cereals. A total of 84 fungal-secreted proteins were discovered using 1DE and 2DE and then MS analysis. Cellulases, glucanosyltransferases, endoglucanases, phospholipases, proteinases and chitinases were among the proteins that were discovered (Phalip et al. 2005). In *F. graminearum* cultivated with hop cells present, it was shown that 45% of the proteins were specifically related to cell wall breakdown and only secondarily connected to carbon and nitrogen absorption. However, the enzyme patterns were completely different when the same fungus was grown on a medium that contained glucose, demonstrating that fungi are able to control their secretion in response to the availability of substrate (Phalip et al. 2005). *Sheath blight*, which is caused by the fungus *Rhizoctonia solani*, was the subject of a recent study on the interactions between rice and fungi. After this fungus had infected the rice sheath leaves, an investigation was conducted. The results showed that six proteins had relative abundances that differed significantly between resistant and susceptible lines, and 11 additional proteins had only been identified in abundance in the response of the resistant line. These proteins have been linked to a similar route for both stress and non-stress plant processes, including antifungal activity, signal transduction, energy metabolism, photosynthesis, protein folding and degradation and anti-oxidation (Lee et al. 2006).

A resistance response of pea to the biotrophic fungus *Erysiphe pisi* was discovered using the proteomic technique (Curto et al. 2006). Infected and uninfected Messire leaves and JI2480 leaves, respectively, displayed 19 and 12 different proteins, according to a MALDI-TOF/TOF investigation and database analysis. Proteins involved in signalling, metabolism, energy production and stress tolerance were discovered. To identify proteins indicating *F. graminearum* tolerance of Fhb1, 72 hours after challenge, comparative protein expression analysis was performed among wheat near-isogenic lines for alleles of Fhb1, an essential FHB resistance gene (Zhang et al. 2013). When compared to un-inoculated Fhb1+ near-isogenic lines, it was found that eight proteins were either induced or upregulated, while nine proteins were either induced or upregulated in the *Fusarium*-infected Fhb1+ near-isogenic line when compared to the *Fusarium*-inoculated Fhb1 near-isogenic line. The main functions of differently expressed proteins were defence against pathogen invasion, energy metabolism and photosynthesis (Zhang et al. 2013).

10.4.3 PLANT–NEMATODE INTERACTION

Phytonematodes, which attack plants continuously and severely harm crops that are vulnerable to them, generate significant economic losses worldwide. *Meloidogyne* spp., *Heterodera* spp. and *Globodera* spp. are obligate sedentary endoparasites that are among the most dangerous plant-parasitic nematodes (Chitwood 2003). The juvenile larvae (J2) of these organisms enter plant roots, and after three moults, they mature into adult forms that repeatedly reproduce. As a result, the root system is severely altered, which significantly reduces nutrient and water intake and results in plant mortality (Curtis 2007). Various nematode-expressed sequence tag (EST) libraries have been created recently, mostly to uncover parasitic nematode-specific genes (Smant et al. 1998). Roughly

100,000 ESTs from *Heterodera*, *Globodera* and *Meloidogyne* species have been sequenced (Gao et al. 2003). Even though there are many ESTs, only a small number of these genes are known to be involved in parasitism, even though many of the transcripts exhibit differential expression at different stages of the parasitic life cycle (Huang et al. 2007). Although to a lesser extent, proteomic methods have also helped to identify potential phytonematode parasitome possibilities (Jauber et al. 2002). Some of the nematode proteins that have been discovered, such as tropomyosin, ATP synthase β-chain, enolase and chaperonin protein HSP-60, are involved in feeding site and cell wall breakdown (Calvo et al. 2005).

Despite the paucity of proteomic investigations, 2DE combined with MS is a potent and quick method for producing peptide sequence tags that can be *in silico* linked to ESTs. In order to access full-length gene sequences, these peptides can also be utilised to construct primers, which helps parasitic genome studies (Ashton et al. 2001). Few studies have attempted to understand the true significance of these sequences in plant–nematode interactions, despite the abundance of experimental and *in silico* evidence. Additionally, EST libraries from oesophageal glands of *Meloidogyne incognita* and *Heterodera glycines* show that the bulk of the genes expressed in these salivary glands encode proteins with unknown activities (89% in *M. incognita* and 72% in *H. glycines*). Taking into account the nematode–plant interaction from the other perspective, some plants have developed defence mechanisms that thwart nematode attraction, penetration and migration, feeding site development, sustenance by digesting, reproduction and survival. In many plants, a number of resistance genes have been discovered (Williamson & Kumar 2006). Studies on the proteome of the relationship between plants and nematodes are still in their early stages. Three proteins, including chitinase, PR protein in *Coffea canephora* and quinone reductase2 in *Gossipium hirsutum*, have been identified in a recent study as being upregulated in response to nematode infection (Mehta et al. 2008).

10.4.4 Plant–Virus Interaction

Due to their non-cellular nature and lack of life outside the host cell, viruses are the ultimate parasites. They have tiny genomes that only encode a small number of proteins (Quirino et al. 2010). For a plant infection to be successful, the virus must first be transferred mechanically or by a vector, reproduce in plant cells, then travel from one cell to another through plasmodesmata and lastly reach the vascular tissue to circulate systemically through the phloem to the host sink tissues (vascular movement). Viral systemic infection is established as viruses are released from the phloem through new replication cycles and cell-to-cell/vascular migration (Whitham et al. 2006). The intricate and varied molecular interactions between host plants and invading viruses can be understood by proteomic research on the differential protein expression during virus infection, proteins implicated in host-viral protein–protein interactions and other proteins (Alexander & Cilia 2016). Numerous methods have been used to investigate the function of viral-encoded proteins, plant–virus interactions and proteins attracted by virus challenge.

Quantitative 2D-DIGE from susceptible and resistant types of sugarcane mosaic virus (SCMV)-challenged plants revealed 93 protein locations that were shown to demonstrate substantial expression variation in a proteomic investigation on the interaction between maize and SCMV. The functional analysis of SCMV-induced proteins revealed the presence of proteins involved in the photosynthetic pathway, stress and defence responses and energy and carbon metabolism (Wu et al. 2013).

A recent analysis of the enhanced expression of nuclear proteins in *Capsicum annuum* cv. *Bugang* (hot pepper) infected with tobacco mosaic tobamovirus used 2DE and then MALDI-TOF MS (TMV) (Lee et al. 2006). *C. annuum* cv. *Bugang* strains are hypersensitive response resistant against TMV-P0, but are susceptible to TMV-P1.2 strains. In hot pepper infected with TMV-P0, a hypothetical protein and five annotated nuclear proteins were found, including four defence-related proteins (messenger RNA [mRNA]-binding protein [which may interact with viral RNA or interfere with plant RNA metabolism], 26S proteasome subunit [RPN7][possibly involved in programmed cell death], 14-3-3 protein [regulator of proteins involved in response to biotic stresses] and Rab11 GTPase [responsible for membrane trafficking]) and a ubiquitin extension protein. The PR-proteins and antioxidant proteins showed differential protein expression in asymptomatic tomato fruits infected and uninfected with TMV, suggesting a potential role for these proteins in the defence against TMV challenge (Casado-Vela et al. 2006). Thus, a conspicuous TMV coat protein spot found in the gel of asymptomatic fruits can be exploited as a key analysis approach for virus detection and characterisation in fruit that has been exposed to the virus. Using 2D-DIGE and MudPIT, it was possible to examine the different ways that cytoplasmic and ER-associated molecular chaperones are regulated in TMV-infected *N. glutinosa* (Caplan et al. 2009). A plasmodesmata protein NbEXPA1, which is unique to *N. benthamiana*, was recently discovered to be downregulated as a result of turnip mosaic virus infection. According to the study, plasmodesmata protein plays a role in encouraging potyviral infection. High-sensitivity MS and 2D-gel electrophoresis techniques can be used to further gain a thorough grasp of how plants and viruses might interact with one another (Park et al. 2017).

Nearly two decades ago, when electrophoretic and protein identification techniques were less advanced than they are now, the field of proteomics of plant–pathogen interactions was established. Ekramoddoullah and Hunt used 2D-PAGE in 1993 to differentiate between the proteins in sugar–pine seedlings that were susceptible to or resistant to the fungus that causes white pine blister rust. Worldwide, plant diseases significantly reduce crop productivity; thus, it is crucial to understand the tactics used by these virulent microbes to spread illness. They have extremely flexible infection mechanisms since they can infect a variety of plants and tissues. Their interaction is dependent on reciprocal recognition, which calls for the invading microbe to release pathogenicity factors and the host to activate defence mechanisms. For efficient disease management, especially when using biocontrol agents, complete knowledge of pathogenic components and the disease cycle is necessary. We will be able to identify the gene product in a specific pathogen and beneficial microorganism for plants with the use of bioinformatic techniques. Additionally, combining an omics analysis with targeted mutagenesis or transgenic research will aid in our understanding of the genetic underpinnings of host–pathogen cross-talks. All live cells contain proteins as structural and functional elements. Studies at the proteome level are perfect for revealing information about the functional molecules of a cell and the changes that various states and stresses cause in the expression patterns of proteins. Proteomic analysis has been used to better understand how plants and microbes interact, how a beneficial microbe might cause a systemic response in plants and how pathogenic and virulence factors contribute to the control of disease. Complex cellular networks are present in many plant–pathogen interactions and plant-beneficial microbe interactions, according to the studies discussed in the previous section. We will be able to identify interesting new targets for the enhancement of disease tolerance by having a system-level understanding of biotic stress reactivity and the impact of applied biocontrol agents. Functional investigation of microbial relationships associated with plants can also be accomplished by the application of proteomics, meta-proteomics and secretomics. Proteomics combined with bioinformatics technologies can offer intriguing insights to build a thorough understanding of the complex interactions between plants and microorganisms.

10.5 CONCLUSION

There is no better way to comprehend the biological mechanisms underlying the relationship between plants and microbes than to examine proteins, which are directly responsible for every cellular action. Protein identification is made considerably easier by technological advancements in MS, bioinformatic techniques and the growing number of sequenced genomes. As more genes are sequenced, protein identification will improve, making it possible to use peptide mass fingerprinting applicable to more plants, diseases and/or beneficial microorganisms. The combination of proteomics with genomics data, traditional plant pathology knowledge and advances in genetic engineering techniques, such as the CRISPR-Cas system, will eventually help to develop new disease-resistant varieties of plants or crops treated with beneficial microorganisms that exhibit enhanced resistance. According to the present plan, comparative proteomics will be the method of choice for researchers all over the world for analysing the interactions between plants and microbes.

10.6 FUTURE PERSPECTIVES

We can now firmly say that proteomics is a mature platform for proteome analysis during plant–pathogen interaction based on the experiments of plant–pathogen interaction that have been done so far. Unknown protein functions are being revealed through proteome-wide functional categorisation utilising bioinformatics techniques. Scientists can better understand the activities of proteins and the intricate regulatory networks that govern the basic biological processes thanks to proteomics. The biological enhancement of agricultural yield requires the identification of proteins that regulate crop architecture and/or stress resistance in a variety of conditions in order to address the current concerns of food insecurity. Modern proteomic techniques are the first choice of the scientists for identifying these proteins. The usage of proteomics is, however, being constrained by a few drawbacks. Proteomic snapshots of proteins can be challenging due to their changeability as dynamic, interacting entities. In addition, a functional connection between two proteins may not always be present. In order to, increase our understanding of protein expression during plant-pathogen interactions, more sensitive analytical tools and efficient methodologies for large-scale data comparison are required. However, new techniques and apparatus are being developed, and proteomic tools are advancing quickly. Future proteomic research, along with functional validation and bioinformatic analysis, may, in our opinion, offer fresh perspectives on plant disease resistance and pathogenicity. A symbiotic link between proteomics technology and systems biology approaches in future will enable new problems to be addressed in the field of plant–pathogen interaction. Proteomics will continue to be one of the fastest expanding fields of study.

REFERENCES

Aggarwal, K., Choe, L., & Lee, K. (2006). Shotgun proteomics using the iTRAQ isobaric tags. *Briefings in Functional Genomics*, 5(2), 112–120. https://doi.org/10.1093/bfgp/ell018

Alexander, M., & Cilia, M. (2016). A molecular tug-of-war: Global plant proteome changes during viral infection. *Current Plant Biology*, 5, 13–24. https://doi.org/10.1016/j.cpb.2015.10.003

Andersson, C.-O. (1958). Mass spectrometric studies on amino acid and peptide derivatives. *Acta Chemica Scandinavica*, 12(6), 1353.

Ashton, P., Curwen, R., & Wilson, R. (2001). Linking proteome and genome: How to identify parasite proteins. *Trends in Parasitology*, 17(4), 198–202. https://doi.org/10.1016/S1471-4922(00)01947-4

Böhmer, M., Colby, T., Bohmer, C., Brautigam, A., Schmidt, J., & Bölker, M. (2007). Proteomic analysis of dimorphic transition in the phytopathogenic fungus *Ustilago maydis*. *Proteomics*, 7(5), 675–685. https://doi.org/10.1002/pmic.200600900

Calvo, E., Flores-Romero, P., Lopez, J., & Navas, A. (2005). Identification of proteins expressing differences among isolates of *Meloidogyne* spp. (Nematoda: Meloidogynidae) by nano-liquid chromatography coupled to ion-trap mass spectrometry. *Journal of Proteome Research*, 4(3), 1017–1021. https://doi.org/10.1021/pr0500298

Caplan, J., Zhu, X., Mamillapalli, P., Marathe, R., Anandalakshmi, R., & Kumar, S. (2009). Induced ER chaperones regulate a receptor-like kinase to mediate antiviral innate immune response in plants. *Cell Host & Microbe*, 6(5), 457–469. https://doi.org/10.1016/j.chom.2009.10.005

Casado-Vela, J., Sellés, S., & Martínez, R. (2006). Proteomic analysis of tobacco mosaic virus-infected tomato (*Lycopersicon esculentum* M.) fruits and detection of viral coat protein. *Proteomics*, 6(S1), S196–S206. https://doi.org/10.1002/pmic.200500317

Chait, B. (2011). Mass spectrometry in the postgenomic era. *Annual Review of Biochemistry*, 80, 239–246. https://doi.org/10.1146/annurev-biochem-110810-095744

Chi, F., Yang, P., Han, F., Jing, Y., & Shen, S. (2010). Proteomic analysis of rice seedlings infected by *Sinorhizobium meliloti* 1021. *Proteomics*, 10(9), 1861–1874. https://doi.org/10.1002/pmic.200900694

Chitwood, D. (2003). Research on plant-parasitic nematode biology conducted by the United States Department of Agriculture–Agricultural Research Service. *Pest Management Science, 59*(6–7), 748–753. https://doi.org/10.1002/ps.684

Costa, T. R., Felisberto-Rodrigues, C., Meir, A., Prevost, M. S., Redzej, A., Trokter, M., & Waksman, G. (2015). Secretion systems in Gram-negative bacteria: Structural and mechanistic insights. *Nature Reviews Microbiology, 13*, 343–359. https://doi.org/10.1038/nrmicro3456

Curtis, R. H. (2007). Plant parasitic nematode proteins and the host–parasite interaction. *Briefings in Functional Genomics, 6*(1), 50–58. https://doi.org/10.1093/bfgp/elm006

Curto, M., Camafeita, E., Lopez, J., Maldonado, A., Rubiales, D., & Jorrín, J. (2006). A proteomic approach to study pea *Pisum sativum* responses to powdery mildew *Erysiphe pisi. Proteomics, 6*(S1), S163–S174. https://doi.org/10.1002/pmic.200500396

Devos, S., Laukens, K., Deckers, P., Straeten, D., Beeckman, T., Inzé, D., ... Prinsen, E. (2006). A hormone and proteome approach to picturing the initial metabolic events during *Plasmodiophora brassicae* infection on *Arabidopsis. Molecular Plant-Microbe Interactions, 19*(12), 1431–1443. https://doi.org/10.1094/MPMI-19-1431

Dubey, H., & Grover, A. (2001). Current initiatives in proteomics research: The plant perspective. *Current Science, 80*(2), 262–269. http://www.jstor.org/stable/24104286

Gao, B., Allen, R., Maier, T., Davis, E., Baum, T., & Hussey, R. (2003). The parasitome of the phytonematode *Heterodera glycines. Molecular Plant-Microbe Interactions, 16*(8), 720–726. https://doi.org/10.1094/MPMI.2003.16.8.720

Grenville-Briggs, L., Avrova, A., Bruce, C., Williams, A., Whisson, S., Birch, P., & West, P. (2005). Elevated amino acid biosynthesis in *Phytophthora infestans* during appressorium formation and potato infection. *Fungal Genetics and Biology, 42*(3), 244–256. https://doi.org/10.1016/j.fgb.2004.11.009

Gygi, S. P., Rist, B., Gerber, S. A., Turecek, F., Gelb, M. H., & Aebersold, R. (1999). Quantitative analysis of complex protein mixtures using isotope-coded affinity tags. *Nature Biotechnology, 17*, 994–999. https://doi.org/10.1038/13690

Hammond-Kosack, K. (2000). Responses to Plant Pathogens. In Jones, R. L., Buchanan, B. B., & Gruissem, W. (Eds.), *Biochemistry and Molecular Biology of Plants*, 2nd ed. (pp. 984–1050). John Wiley & Sons Chichester, West Sussex

Huang, G., Gao, B., Maier, T., Allen, R., Davis, E., Baum, T., & Hussey, R. (2007). A profile of putative parasitism genes expressed in the esophageal gland cells of the root-knot nematode *Meloidogyne incognita. Molecular Plant-Microbe Interactions, 16*(5), 376–381. https://doi.org/10.1094/MPMI.2003.16.5.376

Issaq, H., Chan, K., Janini, G., Conrads, T., & Veenstra, T. (2005). Multidimensional separation of peptides for effective proteomic analysis. *Journal of Chromatography B, 817*(1), 35–47. https://doi.org/10.1016/j.jchromb.2004.07.042

Jain, A., Singh, H. B., & Das, S. (2021). Deciphering plant-microbe crosstalk through proteomics studies. *Microbiological Research, 242*. https://doi.org/10.1016/j.micres.2020.126590

Jauber, S., Laffaire, J. B., Piotte, C., Abad, P., Rosso, M. N., & Ledger, T. N. (2002). Direct identification of stylet secreted proteins from root-knot nematodes by a proteomic approach. *Molecular and Biochemical Parasitology, 121*(2), 205–211. https://doi.org/10.1016/S0166-6851(02)00034-8

Jayaraman, D., Forshey, K., Grimsrud, P., & Ané, J.-M. (2012). Leveraging proteomics to understand plant–microbe interactions. *Frontiers in Plant Science, 3*(44). https://doi.org/10.3389/fpls.2012.00044

Jones, A., Bennett, M., Mansfield, J., & Grant, M. (2006). Analysis of the defence phosphoproteome of *Arabidopsis thaliana* using differential mass tagging. *Proteomics, 6*(14), 4155–4165. https://doi.org/10.1002/pmic.200500172

Kav, N., Srivastava, S., Yajima, W., & Sharma, N. (2007). Application of proteomics to investigate plant-microbe interactions. *Current Proteomics, 4*(1), 28–43. https://doi.org/10.2174/157016407781387357

Larrainzar, E., Wienkoop, S., Weckwerth, W., Ladrera, R., CesarArrese-Igor, C., & González, E. (2007). *Medicago truncatula* root nodule proteome analysis reveals differential plant and bacteroid responses to drought stress. *Plant Physiology, 144*(3), 1495–1507. https://doi.org/10.1104/pp.107.101618

Lee, J., Bricker, T., Lefevre, M., Pinson, S., & Oard, J. (2006). Proteomic and genetic approaches to identifying defence-related proteins in rice challenged with the fungal pathogen *Rhizoctonia solani. Molecular Plant Pathology, 7*(5), 405–416. https://doi.org/10.1111/j.1364-3703.2006.00350.x

Lee, B.-J., Kwon, S. J., Kim, S. K., Kim, K.-J., Park, C.-J., Kim, Y.-J., ... Paek, K.-H. (2006). Functional study of hot pepper 26S proteasome subunit RPN7 induced by *Tobacco mosaic virus* from nuclear proteome analysis. *Biochemical and Biophysical Research Communications, 351*(2), 405–411. https://doi.org/10.1016/j.bbrc.2006.10.071

Lodha, T. D., Hembram, P., Tep, N., & Basak, J. (2013). Proteomics: A successful approach to understand the molecular mechanism of plant-pathogen interaction. *American Journal of Plant Sciences*, 4(6). https://doi.org/10.4236/ajps.2013.46149

Mahmood, T., Jan, A., Kakishima, M., & Komatsu, S. (2006). Proteomic analysis of bacterial-blight defense-responsive proteins in rice leaf blades. *Proteomics*, 6(22), 6053–6065. https://doi.org/10.1002/pmic.200600470

Mallick, P., & Kuster, B. (2010). Proteomics: A pragmatic perspective. *Nature Biotechnology*, 28, 695–709. https://doi.org/10.1038/nbt.1658

Mathesius, U., Imin, N., Natera, S. H., & Rolfe, B. G. (2003). Proteomics as a Functional Genomics Tool. In Ulrike Mathesius, N. I., & Grotewold, E. (Eds.), *Methods in Molecular Biology* (Vol. 236, p. 19). Humana Press. https://doi.org/10.1385/1-59259-413-1:395

Mehta, A., Magalhaes, B. S., Souza, D. S., Vasconcelos, E. A., Silva, L. P., Grossi-de-Sa, M. F., ... Rocha, T. L. (2008). Rooteomics: The challenge of discovering plant defense-related proteins in roots. *Current Protein and Peptide Science*, 9(2), 108–116.https://doi.org/10.2174/138920308783955225

Murad, A., Laumann, R., Lima, T., Sarmento, R., Noronha, E., Rocha, T., ... Franco, O. (2006). Screening of entomopathogenic *Metarhizium anisopliae* isolates and proteomic analysis of secretion synthesized in response to cowpea weevil (*Callosobruchus maculatus*) exoskeleton. *Comparative Biochemistry and Physiology Part C: Toxicology & Pharmacology*, 142(3–4), 365–370. https://doi.org/10.1016/j.cbpc.2005.11.016

Park, S.-H., Li, F., Renaud, J., Shen, W., Li, Y., Guo, L., ... Wang, A. (2017). NbEXPA1, an α-expansin, is plasmodesmata-specific and a novel host factor for potyviral infection. *The Plant Journal*, 92(5), 846–861. https://doi.org/10.1111/tpj.13723

Peck, S., Nuhse, T., Hess, D., Iglesias, A., Meins, F., & Boller, T. (2001). Directed proteomics identifies a plant-specific protein rapidly phosphorylated in response to bacterial and fungal elicitors. *The Plant Cell*, 13(6), 1467–1475. https://doi.org/10.1105/TPC.000543

Phalip, V., Delalande, F., Carapito, C., Goubet, F., Hatsch, D., Leize-Wagner, E., Dupree, P., Dorsselaer, A. V., & Jeltsch, J.-M. (2005). Diversity of the exoproteome of *Fusarium graminearum* grown on plant cell wall. *Current Genetics*, 48, 366–379. https://doi.org/10.1007/s00294-005-0040-3

Philip, L.R., Huang, Y. N.,, P., Marchese, J. N., Williamson, B., Parker, K., Hattan, S., Khainovski, N., Pillai, S., Dey, S., Daniels, S., Purkayastha, S., Juhasz, P., Martin, S., Barlett-Jones, M., He, F., Jacobson, A. & Pappin, D. J. (2004). Multiplexed protein quantitation in *Saccharomyces cerevisiae* using amine-reactive isobaric tagging reagents. *Molecular & Cellular Proteomics*, 3(12), 1154–1169. https://doi.org/10.1074/mcp.M400129-MCP200

Prayitno, J., Imin, N., Rolfe, B., & Mathesius, U. (2006). Identification of ethylene-mediated protein changes during nodulation in *Medicago truncatula* using proteome analysis. *Journal of Proteome Research*, 5(11), 3084–3095. https://doi.org/10.1021/pr0602646

Quirino, B., Candido, E., Campos, P., Franco, O., & Kruger, R. (2010). Proteomic approaches to study plant–pathogen interactions. *Phytochemistry*, 71(4), 351–362. https://doi.org/10.1016/j.phytochem.2009.11.005

Rep, M., Dekker, H., Vossen, J., Boer, A., Houterman, P., Speijer, D., Back, J. W., de Koster, C. G., Cornelissen, B. (2002). Mass spectrometric identification of isoforms of PR proteins in xylem sap of fungus-infected tomato. *Plant Physiology*, 130(2), 904–917. https://doi.org/10.1104/pp.007427

Righetti, P., Castagna, A., Antonucci, F., Piubelli, C., Cecconi, D., Campostrini, N., Antonioli, P., Astner, H., & Mahmoud, H. (2004). Critical survey of quantitative proteomics in two-dimensional electrophoretic approaches. *Journal of Chromatography A*, 1051(1–2), 3–17. https://doi.org/10.1016/j.chroma.2004.05.106

Roe, M., & Griffin, T. (2006). Gel-free mass spectrometry-based high throughput proteomics: Tools for studying biological response of proteins and proteomes. *Proteomics*, 6(17), 4678–4687. https://doi.org/10.1002/pmic.200500876

Rose, J., Bashir, S., Giovannoni, J., Jahn, M., & Saravanan, R. S. (2004). Tackling the plant proteome: Practical approaches, hurdles and experimental tools. *The Plant Journal*, 39(5), 715–733. https://doi.org/10.1111/j.1365-313X.2004.02182.x

Schenk, P. M., Carvalhais, L. C., & Kazan, K. (2012). Unraveling plant–microbe interactions: Can multi-species transcriptomics help? *Trends in Biotechnology*, 30(3), 177–184. https://doi.org/10.1016/j.tibtech.2011.11.002

Schenk, P., Kazan, K., Wilson, I., & Manners, J. (2000). Coordinated plant defense responses in *Arabidopsis* revealed by microarray analysis. *Proceedings of National Academy of Science*, 97(21), 11655–11660. https://doi.org/10.1073/pnas.97.21.11655

Serna-Sanz, A., Parniske, M., & Peck, S. C. (2011). Phosphoproteome analysis of *Lotus japonicus* roots reveals shared and distinct components of symbiosis and defense. *Molecular Plant-Microbe Interactions*, *24*(8), 932–937. https://doi.org/10.1094/MPMI-09-10-0222

Singh, A., & Roychoudhury, A. (2021) Augmenting the Abiotic Stress Tolerance in Plants Through Microbial Association. In: Nath, M., Bhatt, D., Bhargava, P., & Choudhary, D. K. (Eds.), *Microbial Metatranscriptomics Belowground* (pp. 179–198). Springer, Singapore.

Smant, G., Stokkermans, J., Yan, Y., & Bakker, J. (1998). Endogenous cellulases in animals: Isolation of β-1,4-endoglucanase genes from two species of plant-parasitic cyst nematodes. *Proceedings of the National Academy of Sciences of the United States of America*, *95*(9), 4906–4911. https://doi.org/10.1073/pnas.95.9.4906

Ünlü, M., Morgan, M. E., & Minden, J. S. (1997). Difference gel electrophoresis. A single gel method for detecting changes in protein extracts. *Electrophoresis*, *18*(11), 2071–2077. https://doi.org/10.1002/elps.1150181133

Washburn, M., Wolters, D., & Yates, J, III. (2001). Large-scale analysis of the yeast proteome by multi-dimensional protein identification technology. *Nature Biotechnology*, *19*, 242–247. https://doi.org/10.1038/85686

Whitham, S., Yang, C., & Goodin, M. (2006). Global impact: Elucidating plant responses to viral infection. *Molecular Plant-Microbe Interactions*, *19*(11), 1207–1215. https://doi.org/10.1094/MPMI-19-1207

Williamson, V. M., & Kumar, A. (2006). Nematode resistance in plants: The battle underground. *Trends in Genetics*, *22*(7), 396–403. https://doi.org/10.1016/j.tig.2006.05.003

Wittmann-Liebold, B., Graack, H.-R., & Pohl, T. (2007). Two-dimensional gel electrophoresis as tool for proteomics studies in combination with protein identification by mass spectrometry. *Proteomics*, *7*(5), 824. https://doi.org/10.1002/pmic.200790015

Wu, L., Han, Z., Wang, S., Wang, X., Sun, A., Zu, X., & Chen, Y. (2013). Comparative proteomic analysis of the plant–virus interaction in resistant and susceptible ecotypes of maize infected with *sugarcane mosaic virus*. *Journal of Proteomics*, *89*, 124–140. https://doi.org/10.1016/j.jprot.2013.06.005

Xing, T., Ouellet, T., & Miki, B. L. (2002). Towards genomic and proteomic studies of protein phosphory-lation in plant–pathogen interactions. *Trends in Plant Science*, *7*(5), 224–230. https://doi.org/10.1016/S1360-1385(02)02255-0

Zhang, X., Fu, J., Hiromasa, Y., Pan, H., & Bai, G. (2013). Differentially expressed proteins associated with Fusarium head blight resistance in wheat. *PLoS One*, *812*. https://doi.org/10.1371/journal.pone.0082079

Zieske, L. R. (2006). A perspective on the use of iTRAQ™ reagent technology for protein complex and profiling studies. *Journal of Experimental Botany*, *57*(7), 1501–1508. https://doi.org/10.1093/jxb/erj168

11 Deciphering Plant–Microbe Crosstalk through Proteomics Tools

*Sampat Nehra, Aarushi Sachdeva, Jinal Paresh Bhavsar,
Erica Zinnia Nehra, Raj Kumar Gothwal, and Purnendu Ghosh*

11.1 INTRODUCTION

Interaction between plants and microbes is a continuous, dynamic, and intricate process. Plants are frequently invaded by both useful and harmful microorganisms, mostly bacteria and fungus, in both natural and farmed settings. A collection of host and non-host species have developed over millions of years as a result of the interaction of plants and microorganisms, creating an ecological entity known as a "holobiont" (1). Since people started relying heavily on cultivated crops for nourishment, the interactions between microbes and plants unquestionably had a significant impact on the evolution of civilisation. The earliest records of famines, plagues and epidemics reveal that some of the most dangerous plant diseases, such as rusts, smuts and mildews, were identified not long after agriculture first took root (2). The processes underlying the plant–microbe interactions are gaining more attention. Plant development under biotic and abiotic conditions depends critically on the activities of these microbial populations (3). To get a better understanding of how plants react and are regulated when they interact with other plants, researchers are examining plant–microbe interactions (4). This might result in the creation of innovative techniques to increase agricultural yields and stress resistance (3). Proteomic methods offer a distinctive perspective to explain these complex interactions between plants and microbes. A group of proteins under particular circumstances may be defined with the aid of proteomics, making it an important tool for developing plant biology. These studies should provide knowledge that may be used to develop plants with enhanced disease resistance and innovative symbiotic relationships (4, 5). Over the last few years, the study of plant proteomes has expanded significantly. Protein identification has become comparatively simple because of advancements in bioinformatics, software and mass spectrometry (MS) technologies (6). Many heritable traits are not encoded by DNA, as has long been recognised, and gene expression is controlled at several levels. For instance, initial rapid changes in cell activity frequently rely on pre-existing proteins that are either post-translationally modified, have altered their subcellular location, or are degrading. Changes in mRNA abundance are therefore not always matched by matching protein levels. Many biological problems can only be answered at the protein level since existence of a gene or its mRNA does not always indicate a function in cellular activity (6).

11.2 DIFFERENT TECHNIQUES IN VOLVED IN PROTEOMICS

We may now investigate the protein composition of cells whose protein synthesis varies depending on the physiological and stress-related situations they are exposed to, thanks to modern proteomics methods (7). Proteomic studies use a variety of techniques, such as protein extraction and separation which uses either a gel-based approach or a non-gel-based one. Two-dimensional gel electrophoresis (2-DE) and differential gel electrophoresis are two methods of protein separation used in gel-based proteomic techniques (DIGE). Isobaric method for relative and absolute quantification (iTRAQ), stable isotope labelling by amino acid in cell culture (SILAC),

DOI: 10.1201/b23255-11

isotope-coded affinity tag (ICAT)and multidimensional protein identification are examples of non-gel-based approaches (MudPIT). After separation, MS or tandem MS is used to identify or quantify the sample. With the aim of identification of a protein and its function, the databases are compared (8).

With the methods like MS and 2-DE being utilised to quantify cellular virulence, extracellular virulence and pathogenicity, the use of proteomics in plant pathology is becoming increasingly widespread. When pathogenic microbes and their symbiotic counterparts attack plant hosts, proteomics can assist in identifying changes at protein levels in those hosts (5).

When studying at the proteome level, it is possible to simultaneously examine the whole proteome that an organism possesses, along with its qualitative presence, localisation, cataloguing, quantitative abundance and population-level variation. Understanding the function of various proteins in growth and development, changes to the proteome caused by biotic and abiotic stimuli, post-translational modifications, and connections with other proteins and molecules may all be aided by the use of proteomics (9).

11.2.1 Techniques Involved in Protein/Peptide Separation

There are several techniques for separating proteins and peptides, the most of which rely on taking advantage of variations in size, charge and/or hydrophobicity. Since resolved proteins are frequently identifiable and can often be further described, proteomic investigations utilising gel-based separation have yielded a large quantity of information. Combining MS technologies with gel-free separation methods typically results in more effective MS technologies for large-scale proteomics.

To distinguish a specific protein from a mixture of proteins, 2D-PAGE proved to be a great technique which uses the two distinct characteristics of proteins—their molecular size and isoelectric point (pI). On a large-scale 2D-gel, up to 5000 protein spots can be found under the suitable conditions. Primarily, proteins are divided based on their isoelectric point via isoelectric focusing. Subsequently, proteins are divided in the second direction using SDS-PAGE based on their size. By using several staining methods of the gel such as silver staining, SYPRO staining or Coomassie Brilliant Blue staining, the separated protein spots may be seen. Sensitive scanners produce digital pictures of 2D-PAGE, which are then examined using programmes like PDQuest, Phoretix, Melanie, Progenesis, Z3/Z4000 (7). For instance, the root proteome of *Medicago truncatula* alters when the arbuscular mycorrhizal (AM) symbiosis is in its early stages. It was examined utilising 2D gel electrophoresis by evaluating the protein patterns obtained from non-inoculated roots and roots synced for *Glomus intraradices* appressorium formation. This method was applied to the mycorrhizal (*TRV25, DMI3*), autoregulation (*TR122, sunn*) and wild-type (J5) genotypes of *M. truncatula*. Further comparisons amongst mutant and wild-type genotypes of the protein groups that reacted to appressorium development revealed minimal overlaps and considerable changes were found, proving that *DMI3* and *SUNN* mutations altered the appressorium-responsive primary proteome (10).

When comparing control and target samples, DIGE proteomics employs 2D gel electrophoresis to examine the differences in protein regulation. In this method, one of two fluorescent dyes (cy2, cy3, cy5, etc.) that are distinct, but structurally similar, is applied to each of the two samples to be examined. Each dye interacts with amino groups, and as a result, the dye binds to lysine as well as the N-terminal amino group, fluorescently labelling each protein (7, 11). Then, the two proteins for comparison are combined, and a single 2D gel is performed. As a result, every protein in a sample overlays with its exact counterpart that has been differently labelled in another sample. If any particular spot is connected to just one dye molecule instead of two, it may be determined via scanning the gel at two separate wavelengths that can excite the two dye molecules. Software created particularly for 2D-DIGE analysis is then used to analyse the obtained photos (7, 11). By comparing multiple of protein samples run under comparable electrophoresis conditions, DIGE benefits include better sensitivity, linearity, the accuracy of quantification and improved reproducibility. Since the DIGE approach permits running an internal standard as part of every separation, even minor abundance variations may be statistically displayed using the pooled internal standard method, finding real differences in protein expressions.

The pooled internal sample can be tagged with a third fluorophore, Cy2, allowing the separation of the two additional samples, tagged with Cy3 and Cy5, on the same gel. The pooled internal standard can also be utilised with only two dyes, allowing for the running of just one experimental sample per gel. However, the latter strategy adds more gels to the experiment yet again (7, 11). For instance, a 2D-DIGE method was employed to detect protein induction in the roots of *M. truncatula* after inoculation with either symbiotic microorganisms (represented by the AM fungi), *Glomus intraradices* and the rhizobia bacterium *Sinorhizobium meliloti*, or with the pathogenic oomycete (*Aphanomyces euteiches*) root pathogen (4, 7, 12).

Although 2-DE is a fully developed and firmly established technology, it is still hindered by issues with quantitative reproducibility and limits the research of specific protein classes. The majority of development efforts in recent years have been concentrated on promising gel-free proteomics (13). A completely new toolset for quantitative analysis has been accessible with the advent of MS-based proteomics. Various fractionation strategies can be used in shotgun proteomics to resolve complex fractions of peptides produced after proteolytic digestion. This allows for high-throughput studies of the proteome of a particular organelle or cell type and gives a review of the main protein components. Despite the fact that these innovative approaches were first promoted as alternatives to gel-based techniques, it is more likely that they should be considered complements to 2-DE but instead of alternatives (13).

The usual shotgun analyses and 2-DE may be compared and contrasted in various ways, including quantitative statistical power, the degree of proteome coverage, sample consumption and iso-form analysis. The biological topic being addressed frequently dictates whether the platform is used because both are capable of resolving lots of characteristics. There is not one approach available right now that can give quantitative and qualitative data on every protein component of a complicated combination (13).

For large-scale shotgun proteomics, gel-free methods for isolating peptides following sequence-specific degradation have become the norm. The majority of gel-free techniques use liquid chromatography in two complementary dimensions (2D LC). Proteome coverage using MS-based peptide sequencing is considerably increased by thorough pre-fractionation of peptide mixtures. There are presently several gel-free-based proteomics methods available (4).

Particularly for the study of complicated protein combinations, MudPIT is employed. Trypsin and endoproteinase lysis are used in this method to digest protein samples in a sequence-specific manner, and the resulting peptide mixtures are then separated using strong cation exchange (SCX) and reverse phase (RP) high-performance liquid chromatography (HPLC). The mass spectrometer receives peptides from the RP column and the MS data is utilised to analyse the protein databases (4, 7, 14). The benefit of this method is that it produces a complete list of all the proteins contained in a specific protein sample while being quick, sensitive and reproducible. MudPIT is utilised effectively in a variety of proteomics investigations since it combines SCX and RP-HPLC separation (4, 7).

The potential of gel-free proteomic techniques for detecting protein alterations in plant-microbe interaction systems has been shown in a number of researches. Plants of *M. truncatula* "Jemalong A17" planted symbiotically with *Sinorhizobium meliloti* strain were subjected to drought stress in order to understand the proteome modifications that occur in nodules under drought stress (15). To assess alterations at the protein level, a non-gel technique based on MS/MS and liquid chromatography was employed. In order to lessen cross-contamination between the fractions due to the complexity of nodule tissue, the segregation of bacteroid and plant fractions in *M. truncatula* root nodules was originally investigated. Profiling the plant protein fraction of *M. truncatula* nodules led to the finding of 377 plant proteins and the greatest presentation of the plant nodule proteome to date. Multivariate data mining made it feasible to classify protein groups that were exhibiting drought stress responses (15).

11.2.2 Mass Spectrometry for Protein Identification

In the study of proteomics, MS is a crucial analytical method (7). The most popular method for unbiased identification of protein is MS, which has been extensively used in plant-microbe proteomics. The three main steps in an MS analysis are ionisation of protein/peptide, ion separation and detection (4). Test materials are transformed into gaseous ions in the mass spectrometer in MS, which are then separated and identified based on their mass-to-charge ratio (m/z). Trypsin is used to fragment proteins or peptides in the case of protein samples. Liquid chromatography is then used to separate the disrupted proteins. The samples are converted into gaseous form by ionisation. The two main methods for ionising proteins and peptides are matrix-assisted laser desorption/ionisation and electrospray ionisation. Different types of mass analysers are used such as quadrapole, time of flight, orbitrap, ion trap, and Fourier transform ion cyclotron resonance (4, 16). MS underwent a revolution with the development of TOF-MS and comparatively nondestructive techniques to turn proteins into gaseous ions.

In the mass analyser, gaseous ions are separated. Both collision-based such as collision-induced dissociation (CID), heated capillary dissociation (HCD), infrared multiphoton dissociation (IRMPD), sustained off-resonance irradiation CID(SORI-CID) and electron-based such as electron transfer dissociation (ETD) and electron capture dissociation (ECD) techniques are used for fragmentation (16). In order to detect peptides, search methods (such as SEQUEST, Mascot and OMSSA) are utilised to compare measured ions to databases that include known protein sequences after data capture (4). For instance, proteomics of plants and microbes has recently seen a surge due to advances in MS technology (4). To compare the reactions of wild-type *M. truncatula* with the ethylene-insensitive mutant sickle, MALDI-TOF/TOF was used (4).

A comparison between supernodulating and non-nodulating variants to wild-type plants was done with the help of a combination technique in which nano-LC-MS/MS was employed with transcriptome analysis to explore the beginning of symbiosis among soybean and *Bradyrhizobium japonicum*. According to these findings, the inhibition of the auto-regulatory mechanisms in the supernodulating type may be brought about by the negative regulation of processes involved in signal transduction and defence (17).

11.2.3 Techniques Involved in Protein Quantification

Stable-isotope labelling and label-free techniques can be used for proteomics quantification (4). Label-free methods for quantification offer a simple solution for vast biological sample analysis. As opposed to label-based approaches, separate injections of materials of interest are made into MS. Label-free quantification is more advantageous than label-based quantitative approaches in a number of ways since it is less costly and does not demand for expensive labelling agents. Additionally, label-free quantification takes less time than certain label-based approaches, which include time-consuming labelling stages (18).

Label-based quantitation compares samples by assigning them distinct differential mass tag, enabling detection based on particular mass changes. This is a comparison technique which in general involves the incorporation of chemically identical but isotopically distinct labels. Thus, the labelling strategy easily reveals both absolute and relative quantification of proteins from individual samples in the same run (18). Stable isotope labelling includes several techniques like ICAT, iTRAQ, tandem mass tag (TMT)and SILAC.

ICATs technique is an *in vitro* technique; the majority of the proteins in two different cell populations can be identified and quantified using this quantitative proteomics technique, which does not need the gel. Iodoacetamide derivatives that have been biotinylated are known as ICATs, and they both interact with the side chains of cysteine present in denatured proteins. Originally, a linker that was available in two versions, i.e., light form (normal) and heavy (deuterated form), was used to join the reactive group and biotin. Deuterium atoms were used in place of hydrogen atoms in the heavy form. Apart from deuterium (heavy isotope of hydrogen; ^{2}H), ICAT makes use of the isotopes ^{13}C and ^{14}N (7, 16, 18).

When two cell populations are pooled and enzymatically digested to produce peptide fragments, some of them are tagged with two distinct ICAT reagents, i.e., one with light and the other with heavy. Avidin affinity chromatography is used to separate the tagged peptides, and microcapillary liquid chromatography-electrospray tandem MS (LC-MS/MS) is then used for analysis. Combining the results from MS with a further examination of ICAT-labelled peptides allows for the automated determination of the relative amounts of the various components of the protein mix (7).

Another in vitro technique for quantification is iTRAQ (Isobaric tags for relative and absolute quantitation) (4). It is a non-gel technique in which proteins from several sources are quantified together in one run. This method makes use of isotope-coded covalent tags, just as ICAT. The differentiating factor between ICAT and iTRAQ is the tagging of proteins. This approach is based on covalently attaching an isotopic tag with a different mass to the N terminus and amine present in the side chain of peptides produced during protein digestion (7). The tags are composed of three regions: balance, reporter and reactive (16). Nano chromatography is used to extract the peptides from the mixture. MS/MS analysis of the isolated peptides is then performed. The tagged peptides and related proteins are then found via a database search based on the fragmentation data (7).

Stable-isotope labelling with amino acids in cell culture (SILAC) is an in vivo technique for the quantification of proteins (4). In SILAC, one group of cells receives isotopically tagged amino acids (for instance, [15N] lysine) in the media, which is then compared to controls receiving regular lysine. Cultures treated with up to five distinct isotopic variants of arginine can be compared using more ambitious methodologies (16).

A mass normalisation area and a mass reporter region are separated by a linker that is prone to fragmentation in the TMT system. The various label forms are formed in the mass reporter and normalisation areas by differential isotopic substitutions (16).

11.2.4 Plant–Microbe Interactions

Microbes (bacteria, virus, fungi, and protozoa) interact with plants. The interaction could be of two types: positive and negative. Plants and microorganisms interact mostly commensally and mutualistically in the rhizosphere. Plants receive water and minerals from ecto- and endo-mycorrhizal fungus in exchange for photosynthate. This mutualistic relationship can be crucial for plant survival in difficult environment. A critical source of combined nitrogen for crops and ecosystems is provided by the interactions of dinitrogen-fixing bacteria with specific plants. Most common bacteria reside on the aerial parts of plants. On the other side, certain bacteria, viruses, protozoa and fungus produce plant infections that can cause significant economic loss and even serious food shortages due to which it is of ecological and economic importance (19).

There are two types of plant–pathogen interaction: compatible and incompatible. Plants are unable to produce effective anti-infectious defensive responses in encounters that are compatible, allowing pathogens to finish their life cycle and cause disease. However, in incompatible interactions, plants successfully launch a series of intricate defensive reactions against harmful interactions which prevent the development of pathogens (20).

11.2.4.1 Interactions between Plants and Bacteria

There are numerous ways in which bacteria and plants can interact. The contact may be advantageous, damaging, or neutral for the plant, and sometimes effects of bacterium might change depending on the soil conditions. The mechanical aspects of plant–bacterial interactions have been clarified in great depth in recent years, but many fundamental concerns regarding these processes still need to be answered. Using proteomics technology, it is possible to learn more about the specific interactions between soil microorganisms and plants. Proteomic techniques have used to identify key proteins expressed during interaction between bacteria and plants (Table 11.1). In this respect, the

TABLE 11.1

Proteomic Techniques Were Used to Identify Some of the Key Proteins Expressed during Interactions between Plants and Bacteria

S. No.	Host Plant	Bacteria	Protein that Appears or Regulates (Upregulates/Downregulates) When Plant and Bacteria Interact	References
1	*Medicago truncatula*	*Sinorhizobium meliloti*	PR10-1, glutathione reductase, etc.	(25)
2	*Glycine max (Soybean)*	*Bradyrhizobium japonicum*	Chaperonin GroEL4, Elongation factor Tu, Malate dehydrogenase, etc.	(26)
3	*Arabidopsis thaliana*	*Pseudomonas syringae* DC3000	GSTs F2, GSTs F6, GSTs F7, GSTs F8, PrxA, PrxB, and PrxIIE	(27)
4	*Lycopersicon hirsutum*	*Clavibacter michiganensis* subsp. *michiganensis*	Glutathione S-transferase, Ascorbate peroxidase, Alkene oxide cyclase, PR10-1, etc.	(28)
5	Sugarcane (SP70-1143 and Chunee)	*Gluconacetobacter diazotrophicus*	SP70-1143—Aconitatehydratase, Enolase, Glutamine synthetase, etc. Chunee—Phosphoketolase, Catalase, Inositol-3-phosphate synthase, etc.	(29)
6	*Oryzasativa*	*Xanthomonas oryzae* pv. *oryzae*	Ribose-5-phosphate isomerase, Peroxiredoxin, RuBisCo LSU, etc.	(30)
7	*Solanum lycopersicum*	*Ralstonia solanacearum*	Fructokinase-2, nucleoside diphosphate kinase, catalase, enolase, etc.	(31)

symbiotic relationship between legumes and nitrogen-fixing bacteria has been investigated in the greatest depth of all the proteome studies of plant–bacterial interactions. Significant emphasis has also been paid to studies of the connections between the proteome of plants and pathogens (21).

An example of negative interaction between plant and bacteria is observed between *Xanthomonas oryzae* pv. *oryzae* and rice plant. *X. oryzae* pv. *oryzae* is a significant plant pathogen of rice plant that causes a disease known as bacterial leaf blight. Globally, considerable yield losses are caused by Xoo (*X. oryzae* pv. *oryzae*) because it colonises and infects vascular tissue, causing wilting and tissue necrosis. The xylem sap, which was infected with Xoo, was retrieved and its released Xoo protein content was examined. Three different assays identified a total of 324 different proteins. Out of 324, 64 of these proteins, which included a large number of known virulence-related components, were found in all the samples. Additionally, ten of the discovered protein-coding genes were inactivated, and one mutant showed statistically significant reductions in virulence as compared to the wild-type Xoo, indicating the discovery of a new virulence-associated factor. The technique used for the analysis of the xylem sap samples was MS, in which MALDI-TOF/TOF was used for fragmentation and then the peptide digests underwent LC-MALDI. Once the peptides were separated, the analysis by tandem MS of top ten precursor ions from each spot was performed (22).

Yeast two-hybrid (Y2H) tests were used to build an interactome of 100 proteins concentrating on important regulators of biotic and abiotic stress responses of rice plants, and the signalling networks that underlie stress response were better understood. Protein–protein interaction (PPI) tests, transcript co-expression, and phenotypic analyses were used to verify the interactome. Ten novel stress tolerance regulators, including two from protein classes not previously recognised to play a role in stress responses, were found using this interactome-guided prediction and phenotype validation. Cross-talk between biotic and abiotic stress responses is supported by numerous lines of evidence. The combination of focused interactome and systems investigations offered substantial advancement in understanding the molecular basis of characteristics that are crucial for agronomy (23).

Twenty distinct protein spots involved in defence, metabolism, energy and protein synthesis were identified by analysis of samples of proteins from rice leaves infected with either compatible or incompatible Xoo strains. Ten of these proteins, including ATP synthase, the RuBisCO (large subunit), and peroxiredoxin, were identified as being particular to the compatible association (9).

An example of positive interactions between plant and bacteria can be seen between *Kosakonia radicincitans* DSM 16656 and *Arabidopsis thaliana*. Plant physiology gets impacted when endophytic plant growth-promoting bacteria interact with the plant. It is quintessential to understand this kind of interaction at a molecular level as endophytic plant growth-promoting bacteria bolsters sustainable production systems and crop productivity. When *K. radicincitans* DSM 16656 interacted with *A. thaliana*, the molecular processes underpinning plant growth promotion were examined using a proteomics technique. Using 2-DE, the proteome of roots from infected and control plants was examined four weeks after the inoculation, and differently abundant protein spots were found using LC-MS/MS. There were 12 protein sites that responded to the inoculation, with cellular stress responses being the main cause of most of them. *K. radicincitans* was present, and this boosted the protein expression of the 20S proteasome alpha-3 subunit. Out of 12, 2 spots were having multiple proteins and the quantification of such spots is not completely reliable. Therefore, further investigation was not done. However, further investigation was done for the rest of the spots. Out of eight proteins, six spots were related to stress responses and identified as Heat-shock 70 kDa protein 14, Heat-shock 20 kDa like chaperone, glutathione S-transferase F7, a universal stress protein, Heat-shock 70 kDa protein 10 and lipase/lipoxygenase. Furthermore, two distinctly expressed spots were observed to be related with protein metabolism and identified as Elongation factor 1-beta 2 and 20S proteasome alpha-3 subunit, and two remaining spots were identified as fructose bisphosphate aldolase and ferredoxin nitrite reductase (24).

11.2.4.2 Interactions between Plants and Fungi

Plants and fungi interact closely. Fossil data shows that fungus and plants have coexisted in harmony for 400 million years. Almost all plants in natural ecosystems have been discovered to be infected by one or more bacterial and fungal symbionts. Contrary to parasites, which spread illness, endophytes—fungi that grow inside plant tissue—provide diverse fitness benefits to the plants. According to some reports, endophytes can turn into parasites and vice versa depending on the situation. Previously, it was believed that the symbiotic relationships between fungi and other organisms could only take the form of mutualism, commensalism or parasitism. Recent research, however, suggested that the fungi might show various lifestyles in response to signals from inside the host or outside influences (32). Proteomic techniques have utilised to identify key proteins expressed during interaction between bacteria and fungi (Table 11.2).

Further, fungi are divided into three groups: necrotrophic, hemibiotrophic, and biotrophic, depending on the type of parasitism and infection technique. The latter obtains nutrients from the plant but does not kill it, whereas the former gets its nutrients from dead cells. Hemibiotrophs alternate between a necrotrophic and a biotrophic mode of nutrition. Biotrophic species typically show a high degree of specialisation for certain plant species, in contrast to necrotrophic species, which typically attack a wide variety of plant species. The majority of biotrophic fungi are obligate parasites that only have limited saprophytic stages (33).

Biotrophic fungi have evolved to grow in living host tissue with the least amount of induction of host defence. Members of this category must resist or eliminate host defences and alter the metabolism of their hosts in order to complete their life cycles. Some fungi also establish close intracellular connections with living plant cells, frequently through mutualistic relationships, including the control of the host defences and metabolism. When virulence-effector proteins are released into host cells, the physiology of the host is changed, which helps biotrophic fungus flourish and survive in the host environment (34).

For instance, the vascular wilt fungus *Fusarium oxysporum* was discovered to cause a significant alteration in the protein content of tomato (*Lycopersicon esculentum*) xylem sap. The

TABLE 11.2

Proteomic Techniques Were Used to Identify Some of the Key Proteins Expressed during Interactions between Plants and Fungi

S. No.	Host Plant	Fungus	Protein that Appears or Regulates When Plant and Fungi Interact	References
1	Chickpea (*Cicer arietinum* L.)	*Fusarium oxysporum* f. sp. *ciceri* Race 1 (Foc1)	Superoxide dismutase, Pathogenesis-related protein 1, Trypsin protein inhibitor3, Guanine nucleotide binding protein, etc.	(37)
2	*Pinus strobus*	*Cronartium ribicola*	Transmembrane receptor kinase, etc.	(38)
3	*Brassica napus* and *B. carinata*	*Alternaria brassicae*	Cinnamyl alcohol dehydrogenase, Putative dehydroascorbate reductase, Triosephosphate isomerase, etc.	(39)
4	*Pisum sativum*	*Erysiphe pisi*	RuBisCo activase, Isocitrate dehydrogenase, Resistance protein RGC2, Ras-related protein Rab-2-B, etc.	(40)
5	Rice (*Oryzasativa* L.)	*Rhizoctonia solani*	β-1,3-glucanase, RuBisCo large subunit, 20S proteasome beta subunit, Putative 26S proteasome non-ATPase regulatory subunit 14, etc.	(41)
6	Wheat (*Triticum aestivum*)	*Fusarium graminearum*	Chloroplast OEE1, glyceraldehyde-3-phosphate dehydrogenase, 20 kDa chaperonin, superoxide dismutase, etc.	(42)
7	*Gossypium barbadense*	*Verticillium dahlia*	Ribosomal protein, Glycolipid transfer protein 1, Isopentyl-diphosphate isomerase, etc.	(43)

most prevalent proteins that appeared during compatible or incompatible interactions were found using peptide mass fingerprinting and mass spectrometric sequencing. Identification of xylem sap proteins were performed, in this technique that was used is MS, to generate sequence data in order to determine whether the disease-related proteins in the xylem sap are the same as proteins previously found in other tomato-pathogen relationships or are yet unknown proteins produced by either plant or fungus. Trypsin was used to digest proteins in gel, and a matrix-assisted laser desorption ionisation time of flight (MALDI-TOF) mass-spectrometer was used to collect the peptide mass spectra (a peptide mass fingerprint). Following that, databases were searched for connection to anticipated tryptic digests of known proteins using the list of apparent peptide masses. Individual peptides were chosen for tandem MS sequence analysis when sufficient material could be acquired, either to confirm a potential identification or to produce "sequence tags" permitting further database searches. It was discovered that a new PR-5 family member accumulated early in both kinds of interaction. Other pathogenesis-related proteins only showed interactions that were compatible with the development of the disease. This shows how plants and pathogens interact in vascular wilt illnesses by using proteomics to discover novel proteins in xylem sap (35).

Another such example is the interaction between *Oryza sativa* L. (rice plant) and *Magnaporthe grisea*. If a rice plant recognises the host resistance gene of infecting pathogens early, it results in a hypersensitive response (HR), leading to instant and efficient defence responses such as the formation of phytoalexins, pathogenesis-related proteins and oxidative enzymes. This is termed incompatible interaction, whereas the interaction is called compatible when the rice plant is unable to prevent the pathogen entry causing rice blast disease. The molecular identification of host genes involved in the response to defence against *M. grisea* was investigated via the suppression subtractive hybridisation method. Differential modifications in protein levels from leaf blades of rice plants cultivated in nitrogen-starved conditions and infected with *M. grisea* were examined in the fungal pathogen proteome by 2-DE. Forty-five out of 63 proteins were observed to be involved in the fungal pathogen proteomes. PR-5 was the only member of the PR protein family that was stimulated. Because there are plenty of RuBisCo proteins in leaves, it is still quite challenging to isolate rare proteins that may be differentially expressed. To resolve this problem, a PEG-mediated prefractionation technique was used. This method assisted in enhancing the proteins found in leaves that are less abundant. The proteome study of rice suspension-cultured cells inoculated with rice blast fungus or treated with fungal elicitor and jasmonic acid was carried out to discover pathogen-induced proteins. A variety of pathogen-induced proteins, including probenazole-inducible protein, pathogenesis-related class 10, isoflavone reductase and salt-induced protein were discovered using 2-DE and N-terminal protein sequencing. The rice blast fungus and abiotic stressors such as salicylic acid, abscisic acid and jasmonic acid treatment were given to the leaves to distinguishably reveal plant self-defence response proteins following PEG-mediated pre-fractionation. Eight defence-related proteins that were produced differently in rice leaf blades infected with rice blast fungus were discovered using MALDI-TOF analysis (36).

11.2.4.3   Interactions between Plants and Viruses

Many important crop diseases are caused by viruses, which are microscopic pathogens. They cause large economic losses in crop quality and productivity in many regions of the world. Single-stranded ribonucleic acid (ssRNA) is found in the majority of plant viruses; however, single- or double-stranded DNA can also be found. Because they are obligate parasites, they are dependent on host machinery for reproduction. They enter plant cells passively through wounds brought on by physical trauma, environmental conditions, or the vectors themselves, which can include insects, nematodes, fungi, and even mites. Plants also evolve specific defensive systems in response to the viral infection. Because a virus has to be able to discover the right supporting host factors and must be able to avoid the host defensive responses, the majority of plant–virus interactions do not always result in successful infections. Even while a virus has the ability to reproduce, it may not always be harmful. The outcome of the infection is the consequence of several interactions between the virus and host components involved in the reproduction of plant viruses, which can have either positive or negative effects. The host defensive mechanisms against the infectious virus and the results of viral replication in the host are both variables that might result in pathogenesis. As a result, just one viral component may be the primary cause of a pathogenic process (44).

Proteomic research on proteins engaged in host-viral PPIs and differentially expressed proteins during viral infection can shed light on the intricate and varied molecular interactions that occur between plants and viruses. Many methods have been used to investigate the purpose of proteins that are encoded by viruses (9). Proteomic techniques have proved useful to identify key proteins expressed during interaction between bacteria and viruses (Table 11.3).

There are multiple examples where negative plant–virus interactions could be observed. For instance, a significant viral infection that has caused major losses in grain and pasture yield is the sugarcane mosaic virus (SCMV). Proteomic studies on leaf samples from both resistant and susceptible ecotypes of maize were done. Both the samples were infected with SCMV in order to discover potential SCMV resistance proteins and investigate the molecular processes underlying the plant-SCMV interaction. Quantitative 2D-DIGE was used to evaluate proteins, and 93 protein spots following viral inoculation revealed statistically significant variations. SCMV-responsive proteins were predominantly connected to metabolism and energy, photosynthesis, stress and defence reactions, and carbon fixation, according to functional classification (45).

There are countless examples where negative plant–virus interactions could be seen. As an illustrative example, a developing viral pathogen in regions where cucurbits are produced is the zucchini yellow mosaic virus (ZYMV). Infection reduces production significantly in a number of species of Cucurbitaceae. Two cultivars of *Cucurbita pepo* (i.e., Zelena susceptible and Jaguar resistant) were

TABLE 11.3

Proteomic Techniques Were Used to Identify Some of the Key Proteins Expressed during Interactions between Plants and Viruses

S. No.	Host Plant	Virus	Protein that Appears or Regulates When Plant and Fungi Interact	References
1	*Solanum lycopersicum* (Tomato)	*Cucumber mosaic virus*	RuBisCo large subunit, Photosystem II 23 kDa protein, Carbonic anhydrase, RuBisCo activase, etc.	(47)
2	*Lycopersicon esculentum* M.	*Tobacco mosaic virus*	Ascorbate peroxidase, Chitinase, GSH-dependent dehydroascorbate reductase, DIP-1 peptidase, etc	(48)
3	*Nicotiana tabacum*	*T. mosaic virus*	Polyphenol oxidase, Cytosolic ascorbate peroxidase, Protein disulfide isomerase, Inositol-3-phosphate synthase, etc.	(49)
4	Peach leaves	*Plum pox virus*	Thaumatin-like protein, Mandelonitrile lyase, etc.	(50)
5	Rice plant	*Rice stripe virus*	Glutamyl-Trna reductase, Oxygen-evolving enhancer protein 3, Magnesium-chelatase subunit ChlD, Porphobilinogen deaminase, etc.	(51)
6	*Oryza sativa*	*Rice yellow mottle virus*	Chaperonin CPN60-2, mitochondrial precursor, Chaperonin CPN60-1, mitochondrial precursor, fructose-bisphosphate aldolase, etc.	(52)

examined with the help of a 2DE-based proteomic method to identify proteins that may be resistant to infection caused by the severe ZYMV-H strain. The susceptible *C. pepo* cv. Zelena developed the first sign (i.e., clearing veins) on leaves after six to seven days of inoculation (dpi). On the leaves of the partially resistant *C. pepo* cv. Jaguar, similar symptoms arose; however, the only difference was that the symptoms appeared after 15 dpi. Immunoblot analysis, which revealed increased viral protein levels at 6 dpi in the susceptible cultivar, supported this finding. Twenty-eight and 31 sites were found to be differentially abundant amongst cultivars at 6 and 15 dpi, respectively, according to leaf proteome analysis. The early infection variation might result from the quick activation of redox homeostasis-related proteins of the partially resistant cultivar. The cytoskeleton and photosynthesis are altered in the proteome of susceptible cultivar (46).

11.3 CONCLUSION AND FUTURE PERSPECTIVE

There is no better way to comprehend the biological mechanisms behind the relationship between plants and microbes than to examine proteins, which are directly responsible for every cellular function. Protein identification is made considerably easier by technological advancements in MS, techniques of bioinformatics, and the growing number of sequenced genomes. As more genomes are sequenced, protein identification will improve, making it possible to use peptide mass fingerprinting in more plants, diseases and/or beneficial microorganisms. The combination of proteomics with genomics data, traditional plant pathology knowledge, and advances in genetic engineering techniques, such as the CRISPR-Cas system, will eventually help to develop new disease-resistant varieties of plants or crops studied with beneficial microorganisms that exhibit enhanced resistance. According to the current strategy, comparative proteomics will be the method of choice for researchers from all over the world for analysing the interactions between plants and microbes.

Although model plants have been used to make significant progress in plant–microbe proteomics, comparable strategies are now being developed for crops. Improved agricultural techniques, such as proteomics-based fungicides, will result from more knowledge of how these plants react to symbiotic and harmful microbes. The lack of sequencing data and comprehensive protein databases is the main barrier to proteomic application. While methods like "proteogenomics" and "de novo sequencing" make up for this shortcoming, there is still a pressing need to build and maintain plant protein databases. In the future, it is important to extend and combine a number of current databases, such as LegProt, the plant proteome database, PlProt, the post-translational modification database, the *Medicago* phosphoprotein database, and the rice proteome database. Plant defence and symbiosis-induced responses are now well understood because of proteomic investigations of plant–microbe interactions. However, it is noteworthy how little research utilising quantitative and in vivo proteomic approaches has been published. Additionally, in investigations that do include quantification, additional biological duplicates should be carried out rather than simply technical ones. Furthermore, even though most proteomic studies identify proteins and anticipate their functions, most of them do not verify their ideas using genetics even when the techniques and tools are available. Although it is yet uncommon, coupling proteomic analysis with genomics and other omics methods would increase the biological importance of many investigations. It will be possible to predict and manipulate response of plants to symbiotic and harmful bacteria more accurately with the use of beneficial information that results from the more implementation of these complementary techniques.

REFERENCES

1. Dolatabadian, A. (2020). Plant-microbe interaction. *Biology*, *10*(1), 15. https://doi.org/10.3390/biology10010015
2. Jackson, A. O., & Taylor, C. B. (1996). Plant-microbe interactions: Life and death at the interface. *The Plant Cell*, *8*(10), 1651–1668. https://doi.org/10.2307/3870220
3. Kafle, A., Garcia, K., Peta, V., Yakha, J., & Bücking, A. S. H. (2018). Beneficial plant microbe interactions and their effect on nutrient uptake, yield, and stress resistance of soybeans. In M. Kasai (Ed.), *Soybean - Biomass, Yield and Productivity*. IntechOpen. https://doi.org/10.5772/intechopen.81396
4. Jayaraman, D., Forshey, K. L., Grimsrud, P. A., & Ané, J. M. (2012). Leveraging proteomics to understand plant-microbe interactions. *Frontiers in Plant Science*, *3*, 44. https://doi.org/10.3389/fpls.2012.00044
5. Kav, N. N., Srivastava, S., Yajima, W., & Sharma, N. (2007). Application of proteomics to investigate plant-microbe interactions. *Current Proteomics*, *4*(1), 28–43.
6. Quirino, B. F., Candido, E. S., Campos, P. F., Franco, O. L., & Krüger, R. H. (2010). Proteomic approaches to study plant–pathogen interactions. *Phytochemistry*, *71*(4), 351–362.
7. Lodha, T. D., Hembram, P., Tep, N., & Basak, J.. (2013). Proteomics: A successful approach to understand the molecular mechanism of plant-pathogen interaction. *American Journal of Plant Sciences*, 4(6). https://doi.org/10.4236/ajps.2013.46149
8. Singh, P., Pitambara, R. R., Dev, D., & Maharshi, A. (2018). Proteomics approaches to study host pathogen interaction. *Journal of Pharmacognosy and Phytochemistry*, *7*(4), 1649–1654.
9. Jain, A., Singh, H. B., & Das, S. (2021). Deciphering plant-microbe crosstalk through proteomics studies. *Microbiological Research*, *242*, 126590.
10. Amiour, N., Recorbet, G., Robert, F., Gianinazzi, S., & Dumas-Gaudot, E. (2006). Mutations in DMI3 and SUNN modify the appressorium-responsive root proteome in arbuscular mycorrhiza. *Molecular Plant-Microbe Interactions*, *19*(9), 988–997.
11. Van den Bergh, G., & Arckens, L. (2005). Recent advances in 2D electrophoresis: An array of possibilities. *Expert Review of Proteomics*, *2*(2), 243–252.
12. Schenkluhn, L., Hohnjec, N., Niehaus, K., Schmitz, U., & Colditz, F. (2010). Differential gel electrophoresis (DIGE) to quantitatively monitor early symbiosis-and pathogenesis-induced changes of the *Medicago truncatula* root proteome. *Journal of Proteomics*, *73*(4), 753–768.
13. Abdallah, C., Dumas-Gaudot, E., Renaut, J., & Sergeant, K. (2012). Gel-based and gel-free quantitative proteomics approaches at a glance. *International Journal of Plant Genomics*, *2012*, 494572. https://doi.org/10.1155/2012/494572
14. Baggerman, G., Vierstraete, E., De Loof, A., & Schoofs, L. (2005). Gel-based versus gel-free proteomics: A review. *Combinatorial Chemistry & High Throughput Screening*, *8*(8), 669–677.

15. Larrainzar, E., Wienkoop, S., Weckwerth, W., Ladrera, R., Arrese-Igor, C., & González, E. M. (2007). *Medicago truncatula* root nodule proteome analysis reveals differential plant and bacteroid responses to drought stress. *Plant Physiology, 144*(3), 1495–1507.
16. Twyman, R. (2013). *Principles of Proteomics.* Taylor & Francis, Oxfordshire, UK
17. Salavati, A., Bushehri, A. A., Taleei, A., Hiraga, S., & Komatsu, S. (2012). A comparative proteomic analysis of the early response to compatible symbiotic bacteria in the roots of a supernodulating soybean variety. *Journal of Proteomics, 75*(3), 819–832. https://doi.org/10.1016/j.jprot.2011.09.022
18. Anand, S., Samuel, M., Ang, C. S., Keerthikumar, S., & Mathivanan, S. (2017). Label-based and label-free strategies for protein quantitation. In *Proteome Bioinformatics* (pp. 31–43). Springer Science+Business Media, New York city, USA.
19. Atlas, R. M. (1998). *Microbial Ecology: Fundamentals and Applications.* Pearson Education India, Bengaluru, Karnataka, India.
20. Ponzio, C., Weldegergis, B. T., Dicke, M., & Gols, R. (2016). Compatible and incompatible pathogen–plant interactions differentially affect plant volatile emissions and the attraction of parasitoid wasps. *Functional Ecology, 30*(11), 1779–1789.
21. Cheng, Z., McConkey, B. J., & Glick, B. R. (2010). Proteomic studies of plant–bacterial interactions. *Soil Biology and Biochemistry, 42*(10), 1673–1684.
22. González, J. F., Degrassi, G., Devescovi, G., De Vleesschauwer, D., Höfte, M., Myers, M. P., & Venturi, V. (2012). A proteomic study of *Xanthomonas oryzae* pv. *oryzae* in rice xylem sap. *Journal of Proteomics, 75*(18), 5911–5919. https://doi.org/10.1016/j.jprot.2012.07.019
23. Seo, Y. S., Chern, M., Bartley, L. E., Han, M., Jung, K. H., Lee, I., Walia, H., Richter, T., Xu, Xia., Cao, P., Bai, Wei., Ramanan, R., Amonpant, F., Arul, L., Canlas, P. E., Ruan, R., Park, C.-J., Chen, X., Hwang, S., Jeon, J.-S., & Ronald, P. C. (2011). Towards establishment of a rice stress response interactome. *PLoS Genetics, 7*(4), e1002020. https://doi.org/10.1371/journal.pgen.1002020
24. Witzel, K., Üstün, S., Schreiner, M., Grosch, R., Börnke, F., & Ruppel, S. (2017). A proteomic approach suggests unbalanced proteasome functioning induced by the growth-promoting bacterium *Kosakonia radicincitans* in Arabidopsis. *Frontiers in Plant Science, 8*, 661.
25. Larrainzar, E., Wienkoop, S., Weckwerth, W., Ladrera, R., Arrese-Igor, C., & González, E. M. (2007). *Medicago truncatula* root nodule proteome analysis reveals differential plant and bacteroid responses to drought stress. *Plant Physiology, 144*(3), 1495–1507.
26. Nomura, M., Arunothayanan, H., Van Dao, T., LE, H. T. P., Kaneko, T., Sato, S., & Tajima, S. (2010). Differential protein profiles of *Bradyrhizobium japonicum* USDA110 bacteroid during soybean nodule development. *Soil Science & Plant Nutrition, 56*(4), 579–590.
27. Jones, A. M., Thomas, V., Truman, B., Lilley, K., Mansfield, J., & Grant, M. (2004). Specific changes in the Arabidopsis proteome in response to bacterial challenge: Differentiating basal and R-gene mediated resistance. *Phytochemistry, 65*(12), 1805–1816.
28. Coaker, G. L., Willard, B., Kinter, M., Stockinger, E. J., & Francis, D. M. (2004). Proteomic analysis of resistance mediated by Rcm 2.0 and Rcm 5.1, two loci controlling resistance to bacterial canker of tomato. *Molecular Plant-Microbe Interactions, 17*(9), 1019–1028.
29. Lery, L. M., Hemerly, A. S., Nogueira, E. M., von Krüger, W. M., & Bisch, P. M. (2011). Quantitative proteomic analysis of the interaction between the endophytic plant-growth-promoting bacterium *Gluconacetobacter diazotrophicus* and sugarcane. *Molecular Plant-Microbe Interactions, 24*(5), 562–576.
30. Mahmood, T., Jan, A., Kakishima, M., & Komatsu, S. (2006). Proteomic analysis of bacterial-blight defense-responsive proteins in rice leaf blades. *Proteomics, 6*(22), 6053–6065.
31. Dahal, D., Pich, A., Braun, H. P., & Wydra, K. (2010). Analysis of cell wall proteins regulated in stem of susceptible and resistant tomato species after inoculation with *Ralstonia solanacearum*: A proteomic approach. *Plant Molecular Biology, 73*(6), 643–658.
32. Rai, M., & Agarkar, G. (2016). Plant–fungal interactions: What triggers the fungi to switch among life-styles? *Critical Reviews in Microbiology, 42*(3), 428–438.
33. Mathesius, U. (2009). Comparative proteomic studies of root–microbe interactions. *Journal of Proteomics, 72*(3), 353–366.
34. Rafiqi, M., Ellis, J. G., Ludowici, V. A., Hardham, A. R., & Dodds, P. N. (2012). Challenges and progress towards understanding the role of effectors in plant–fungal interactions. *Current Opinion in Plant Biology, 15*(4), 477–482.
35. Rep, M., Dekker, H. L., Vossen, J. H., de Boer, A. D., Houterman, P. M., Speijer, D., Back, J. W., Koster C. G. De, & Cornelissen, B. J. C. (2002). Mass spectrometric identification of isoforms of PR proteins in xylem sap of fungus-infected tomato. *Plant Physiology, 130*(2), 904–917.

36. Kim, S. T., Kim, S. G., Hwang, D. H., Kang, S. Y., Kim, H. J., Lee, B. H., Lee, J. J., & Kang, K. Y. (2004). Proteomic analysis of pathogen-responsive proteins from rice leaves induced by rice blast fungus, *Magnaporthe grisea*. *Proteomics, 4*(11), 3569–357.

37. Chatterjee, M., Gupta, S., Bhar, A., Chakraborti, D., Basu, D., & Das, S. (2014). Analysis of root proteome unravels differential molecular responses during compatible and incompatible interaction between chickpea (*Cicer arietinum* L.) and *Fusarium oxysporum* f. sp. *ciceri* Race1 (Foc1). *BMC Genomics, 15*(1), 1–19.

38. Smith, J. A., Blanchette, R. A., Burnes, T. A., Jacobs, J. J., Higgins, L., Witthuhn, B. A., & Gillman, J. H. (2006). Proteomic comparison of needles from blister rust-resistant and susceptible *Pinus strobus* seedlings reveals upregulation of putative disease resistance proteins. *Molecular Plant-Microbe Interactions, 19*(2), 150–160.

39. Sharma, N., Rahman, M. H., Strelkov, S., Thiagarajah, M., Bansal, V. K., & Kav, N. N. (2007). Proteome-level changes in two Brassica napus lines exhibiting differential responses to the fungal pathogen *Alternaria brassicae*. *Plant Science, 172*(1), 95–110.

40. Curto, M., Camafeita, E., Lopez, J. A., Maldonado, A. M., Rubiales, D., & Jorrín, J. V. (2006). A proteomic approach to study pea (*Pisum sativum*) responses to powdery mildew (*Erysiphe pisi*). *Proteomics, 6*(S1), S163–S174.

41. Lee, J., Bricker, T. M., Lefevre, M., Pinson, S. R., & Oard, J. H. (2006). Proteomic and genetic approaches to identifying defence-related proteins in rice challenged with the fungal pathogen *Rhizoctonia solani*. *Molecular Plant Pathology, 7*(5), 405–416.

42. Zhang, X., Fu, J., Hiromasa, Y., Pan, H., & Bai, G. (2013). Differentially expressed proteins associated with *Fusarium* head blight resistance in wheat. *PLoS One, 8*(12), e82079.

43. Wang, F.-X., Ma, Y.-P., Yang, C.-L., Zhao, PI-M., Yao, Y., Jian, G.-L., Luo, Y.-M., & Xia, G.-X. (2011). Proteomic analysis of the sea-island cotton roots infected by wilt pathogen *Verticillium dahliae*. *Proteomics, 11*(22), 4296–4309.

44. Yadav, S., & Chhibbar, A. K. (2018). Plant–virus interactions. In *Molecular Aspects of Plant-Pathogen Interaction* (pp. 43–77). Springer, Singapore.

45. Wu, L., Han, Z., Wang, S., Wang, X., Sun, A., Zu, X., & Chen, Y. (2013). Comparative proteomic analysis of the plant–virus interaction in resistant and susceptible ecotypes of maize infected with sugarcane mosaic virus. *Journal of Proteomics, 89*, 124–140.

46. Nováková, S., Flores-Ramírez, G., Glasa, M., Danchenko, M., Fiala, R., & Skultety, L. (2015). Partially resistant *Cucurbita pepo* showed late onset of the Zucchini yellow mosaic virus infection due to rapid activation of defense mechanisms as compared to susceptible cultivar. *Frontiers in Plant Science, 6*, 263.

47. Di Carli, M., Villani, M. E., Bianco, L., Lombardi, R., Perrotta, G., Benvenuto, E., & Donini, M. (2010). Proteomic analysis of the plant–virus interaction in cucumber mosaic virus (CMV) resistant transgenic tomato. *Journal of Proteome Research, 9*(11), 5684–5697.

48. Casado-Vela, J., Sellés, S., & Martínez, R. B. (2006). Proteomic analysis of tobacco mosaic virus-infected tomato (*Lycopersicon esculentum* M.) fruits and detection of viral coat protein. *Proteomics, 6*(S1), S196–S206.

49. Caplan, J. L., Zhu, X., Mamillapalli, P., Marathe, R., Anandalakshmi, R., & Dinesh-Kumar, S. P. (2009). Induced ER chaperones regulate a receptor-like kinase to mediate antiviral innate immune response in plants. *Cell Host& Microbe, 6*(5), 457–469.

50. Díaz-Vivancos, P., Rubio, M., Mesonero, V., Periago, P. M., Ros Barceló, A., Martínez-Gómez, P., & Hernández, J. A. (2006). The apoplastic antioxidant system in Prunus: Response to long-term plum pox virus infection. *Journal of Experimental Botany, 57*(14), 3813–3824.

51. Wang, B., Hajano, J. U., Ren, Y., Lu, C., & Wang, X. (2015). iTRAQ-based quantitative proteomics analysis of rice leaves infected by rice stripe virus reveals several proteins involved in symptom formation. *Virology Journal, 12*(1), 1–21.

52. Delalande, F., Carapito, C., Brizard, J. P., Brugidou, C., & Van Dorsselaer, A. (2005). Multigenic families and proteomics: Extended protein characterization as a tool for paralog gene identification. *Proteomics, 5*(2), 450–460.

12 Dissecting Plant–Virus Interactions through Proteomics

Nidhi Choudhary, Sampat Nehra, and Raj Kumar Gothwal

12.1 INTRODUCTION

Plant viruses are able to interact and code only with specific host proteins of the plant to infect it. Hence, they are dependent on host plant cells by co-opting such host components during the infection cycle. The host components which get subverted during the viral infection include subcellular membrane host proteins and metabolites. To cite an example, the translation of the viral genomic RNA of plant RNA viruses is performed by the host ribosomes (Wang et al., 2009). The virus-infected host proteins also facilitate the multiplication of the viral genome and transportation of viral components in host plant cells to neighboring tissues, as well as RNA recombination (Barajas et al., 2008). Therefore, to fully understand the mechanism of virus replication and its responses by the defense system of the host cell against the infecting virus, it is very important to dissect virus–host interactions at the molecular level. Researching and understanding this particular aspect will help in the efficient development of new more potent antiviral strategies to overcome the damage caused by invading viruses on the plant cells.

There are a wide range of methods that can be applied for either monitoring the changes caused in a plant cell due to virus infection or to identify certain specific host proteins that establish interaction with the viral protein or with the viral RNAs. For this monitoring, the proteomics method has more advantages than the genomics approach as they can directly measure protein levels in cells which may not be reflected by mRNA levels because of various regulations of translation, different half-lives of post-translational modifications (PTMs) of proteins, and subcellular localization of proteins in the cell (Irish et al., 2006). However, the proteomics method has certain disadvantages like the difficulty faced in identifying minor protein components due to limited sensitivity of the detection methods and underestimation of those proteins that interact with other components only transiently. However, these limitations are becoming less prevalent with the development of newer, more sophisticated and sensitive proteomics approaches with every passing year (Newman et al., 2006).

Till two decades ago, the application of proteomics approaches in plant virus research used two-dimensional gel electrophoresis (2-DE) (Zhang et al., 2010) and a small antibody library of pathogenesis-related proteins (PRs) to characterize infected plants, to dissect virus–plant interactions. However, later on, genome sequencing of different organisms became available (Yu et al., 2002) in combination with genetic, biochemical and bioinformatics tools, which led to the creation of more advanced and sophisticated host protein functional interaction databases (Jensen et al., 2009). Further advances have also been achieved with the introduction of mass-spectrometric analysis (Casado-Vela et al., 2006), and protein array approaches to screen for host proteins that bind to viral components (Zhu et al., 2007). Proteomics-based studies on the host plants can give us a clearer understanding of how a particular viral infection affects the expression profile of the host proteome. In return, these data sets could also be useful to identify proteins involved in defensive responses as well as damage control to protect the cells. Overall, these advances using proteomics have already led to a greatly improved understanding of virus–plant interactions, as described in this chapter.

DOI: 10.1201/b23255-12

12.2 VARIOUS SEPARATION TECHNIQUE   .

Proteomics field is a collection of technologies for the purpose of identification and quantification of overall proteins present in a cell, tissue or an organism. There are five main technologies which are used under proteomics; they are conventional techniques, advanced techniques, quantitative techniques, high-throughput techniques and bioinformatics analysis technologies.

12.3 THE CONVENTIONAL TECHNOLOGIES OF PROTEOMICS

The conventional techniques for identifying and quantifying different protein structures work by performing purification of proteins using chromatography-based separation such as ion exchange chromatography (IEX), size exclusion chromatography (SEC) and affinity chromatography. Apart from this, there are also two other conventional techniques which are enzyme-linked immunosorbent assay (ELISA) and Western blotting techniques.

The ion exchange chromatography (IEX) is a protein separation method on the basis of the net charge of the protein as proteins carry a net charge depending on amino acid composition and any covalently attached modifications. IEX works by identifying two types using cation exchange and anion exchange chromatography. The principle of separation is to allow reversible exchange of ions between the cationic or anionic ions present in ion exchangers and the target ions present in the sample solution. The SEC identifies target proteins by separating the proteins using a porous carrier matrix with different pore size on the basis of permeation. The proteins are then separated on the basis of their differing molecular size. The last chromatography technique is the affinity chromatography which involves separation and identification of proteins through reversible interaction between the affinity ligand of chromatographic matrix and the proteins to be purified.

Another commonly used conventional technique is ELISA which is performed to detect and measure peptides and proteins in the plant cells. The plant defense mechanism produces antibodies which are proteins produced in response to any particular virus. ELISA techniques work by complexing antibodies and viruses together to produce a measurable result of protein expression levels.

The last conventional technique is the Western blotting procedure. This process works by detecting a single, specific protein within the complex mixture extracted as a sample from cells or tissues. Western blotting method not only detects the presence or absence of a protein but also indicates whether proteins are being upregulated or downregulated in a system. It also detects protein modifications and quantifies protein expression levels relative to standard. (https://academic.oup.com/chromsci/article/55/2/182/2333796).

12.4 ADVANCED TECHNOLOGIES OF PROTEOMICS

The advanced technologies of proteomics involve methods like microarray, gel-based approaches, mass spectrometer and Edman sequencing.

12.4.1 PROTEIN MICROARRAY METHOD

The protein microarray is a proteomics technique which is capable of high-throughput detection from a very small sample size. It can simultaneously detect multiple pathogens, infecting plants in a single reaction. Microarrays are further divided into three categories: analytical protein, functional protein and reverse-phase protein.

In the analytical protein microarray method, small amounts of samples are applied to a membrane or a "chip" (a chemically modified glass slide surface) for analysis. Next, certain virus-specific antibodies are immobilized to the membrane/chip surface and are used to capture target proteins in a complex sample. After capturing antibodies, proteins are detected by direct protein labeling. This method of microarray is used to measure the expression levels and binding affinities of proteins in a sample. The functional protein microarray method involves constructing an array by means of purified proteins and characterizing their functionalities like RNA interactions and enzyme-substrate turnover. This method characterizes the functions of thousands of proteins for easy detection and identification. The last microarray method is the reverse-phase protein microarray. In this method, proteins from two different samples like healthy vs. virus-infected tissues or treated vs. untreated cells are bound to the chip, and then specific antibodies are used to probe the target proteins. The antibodies are identified using fluorescent, chemiluminescent and colorimetric assays. The quantification of proteins is done on the basis of reference peptides already printed on slides. These microarrays are then used to detect the altered or dysfunctional protein which indicates presence of a certain disease.

12.5 GEL-BASED PROTEOMICS TECHNOLOGY

12.5.1 SDS-PAGE Method

The gel-based proteomics approach involves three methods called sodium dodecyl sulfate-polyacrylamide gel electrophoresis (SDS-PAGE), 2D PAGE and Two-dimensional difference gel electrophoresis (2D-DIGE). The first one, SDS-PAGE, involves separation of complex mixtures of proteins on the basis of their size and molecular weight. The procedure works by initially denaturing proteins in the complex protein mixture sample with an anionic detergent that also binds to them. This gives a negative charge to all proteins in proportion to their respective molecular mass. Proteins have a tendency of moving with an electric field in a medium which has a different pH from their isoelectric point. As such, different proteins in the mixture move with different velocities depending upon the ratio between their charge and mass. Next, by performing electrophoresis through a porous acrylamide gel matrix, the proteins are separated according to their molecular mass.

12.5.2 2D-PAGE Method

The two-dimensional polyacrylamide gel electrophoresis (2D-PAGE) is another gel-based method for separation of proteins dependent upon their mass and charge. The 2D-PAGE method can resolve up to 5,000 different proteins successively, depending on the size of gel. Unlike SDS-PAGE where proteins are separated on the basis of their charge, in the 2D-PAGE method separation occurs on the basis of their varying mass.

12.5.3 2D-DIGE Method

The third gel-based protein separation method is the 2D-DIGE in which the proteins are covalently labeled using fluorescent dyes (Cy3 and Cy5) before performing electrophoresis. Next, the dye is exposed to specific wavelengths to excite and emit and then they are separated on the basis of their charge and mass.

12.5.4 Mass Spectrometry (MS)

In proteomics, mass spectrometry (MS) technique is applied to measure the mass-to-charge ratio (m/z) of one or more molecules present in a protein sample. This also helps in determining the molecular weight of the proteins. The MS process involves three steps. In the first step, the molecules are transformed into gas-phase ions which however is quite challenging if biomolecules are in a liquid or solid phase. In the second step, the ions are separated on the basis of m/z values in the presence of electric or magnetic fields. This separation process is done in a compartment known as a mass analyzer. In the third and final step, the separated ions and the amount of each type of ions with a particular m/z value are measured. In MS method, the methods used for ionization include matrix-assisted laser desorption ionization (MALDI), surface-enhanced laser desorption/Ionization (SELDI) and electrospray ionization (ESI).

12.5.4.1 Edman Sequencing

Edman sequencing procedure is used to determine the amino-acid sequence in peptides or proteins. The method involves introducing chemical reactions that eliminate and identify amino acid residue that is present at the N-terminus of the polypeptide chain. Edman sequencing has played a very crucial role in development of therapeutic proteins and quality assurance of biopharmaceuticals (https://www.news-medical.net/life-sciences/Proteomics-Methods/).

12.6 QUANTITATIVE TECHNOLOGIES OF PROTEOMICS

12.6.1 The Isotope-coded Affinity Tag (ICAT)

The isotope-coded affinity tag (ICAT) technique is a method for protein quantification and isotopic labeling method uses chemical labeling reagents for quantification. In this pairwise changes in protein, expression is measured through differential stable isotopic labeling of proteins or peptides, which is then identified and quantified using a mass spectrometer. The procedure is to take two identical peptides from the two biological conditions and observe changes in protein expression. The peptides labeled with heavy isotope are compared with the one having a normal isotope. This approach allows the simultaneous comparison of the expression of many proteins between two different biological states.

12.6.2 Stable Isotopic Labeling with Amino Acids in Cell Culture (SILAC)

Stable isotope labeling with amino acids in cell culture (SILAC) is another MS-based approach for quantitative target protein molecules. It is a multiplexing quantitative proteomic measuring method that uses labeled isotopically heavy amino acids, for example $^{13}C6,^{15}N2$-lysine and 13C6,15N4-arginine which are incorporated metabolically into the whole proteome. The proteomes of different cells grown in cell culture are labeled with "light" or "heavy" forms of amino acids and differentiated through MS.

12.6.3 Isobaric Tag for Relative and Absolute Quantitation (iTRAQ)

iTRAQ is another protein quantification method and protein labeling technique which is based on tandem MS (TANDEM MS). This technique depends on labeling the protein with isobaric tags (8-plex and 4-plex) for relative and absolute quantitation. The technique involves labeling of the N-terminus and side chain amine groups of proteins which are then fragmented using liquid chromatography and finally analyzed through MS (https://www.ncbi.nlm.nih.gov/books/NBK56015/).

12.7 HIGH-THROUGHPUT TECHNOLOGIES OF PROTEOMICS

12.7.1 X-Ray Crystallography

When it comes to determining three-dimensional(3D) structure of proteins, the most commonly used technique is the X-ray crystallography method. In this procedure, highly purified crystallized samples are exposed to X-rays. By observing and processing diffraction patterns produced by exposure, information about the size of the repeating unit that forms the crystal and crystal packing symmetry are obtained. X-ray crystallography has a wide range of applications in the virus system, protein–nucleic acid complexes and immune complexes.

12.7.2 NMR – Spectroscopy

The NMR is used in the investigation of molecular structure, folding and in understanding protein behavior. The determination of protein structure through NMR spectroscopy involves different phases where each phase uses a discrete set of extremely specific techniques. The samples of target proteins are prepared, and after making measurements, its structure is confirmed using various interpretative approaches.

12.8 BIOINFORMATICS ANALYSIS

Bioinformatics is an essential component of proteomics; therefore, its implications have been progressively increasing with the advent of high-throughput methods that are dependent on powerful data analysis. This new and emergent field is presenting novel algorithms to manage huge and heterogeneous proteomics data and headway toward the discovery procedure (https://www.technologynetworks.com/proteomics/articles/proteomics-principles-techniques-and-applications-343804).

12.8.1 Two-dimensional Polyacrylamide Gel Electrophoresis (2D-PAGE)

The 2-D electrophoresis method is the commonly used method for the analysis and separation of complex protein mixtures extracted from cells, tissues, etc. This technique separates protein in two steps – first-dimension IEF (isoelectric focusing) and second-dimension SDS-PAGE. First-dimension IEF separates proteins according to their isoelectric points (pI) or the charge that they carry, whereas the second-dimension SDS-PAGE separates protein according to molecular weight and size (Dubey & Grover, 2001).

The advantage of 2D PAGE is that it provides a useful platform for analyzing the protein mixture as on a large format 2D gel; up to 5,000 protein spots can be isolated under right conditions (Righetti et al., 2004). These protein spots can be seen by using many staining methods of the gel such as Coomassie Brilliant Blue staining and silver staining or SYPRO staining. The SYPRO staining is a set of molecular probes which are used for the identification of proteins. It gets discolored in case of luminous protein separation analysis using PAGE detection. Another advantage is that not only does it resolve large numbers of proteins, but also, staining these proteins enables the relative abundances of the proteins to be quantified (Rose et al., 2004).

12.8.2 Objective of 2D-PAGE

The objective of separating proteins using 2D-PAGE includes identification of new proteins and measuring their relative abundance between comparative samples. It is almost impossible to detect the appearance of a few new protein spots or the disappearance of single spots in studies with several thousand spots. Moreover, evaluation of two gels by manual comparison is also impossible. As such, it is necessary to detect differences and obtain information from gels by the usage of image collection hardware and image evaluation software. 2D gel analysis software includes Melanie, PDQuest, Progenesis, REDFIN. The gels can be used for the identification of proteins and other applications by MS (Wittmann-Liebold et al., 2007).

12.8.3 Application of 2D-PAGE

Two-dimensional electrophoresis is an effective and widely used method for the analysis of complex protein mixtures and it has the exceptional ability to separate thousands of proteins at once. It establishes direct visual confirmation of changes in protein/PTMs abundance and provides early justification for downstream analytical steps through detecting post- and co-translational modifications, which otherwise cannot be predicted from the genomic sequence. Other applications of 2-DE include whole proteome analysis, cell differentiation, detection of biomarkers and disease markers, drug discovery, cancer research, bacterial pathogenesis, purity checks, microscale protein purification and product (characterization https://www.future-science.com/doi/10.2144/000112823).

12.9 FLUORESCENT 2D-DIGE

Two-dimensional-DIGE and fluorescence DIGE were designed for more quantitative variation of protein separation. In the 2D-DIGE method, protein mixtures are first prelabeled on the basis of their fluorescence before performing 2D Gel electrophoresis (Ünlü et al., 1997). The samples are selectively and covalently tagged with fluorophores to distinguish separate proteins resolved on the same gel. One of the two fluorescent dyes that are distinct but structurally similar is applied to each sample being compared (cy2, cy3, cy5, etc.). Each dye reacts with amino groups in samples, and the dye binds to lysine residues and the N-terminal amino group, fluorescently labeling each protein. As such, after the 2D gel electrophoresis, the distinctly prelabeled protein extracts can now be visualized separately. The spot pattern in-gel as well as inter-gel comparison reveals differentially expressed proteins (Ünlü et al., 1997).

The gel is scanned at two different wavelengths which excite the two dye molecules for determining whether a specific location is connected to one dye molecule instead of two. By directly comparing hundreds of protein samples examined under similar electrophoresis conditions, the DIGE method demonstrates better sensitivity, linearity, accuracy of measurement and enhanced reproducibility (Mallick & Kuster, 2010).

The DIGE approach allows the quantitative and repeatable comparison of hundreds of proteins in a single experiment. As the DIGE approach allows running an internal strand as part of each separation, more minor abundance variations can be statistically shown with the pooled internal standard methodology, thus identifying real differences in protein expressions of the used samples (Larrainzar et al., 2007).

12.9.1 OBJECTIVE OF 2D-DIGE

The main objective of the fluorescent 2D-DIGE system is to detect changes in the expressed levels of many proteins in two different samples. This method has become popular for validating the generation of fusion proteins, analyzing protein–protein interactions and electrophoretic mobility shift assays (Jayaraman et al., 2012).

12.9.2 APPLICATION OF 2D-DIGE

Two-dimensional DIGE has been used extensively in most areas of biology and medicine. Its application has been crucial amongst various proteomics technologies for studying the mechanisms of diseases, precisely identifying new therapeutic targets or finding potential biomarkers. Apart from this, DIGE has been applied in the fields of cancer research, renal physiology, plant biology and the elucidation of signal transduction pathways and many more fields. Its application has also benefited neurology and neuroscience, especially in neurotoxicology and neurometabolism. In neurodegeneration, it has been used in the determination of specific proteomic aspects of individual brain areas and body fluids to identify biomarkers. The associated detection of several hundred proteins in a gel sample provides comprehensive data to elucidate a physiological protein network and its peripheral representatives.

DIGE-based proteomics has been used to identify molecular bases of neurological processes in research projects. The application is used in many neurological aspects like analyzing protein composition in cerebrospinal fluid of traumatic brain injury patients, understanding molecular mechanisms of sleep deprivation and aging, Western Pacific amyotrophic lateral sclerosis-Parkinsonism–dementia complex (ALS/PDC), etc. (Hariharan et al., 2010).

12.10 MULTIDIMENSIONAL PROTEIN IDENTIFICATION TECHNOLOGY (MUDPIT)

Multidimensional protein identification technology (MudPIT) technique is primarily applied for the study of intricate protein combinations. In this method, trypsin and endo-proteinase lysis are used to degrade protein samples enzymatically in a sequence-specific manner. This forms peptide mixtures which are then separated using strong cation exchange (SCX) and reverse phase (RP) high-performance liquid chromatography (HPLC) (Issaq et al., 2005). The peptides are received by a mass spectrometer from the RP column, and the MS data is utilized to search protein databases. The advantage of this technique is that it produces a complete list of all the proteins contained in a specific protein sample and it is also quick, sensitive and reproducible. MudPIT method combines SCX and RP-HPLC separation and is successfully used in a wide variety of proteomics experiments, including large-scale protein cataloging in cell and organism, membrane and organelle protein profiling, and identification of protein complexes, determination of PTMs and quantitative analysis of protein expression (Washburn et al., 2001).

12.10.1 OBJECTIVE OF MUDPIT

Gel-based proteomics technologies like 2D PAGE have certain drawbacks in identification and separation of proteins. These drawbacks include a limited molecular mass range, improper separation of highly acidic or basic proteins, and inability to test membrane proteins in the analysis. However, membrane proteins like receptors, ion transporters, signal transducers and cell adhesion proteins play important functions so that it is important to use a method like MudPIT. This method provides a gel-free alternative to 2D-PAGE as it can perform analysis of both membrane and soluble proteins. The main objective of MudPIT is to offer a non-gel technology for separating and identifying a very wide range of individual components of complex protein and peptide mixtures, which may be excluded in other protein separation techniques like 2D PAGE (Mauri et al., 2005).

12.10.2 APPLICATION OF MUDPIT

MudPIT has a great applicability in a wide range of proteomics-based investigations of plant–virus interactions. It provides a large-scale catalog of proteins in target cells and conducts more accurate profiling of organelle and membrane proteins which are not covered under gel-based proteomics technologies like SDS-PAGE, discovering plant–virus interaction protein complexes, capturing of PTMs and quantification of protein expression. Due to its unique advantages, MudPIT is a bonafide alternative to traditional methods for quick and complete identification of protein–protein interactions for stoichiometric and substoichiometric partners. MudPIT can also be applied to separate complex sets of proteins related by a common property (Wadhwa et al., 2005).

12.11 ISOTOPE-CODED AFFINITY TAGS (ICAT)

ICAT is a quantitative proteomics technique that uses chemical labeling reagents, instead of gels, to locate and count the respective quantity of the proteins in two cell types and observe pairwise changes in protein expression which have occurred. The ICAT technique uses isotope ^{13}C for labeling. To achieve this, two populations are labeled with two different ICAT reagents, for example one having light reagent and the other one dark. Next, they are mixed and enzymatically cleaved to produce peptide fragments. Some of these are tagged. Then with the avidin affinity chromatography technique, the tagged peptides are isolated, which is followed by their identification and quantification using a mass spectrometer (Gygi et al., 1999a,b). Mass spectrometer conducts analysis with the help of micro-capillary liquid chromatography and electrospray technique. Thus, a quantitative and qualitative assessment of changes in protein levels is observed and this helps in identifying which protein structures can prevent viral propagation. This approach makes it possible to simultaneously observe and compare expression of many proteins in two different biological states; e.g., yeast grown on galactose versus glucose, or normal cells versus cancer cells (Kav et al., 2007).

12.11.1 OBJECTIVE OF ICAT

The objective of using ICAT technology is to perform target protein labeling with ICAT. Another objective is to establish sequence identification and accurate quantification of proteins in complex mixtures. This is achieved using TANDEM MS and it allows analysis of protein expression alterations in affected plant tissue. The advantage of the ICAT method is that it is applicable both in large-scale analysis of complex protein samples including whole proteomes and small-scale analysis of subproteomes. This feature allows easy quantitative analysis of proteins, including those that are difficult to analyze by gel-based proteomics technologies (Gygi et al., 1999a,b).

12.11.2 APPLICATION OF ICAT

The use of ICAT reagent-labeled technique in quantitative MS has applications in a large number of different fields of biochemistry. High-throughput approaches to the quantification and identification of proteins are widely applied in the industrial synthesis of therapeutic enzymes. Proteomic analysis on most recombinant proteins results in very low yield and has poor solubility. This majorly affects the ability to achieve high-throughput protein purification. On the other hand, quantitative methods like ICAT which use isotope-coded affinity tags have been proven to achieve comparatively much high-throughput protein purification and also resulting in increased yield, and greater solubility and folding of the recombinant protein, during the process. This purification results in yields of over 90%, making it economically viable. For achieving high-throughput screening, combinations of two or more isotopic tags are usually needed (Shiio & Aebersold, 2006).

12.12 ISOBARIC TAGS FOR RELATIVE AND ABSOLUTE QUANTITATION (iTRAQ)

iTRAQ is also a non-gel-based technique for quantifying protein which is acquired from different sources and measured in a single experiment. In this technique, like ICAT, isotope-coded covalent tags are used (Rose et al., 2004). How a protein is tagged makes a great difference? As evident from the name itself, this method attaches an isotopic tag covalently with a different mass to the N-terminus and side chain amine of peptides that have been produced during protein digestion (Zieske, 2006). The iTRAQ reagents normally consist of an N-methyl piperazine reporter group, a balance group and an N-hydroxysuccinimide ester group that is reactive with the primary amines of peptides. As of now, mainly two types of reagents are used (4- and 8-plex) which are used for labeling all peptides from different samples (Aggarwal et al., 2006). The peptides need to be extracted from the mixtures, which is done using nanochromatography. Next a TANDEM MS, also named MS/MS analysis of the isolated peptides, is performed. The tagged peptides and the associated protein are then found with the help of a database search based on the fragmentation data. As the applied tag breaks up, it creates a low molecular mass reporter ion. This ion is then used by the software (e.g., i-tracker) to approximately quantify the peptides and proteins that make up those peptides. The benefits of iTRAQ include multiplexing numerous samples in a single experiment (up to 8), easy and quick quantification, a streamlined analysis and more precise analytical results (Lodha et al., 2013).

12.12.1 OBJECTIVE OF iTRAQ

Like any other proteomics technology, iTRAQs aim to identify and quantify differential expression of proteins in perturbed systems. iTRAQ is also useful for comparing normal, virus-infected and drug-treated protein samples. It is also applied in time course studies, developing biological replicates and for assessing relative quantitation. iTRAQ, in combination with TANDEM MS, helps in determining the quantity of proteins from different sources in a single experiment. It allows identification of potential biomarkers and drug targets for resolving virus-infected crops. The development of iTRAQ has made possible multiplexing of samples to improve the accuracy and throughput of experiments as well as reduction of costs. Another objective is to obtain results without any loss of information from samples involving PTMs. Apart from this, iTRAQ and MS technology are used in precise isolation and measurements of non-tryptic peptides (Tian et al., 2019).

12.12.2 APPLICATION OF iTRAQ

ITRAQ procedure is applied to identify and quantify a greater range of altered protein structures, to quantify the amount of proteins from different sources in a single experiment, to locate potential drug targets, to analyze the druggability of selected protein targets and to develop drugs which can be aimed at damage causing protein targets. The iTRAQ protein quantification procedure is suited for unbiased non-targeted biomarker discovery. iTRAQ is also helpful in comparing normal, diseased and drug-treated samples, preparing biological replicates and providing relative quantitation (Tian et al., 2019).

12.13 MASS SPECTROMETRY

MS analysis works in three steps which are ionization of the protein or peptide, ion separation and detection. Ionization of protein or peptides can be done by matrix-assisted laser desorption or ionization (MALDI) method or electrospray ionization (ESI) method, whereas mass analyzers include time-of-flight (TOF), quadrupole, ion trap, orbitrap and Fourier transform ion cyclotron resonance (FT-ICR) methods (Mathesius et al., 2003). In this procedure, the test samples are first transformed into gaseous ions in the mass spectrometer during MS. In the next step, they are separated and identified based on their mass-to-charge ratio(m/z). Trypsin method is used to fragment proteins or peptides of the sample. The fragmented proteins are then separated using the liquid chromatography method. The samples are then changed into the gaseous phase by being ionized. With the development of TOF-MS and comparatively non-destructive techniques to turn proteins into volatile ions, MS has undergone a revolution. In the mass analyzer, gaseous ions are separated. Peptides are fragmented by employing collision-induced dissociation with a certain mass. Then they are subjected to a second MS, which provides a series of fragment peaks from which the amino acid sequence of the peptide may be deduced. Next algorithms of protein identification are used to compare the finding to known benchmarks. These algorithms can be divided into two groups: de novo search algorithms and database search algorithms (Prayitno et al., 2006).

12.13.1 OBJECTIVE OF MASS SPECTROMETRY

The main objective of mass spectrometers is to identify unknown compounds through determining molecular weight and separate to quantify known compounds from unknown. Another objective is to determine structural and chemical properties of protein molecules of a given sample. This is done by ionizing and fragmenting sample molecules in the gas phase. From the resulting pattern of the ion fragmentation, structural information of a given molecule is obtained. Mass spectrometer coupled with chromatography provides detailed information for each separated compound, both of which can be used for comparison with suitable reference standard (https://www.britannica.com/science/mass-spectrometry).

12.13.2 APPLICATION OF MASS SPECTROMETRY

MS and Tandem MS are a powerful tool in the field of proteomics research because they can precisely measure the molecular mass of peptides and proteins and their sequences. The advantage of TANDEM MS is that it gives precise sequence information of sample proteins using the technique of fragmentation of peptides and proteins. Tandem MS also helps in identification and isolation of post-translational and other covalent modifications.

12.13.3 PROTEIN IDENTIFICATION BY MASS SPECTROMETRY

The "de novo" sequencing method and the peptide mass fingerprinting (PMF) database searching are the two main MS-based protein identification methods used. In this procedure, computational methods are used for the identification of proteins on the basis of the peaks of the captured mass spectra. Each peak in the MS results represents a peptide fragment ion.

12.13.4 PROTEIN QUANTIFICATION BY MASS SPECTROMETRY

Proteomics-based research focuses on finding out the level or quantity of protein expression. MS technique proves very effective as it has the ability to provide a more reliable and quantifiable result of the sample proteins and peptides and verifiably highlights their different expressions. Quantitative proteomics methods can be further divided into relative quantification and absolute quantification methods. The relative quantification method with stable isotope labeling aims at ascertaining or identifying the difference of proteome expression amongst the chosen samples, while the absolute quantification quantifies the specific expression level of the protein. Some of the commonly applied methods for protein quantification are iTRAQ, SILAC, ICAT label-free quantification, etc.

12.13.5 POST-TRANSLATIONAL MODIFICATION ANALYSIS BY MASS SPECTROMETRY

PTMs method works by making chemical alterations to the structure of protein using proteolytic cleavage and adding modifying groups like acetyl, phosphoryl, glycosyl and methyl, to one or more amino acids. Different types of PTM include phosphorylation, acetylation, acetylation and glycosylation. MS is a very vital technology of proteomics and measuring modifications in protein samples as it provides universal information about protein modifications without knowing the location of the modifying sites. The most commonly applied MS-based PTM analysis uses approaches like top–down, middle-down and bottom-up strategies of PTM analysis using MS (https://www.creative-proteomics.com/support/applications-of-mass-spectrometry.htm)

12.14 PLANT–VIRUS INTERACTION

As biotrophic pathogens, plant viruses attack a wide range of plant species utilizing cellular environment of host plants for protein synthesis, their genome replication and intercellular movement to make their propagation and proliferation possible (Quirino et al., 2010). Virus infection operates by causing morphological and physiological alterations in the infected plant hosts, which always impedes plant performance such as the decreased host biomass and loss of crop yield.

Plant viruses bring about alterations and deformations inside a plant host cell in various ways. Their genome assumes many forms like dsDNA, ssDNA, dsRNA or ssRNA. It is surrounded by a capsid consisting of coat protein molecules. The viral genome encodes and alters genetic information about proteins, which helps the virus infection spread through local or systemic transport. Plant viruses thus induce multilevel changes in their internal system during infection by being active only inside the host cell and using their cellular mechanism (Hammond-Kosack, 2000).

To overcome these viral or pathogen infections, plants have evolved both constitutive and inducible defense mechanisms. Constitutive defense system works by introducing physical barriers for pathogens, such as waxy epidermal cuticles or cell wall on the surface. This prevents the penetration of pathogens into the plant cellular environment. On the other hand, inducible defense responses work by creating surveillance mechanisms when a plant recognizes any potential pathogens (Alexander & Cilia, 2016). Therefore, it shows that plants do not remain as "static components" during their interaction or exposure to a viral pathogen. Instead, they proactively introduce complex systems of defense response to neutralize or deter the invading virus. How well the host plant will overcome a viral attack or how susceptible it is to infection depends upon the speed, strength and level of effectiveness of this defensive response (Schenk et al., 2000).

Viruses can be called ultimate parasites as they do not show any life process outside the host cell. Their small genomes manage to encode a small number of host proteins. After entering a host cell mechanically or through a vector, they control the cellular environment and reproduce and replicate themselves and move in plant cells, then travel from one cell to another through plasmodesmata (Wu et al., 2013). Finally, they reach the vascular tissue and from there circulate systemically through the phloem to the sink tissues of the host plant (Lee et al., 2006). These complex and varied forms of molecular interactions between host plants and invading pathogens like viruses can be understood with the help of proteomic research done on the differential protein expressions during the occurrence of plant viral infection as well as by observing how proteins are implicated or altered in host proteins. Many methods are used for investigating the function of viral-encoded proteins, plant–virus interactions and susceptibility of certain proteins toward virus attack. There are many proteomics-based methods which can be used to identify the specific host factors which are involved in plant virus–host interactions.

12.15 PLANT CELL WALL AND THE PLANT–VIRUS RESISTANCE RESPONSE

The innate plant immunity system gets activated by receiving exogenous or endogenous signals. The two kinds of signals are – pathogen-associated molecular pattern molecules (PAMPs) and damage-associated molecular pattern molecules (DAMPs). For a plant cell to survive against any viral infection, it is critical to first successfully undergo damage detection (Casado-Vela et al., 2006). Any mechanical damage for pathogen infection disturbs the homeostatic cellular processes within a plant. The plant perceives this damage to its structure independently of the infecting organism. As such, the response that is triggered is not specific to a given pathogen. DAMP signal may be perceived by the plant from the damage caused to cell structures by physical injuries that may result in developmental breaks (Caplan et al., 2009). The response to DAMPs protects the plant not only against any pathogen infection but it also helps in repairing tissue damage due to mechanical injury.

The study of plant response to PAMP and DAMP signals and its effect on the cell wall indicates that the cell wall is actively modified when a plant initiates a defensive response against viral pathogens. Long periods of plant–virus coevolution did not only lead to susceptibility of the plants toward certain pathogens but also led to the development of effective resistance (Park et al., 2017). The way a plant defense mechanism responds along with its associated cellular and physiological features after identifying a specific pathogen or virus is stereotypical.

Normally, plants treated with cell wall biosynthesis inhibitors and a large number of cell wall mutants have been seen to show a wide range of immune responses, which include the development of defense genes and the accumulation of defense compounds. However, in the case of plant–virus interactions, they are rather limited. The earliest investigations of plant–virus interaction responses were based on plasmodesmata. It was observed that inhibition of β-1,3 glucanase enzyme caused an increased deposition of callose in certain parts of the cell wall, resulting in a reduction in the size exclusion limit. This caused reduction in both short- and long-distance transport of CMV, TMV, PVX and tobacco necrosis virus. On the other hand, plants with suppressed β-1,3 glucanase restricted the plant virus to a greatest extent allowing access only to a limited number of cells. This made it possible to induce programmed cell death, which helped in repressing and even stopping virus infection.

12.16   CONCLUSION

In the near past, a couple of decades ago, the electrophoretic and protein identification techniques were not as advanced as they are now. So then, the field of proteomics of plant–pathogen interactions was established. In 1993, Ekramoddoullah and Hunt researched on sugar-pine seedlings that were susceptible to or were resistant to the fungus that caused white pine blister rust. They used the 2D PAGE method to differentiate between the proteins in sugar-pine seedlings to observe which proteins were susceptible and which were resistant to the white pine blister rust fungus.

Plant diseases greatly reduce crop productivity globally; thus, it is very important to understand the tactics used by these virulent plant-infecting microbes to spread infection. They have extremely adaptive infection mechanisms as they are able to infect a wide range of plants and plant tissues. Their interaction is dependent on mutual two-way recognition and response. This involves invading microbes to release pathogenic factors and the host cell to identify threats and trigger necessary and specific defense mechanisms. For effectively dealing with plant diseases particularly when using biocontrol agents, it is crucial to have complete knowledge of pathogenic components and the disease cycle. It will help us to identify the gene product in a specific pathogen and beneficial microorganism for plants with the use of bioinformatic techniques.

We know that proteins are the building blocks of structural and functional elements of all living cells. Therefore, research and investigations at the proteome level are pertinent for accessing information about the functional molecules of a cell and the alterations that occur in the protein expression patterns due to various states and stresses. Proteomic analysis has been applied to closely understand the plant–microbe interaction and establish how a beneficial microbe might cause a systemic response in plants, and how understanding of pathogenic and virulence factors can contribute to the control of plant diseases. So, there is a need to identify new targets in plant defense mechanism so that its disease tolerance abilities can be enhanced by having a system-level understanding of biotic stress reactivity and the impact of applied biocontrol agents. Application of proteomics, meta-proteomics and secret-omics can also be fruitful in investigating the functional aspect of plant–microbe interaction.

Specialized proteomics-based studies have helped in identifying and isolating a large number of host factors involved in plant virus infections. Recording and comparing proteomics datasets with extensive genomics datasets allow bioinformatics analysis which provides a broad view of virus–plant interactions and in creating testable models. However, the proteomics approach is not able to identify the molecular mechanism of the specific host factors during viral infections or how viruses manage to hijack host plant cellular functions so that the pathogen can destroy the host cell defense mechanism and use it for its own propagation. Therefore, to overcome this drawback of proteomics studies, additional experiments using genetics, biochemistry, cell biology and other approaches need to be performed to understand and analyze the functional roles of specific host proteins during plant virus infections. These future studies and results augur for a better understanding of structural and functional alterations during a virus–host interaction event and can allow us to develop new antiviral approaches to inhibit damage caused by disease-causing pathogens in plants and the same knowledge can later also help us in combating virus attacks on humans and animals.

12.17   FUTURE PERSPECTIVE

There is an unlimited potential in proteomics as it is yet in quite early stages of development. Unlike genome sequencing, in which sequence analysis and its mutations are obvious endpoints, there is no such endpoint in proteomics as the functional proteome is applicable on several analytical dimensions. With deeper studies, it is possible to learn more precisely about the function of protein alterations, different proteoforms and protein–protein interactions. Future advancements will depend upon improved sample preparation, better chromatography and more calibered instrumentation. However, the greatest untapped potential of proteomics lies in the data analysis. Protein modifications occur in a vast number of ways and so far we have gathered, identified, labeled and understood only a fraction of the information hidden in the peptide or even peptide-fragment level data. A very sustained and deeper research in data analysis and its comprehensive classification is the key to success of proteomics in future.

However, there are certain bottlenecks in the proteomics field. Proteomics cannot reach its true potential unless the research tools are more efficient at addressing its complexity in both the data collection and data analysis stages. Processing samples which provide a deep, unbiased look at the proteome is not only very laborious but also time-consuming. After proteomic information has been procured, the challenge as of now is to derive meaningful endpoints from that complex data.

There are three main approaches in proteomics in resolving protein alterations by viruses that involve identification and grouping protein molecules, studying them individually and lastly, to try to quantify them using indirect measurements. Future researchers need to improve the usability and success rate of proteomics by upgrading technology and making more efficient recording and analysis of collected data.

The first approach is to work toward identifying and grouping similar molecules. If individual proteins and peptides can be separated because of their distinct structure or functionality and the similar ones grouped together, it would be easy to use a detector, such as a mass spectrometer, with those proteins and peptides at the same time. This would make results faster and more accurate.

The second approach is to direct the focus of research on the study of each protein molecule individually. This can be technologically very challenging, but the development of such technology will go a long way in making proteomics very effective in the field of pathogen resistance.

The third approach involves indirect measurements to study proteins, achieved by tagging a protein with an antibody, and then measuring the abundance of the antibody instead of the protein itself. This method poses a challenge as it does not focus on a direct measurement of the protein itself. Moreover, because of the magnitude of huge varieties of the proteome, it is not possible to tag new and so far, unknown individual proteoforms. The future need is to find ways to conduct proteomic analysis at the proteoform level instead of just the protein level. Therefore, there is a need for a deep, comprehensive and unbiased approach to identify proteoforms at the peptide level to tackle the challenges faced by present proteomics.

REFERENCES

Aggarwal, K., Choe, L., & Lee, K. (2006). Shotgun proteomics using the iTRAQ isobaric tags. Briefings in Functional Genomics, 5(2), 112–120.

Alexander, M., & Cilia, M. (2016). A molecular tug-of-war: Global plant proteome changes during viral infection. Current Plant Biology, 5, 13–24.

Barajas, D., Li, Z., Panavas, T., Herbst, D. A., & Nagy, P. D. (2008). Cdc34p ubiquitin-conjugating enzyme is a component of the Tombusvirus replicase complex and ubiquitinates p33 replication protein. Journal of Virology, 82, 6911–6926.

Caplan, J., Zhu, X., Mamillapalli, P., Marathe, R., Anandalakshmi, R., & Kumar, S. (2009). Induced ER chaperones regulate a receptor-like kinase to mediate antiviral innate immune response in plants. Cell Host & Microbe, 6(5), 457–469.

Casado-Vela, J., Selles, S., & Martinez, R. B. (2006). Proteomic analysis of tobacco mosaic virus-infected tomato (*Lycopersicon esculentum* M) fruits and detection of viral coat protein. Proteomics, 6(suppl. 1), S196–S206.

Dubey, H., & Grover, A. (2001). Current initiatives in proteomics research: The plant perspective. Current Science, 80(2), 262–269.

Gygi, S. P., Rist, B., Gerber, S. A., Turecek, F., Gelb, M. H., & Aebersold, R. (1999a). Quantitative analysis of complex protein mixtures using isotope-coded affinity tags. Nature Biotechnol Oct, 17(10), 994–999.

Gygi, S. P., Rist, B., Gerber, S. A., Turecek, F., Gelb, M. H., &Aebersold, R. (1999b). Quantitative analysis of complex protein mixtures using isotope-coded affinity tags. Nature Biotechnology, 17, 994–999.

Hammond-Kosack, K. (2000). Responses to plant pathogens. In W. G. B. B. Buchanan (Ed.), Biochemistry and Molecular Biology of Plants (pp. 1102–1109). Wiley.

Hariharan, D., Weeks, M. E., & Crnogorac-Jurcevic, T. (2010). Application of proteomics in cancer gene profiling: Two-dimensional difference in gel electrophoresis (2D-DIGE). Methods in Molecular Biology, 576, 197–211.

Irish, J. M., Kotecha, N., & Nolan, G. P. (2006). Mapping normal and cancer cell signaling networks: Towards single-cell proteomics. Nature Reviews Cancer, 6, 146–155.

Issaq, H., Chan, K., Janini, G., Conrads, T., & Veenstra, T. (2005). Multidimensional separation of peptides for effective proteomic analysis. Journal of Chromatography B, 817(1), 35–47.

Jayaraman, D., Forshey, K., Grimsrud, P., & Ané, J.-M. (2012). Leveraging proteomics to understand plant–microbe interactions. Frontiers in Plant Science, 3(44), 1–6.

Jensen, L. J., Kuhn, M., Stark, M., Chaffron, S., Creevey, C., Muller, J., Doerks, T., Julien, P., Roth, A., Simonovic, M., eBork, P., & Mering, C. V. (2009). STRING 8—A global view on proteins and their functional interactions in 630 organisms. Nucleic Acids Research.

Kav, N., Srivastava, S., Yajima, W., & Sharma, N. (2007). Application of proteomics to investigate plant-microbe interactions. Current Proteomics, 4(1), 28–43.

Larrainzar, E., Wienkoop, S., Weckwerth, W., Ladrera, R., CesarArrese-Igor, C., & González, E. (2007). Medicagotruncatula root nodule proteome analysis reveals differential plant and bacteroid responses to drought stress. Plant Physiology, 144(3), 1495–1507.

Lee, J., Bricker, T., Lefevre, M., Pinson, S., & Oard, J. (2006). Proteomic and genetic approaches to identifying defense-related proteins in rice challenged with the fungal pathogen *Rhizoctonia solani*. Molecular Plant Pathology, 7(5), 405–416.

Lodha, T. D., Hembram, P., Tep, N., & Basak, J. (2013). Proteomics: A successful approach to understand the molecular mechanism of plant-pathogen interaction. American Journal of Plant Sciences, 4(6).

Mallick, P., & Kuster, B. (2010). Proteomics: A pragmatic perspective. Nature Biotechnology, 28, 695–709.

Mathesius, U., Imin, N., Natera, S. H., & Rolfe, B. G. (2003). Proteomics as a functional genomics tool. In N. I. Ulrike Mathesius, & E. Grotewold (Eds.), Methods in Molecular Biology (Vol. 236, p. 19). Humana Press, Springer Science + Business Media, NY, USA.

Mauri, P., Scarpa, A., Nascimbeni, A. C., Benazzi, L., Parmagnani, E., Mafficini, A., Peruta, D.M., Bassi, C., Miyazaki, K., & Sorio, C. (2005). Identification of proteins released by pancreatic cancer cells by multidimensional protein identification technology: A strategy for identification of novel cancer markers. FASEB Journal, 19(9), 1125–1127.

Newman, J. R., Ghaemmaghami, S., Ihmels, J., Breslow, D. K., Noble, M., DeRisi, J. L., & Weissman, J. S. (2006). Single-cell proteomic analysis of *Saccharomyces cerevisiae* reveals the architecture of biological noise. Nature, 441, 840–846.

Park, S.-H., Li, F., Renaud, J., Shen, W., Li, Y., Guo, L., Cui, H., Sumarah, M., & Wang, A. (2017). NbEXPA1, an α-expansion, is plasmodesmata-specific and a novel host factor for potyvirus infection. The Plant Journal, 92(5), 846–861.

Prayitno, J., Imin, N., Rolfe, B., & Mathesius, U. (2006). Identification of ethylene-mediated protein changes during nodulation in *Medicagotruncatula* using proteome analysis. Journal of Proteome Research, 5(11), 3084–3095.

Quirino, B., Candido, E., Campos, P., Franco, O., & Kruger, R. (2010). Proteomic approaches to study plant–pathogen interactions. Phytochemistry, 71(4), 351–362.

Righetti, P., Castagna, A., Antonucci, F., Piubelli, C., Cecconi, D., Campostrini, N., Antonioli, P., Astner, H., & Mahmoud, H. (2004). Critical survey of quantitative proteomics in two-dimensional electrophoretic approaches. Journal of Chromatography A, 1051(1–2), 3–17.

Rose, J., Bashir, S., Giovannoni, J., Jahn, M., & Saravanan, R. S. (2004). Tackling the plant proteome: Practical approaches, hurdles and experimental tools. The Plant Journal, 39(5), 715–733.

Schenk, P., Kazan, K., Wilson, I., & Manners, J. (2000). Coordinated plant defense responses in Arabidopsis revealed by microarray analysis. Proceedings of National Academy of Science, 97(21), 11655–11660.

Shiio, Y., & Aebersold, R. (2006). Quantitative proteome analysis using isotope-coded affinity tags and mass spectrometry. Nature Protocols, 1(1), 139–145.

Tian, L., You, H. Z., & Wu, H. et al. (2019). iTRAQ-based quantitative proteomic analysis provides insight for the molecular mechanism of neuroticism. Clinical Proteomics, 16, 38.

Ünlü, M., Morgan, M. E., & Minden, J. S. (1997). Difference gel electrophoresis. A single gel method for detecting changes in protein extracts. Electrophoresis, 18(11), 2071–2077.

Wadhwa, R., Takano, S., Kaur, K., Aida, S., Yaguchi, T., Kaul, Z., Hirano, T., Taira, K., & Kaul, S. C. (2005). Identification and characterization of molecular interactions between mortalin/mtHsp70 and HSP60. Biochemical Journal, Oct 15, 391(Pt 2), 185–190.

Wang, Z., Treder, K., & Miller, W. A. (2009). Structure of a viral cap-independent translation element that functions via high affinity binding to the eIF4E subunit of eIF4F. Journal of Biological Chemistry, 284, 14189–14202.

Washburn, M., Wolters, D., & Yates, J.III. (2001). Large-scale analysis of the yeast proteome by multidimensional protein identification technology. Nature Biotechnology, 19, 242–247.

Wittmann-Liebold, B., Graack, H.-R., & Pohl, T. (2007). Two-dimensional gel electrophoresis as a tool for proteomics studies in combination with protein identification by mass spectrometry. Proteomics, 7(5), 824.

Wu, L., Han, Z., Wang, S., Wang, X., Sun, A., Zu, X., & Chen, Y. (2013). Comparative proteomic analysis of the plant–virus interaction in resistant and susceptible ecotypes of maize infected with sugarcane mosaic virus. Journal of Proteomics, 89, 124–140.

Yu, J., Hu, S., Wang, J., Wong, G. K., Li, S., Liu, B., Deng, Y., Dai, L., Zhou, Y., & Zhang, X. et al. (2002). A draft sequence of the rice genome (Oryza sativa L. ssp. indica) Science, 296, 79–92.

Zhang, F., Dulneva, A., Bailes, J., & Soloviev, M. (2010). Affinity peptidomics: Peptide selection and affinity capture on hydrogels and microarrays. Methods in Molecular Biology, 615, 313–344.

Zhu, J., Gopinath, K., Murali, A., Yi, G., Hayward, S. D., Zhu, H., & Kao, C. (2007). RNA-binding proteins that inhibit RNA virus infection. Proceedings of the National Academy of Sciences, 104(9), 3129–3134.

Zieske, L. R. (2006). A perspective on the use of iTRAQ™ reagent technology for protein complex and profiling studies. Journal of Experimental Botany, 57(7), 1501–1508.

13 Plant Proteomics Research for Enhancing Resistance against Insect Pests

Suvarna and Shivaleela

13.1 INTRODUCTION

Biotic stresses represent a serious threat to crop production to meet global food exigency and thus create a major challenge for scientists, who ought to understand the intricate defence mechanisms. Insect pest infestation poses a major problem among biotic stresses. Many crops are affected by different insect pests causing a huge yield loss globally. Insecticides are being used to control insect pest population to avoid crop losses. However, using insecticides for pest control in crop production has many disadvantages, and the first and foremost is that they are harmful to the environment and humans, also kill parasites and predators which are natural enemies of pests and thus adversely impact biodiversity. Meanwhile, there are other strategies such as cultivating insect-resistant varieties, following cultural practices, using biocontrol agents, solar insect traps, *etc.*, which appear to be quite nature-friendly. To develop and use insect-resistant varieties of crop plants for sustainable insect pest management, there is a demand to study and understand the insect pest-plant interactions and various mechanisms of insect pest resistance. Further, some morphological, physiological and biochemical characters will also contribute to the resistance mechanism of plants through non-preference and antibiosis, but such characters need to be thoroughly studied and introgressed into the high yielding varieties. Understanding the resistance mechanisms at the molecular level by studying proteomics and metabolomics, genomics and transcriptomics helps in the development of high yielding cultivars with insect pest resistance. Studies are being conducted in crops against insect pest resistance using different omics approaches: genomics, proteomics, metabolomics and transcriptomics, *etc.* (Barah and Bones, 2015; Zogli et al., 2020). The study of proteins expressed in response to insect attack in plants and its quantification and identification through proteomic approaches is the best method to understand the defence mechanism of crops against insect pests.

13.2 SIGNIFICANCE FOR ENHANCING INSECT RESISTANCE

Augmentation of insect pest resistance in plants is very much essential. Food and Agricultural Organization (FAO) estimates that annually up to 40% of global crop production is lost due to pests. Each year, invasive insects cost the global economy at least $70 billion (FAO, 2022). More than 1000 pesticides are being used to control insects, fungi, weeds and other pests to guard the crops from acute damage. In India, 6668.73 MT technical grade insecticides are used to control insect pests during 2020–2021 (Indiastat, 2022). More chemical pesticides are used to control pests in cereals (17,149) followed by cash crops (6090), vegetables (4773), oilseeds (4723), pulses (4567) and fibre (4006) and less in other crops during 2020–2021 (FAO, 2022). Each pesticide has different properties and toxicological effects. Spraying of chemical insecticides has disadvantages and is harmful to human health and other organisms and to the environment (Tudi et al., 2021). So integrated pest management is advocated by the FAO that includes the usage of

DOI: 10.1201/b23255-13

natural predators and biopesticides in conjunction with mixed cropping and rotation where possible and the use of insect pest-resistant varieties. However, to develop insect-resistant varieties, one should know the molecular mechanism of defence against an insect pest through various molecular approaches. Proteomics is one such strategy which emphasizes different types of proteins synthesized by plants. The detailed study of each specific protein in plants under proteomics may help in the development of insect-resistant crop plants through various breeding techniques.

13.3 PROTEOMICS: APPROACHES AND PROTEIN ANALYSIS

The term protein was given by Berzelius in 1838. It was derived from a Greek word "proteios", which means "the first rank" (Cristea et al., 2004). The total protein content of a cell that is depicted with respect to their localization, post-translational modifications, interactions and turnover at a particular time is called "proteome". The expression "proteomics" was initially used by Marc Wilkins in 1996 to indicate the "PROTein complement of a genOME" (Wilkins et al., 1996). Most of the functional information of genes is obtained by proteome characterization (Roychoudhury et al., 2011). The study of proteins, its characterization, including structure, functions, expression, interactions and modifications of proteins at any particular stage is called proteomics (Domon and Aebersold, 2006).

Protein analysis involves the steps, *viz.*, protein extraction, purification, analysis, characterization, sequence analysis, quantification and structural analysis. Bioinformatics tool can be used in the detection of proteins (Figures 13.1 and 13.2).

There are several proteomic approaches that have been used for studying the proteins (Table 13.1). Among these, the conventional techniques include chromatography-based approaches, enzyme-linked immunosorbent assay (ELISA), Western blotting and Edman sequencing. Advanced techniques include microarrays, gel-based approaches, quantitative approaches and X-ray crystallography. High-throughput techniques include mass spectrometry (MS) and NMR spectrometry (Aslam et al., 2017).

Among these, the proteomic approaches used for insect pest resistance are 2-DE (two-dimensional gel electrophoresis), 2D-DIGE (two-dimensional differential gel electrophoresis), isobaric tags for relative and absolute quantification (iTRAQ), MS and tandem mass tag (TMT) (Table 13.2).

13.3.1 Two-dimensional Polyacrylamide Gel Electrophoresis (2D-PAGE)

This is an effective and reliable method for separating proteins based on their mass and charge. It has the capacity of resolving ~5000 different proteins based on the gel size successively. In the former dimension, the proteins are separated based on their charge, and in the latter dimension, the proteins are separated based on differences between their mass (Issaq and Veenstra, 2008).

13.3.2 Two-dimensional Differential Gel Electrophoresis (2D-DIGE)

This technique utilizes the proteins, which are labelled with CyDye, and then they can be visualized easily by exciting the dye at a particular wavelength (Marouga et al., 2005).

13.3.3 Isobaric Tags for Relative and Absolute Quantification (iTRAQ)

iTRAQ is a multiplex protein labelling method for protein quantification, which is centred on tandem MS. The proteins are marked with isobaric tags (8-plex and 4-plex) for its quantification. The N-terminus of proteins and side chain amine groups are labelled and fractionated *via* liquid

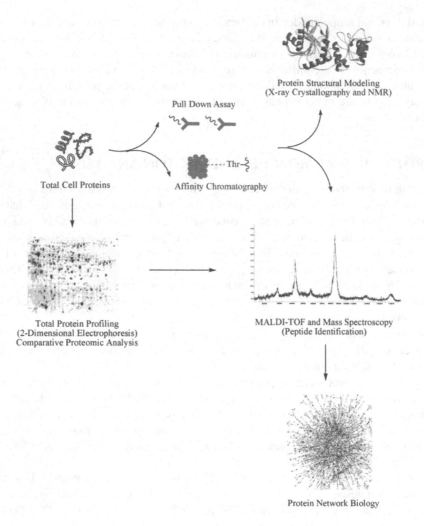

FIGURE 13.1   Schematic representation of protein analysis.

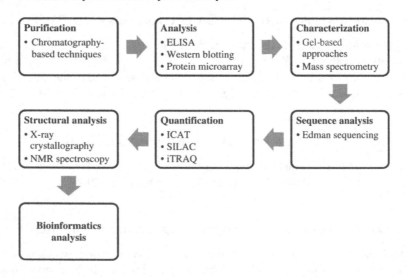

FIGURE 13.2   Applications of proteomics.

TABLE 13.1

Various Procteomic Approaches Used to Study Proteins

Sl. No.	Proteomic Approaches
I	**Conventional Approaches**
	A Chromatography-based approaches
	1 Ion-exchange chromatography (IEX)
	2 Size-exclusion chromatography (SEC)
	3 Affinity chromatography
	B Enzyme-linked immunosorbent assay (ELISA)
	C Western blotting
	D Edman sequencing
II	**Advanced Techniques**
	A Protein microarrays
	1 Analytical protein microarray
	2 Functional protein microarray
	3 Reverse-phase protein microarray
	B Gel-based approaches
	1 Sodium Dodecyl Sulphate-Polyacrylamide Gel Electrophoresis (SDS-PAGE)
	2 Two-dimensional polyacrylamide gel electrophoresis (2D PAGE)
	3 Two-dimensional differential gel electrophoresis (2D-DIGE)
	C Quantitative techniques
	1 ICAT labelling (isotope-coded affinity tags)
	2 Stable isotopic labelling with amino acids in cell culture (SILAC)
	3 Isobaric tag for relative and absolute quantitation (iTRAQ)
	D X-ray crystallography
III	**High-Throughput Techniques**
	A Mass spectrometry
	B NMR spectroscopy
IV	**Bioinformatics**

chromatography. It is ultimately analysed through mass spectrometry. It is a suitable method which helps to ascertain and quantify the proteins simultaneously (Wiese et al., 2007).

13.3.4 MASS SPECTROMETRY (MS)

The molecular weight of the proteins is determined using MS which measures mass-to-charge (m/z) ratio. Three steps are involved in MS. In the preliminary step, molecules have to be transformed togas-phase. The second step comprises the separation of ions based on values of m/z in the presence of mass analyser. Finally, in the third step, the separated ions and the amount of each species with a specific m/z value are measured. Generally used ionization methods comprise surface-enhanced laser desorption/ionization (SELDI), matrix-assisted laser desorption ionization (MALDI) and electrospray ionization (ESI) (Yates, 2011).

13.3.5 TANDEM MASS TAG (TMT)

TMT is sensitive, quantitative proteomic techniques that simultaneously identify and determine the relative abundances of peptides from various samples utilizing isobaric labelling centred on tandem MS (MS/MS) (Thompson et al., 2003).

TABLE 13.2

Proteomics Study in Different Crops against Resistance to Insect Pests

Crop	Proteomic Approach	Insect Pest	Defence Response Mechanisms	Reference
Rice	iTRAQ	*Nilaparvata lugens*	No change in proteins and genes related to callose metabolism and up-regulation of glycine cleavage system protein	Wei et al. (2009)
	2-DE	*N. lugens*	S-like RNase, glyoxalase I, EFTu1 and salt stress root protein RS1	Sangha et al. (2013)
	iTRAQ-based quantitative proteomics approach	*N. lugens*	Proteins involved in defence signal transduction, redox regulation, carbohydrate and protein metabolism and cell structural proteins	Du et al. (2015)
	2-DE, 2D-DIGE	*Laodelphax striatellus*	Activation of salicylic acid pathway	Dong et al. (2017)
	iTRAQ	*N. lugens*	LOXs, DIRs andOsDTC1 are the key enzymes, HSP 20	Zhang et al. (2019)
	SWATH-MS-based quantitative approach	*Cnaphalocrocis medinalis*	Jasmonic acid signalling, carbon remobilization and the production of flavonoids and glutathione	Cheah et al. (2020)
	iTRAQ	*N. lugens*	Changes in the protein level of two virulent biotypes	Zha and You (2020)
Wheat	2-DE, MALDI-TOF MS	*Sitobion avenae*	Proteins which are involved in metabolic processes and photosynthesis, signal transduction, stress and defence, antioxidant activity, regulatory processes and hormone responses.	Ferry et al. (2011)
	Isoelectric focussing (IEF) and SDS-PAGE, MALDI-TOF MS, MS/MS	*S. avenae*	Stress response proteins (including NBS-LRR-like proteins) and oxidative stress response proteins	Guan et al. (2015)
	Isoelectric focusing (IEF) and SDS-PAGE	Sunn pest	Proteins – serpin, β-amylase, α-amylase inhibitor, dehydroascorbate reductase, triticin and α-L arabinofuranidase	Saadati and Toorchi (2017)
	Nano-high performance liquid chromatography-tandem MS (nHPLC−MS/MS)	*Cephus cinctus*	Metabolic pathways involving enzymatic detoxification, proteinase inhibition and antiherbivory compound production	Lavergne et al. (2020)
Sorghum	LC-MS/MS	*Chilo partellus*	Proteins which are involved in stress and defence, small molecule biosynthesis, amino acid metabolism, catabolism and translation-regulated activities	Tamhane et al. (2021)
Maize	2-DE PAGE	*Spodoptera littoralis* and *Busseola fusca*	Protein spots were unidentified	George et al. (2011)
	2-DE, MS	*Ostrinia furnacalis*	Methyl jasmonate (MeJA)-induced mechanism	Zhang et al. (2015)
Pigeonpea	MS	*Helicoverpa armigera*	Reallocation of resources and diversion of metabolic flux to support the production of secondary metabolites	Rathinam et al. (2020)
	TMT quantitative proteomic	*H. armigera*	Secondary metabolite precursors, antioxidants and the phenylpropanoid pathway and reactive oxygen species (ROS)	Dawit et al. (2021)

(Continued)

TABLE 13.2 *(Continued)*
Proteomics Study in Different Crops against Resistance to Insect Pests

Crop	Proteomic Approach	Insect Pest	Defence Response Mechanisms	Reference
Soybean	2-DE, MALDI-TOF MS	*Prodenia litura*	PAL and SAMS were up-regulated at both protein and mRNA levels	Fan et al. (2012)
	iTRAQ-MRM	*Lamprosema indicata*	Induction of the synthesis of anti-digestive proteins, ROS scavenging, signalling pathways and secondary metabolites synthesis	Zeng et al. (2017)
Potato	2-DE and MS	*Leptinotarsa decemlineata* and *Macrosiphum euphorbiae*	Perception of distinct physical and chemical cues in plants	Duceppe et al. (2012)
Chilli/Hot pepper	iTRAQ	*Bemisia tabaci*	Regulation, stress response, protein metabolism, lipid metabolism and carbon metabolism	Wu et al. (2019)
Arabidopsis	2D PAGE	*Plutella xylostella*	Enhanced production of ROS	Collins et al. (2010)
	2-DE and IEF and MALDI-TOF or LC-LTQ orbitrap MS/MS	*B. tabaci*	Proteins involved in protein folding and regulation, redox regulation, primary metabolism RuBisCo activase and secondary metabolism	Yin et al. (2012)
	2-DE, MALDI TOF-MS/MS LC-ESI-MS/MS	*Myzus persicae*	Proteins involved in amino acid synthesis, carbohydrate synthesis, energy metabolism and defence response and translation	Truong et al. (2015)
	MALDI TOF/MS and LS-ESI-MS/MS	*P. xylostella*	Proteins involved in amino acid synthesis, carbohydrate synthesis, lipid metabolism and photosynthesis	Truong et al. (2018)

13.4 PROTEOMICS APPROACHES IN DIFFERENT CROPS AGAINST INSECT RESISTANCE

Many studies have disclosed that expression levels of proteins differed in susceptible and resistant genotypes after insect pest infestation. Further analysis of proteins revealed the participation of these proteins in different physiological processes, activation of stress signalling pathways and different metabolism (Table 13.2). Therefore, assessing the defence mechanisms in plants by using a proteomic approach is helpful in the development of insect-resistant crop varieties and pest control. The research studies on proteomic approaches in different crops for resistance to pests are reviewed and discussed.

13.4.1 RICE

Rice (*Oryza sativa*) is the main staple food crop in the world. Many pests attack the crop and cause yield losses. All portions of the plant are subjected to attack by insect pests at any stages of plant growth. It is reported that around 300 species of insect pests affect rice crop in India. Out of this, 20 species have been found to cause major damages (Arora and Dhaliwal, 1996). Some of these are plant and leaf hoppers, stem borers, defoliators, root feeders and grain sucking insects. Insects can attack rice grains during storage also. Brown plant hopper (BPH), whitebacked BPH (WBPH) and small BPH (SBPH) are typical plant hoppers and major piercing sucking pests in rice.

BPH (*Nilaparvata lugens* Stal) is a major destructive pest of rice. A quantitative MS proteomic approach was used for the comparative analysis of expression profiles of proteins in leaf sheaths of rice in response to the infestation by the BPH (*N. lugens* Stal) in susceptible and resistant genotypes of rice (Wei et al., 2009). A significant change in the expression levels of proteins in response to BPH infestation was reported. The proteins were found to be involved in multiple pathways like jasmonic acid (JA) synthesis, beta-glucanases, oxidative stress response, clathrin protein, protein kinases, photosynthesis, glycine cleavage system and aquaporins. However, some proteins related to callose metabolism remained unchanged.

In another study of resistance to BPH, a proteomics methodology was used in wild type IR 64 genotype and near isogenic rice mutants (Sangha et al., 2013). In response to the BPH infestation, 65 proteins were differentially expressed in wild type IR 64 when compared to near isogenic mutants. Based on MS, 52 proteins identified were involved in 11 functional categories. Expression of proteins differed in terms of initial days of infestation (less altered) and 28 days after of infestation (expressed more). A total of 22 proteins were associated with resistance to BPH. In the near isogenic dominant (D 518) and susceptible (D 1131) mutants also, there was differential expression of proteins along with S-like RNase, EFTu1, glyoxalase I and salt stress root protein RS1.

iTRAQ-based quantitative proteomics approach was used to study the protein expression profiles in the phloem exudates of BPH-resistant and susceptible rice plants after BPH infestation. A total of 238proteins were identified in phloem sap of rice genotypes, which were involved in multiple pathways, including redox regulation, defence signal transduction, protein and carbohydrate metabolism and cell structural proteins (Du et al., 2015).

Using the same iTRAQ proteomics approach, three cultivars of rice, resistant, susceptible and hybrid of these were studied for expression levels of proteins in response to BPH infestation (Zhang et al., 2019) to understand the mechanisms of BPH resistance in rice. Differentially expressed proteins (DEPs) were detected after BPH infestation from a resistant (414), susceptible (425) and hybrid cultivars (470). These identified DEPs are mainly engaged in the biosynthesis of secondary metabolites, glyoxylate, and carbon and dicarboxylate metabolism. Six DEPs, viz., lipoxygenases (LOXs) (two), a lipase, dirigent proteins (DIRs) and an Ent-cassa-12,15-dienesynthase (OsDTC1) and a heat shock protein (HSP20) were reported to be associated with BPH resistance in resistant cultivar of rice (Zhang et al., 2019).

BPH has different biotypes. Studying the biotypes helps in understanding the adjustment of BPH biotypes to rice plants and also helps in framing the better design of new management strategies for host defence against BPH. The protein profiles of the two virulent populations (Y and I) were studied using iTRAQ proteomics approach (Zha and You, 2020). Biotypes Y and I can survive on moderately resistant rice cultivar (YHY15) and on the susceptible rice cultivar (TN1). A total of 258 DEPs were detected. Out of these, 151 DEPs were up-regulated, while 107 were down-regulated.

SBPH (*Laodelphax striatellus*) is another sucking pest of rice. The comparative protein profiles analysis of two contrasting genotypes of rice, *i.e.*, SBPH resistant genotype (Pf 9279–4) and SBPH susceptible genotype (02428) in the leaf sheaths of these plants after infestation with SBPH was analysed (Dong et al., 2017). A total of protein spots were differentially expressed between the SBPH resistant and susceptible genotypes after infestation. Out of these, 24 DEPs were reported. These proteins were involved mainly in stress response, transcriptional regulation, protein metabolism, carbohydrate energy metabolism and amino acid metabolism, photosynthesis and were cell wall-related proteins.

Cnaphalocrocis medinalis is the main insect pest of rice in Asian continent. Proteomic analysis was carried out by SWATH-MS to detect DEPs between the two rice cultivars, resistant (Qingliu) and susceptible (TN1), after four time points of leaf folder herbivory feeding (Cheah et al., 2020). The DEPs expressed were identified to be participative in various physiological processes which were photosynthesis, amino acid, metabolic and secondary metabolic processes. The resistant mechanisms reported to be involved are JA signalling, production of flavonoids and glutathione and carbon remobilization in the resistant genotype of rice.

13.4.2 WHEAT

Wheat (*Triticum aestivum* L.) is another cereal grain crop that is cultivated as staple food around the world. Presently, it is considered the second main staple food crop after rice. *Triticum* genus contains different species which include diploids, tetraploids and hexaploids. Among these, common wheat or bread wheat (*T. aestivum*) is widely cultivated. The insect pests which attack wheat are aphids, brown mite, wheat thrips, shoot fly, American pod borer, army worm or cutworm and pink stem borer.

The grain aphid (*Sitobion avenae*) (Fabricius) is an alarming pest of wheat crop. Aphids differ from other insect pests in their manner of feeding. They feed from phloem tissue by inserting its stylet between the cells (Farmer and Ryan, 1992; Gatehouse, 2002).

Resistance against aphid (*S. avenae*) infestation was studied by using proteomic approaches in wheat (Ferry et al., 2011; Guan et al., 2015). Around 500 protein spots were identified from leaf extracts of seedlings using 2-DE approach. DEPs were reported between control and affected plants after 24 h and eight days after aphid feeding. Some were up-regulated and some were down-regulated in local or systemic tissues. DEPs were also identified after the external application of salicylic acid (SA) or methyl jasmonate in control plants. These proteins were involved in metabolic processes and photosynthesis, in signal transduction, stress and defence, antioxidant activity, regulatory processes and hormone responses. In another study, resistance to aphid infestation was studied using 2D gel electrophoresis and image comparison software in two accessions of wild Einkorn wheat, susceptible line (ACC5 PGR1735) and tolerant line (ACC20 PGR1755) (Guan et al., 2015). Among 200 protein spots identified, 24 spots were considerably up-regulated in the resistant line after 24 h of aphid feeding. The bulk of proteins up-regulated by aphid infestation were entangled in metabolic processes, including photosynthesis and transcriptional regulation.

The wheat stem sawfly (*Cephus cinctus*, WSS) is another pest of wheat, and partial resistance has been deployed by breeding for a solid-stem trait, but this trait is affected by the environment. The molecular response to WSS was studied by using proteomics on four wheat cultivars (Lavergne et al., 2020) differing in the resistance reaction to insect pest. A whole set of 1830 proteins identified were reported to be associated in five major biological processes. All four varieties had a molecular response to WSS following infestation. The most clear-cut changes were observed for the resistant cultivar in which the most expressed proteins were involved in proteinase inhibition, enzymatic detoxification and anti-herbivory compound production.

The usage of the proteomics method to look into plant protein behaviour in insect gut is a new method in insect–plant interaction studies. Sunn pest (*Eurygaster integriceps*) is a major pest of barley and wheat in Iran. The main part of digestive system of insects includes the gut and the salivary glands. The functioning of these organs with respect to digestion is associated with the expressed proteins. Some protein spots were traced in gut of an adult in sunn pest using 2-DE, MS and NCBI database. Six proteins, serpin, β-amylase, dehydroascorbate reductase, α-amylase inhibitor, triticin and α-L arabinofuranidase, were identified using plant database (Saadati and Toorchi, 2017).

13.4.3 MAIZE

Maize (*Zea mays*), well-known as corn, is a cereal grain native to Southern Mexico. It is a staple food in many areas of the world, with the total production of maize exceeding that of rice or wheat. In addition, being used directly by humans, maize is also used for feed, corn ethanol and other maize products, such as corn starch and corn syrup. The percent loss caused by insect pests in maize ranges from 5 to 15%. Among the pests, the four key pests of maize prevalent in India are pink stem borer, spotted stem borer, shoot fly and fall army worm.

George et al. (2011) studied the differential protein expression after infestation of maize plants with two insects having different feeding strategies (*Spodoptera littoralis*, chewing insect, and

Busseola fusca, stem borer). Infestation of maize with 3rd instar larvae belonging to *S. littoralis* and *B. fusca* resulted in differential up-regulation and down-regulation of protein spots (14 and 7 in *S. littoralis* and 12 and 9 in *B. fusca*). Among these, only nine and four were common in up-regulated proteins and down-regulated proteins, respectively.

JA plant hormone is involved in different growth and developmental activities and defence responses to abiotic and biotic stresses. Methyl Jasmonate (MeJA), a volatile form of JA, has been widely used to study jasmonate signalling pathway and mechanism of plant defence.

MeJA activates signalling cascade of JA and defensive proteins such as pathogen-related (PR) proteins which are involved in plant defence and immunity response (Zhang et al., 2015). A total of 62 MeJA response proteins were discovered in maize after MeJA treatment against corn borer. Out of 62 proteins, 43 protein levels were enhanced, and the levels of 11 proteins were decreased. Eight proteins were distinctively expressed in response to MeJA treatments. These identified proteins are important in biological functions, including photosynthesis, energy-related proteins, degradation, protein folding and regulated proteins, defence and stress-regulated proteins and redox responsive proteins, transcription-related protein metabolism, protein synthesis, cell structure and secondary metabolism.

MeJA not only induced plant defence mechanism to insects, but it also enhanced toxic protein production that can potentially be used for biocontrol of Asian corn borer (ACB) (Zhang et al., 2015). ACB larvae fed on leaves treated with altered concentrations of MeJA affected the differential growth stages of insect, *i.e.*, larval mortality, developmental period of the different larval stages of the ACB, adult stage life span and average number of eggs produced by the female per day.

13.4.4 SORGHUM

Sorghum (*Sorghum bicolor* L. Moench) is also an important cereal food, forage and biofuel crop cultivated in the world. Around 150 species of insects are known to affects orghum. Out of these, shoot fly, spotted stem borer, midge and head bugs are the key pests. The sucking pests, *viz.*, shoot bugs and aphids are sporadic pests and seldom cause economic damage. The shoot fly is the key pest in India which leads to heavy yield loss.

An effort was made to study the proteomics in spotted stem borer infestation in sorghum using three genotypes which were varied for susceptibility to spotted stem borer. ICSV 700 and IS 2205 showed variable degree of resistance, and Swarna variety was susceptible to the insect infestation (Tamhane et al., 2021). Insect infestation procedure was carried out with the Bazooka applicator, and leaves were taken five days post infestation and applied label-free quantitative proteomic approach. A total of 967 proteins identified were involved in photosynthesis and stress responses. At steady-state, the resistant genotypes of *S. bicolor* exhibited at least two-fold higher number of unique proteins than that of the susceptible genotype Swarna.

13.4.5 PIGEON PEA

Pigeon pea [*Cajanus cajan* (L.) Millspaugh] is an economically important legume crop in semi-arid tropics. The insect pests which attack pigeon pea are spotted pod borer, gram pod borer, pod fly, leaf webber, blister beetle and plume moth. The crop is susceptible to *Helicoverpa armigera* (Hubner), which causes devastating yield losses. This pest develops resistance to many commercially available insecticides. Therefore, wild relatives of pigeon pea are being considered potential sources of genes to expand the genetic base of cultivated pigeon pea to improve traits such as host plant resistance to pests and pathogens. The wild relative *Cajanus platycarpus* is resistant to *H. armigera*. The resistance mechanism(s) in *C. platycarpus* versus *C. cajan* was studied using proteomic approach after continued herbivory (up to 96 h). DEPs were observed in the wild relative *C. platycarpus* (Rathinam et al., 2020). The up-regulated proteins played a vital role in the phenylpropanoid pathway, microtubule assembly and synthesis of lignins. Some proteins were up-regulated which were involved in polyamine pathway. It was observed that the production of secondary metabolites by reallocation of resources and diversion of metabolic flux could be the probable approach in the wild relative against herbivory. Pod borer resistance mechanism which is present in wild relative *C. platycarpus* can also be used in crop improvement.

Quantitative proteomic analysis was conducted using the TMT platform to detect differentially abundant proteins between tolerant (IBS 3471) and susceptible variety (ICPL 87) accession to *H. armigera* (Dawit et al., 2021). Leaf proteomes were analysed at the vegetative and flowering/podding growth stages. *H. armigera* tolerance in IBS 3471 appeared to be related to enhanced defence responses, such as changes in secondary metabolite precursors, antioxidants and the phenylpropanoid pathway. A rapid accumulation of reactive oxygen species (ROS) was reported in the IBS 3471, when the insect larvae were fed on artificial media with lyophilized leaves of resistance genotype.

13.4.6 SOYBEAN

Soybean (*Glycine max*) is the unique grain legume globally accepted for its dual-purpose use as pulse and oilseed containing 38%–44% protein and 18%–22% oil. It contains proteins, vitamins, fatty acids, minerals and other nutrients for animals and humans, and it also has non-food uses, such as the production of combustible fuels and industrial feedstocks. However, pests can poorly affect the yield and quality of soybean. India is the third largest importer of soya oil in the world and one of the major exporters of soya meal to the other Asian countries. Soybean offers substratum for about 275 species of pests in India. Out of these, only a dozen of species, like tobacco caterpillar, girdle beetle, Bihar hairy caterpillar, green semilooper, stemfly, jassids, aphids and white fly, attain the major pest status. A total of 380 species are reported in soybean crop in many parts of the world.

Cotton worm (*Prodenia litura*, fabricius) is one of the key insect pests of soybean. Soybean leaves were fed with this worm, and proteomic analysis was done using 2D GE and MALDI-TOF MS (Fan et al., 2012). A total of 11unique proteins were detected out of which 10 protein spots were expressed in response to worm feeding. These proteins were mainly involved in physiological processes, including defence signal transduction, reactive oxygen removal and metabolism regulation.

Lamprosema indicata (Fabricius) is a major leaf feeding pest of soybean, whose larvae hide in soybean leaves, induce leaf curling and feed on leaf tissues. This feeding impacts the plant photosynthesis and causes abnormal plant growth. It belongs to the Lepidoptera and Pyralidae groups. To understand the defence mechanism of resistance to *L. indicata* in soybean, differential proteomic analysis by iTRAQ-MRM was done using a highly resistant line (Gantai-2–2) and a highly susceptible line (Wan 82–178) after larval feedings for 0 and 48 h (Zeng et al., 2017). DEPs were reported 0 and 48 h after feeding. Most of the DEPs were coupled with ribosome, peroxisome, flavonoid biosynthesis, linoleic acid metabolism, phenylpropanoid biosynthesis, stilbenoid, diarylheptanoid and gingerol biosynthesis, glutathione metabolism, plant hormone signal transduction, and flavonol and flavone biosynthesis, as well as other resistance-related metabolic pathways. Resistance to this insect was due to the production of anti-digestive proteins which hinder the growth and development of insects, signalling pathways, secondary metabolites synthesis, ROS scavenging and so on.

13.4.7 COTTON

Cotton (*Gossypium hirsutum*) is one of the world's leading fibre crops and is abundant and economically produced, making cotton relatively inexpensive. In India, it is a major important commercial crop. The crop is infested with different kinds of insects in the production cycle. Jassids, thrips, aphids, mealy bugs and whiteflies are among sucking pests, and red cotton bug, bollworm (American, pink and spotted) and *Spodoptera* caterpillars attack leaves and bolls.

Among these pests, cotton bollworm (*Helicoverpa armigera*) is a key pest that feeds on bolls caus-ing widespread damage leading to productivity loss. The genome wide response of bolls infested with bollworm was studied using proteomic approach (Kumar et al., 2015). About 35% of proteome was regulated differentially during bollworm infestation, and 45% of differentially expressed proteins were found to be associated in signalling pathways followed by redox regulation. They suggested that defence against insect *H. armigera* is associated with stress-responsive hormone regulation.

13.4.8 POTATO

Potato (*Solanum tuberosum*) is an important vegetable crop of the world. It is a temperate crop grown under subtropical conditions in India. Potatoes are used in numerous industrial purposes such as for the production of alcohol, starch, dextrin and glucose. More than 100 insect species infect potato plants. These pests cause damage to potato by feeding on leaves, reducing photosynthetic efficiency by attacking stems, thereby weakening plants and impeding nutrient transport. Potato tubers are heav-ily damaged in the field and during storage, causing direct losses to farmers. Insect pests that damage tubers include white grubs, cutworms, potato tuber moth, termites, red ants and mole crickets. Sap feeding insects such as aphids, leafhoppers, thrips and whiteflies inflict damage by directly feeding on different parts of a plant and as vectors of plant viruses. The important leaf feeding caterpillars are *Spodoptera* spp., *Helicoverpa armigera*, *Thysanoplusia orichalcea* and *Spilosoma obliqua*. Among coleopterans, the most destructive pests are hadda beetles, flea beetles and blister beetles.

The response of potato plants to herbivorous insects was studied by using proteomic approaches. The plants were subjected to mechanical wounding, defoliation by the Colorado potato beetle *Leptinotarsa decemlineata* Say, or phloem sap feeding by the potato aphid *Macrosiphum euphor-biae* Thomas (Duceppe et al., 2012). Approximately 500 leaf proteins were monitored by herbivory compared to healthy control plants. A total of 31 were up-regulated in the herbivory treatment. Some proteins were up- or down-regulated in response to different treatments. The up-regulated proteins were identified by MS as typical defence proteins, including wound-inducible protease inhibitors and pathogenesis-related proteins. Some proteins were involved in photosynthesis.

13.4.9 TOMATO

Tomato (*Solanum lycopersicum*) is a herbaceous, sprawling plant with a weak woody stem that grows 1–3 m in height. Tomato is a Peruvian and Mexican native and third important crop in India following potato and onion. India is the second-largest producer of tomato in the world. The pests which can attack tomato are aphids, tobacco caterpillar, whitefly, serpentine leaf miner, gram pod borer, spider mites and root-knot nematode.

Aphids are among the most devastating pests in temperate climates, causing substantial damage in several crops including tomato. The proteomic analysis was carried out to study the molecular mechanism of *Macrosiphum euphorbiae* aphid and response of tomato. Protein profiles were esti-mated 24, 48 and 96 h after aphid infestation (Coppola et al., 2013). A total of 899 differentially expressed genes and 57 identified proteins indicated that the tomato response is characterized by an enhanced oxidative stress, which was counteracted by the induction of the proteins involved in the detoxification of oxygen radicals. Aphids elicit a defence reaction based on cross-communication of various hormone-related signalling pathways such as those related to the SA, ethylene, JA and brassinosteroids. Among them, stress-responsive SA-dependent genes and the SA-signalling path-way play a dominant role.

13.4.10 ARABIDOPSIS

Arabidopsis thaliana, mouse-ear cress, the thale cress or Arabidopsis, is a small flowering plant native to Eurasia and Africa. It is considered a weed and is found along the roads and in disturbed land.

Arabidopsis is a winter annual with relatively short lifecycle. *A. thaliana* is a popular model organism in genetics and plant biology. It was the first plant whose genome has been sequenced and is a popular specimen for understanding the molecular biology of many plant traits, comprising flower development and light sensing. It is a model crop of the Brassicaceae family and is used to study insect resistance mechanism in crops belonging to this family.

Diamondback moth (*Plutella xylostella*) is a specialist herbivore that feeds on species belongs to the family Brassicaceae, including *Arabidopsis thaliana* (Stotz et al., 2000). This pest is regarded as the most destructive insect pest throughout the world (Mohan and Gujar, 2001). *A. thaliana-P. xylostella* interaction is a model system used to investigate insect resistance and the analysis of inducible defence mechanisms.

A proteomic approach (2D-PAGE) was used to study the physiological factors affecting feeding behaviour by larvae of the insect, *P. xylostella*, in herbivore-susceptible and herbivore-resistant *Arabidopsis thaliana* (Collins et al., 2010) of 162 recombinant inbred lines (RILS) which revealed significant differences in the proteomes between the resistant and susceptible RILS. Increased production of hydrogen peroxide in resistant RILs was reported suggesting that enhanced production of ROS may be a major pre-existing mechanism of *Plutella* resistance in *Arabidopsis*. In another study of resistance to *P. xylostella*, a 2-DE proteomic approach was used (Truong et al., 2018). Approximately 450 protein spots were reproducibly detected. Out of these, 18 DEPs were identified between healthy and infested leaves. The identified proteins were associated with amino acid, carbohydrate, energy, lipid metabolism and photosynthesis.

The defence response of green peach aphid (*Myzus persicae* Sulzer), a polyphagous pest feeding on *Arabidopsis thaliana* (L.), was investigated by using the 2-DE proteomic approaches following aphid colony development after three days on the host plants (Truong et al., 2015). A total of 31 differentially expressed protein spots were reported. The selected proteins were identified to be associated in different metabolic pathways: carbohydrate, amino acid and energy metabolism, photosynthesis, defence response and translation after aphid feeding.

13.4.11 CHILLI/HOT PEPPER

Chilli (*Capsicum annuum* L.) is an important economic and commercial cash crop and is primarily used as a spice and vegetable in various cuisines globally. The production of chilli suffers from many insect pest damage, *viz.*, sucking pests like aphids, thrips, whiteflies, mealy bugs and mites, leaf feeders, leaf eating caterpillars, fruit borers and root grubs.

The *Bemisia tabaci* is a major leaf feeding insect pest to pepper causing serious damage to growth and yield. iTRAQ proteomics approach was used to study the molecular mechanism of resistance to *B. tabaci* after feeding for 48 h in highly resistant genotype and a highly susceptible genotype (Wu et al., 2019). A total of 37 differential abundance proteins (DAPs) and 17 DAPs were identified in the resistant genotype and susceptible genotype, respectively, at 48 h after *B. tabaci* feeding. These DAPs, identified to be involved in different pathways, were connected with redox regulation, stress response, protein metabolism, lipid metabolism and carbon metabolism. Among DAPs, some candidate DAPs are closely related to *B. tabaci* resistance such as annexin D4-like (ANN4), calreticulin-3 (CRT3), heme-binding protein 2-like (HBP1), acidic endochitinase pcht28-like (PR3) and lipoxygenase 2 (LOX2).

13.5 CONCLUSION

Understanding the molecular mechanisms of insect resistance in crop plants helps in developing the resistant varieties to a particular insect pest. Different omics approaches have been used to decipher the molecular mechanisms of defence against insect pests, *viz.*, genomics, transcriptomics, QTLomics, proteomics, etc. Proteomics is an ideal approach to study the expression of different proteins in response to insect feeding in susceptible and resistant genotypes which helps in utilizing the resistance mechanism in crop improvement against insect pests.

REFERENCES

Arora, R. and Dhaliwal, G.S. (1996), "Agro-ecological changes and insect pest problems in Indian agriculture", *Indian Journal of Ecology*, 23, 109–122.

Aslam, B., Basit, M., Nisar, M.A., Khurshid, M. and Rasool, M.H. (2017), "Proteomics: technologies and their applications", *Journal of Chromatographic Science*, 55(2), 182–196.

Barah, P. and Bones, A.M. (2015), "Multidimensional approaches for studying plant defence against insects: from ecology to omics and synthetic biology", *Journal of Experimental Botany*, 66(2), 479–493.

Cheah, B.H., Lin, H.H., Chien, H.J., Liao, C.T., Liu, L.Y.D., Lai, C.C., Lin, Y.F. and Chuang, W.P. (2020), "SWATH-MS-based quantitative proteomics reveals a uniquely intricate defense response in *Cnaphalocrocis medinalis*-resistant rice", *Scientific Reports*, 10(1), 1–11.

Collins, R.M., Afzal, M., Ward, D.A., Prescott, M.C., Sait, S.M., Rees, H.H. and Tomsett, A.B. (2010), "Differential proteomic analysis of *Arabidopsis thaliana* genotypes exhibiting resistance or susceptibility to the insect herbivore", *Plutella xylostella*", *PLoS One*, 5(4), e10103.

Coppola, V., Coppola, M., Rocco, M., Digilio, M.C., Ambrosio, D., Renzone, C., Martinelli, G., Scaloni, R., Pennacchio, A., Rao, F. and Corrado, R. (2013), "Transcriptomic and proteomic analysis of a compatible tomato-aphid interaction reveals a predominant salicylic acid-dependent plant response", *BMC Genomics*, 14(1), 1–18.

Cristea, I.M., Gaskell, S.J. and Whetton, A.D. (2004), "Proteomics techniques and their application to hematology", *Blood*, 103(10), 3624–3634.

Dawit, N.A., Njaci, I., Higgins, T.J., Williams, B., Ghimire, S.R., Mundree, S.G. and Hoang, L.T.M. (2021), "Comparative TMT proteomic analysis unveils unique insights into *Helicoverpa armigera* (Hübner) resistance in *Cajanus scarabaeoides* (L.) Thouars", *International Journal of Molecular Sciences*, 22(11), 5941.

Domon, B. and Aebersold, R. (2006), "Mass spectrometry and protein analysis", *Science* (New York, NY), 312(5771), 212–217.

Dong, Y., Fang, X., Yang, Y., Xue, G.P., Chen, X., Zhang, W., Wang, X., Yu, C., Zhou, J., Mei, Q. and Fang, W. (2017), "Comparative proteomic analysis of susceptible and resistant rice plants during early infestation by small brown planthopper", *Frontiers in Plant Science*, 8, 1744.

Du, B., Wei, Z., Wang, Z., Wang, X., Peng, X., Du, B., Chen, R., Zhu, L. and He, G. (2015), "Phloem-exudate proteome analysis of response to insect brown plant-hopper in rice", *Journal of Plant Physiology*, 183, 13–22.

Duceppe, M.O., Cloutier, C. and Michaud, D. (2012), "Wounding, insect chewing and phloem sap feeding differentially alter the leaf proteome of potato, *Solanum tuberosum* L", *Proteome Science*, 73(10).

FAO. (2022). www.fao.org.

Fan, R., Wang, H., Wang, Y. and Yu, D. (2012), "Proteomic analysis of soybean defense response induced by cotton worm (*Prodenialitura*, fabricius) feeding", *Proteome Science*, 10(1), 1–11.

Farmer, E.E. and Ryan, C.A. (1992), "Octadecanoid precursors of jasmonic acid activate the synthesis of wound-inducible proteinase inhibitors", *Plant Cell*, 4, 129–134.

Ferry, N., Stavroulakis, S., Guan, W., Davison, G.M., Bell, H.A., Weaver, R.J., Down, R. E., Gatehouse, J. A., and Gatehouse, A. M. R. (2011), Molecular interactions between wheat and cereal aphid (*Sitobion avenae*): Analysis of changes to the wheat proteome. *Proteomics*. 11, 1985–2002. doi: 10.1002/pmic.200900801

Gatehouse, J.A. (2002), "Plant resistance towards insect herbivores: a dynamic interaction", *New Phytology*, 156, 145–169.

George, D., Babalola, O.O. and Gatehouse, A.M.R. (2011), "Differential protein expression in maize (*Zea mays*) in response to insect attack", *African Journal of Biotechnology*, 10(39), 7700–7709.

Guan, W., Ferry, N., Edwards, M.G., Bell, H.A., Othman, H., Gatehouse, J.A. and Gatehouse, A.M. (2015), "Proteomic analysis shows that stress response proteins are significantly up-regulated in resistant diploid wheat (*Triticum monococcum*) in response to attack by the grain aphid (*Sitobionavenae*)", *Molecular Breeding*, 35(2), 1–22.

Indiastat. (2022). www.indiastat.com

Issaq, H. and Veenstra, T. (2008), "Two-dimensional polyacrylamide gel electrophoresis (2D-PAGE): advances and perspectives", *Biotechniques*, 44(4), 697–700.

Kumar, S., Kanakachari, M., Gurusamy, D., Kumar, K., Narayanasamy, P., Venkata, P.K., Solanke, A., Gamanagatti, S., Hiremath, V., Katageri, I.S., Leelavathi, S., Kumar, P.A. and Reddy, V.S. (2015),"Genome-wide transcriptomic and proteomic analyses of bollworm-infested developing cotton bolls revealed the genes and pathways involved in the insect pest defence mechanism", *Plant Biotechnology Journal*, 1–18. doi: 10.1111/pbi.12508

Lavergne, F.D., Broeckling, C.D., Brown, K.J., Cockrell, D.M., Haley, S.D., Peairs, F.B., Pearce, S., Wolfe, L.M., Jahn, C.E. and Heuberger, A.L. (2020), "Differential stem proteomics and metabolomics profiles for four wheat cultivars in response to the insect pest wheat stem sawfly", *Journal of Proteome Research*, 19(3), 1037–1051.

Marouga, R., David, S. and Hawkins, E. (2005), "The development of the DIGE system:2D fluorescence difference gel analysis technology", *Analytical and Bioanalytical Chemistry*, 382(3), 669–678.

Mohan, M. and Gujar, G.T. (2001), "Toxicity of *Bacillus thuringiensis* strains and commercial formulations to the diamond back moth, *Plutella xylostella* (L.)", *Crop Protection*, 20, 311–316.

Rathinam, M., Roschitzki, B., Grossmann, J., Mishra, P., Kunz, L., Wolski, W., Panse, C., Tyagi, S., Rao, U. and Schlapbach, R. (2020), "Unraveling the proteomic changes involved in the resistance response of *Cajanus platycarpus* to herbivory by *Helicoverpa armigera*", *Applied Microbiology and Biotechnology*, 104, 7603–7618.

Roychoudhury, A., Datta, K. and Datta, S.K. (2011), Abiotic stress in plants: From genomics to metabolomics. In: Tuteja, N., Gill, S.S. and Tuteja, R. (Eds.). Omics and Plant Abiotic Stress Tolerance, Bentham Science Publishers, Pp. 91–120.

Saadati, M. and Toorchi, M. (2017), "The study of plant protein accumulation in gut of insect using proteomics technique: Wheat–sunn pest interaction", *Journal of the Saudi Society of Agricultural Sciences*, 16(3), 205–209.

Sangha, J.S., Yolanda, H.C., Kaur, J., Khan, W., Abduljaleel, Z., Alanazi, M.S., Mills, A., Adalla, C.B., Bennett, J., Prithiviraj, B. and Jahn, G.C. (2013), "Proteome analysis of rice (*Oryza sativa* L.) mutants reveals differentially induced proteins during brown planthopper (*Nilaparvata lugens*) infestation", *International Journal of Molecular Sciences*, 14(2), 3921–3945.

Stotz, H.U., Pittendrigh, B.R., Kroymann, J., Weniger, K., Fritsche, J., Bauke, A. and Mitchell-Olds, A. (2000), "Induced plant defence responses against chewing insects. Ethylene signalling reduces resistance of *Arabidopsis* against Egyptian cotton worm but not diamond back moth", *Plant Physiology*, 124, 1007–1017.

Tamhane, V.A., Sant, S.S., Jadhav, A.R., War, A.R., Sharma, H.C., Jaleel, A. and Kashikar, A.S. (2021), "Label-free quantitative proteomics of *Sorghum bicolor* reveals the proteins strengthening plant defense against insect pest *Chilopartellus*", *Proteome Science*, 19(1), 1–25.

Thompson, A., Schafer, J., Kuhn, K., Kienle, S., Schwarz, J., Schmidt, G., Neumann, A.T. and Hamon, C. (2003), "Tandem mass tags: A novel quantification strategy for comparative analysis of complex protein mixtures by MS/MS", *Analytical Chemistry*, 75, 1895–1904.

Truong, D.H., Bauwens, J., Delaplace, P., Mazzucchelli, G., Lognay, G., Francis, F. and Leiss, K. (2015), "Proteomic analysis of *Arabidopsis thaliana* (L.) Heynh responses to a generalist sucking pest (*Myzus persicae* Sulzer)", *Plant Biology (Stuttgart, Germany)*, 17, 1210–1217.

Truong, D.H., Nguyen, H.C., Bauwens, J., Mazzucchelli, G., Lognay, G. and Francis, F. (2018), "Plant defense in response to chewing insects: proteome analysis of *Arabidopsis thaliana* damaged by *Plutella xylostella*", *Journal of Plant Interactions*, 13(1), 30–36.

Tudi, M., Ruan, H.D., Wang, L., Jia Lyu, j, Sadler, R., Connell, D., Chu, C. and Phung, D.T. (2021), "Agriculture development, pesticide", *International Journal of Environmental Research and Public Health*, 18, 1112.

Wei, Z., Hu, W., Lin, Q., Cheng, X., Tong, M., Zhu, L., Chen, R. and He, G. (2009), "Understanding rice plant resistance to the brown planthopper (*Nilaparva talugens*): a proteomic approach", *Proteomics*, 9, 2798–2808.

Wiese, S., Reidegeld, K.A., Meyer, H.E. and Warscheid, B. (2007), "Protein labelling by iTRAQ: a new tool for quantitative mass spectrometry in proteome research", *Proteomics*, 7(3), 340–350.

Wilkins, M.R., Sanchez, J.C., Gooley, A.A., Appel, R.D., Humphery-Smith, I. and Hochstrasser, D.F. (1996), "Progress with proteome projects: why all proteins expressed by a genome should be identified and how to do it", *Biotechnology and Genetic Engineering Reviews*, 13(1), 19–50.

Wu, X., Yan, J., Wu, Y., Zhang, H., Mo, S., Xu, X., Zhou, F. and Ding, H. (2019), "Proteomic analysis by iTRAQ-PRM provides integrated insight into mechanisms of resistance in pepper to *Bemisia tabaci* (Gennadius)", *BMC Plant Biology*, 19, 270.

Yates, J.R. III (2011), "A century of mass spectrometry: from atoms to proteomes", *Nature Methods*, 8(8), 633–637.

Yin, H., Yan, F., Jianguo, J., Li, Y., Wang and Xu, C. (2012), Proteomic analysis of *Arabidopsis thaliana* leaves infested by tobacco whitefly *Bemisia tabaci* (Gennadius) B biotype. Plant Molecular Biology Reports, 30, 379–390. doi: 10.1007/s11105-011-0351-0

Zeng, W., Sun, Z., Cai, Z., Chen, H., Lai, Z., Yang, S. and Tang, X. (2017), "Proteomic analysis by iTRAQ-MRM of soybean resistance to *Lamprosema indicata*", *BMC Genomics*, 18(1), 1–22.

Zha, W. and You, A. (2020), "Comparative iTRAQ proteomic profiling of proteins associated with the adaptation of brown plant hopper to moderately resistant vs. susceptible rice varieties", *PLoS ONE*, 15(9), e0238549.

Zhang, X., Yin, F., Xiao, S., Jiang, C., Yu, T., Chen, L., Ke, X., Zhong, Q., Cheng, Z. and Li, W. (2019), "Proteomic analysis of the rice (*Oryza officinalis*) provides clues on molecular tagging of proteins for brown planthopper resistance", *BMC Plant Biology*, 19, 30.

Zhang, Y.T., Zhang, Y.L., Chen, S.X., Yin, G.H., Yang, Z.Z., Lee, S., Liu, C.G., Zhao, D.D., Ma, Y.K., Song, F.Q. and Bennett, J.W. (2015), "Proteomics of methyl jasmonate induced defence response in maize leaves against Asian corn borer", *BMC Genomics*, 16(1), 1–16.

Zogli, P., Pingault, L., Grover, S. and Louis, J. (2020), "Ento(o)mics: the intersection of 'omic' approaches to decipher plant defense against sap-sucking insect pests", *Current Opinion in Plant Biology*, 56, 153–161.

14 Proteomics Approaches in Medicinal Plant Research and Pharmacological Studies

Riya Jain, Shruti Rohatgi, Deeksha Singh,
Shivangi Mathur, and Rajiv Ranjan

14.1 INTRODUCTION

Medicinal plants have been used for healing since the dawn of civilisation. The use of medicinal plants as herbal medicines was discovered after extensive research on the effectiveness of using different parts of the plant, such as leaves, roots, fruits, and barks, to cure many diseases. It is evident from many sources that humans have been searching for herbal medicines from ancient times. The various evidences include written records, historical monuments, and existing herbal medicines. These herbs were used to treat contagious diseases of humans as well as animals (Petrovska, 2012; Roychoudhury and Bhowmik, 2021a; Stojanoski, 1999).The health care systems of China, India, and Greece depend on the plants as an essential part of their pharmacopoeia along with Africa, America, Persia, and Rome (Pandey et al., 2013; Petrovska, 2012; Wilkins, 2014). Plant-based remedies are still used for the medications of several diseases. For instance, various manuals of phytotherapy report the use of cranberry juice (*Vaccinium macrocarpon*) and bearberry (*Arctostaphylos uva-ursi*) to diagnose cystitis infections, while plants like garlic (*Allium sativum*), lemon balm (*Melissa officinalis*), and tea tree (*Melaleuca alternifolia*) are defined as wide-ranging antimicrobial agents (Heinrich et al., 2017).

Medicinal plants play a significant role in the biomedical invention and served as the global foundation of naturopathy for a long period of time. By combining traditional knowledge of medicinal plants with interdisciplinary science, a number of therapeutic medications used in traditional medicine were developed later (Banerjee and Roychoudhury, 2017; Cragg et al., 1994). The inspiration to use traditional ideas to cure health has been proved by scientific data (Pandey et al., 2016). A growing body of knowledge is emerging about the cellular and molecular mechanisms of plants, as well as how genes and proteins work in various metabolic processes of plants, only with the advancement of current molecular biology and biotechnological tools, referred to as omics science (Van Emon., 2016). Predicting the gene networks that bioactive components of medicinal plants are regulating is one of the novel techniques to learn more about how active chemicals accomplish their therapeutic impact (Roychoudhury and Bhowmik, 2021b; Shao and Zhang, 2013).When compared to single-constituent doses, herbal medication combinations frequently have synergistic therapeutic effects. They can also increase the cytotoxicity caused by chemotherapy agents.

With the development of omics technology, omics research is now being used in almost all sectors of biomedicine. Omic approaches offer simultaneous evaluation of molecular classes, which are the primary experimental focus for such an approach (Gowda et al., 2008; Hashiguchi et al., 2017; Roychoudhury et al., 2011; Zaynab et al., 2018). Proteomic techniques have been used to study the physiology of therapeutic plants and their effects on animals in order to investigate the mechanisms behind the pharmacological activity of herbs. Analysis of the production of bioactive compounds that confer medicinal plants their health-promoting properties has been done using proteomic methods, which allow the assessment of systemic changes throughout the cellular metabolism. In comparison to metabolomics, proteomics is an effective method for dealing with analysing various biomarkers and systematic protein expression analysis during pharmacological treatment. Proteomic approaches are valuable in the study of medicinal plants because they may be used to isolate important enzymes involved in the production of beneficial compounds in addition to elucidating systemic physiologic changes (Kim et al., 2016; Mumtaz et al., 2017; Milani and Sparavigna, 2017).

Proteomics is the study of the proteome, which includes information about the expression, functions, structure, interactions, and protein modifications at every stage (Domon and Aebersold, 2006). It is the large-scale, high-efficiency systematic analysis of the proteome of a particular kind of cell, tissue, or body fluid, with a focus on structures and functions (Anderson and Anderson, 1998; Blackstock and Weir, 1999). Additionally, the proteome alters on a regular basis from cell to cell, from time to time, and in response to external stimuli. The complexity of proteomics in eukaryotic cells is caused by post-translational modifications, which can occur at various sites in a number of ways (Krishna and Wold, 1993).The total amount of proteins in a cell determined by their distribution, post-translational modifications, interactions, and turnover at a given moment is known as the "proteome" (Wilkins et al., 1996). For early illness diagnosis, prognosis, and disease development monitoring, proteomics is essential. Additionally, it plays a crucial role in the development of new drugs as target molecules (Domon and Aebersold, 2006).

Proteomics have quickly advanced in several research fields due to the development of related mass spectrometry (MS) and high-throughput analytical methods. A classical method for studying proteomics that is still used is two-dimensional gel electrophoresis (2-DE). The most significant property of 2-DE is the ability to increase the capacity, sensibility, resolution, and accuracy rate of the proteome (Marouga et al., 2005; Tannu and Hemby, 2006). MS is one of the proteomic techniques that has evolved to evaluate complicated protein mixtures more sensitively (Yates, 2011). Furthermore, Edman degradation is used to identify the amino acid sequence of a specific protein (Smith, 2001). Recent developments in quantitative proteomic approaches include stable isotope labelling with amino acids in cell culture (SILAC), Isotope-Coded Affinity Tag (ICAT), and Isobaric Tag for Relative and Absolute Quantification (iTRAQ) (Kroksveen et al., 2015; Ong et al., 2006; Shiio and Aebersold, 2006; Wiese et al., 2007). The three-dimensional (3D) structure of proteins is detected by two important high-throughput methods, X-ray crystallography and nuclear magnetic resonance (NMR) spectroscopy which may be useful in understanding their biological functions (Smyth and Martin, 2000; Wiese et al., 2007). A more effective proteomics methodology is 2-D fluorescence difference gel electrophoresis (2-D DIGE), which combines high-sensitivity protein-labelling methods with a narrow pH gradient gel separation (Marouga et al., 2005; Tannu and Hemby, 2006). Analysis of the transcriptome or proteome can help differentiate between two biological phases of the cell, allowing for the determination of variations in gene expression levels. For a comprehensive transcriptome study on a wide scale, microarray chips were created. Microarrays, however, cannot be used directly to assess an increase in mRNA production (Canales et al., 2006).

The capacity of proteomics to identify various species is among the most remarkable uses of this method in herbal therapy. These applications could be highly useful tools for the standardisation of herbal preparations, toxicity investigations, and quality control (Wang et al., 2011c). Genomic, transcriptomic, proteomic, metabolomic, and even a combination of various omics technologies are being used more and more in pharmaceutical research. The most crucial component of network biology and systems biology is omics research, which enables us to completely comprehend the pathological processes leading to illness, as well as to identify the major pathways and potential mechanisms underlying pharmaceutical discovery and drug therapy. Additionally, omics research may identify prospective drug development targets, facilitating effective safety evaluation and personalised therapy. These days, omics is acknowledged as a potent methodology for pharmaceutical research, particularly in investigations of target discovery, personalised medicine, toxicity, and classic Chinese medicine (Yan et al., 2015).

14.2   HISTORY OF PROTEOMICS IN MEDICINAL PLANT RESEARCH

In 1994, Marc Wilkins coined the word "proteomics," referring to the "PROTein complement of a genOME" (Wilkins et al., 1996). The foundation of herbal medicine and biomedical innovation worldwide is provided by medicinal plants. Traditional knowledge of medicinal plants was important for the discovery of many common therapeutic drugs, which were then validated by scientific discovery. "Paclitaxel," an anticancer drug made from the bark of *Taxus brevifolia*, is one of the amazing discoveries in the field of herbal medicines (Weaver, 2014). Scientific investigations on medicinal plants should focus on physiological studies of the generation of secondary metabolites and the pharmacological effects of active substances on animals to ensure proper usage of plant metabolic products and genetic resources. Proteomics research can be used to gather detailed physiological data on the therapeutic plants and the animals that consume them. Proteomic studies show how external factors can alter physiological processes and metabolic pathways (Zaynab et al., 2018).

The Human Genome Project is now complete, and the emphasis is laid on the protein content of humans. This has led to the development of proteomics, which examines all proteins generated by cells and organisms. This involves identifying proteins in the body and determining how they contribute to both physiological and pathological processes (Kosak and Groudine, 2004). The proteome identifies most of the functional information of a gene. Eukaryotic cell proteomes are relatively complicated and have a dynamic range. Additionally, prokaryotic proteins are responsible for the pathogenic mechanisms; however, because of the huge diversity in their characteristics, such as a dynamic range in concentration, molecular size, hydrophobicity and hydrophilicity, and examination of these proteins is difficult (Pandey and Mann, 2000).

Modern pharmacological sciences take a new approach to the equilibrium of biological systems that are simultaneously perturbed by primary and secondary molecular targets from a holistic systems biology perspective that goes beyond target specificity and single molecule pharmacology (Auffray et al., 2009). To employ novel medicines most effectively in clinical situations, it is essential to have a thorough understanding of their pharmacological features. Publishing scientific proof of the efficacy of natural products produced from plants has received a lot of attention (Efferth, 2017). Proteomic methods allow the assessment of systemic alterations during cellular metabolism and have been employed to examine the synthesis of the bioactive chemicals that give medicinal plants their health-promoting characteristics (Kim et al., 2016). Employing comparative proteomics, specific tissue-expressed proteins of *Cannabis sativa* were identified since the major active compound, viz., cannabinoids, are present in different organs including glands, flowers, and leaves (Raharjo et al., 2004).

14.3   TYPES OF PROTEOMICS APPROACHES FOR MEDICINAL PLANTS

Proteomics is a great technique for studying how metabolism alters in response to various stresses (Patel et al., 2009). It is categorised into various groups based on protein response towards the stress as shown in Figure 14.1.

14.3.1   FUNCTIONAL PROTEOMICS

Functional proteomics approach involves identifying proteins that interact with one another to understand functions and molecular mechanisms within a cell (Kolchand Pitt, 2010). The interaction of an unidentified protein with members of a particular protein complex involved in a given process would be very suggestive of its biological function (Chandrasekhar et al., 2014). The two primary objectives of functional proteomics techniques are the description of cellular mechanisms at the molecular level and the clarification of the biological roles of unknown proteins. Many proteins exhibit their biological properties in cells by quickly and briefly adhering together to form massive protein complexes (Böttcher et al., 2010). Additionally, the understanding of protein-protein

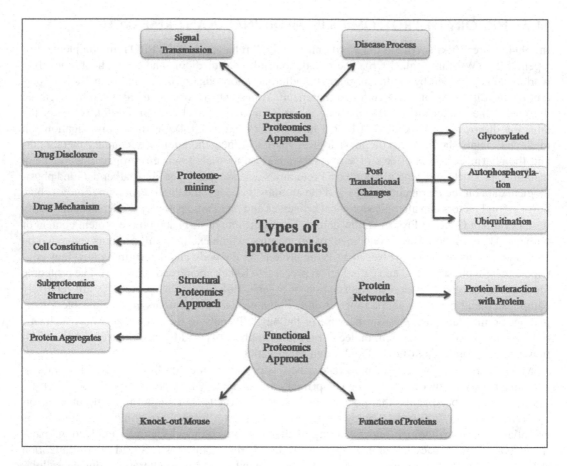

FIGURE 14.1 Types of proteomics approaches.

interactions *in vivo* may substantially aid in the precise description of cellular signalling networks (Monti et al., 2009).

Functional proteomics is a new area of study in the field of proteomics that is "focused on monitoring and analysing the spatial and temporal features of the molecular pathways and dynamics involved in living cells" (Godovac-Zimmermann and Brown, 2001).With a growing number of genomic initiatives, there is an exponential increase in protein sequences with unidentified functions. The biological function of the proteins in the cell is still completely unknown, despite advances in our understanding of the protein sequences, associated coding genes, chromosomal distribution, and even regulatory processes (Monti et al., 2005).

14.3.2 EXPRESSION PROTEOMICS

Expression proteomics is an approach used to analyse both quantitative and qualitative expressions of protein content under two distinct states. The protein that causes stress or a diseased condition, and the proteins that are expressed as a result of a disease, can be understood by comparing normal cells to treated or diseased cells (Hurwitz et al., 2016). Expression proteomics is typically focused on studying the protein expression patterns in diseased cells (Monti et al., 2009). For example, the differential protein expression between a sample of normal and tumour tissue can be examined (Chandrasekhar et al., 2014).

The protein expressional alterations that are present and absent in tumour tissue as compared to normal tissue were observed using the 2-D gel electrophoresis and MS techniques (Bauer and

Kuster, 2003; Liao et al., 2009). Overexpression and under expression of specific protein functions, signalling pathways, and multi-protein complexes can be identified and defined. The discovery of such proteins will provide important insights into the molecular biology of tumour development and disease-specific methods for usage as diagnostic indicators or therapeutic targets (McConkey et al., 2010).

14.3.3 STRUCTURAL PROTEOMICS

The three-dimensional shape and structural complexity of functional proteins are made possible by structural proteomics. Homology modelling is a technique for predicting the structure of a protein whose amino acid sequence has been known *via* sequence analysis or from the gene (Bauer and Kuster, 2003). The structure and functions of protein complexes that are present in a particular cellular organelle can be thoroughly explained by structural proteomics (Reisinger and Eichacker, 2007). It is feasible to recognise every protein found in a complex system, such as membrane surface, ribosomes, and cell organelles and to describe every protein interaction that might exist between proteins and their complexes. The main techniques for determining structure include X-ray crystallography and NMR spectroscopy (Yokoyama et al., 2000).

Structural proteomics has been used to determine the value of approximately 25–30% of proteins from eukaryotic organisms (Christendat et al., 2000). Determining how many three-dimensional structures are required to establish a "basic parts list" of protein folds is one such related application of structural proteomics data. The majority of other structures might be then represented using computational approaches from this basic set (Šali., 1998; Sanchez and Sali, 1998) The long-term objective is to discover the experimental structures of every protein, as current methodologies are not yet precise enough to reveal the minute changes in protein structures that contribute to the complexity and diversity of life (Moult et al., 2014).

14.4 METHODS OF PROTEOMICS FOR PROTEIN SEPARATION

14.4.1 PROTEOMICS SAMPLE PREPARATION

In proteomics research, sample preparation is the most important stage having a significant impact on an experimental result. For this, choosing an adequate experimental framework and sample preparation technique is crucial for producing data that is reliable, specifically in comparison to proteomics where small differences between experimental and control samples are considered (Freeman and Hemby, 2004). The large range of protein abundance is one of the major challenges in the study of complex biological materials. A protein may only exist in a small number of copies in a given cell, but an abundant protein may exist in up to a million copies; hence, these ample proteins must be excluded from the majority of proteome analyses. This might be accomplished by pre-analysing materials using various techniques as shown in Figure 14.2 for fractionation and protein enrichment (Bodzon-Kulakowska et al., 2007).

14.4.2 TRADITIONAL METHODS

The traditional methods for purifying proteins include chromatographic methods and western blotting. These methods may only be able to analyse a small number of proteins individually, and they are also unable to determine the degree of protein expression (Kurien and Scofield, 2006; Lequin, 2005).

14.4.2.1 Chromatographic Methods

Chromatography is a crucial biophysical method that helps to separate, identify and purify constituents of a mixture for quantitative and qualitative study. Proteins may be purified using factors, including their shape and size, total charge, the presence of hydrophobic groups on their surface, and their ability to bind to the stationary phase (Coskun, 2016). According to the basic

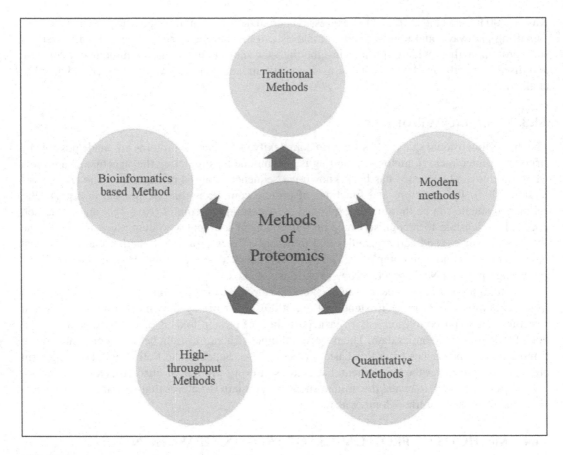

FIGURE 14.2   Methods of proteomics studies.

principle of chromatography, molecules in a mixture applied to a surface or a solid, and a liquid stationary phase (stable phase), separate from one another while moving with the assistance of a mobile phase. Molecular properties linked to partition (liquid-solid) or adsorption (liquid-solid), chromatography, or discrepancies between their molecular weights are among the elements that have an impact on this separation process. Due to these variations, certain mixture components flow quickly into the mobile phase and leave the chromatographic system faster, while others pass slowly into the stationary phase and move slowly through the system (Cuatrecasas et al., 1968; Harris, 2012; Porath, 1997).

14.4.2.1.1   Affinity Chromatography

A significant advancement in protein purification was made using affinity chromatography, which allows scientists to study protein breakdown, protein-protein interactions, and post-translational modifications. Affinity chromatography works on the principle of reversible interaction of the purifying proteins with the affinity ligand of the chromatographic matrix (Hage et al., 2012).With affinity chromatography, biological macromolecules are selectively isolated from a crude mixture based on their extremely specific interactions with tag proteins. The interaction is reversible, and the crude mixture is purified by keeping the fusion tag or affinity ligand stable to a support matrix, while keeping the target protein in a mobile phase (Dutta and Bose, 2022).

There are many applications for affinity chromatography in identifying pathogenic enzymes in microorganisms. Different amyloidogenic proteins and peptides interact with a set of amyloid-associated proteins, modifying the pathogenic and physical effects of these proteins. Purification of the amyloid peptide of Alzheimer's from the plasma of humans might be used to diagnose Alzheimer's disease using affinity chromatography (Aslam et al., 2017; Calero et al., 2012).

14.4.2.1.2 Ion Exchange Chromatography

The Ion Exchange Chromatography (IEX) is a versatile instrument for protein purification based on ionised groups found on its surface. The sequence of amino acids in different proteins varies; some amino acids are cationic and others are anionic. A balance between these charges is used to determine the net charge that a protein contains at a physiological pH level. It divides the protein first according to the kind of charge (cationic or anionic) and then further according to the relative charge strength. The IEX is extremely useful since it is cost-effective and can sustain in buffer conditions (Jungbauer and Hahn, 2009). Negatively charged proteins are adsorbed by anion-exchange matrices, which are positively charged ion-exchange materials. Cation-exchange matrices, on the other hand, absorb positively charged proteins and are known to be coupled with negatively charged groups (Biosciences, 2002).

Pharmaceutical goods use proteins produced in transgenic plants commercially. For instance, take the purified and expressed serine protease inhibitor aprotinin from maize seeds (Azzoni et al., 2005). Proteins of *Nigella sativa* having immunomodulating activity are separated using IEX; four peaks were obtained after the complete separation of the protein (Haq et al., 1999). Strong cation exchange chromatography may be used to purify the serum, which is made up of numerous cytokines, chemokines, proteolytic fragmentation, and peptide hormones of larger proteins (Tirumalai et al., 2003).

14.4.2.1.3 Gel-Permeation (Molecular Sieve) Chromatography

The basic idea behind this technique is to employ dextran-containing materials to divide macromolecules according to their disparities in molecular size. This process is used to calculate the molecular weight of proteins and reduce the concentration of salt in protein solutions. The stationary phase of a gel-permeation column is made up of inactive molecules with tiny pores. A steady flow rate is used to continually circulate the solution, which contains molecules of various sizes, across the column. Molecules bigger than the pores are unable to penetrate gel particles; instead, they are trapped between particles in small spaces. Larger molecules can travel quickly through the column because they can pass through the gaps between porous particles. As molecules become smaller, they diffuse into the pores and have proportionately longer retention durations when they exit the column (Determann, 2012).

14.4.2.1.4 Size Exclusion Chromatography

In liquid phase chromatography, a method known as size exclusion chromatography (SEC) is used to separate molecules, macromolecules, proteins, and nanoparticles (less than 100 nm) entirely based on their hydrodynamic capacity in solution (Peukert et al., 2022). Proteins are separated by SEC using a highly permeable carrier matrix with various pore sizes based on permeation; as a result, the proteins can be separated according to their molecular size. The SEC is a reliable method that can handle proteins in a variety of physiological circumstances, including those involving the presence of surfactants, co-factors, ions, or different temperatures. Under biological circumstances, the SEC is an effective method for purifying multimeric non-covalent protein complexes and is utilised to separate proteins of low molecular weight (Voedisch & Thie, 2010).

Through the use of SEC, the marine bacteria *Pseuodoalteromonas* produces antimicrobial peptides that have a potent inhibitory impact on pathogens responsible for skin diseases (Longeon et al., 2004).Various cytosolic proteins from *Arabidopsis thaliana* have also been isolated to understand how cells coordinate metabolic, mechanical, and developmental functions (Aryal et al., 2014).

14.4.2.1.5   High-Performance Liquid Chromatography

The identification and separation of carbohydrates, amino acids, nucleic acids, proteins, lipids, steroids, and some other biologically active compounds can be done quickly and accurately using this chromatography technique. It allows the purification and functional and structural analysis of numerous molecules. The mobile phase flows through columns in HPLC at a high flow rate of 0.1–5 cm/s while being subjected to ambient pressures of 10–400 atm. With this method, HPLC separation power is increased by the use of tiny particles and high pressure applied at the rate of solvent flow, and analysis is completed rapidly (Coskun, 2016). An HPLC device must have a solvent store, high-pressure pump, commercially manufactured column, sensor, and recorder. A computerised system is used to monitor the duration of separation and accumulation of material (Regnier, 1983). The primary chemicals accountable for the antioxidant function of indigenous Chinese medicinal plants have been identified in the first study from the twigs of *Morus alba* using HPLC-HRMS and HPLC-CD (Thomas Pannakal et al., 2022).

14.4.2.2   Western Blotting

Proteins are separated through electrophoresis, transferred to a nitrocellulose membrane, and then precisely detected by enzyme-conjugated antibodies in the western blotting process, which is a significant and effective method for detecting low-abundance proteins (Kurien and Scofield, 2006). Western blotting is a prominent method for identifying antigens from a variety of microorganisms and is useful for identifying infectious disorders (Aslam et al., 2017). Li et al. used western blotting to determine and validate the identities of ten reference proteins of rice (Li et al., 2011). The two proteins that were highly expressed in rice were heat-shock proteins and elongation factor 1-α (Kollerová et al., 2008).

14.4.3   MODERN METHODS

14.4.3.1   Electrophoresis (Gel-Based Methods)

14.4.3.1.1   2-DE

The two-dimensional gel electrophoresis (2-DE) served as the foundation for proteomics. The technique of analysing various proteins in parallel arose from 2D gel maps (Anderson et al., 2001; Blackstock and Weir, 1999; Wasinger et al., 1995). 2D gels have delivered a wealth of useful information and will remain an important aspect of proteomics study in the conceivable future (Harry et al., 2000; Rabilloud, 2002). Two-dimensional polyacrylamide gel electrophoresis (2D-PAGE) is an effective and dependable technique for separating proteins based on their charge and mass. According to the size of the gel, 2D-PAGE may detect around 5,000 distinct proteins sequentially. In the first of the two dimensions, the proteins are separated based on their charge, and in the second, based on the variations in their mass. The two-dimensional gel electrophoresis is effectively used for the study of metabolic pathways, post-translational modifications, mutant proteins, and other biological processes (Aslam et al., 2017).

Extraction of proteins from grapes is difficult because of the high protease activity, low protein concentration, and elevated levels of interfering substances like polyphenols, flavonoids, terpenes, lignans, and tannins. However, Marsoni et al. (2005) separated the proteins through 2-DE from the tissue of grapes, and Islam et al. (2004) as well separated the proteins from the fully developed leaf tissue of rice plants and used them in the analysis of proteome.

14.4.3.1.2   2 D-DIGE

Proteomics research is a potent method that may completely analyse the protein expressions. The 2D-DIGE (two-dimensional differential gel electrophoresis) technology allows the researcher to examine the difference between various samples (approximately two) in a single experiment utilising fluorescent dyes that have already been pre-labelled (Ciordia et al., 2006).This makes it easier to find proteins that are differently expressed in conditions like cancer. Compared to the conventional 2-D electrophoresis (2-DE), the 2D-DIGE method takes a shorter time and is more sensitive (Qi et al., 2008). In 2D-DIGE, proteins that have been tagged with cyanine dye are used because the dye can be easily seen when excited at a certain wavelength (Marouga et al., 2005). Multiple proteins were initially isolated in prophase and anaphase and afterwards separated using 2D-DIGE to study the crucial process of cellularisation in female gametophytes of *Pinus tabuliformis* during ovule development (Lv et al., 2015).

14.4.3.1.3   SDS-PAGE

Sodium dodecyl sulphate-polyacrylamide gel electrophoresis (SDS-PAGE) is a high-resolution method for separating proteins based on their size, which makes it easier to estimate their molecular weight. Proteins can move in an electrical field of media containing the pH that differs from the isoelectric point of the protein. Depending on the ratio of mass and charge of the protein, various proteins in a mixture move at different speeds. Proteins get denatured by the addition of sodium dodecyl sulphate; thus, they must be separated according to their molecular weight (Dunn and Burghes, 1986).

SDS-PAGE analysis revealed that the combination of reduced mineral milk and heat treatment led to the formation of complexes with high molecular weight (Jovanovic et al., 2007). Species of *Cleome* are particularly beneficial in the treatment of asthma, cough, rheumatism, fever, and many other ailments. They are used as green veggies in African nations. SDS-PAGE was used to compare the proteins in the leaves and seeds of *Cleome* species (Aparadh et al., 2012). SDS-PAGE is the preliminary stage of immunoblotting, which transfers protein on the membrane to study with protein-specific antibodies that are separated electrophoretically. Two-dimensional SDS-PAGE is the second step which is used for the high-resolution extraction of many proteins on a single gel (Gallagher, 2014).

14.4.3.1.4   Phos-Tag SDS-PAGE

SDS-PAGE using Phos-tag (tags that bind to phosphate) is a significant advancement that can assess the level of protein phosphorylation. Initially, the proteins that are phosphorylated are separated through Phos-tag SDS-PAGE and SDS-PAGE using gels of different compositions, although sometimes it is difficult to compare the electrophoretic mobility of proteins in different gels that were used. This issue has been resolved by the newly invented diagonal Phos-tag electrophoresis. Proteins that have been phosphorylated may be easily separated from proteins that have not been phosphorylated because it can show the pattern of both Phos-tag SDS-PAGE and SDS-PAGE on an individual gel. In the proteomics study, phospho-tag SDS-PAGE provides important information about the phosphorylation status of proteins.

Despite the fact that shotgun scanning of proteome enables the detection of several phosphorylated sites, it is difficult to distinguish between the various phosphorylated states of protein compounds utilising the method. The phosphorylation status of proteins is therefore commonly determined using Phos-tag electrophoresis. With the recent discovery of Phos-tag diagonal electrophoresis, this method has grown more potent (Hirano and Shirakawa, 2022).The described issues are resolved by Phos-tag, a phosphate-affinity SDS-PAGE method, in which proteins that are phosphorylated are sorted, based on the quantity and locations of phosphorylation of proteins during electrophoresis (Hisanaga et al., 2022).

14.4.3.1.5   μFFE

One of the most effective proteomics technologies involves microfluidic free-flow electrophoresis. μFFE (microfluidic free-flow electrophoresis) is a method that carries out the electrophoresis-based separation of proteins in real time on a continual flow of analytes as it passes through the buffer channels(Lee and Kwon, 2022). The electric field is applied perpendicularly to the flow to deflect

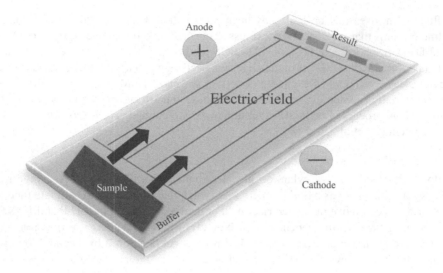

FIGURE 14.3 Microfluidic free-flow electrophoresis.

the analytes laterally according to their mobility as they flow through the separation buffer channel as shown in Figure 14.3.

Microfluidic free-flow electrophoresis permits the preparatory and systematic separation of low-volume proteins over 15 years. As a result of advancements in chip design, bubbles that are produced by electrolysis are less likely to interfere with separation. To increase the resolving power in µFFE, principles of band widening have been studied empirically and theoretically (Turgeon and Bowser, 2009).Numerous applications for the enrichment, fractionation, and purification of target molecules of proteins in natural systems have been stimulated by the capacity of µFFE.

14.4.3.2 Protein Sequencing

The main components of life are proteins. Knowledge of biological systems and diseases relies heavily on the protein composition of cells and organisms, although the significance of protein assay includes a limited number of methods existing to identify sequences of protein, and these approaches have drawbacks, such as the need for a large amount of material. The discovery of unknown proteins and the implementation of single-cell proteomics would be made possible by single-molecule methods, which would revolutionise the field of proteomics research. For individual molecules of proteins, sequencing techniques have been developed in current history that make use of tunnelling currents, fluorescence, nanopores, and Edman degradation (Bradley et al., 1982; Miyashita et al., 2001; Restrepo-Pérez et al., 2018; Shimonishi et al., 1980).

14.4.3.2.1 Edman Degradation Sequencing of Protein

Edman degradation technique was developed in 1950 by Pehr Edman who has included the orderly characterisation from the N-terminal to the C-terminal of the protein amino acid sequences (Edman, 1950).This method of sequencing includes periodic chemical processes that identify, cleave, and label the sequence of amino acids at the N-terminus of the protein, consecutively as shown in Figure 14.4. The initial step of the process includes the reaction of the N-terminal amino acids of protein with the phenyl-isothiocyanate (PITC), an Edman reagent under alkaline buffer conditions. In the acidic conditions, the altered amino acid of the N-terminus is separated as derivatives of thia-zolinone and detected through the chromatographic technique (Restrepo-Pérez et al., 2018).

Edman degradation is a useful tool for sequencing, but it is limited to the analysis of purified peptides that are shorter than ~50 amino acids. This technique cannot be used to analyse the mixture of

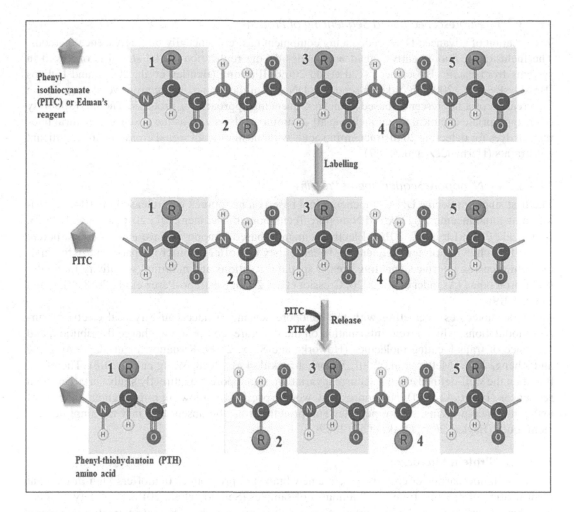

FIGURE 14.4   N-terminal sequencing of protein through Edman degradation.

complicated proteins, such as the ones seen in the majority of biological samples. Furthermore, each cycle of protein degradation might take ~45 minutes (Li and Geraerts, 1997), making the process of longer duration. Modifications of the N-terminal may obstruct the process as well. For instance, protein sequencing is impossible when the peptides of the N-terminus are acetylated (Restrepo-Pérez et al., 2018).

14.4.3.2.2   Tunnelling Currents-based Sequencing of Proteins

In the 1970s, the concept of measuring individual molecules with tunnelling currents was initially proposed (Aviram and Ratner, 1974). Measurements of tunnelling currents are carried out between two electrodes of metal that are spaced apart by some gap that varies in size ranging from angstroms to nanometres. The tunnelling current changes are determined when molecules move across the gap. The molecule that is momentarily dwelling in the gap may be identified in real time using these current variations. The potential to fulfil this theory became apparent with the introduction of the scanning tunnelling microscope (STM) in the late 1970s, and this resulted in the creation of a new discipline called molecular electronics (Dekker et al., 1997; Ratner, 2013; Reed et al., 1997).

14.4.3.2.3   Fluorescence-based Sequencing of Proteins

The creation of advanced DNA sequencing equipment has relied heavily on fluorescence methods. The inclusion of fluorescently tagged nucleotides during replication of the strand is observed in systems from Pacific Biosciences (Eid et al., 2009), Illumina (Bentley et al., 2008), and Helicos (Braslavsky et al., 2003) in order to sequence DNA from the strand. There are several obstacles in the creation of a fluorescence-based de novo sequencing approach for proteins. The unavailability of an appropriate chemical to uniquely tag all 20 amino acids as well as the absence of natural fluorescent dyes for detecting 20 distinct amino acids without significant signal crosstalk are significant constraints (Hernandez et al., 2017).

14.4.3.2.4   Nanopore Sequencing of Proteins

The first single-molecule DNA sequencing tool based on nanopores was released in 2014, according to an announcement by Oxford Nanopore Technologies (Deamer et al., 2016; Jain et al., 2015; Jain et al., 2016; Lu et al., 2016). Electrochemically based nanopores have emerged as a potent option for achieving protein sequencing because they can offer submicron space and individual-molecule sensing interface, enabling greater spatial resolutions and maximal sensitivity for analysis of proteomes (Asandei et al., 2020; Cressiot et al., 2020; Restrepo-Pérez et al., 2018; Ying and Long, 2019).

Single molecules interacting with the nanopore sensing surface cause typical electrical current modulations, which reveal information on the structure, composition, charge distribution, and sequence of trans-locating molecules (Howorka and Siwy, 2009; Kasianowicz et al., 1996; Long and Zhang, 2009; Meller et al., 2001; Varongchayakul et al., 2018; Wang et al., 2018). Therefore, based on the well-defined electrical current variation, a nanopore may directly study certain protein sequences (Hu et al., 2021). Nanopore array would specifically allow high-throughput analysis of low-abundance proteins, saving peptides and proteins from the absence of in vitro amplification techniques (Bayley, 2015; Osaki et al., 2009).

14.4.3.3   Protein Microarray

The protein microarray, or protein chip, is a new branch of proteomics that offers high-throughput identification of proteins from small amounts of samples (Sutandy et al., 2013). A variety of functions have been performed with protein chips, such as protein-phospholipid interactions, protein-protein interactions, and protein kinase component detection (Hall et al., 2007). They may also be employed to monitor pathological conditions and perform clinical diagnostics. Functional and experimental protein microarrays are the two categories into which protein microarrays may be divided (Sutandy et al., 2013).

14.4.3.3.1   Functional Microarray

Functional protein microarrays are constructed using purified proteins. It is possible to study interactions between proteins and DNA, RNA, and other proteins, as well as those between proteins and drugs and lipids, using these techniques. Furthermore, they facilitate the interaction between enzymes and substrates. The activities of thousands of proteins were identified via functional protein microarray. *Arabidopsis thaliana* protein-protein interactions were investigated, and calmodulin substrates (CaM) and calmodulin-like proteins (CML) were discovered (Popescu et al., 2007; Sutandy et al., 2013).

14.4.3.3.2   Experimental Microarray

The most typical category of analytical protein microarrays is the antibodies microarray. Directly protein labelling is used to find proteins that have been captured by antibodies. These are often employed to assess the degree of protein expression and their binding affinity (Ebhardt et al., 2015; Rosenberg and Utz, 2015; Sutandy et al., 2013).

To describe the kinase activity of plants using protein microarrays and identify cellular signalling cascades, experimental and analytical methods have been established (Brauer et al., 2014). It has been reported that the *A. thaliana* mitogen-activated protein kinases (MAPKs) have been described using experimental microarray. Plants that respond to a variety of external stimuli use molecules called MAPKs, which are highly conserved signal transduction molecules (Feilner et al., 2005).

14.4.4   QUANTITATIVE METHODS

A method that enables the detection of protein quantity using different quantity measurement techniques such as MS works on the principles of analysing the quantity of protein based on their mass-to-charge ratio and other techniques like iTRAQ, SILAC, and ICAT labelling.

14.4.4.1   Mass Spectrometry (MS)

Proteomics based on MS have made it easier to explore signalling cascades across biology during the past 20 years. MS is needed in plants where a developmental history of genome doubling has produced enormous gene families engaged in regulatory pathways and post-translational alterations (Blackburn et al., 2022).MS is mainly used to compute the mass-to-charge ratio of proteins by which the molecular mass of protein can be determined. There are three steps in the entire procedure. Bio-molecules in a solid or liquid phase encounter difficulty since the molecules should be converted to ions in the gas phase in the initial step. In the second step, ions are separated based on their values of mass-to-charge ratio while being exposed to magnetic or electrical fields in a region called a mass analyser. At last, the isolated ions are measured, together with the quantity of every entity with a certain mass-to-charge value as shown in Figure 14.5.

Typical ionisation techniques include electrospray ionisation (ESI), surface-enhanced laser ionisation (SELDI), and matrix-assisted laser desorption ionisation (MALDI) (Yates, 2011). Through the use of MS, post-translational modifications within plants, such as protein phosphorylation, have been identified (Novakova et al., 2011). Proteins found in *A. thaliana* chloroplasts were characterised using Fourier transform MS (Zabrouskov et al., 2003).

14.4.4.2   iTRAQ

In order to analyse quantitative variations in the proteome using tandem MS, the method known as "isobaric tags for relative and absolute quantification" (iTRAQ) uses isobaric agents to mark the basic amines of proteins and peptides (Vélez-Bermúdez et al., 2016). iTRAQ is a tandem MS-based multiplex protein tagging method for protein quantitation. For absolute and relative quantification, this method uses isobaric labels (4-plex and 8-plex) to mark the protein. Liquid chromatography (LC) is used to separate the proteins and amine residues of the side chain and N-terminus, and then MS is used to assess the results. Finding the regulation of genes is crucial to understand how the disease manifests; hence, protein characterisation with iTRAQ is a suitable technique that aids in both protein identification and quantification together (Wiese et al., 2007).

The growth of crops is limited by the presence of soil-dissolving aluminium (Al^{3+}) ions, although *Oryza sativa* is extremely aluminium resistant; thus, a quantitative study of proteome was conducted to analyse the effect of Al^{3+} on the roots of *O. sativa* during the initial stages of growth. A total of 106 proteins exhibited distinct expression patterns between cultivars that were Al^{3+} resistant and Al^{3+} susceptible (Wang et al., 2014).

An iTRAQ-based quantitative technique was used to determine the function of hydrogen peroxide (H_2O_2) in the development of wheat, and the results revealed that a rise in H_2O_2 concentration inhibited the development of wheat seedlings and roots. Despite ~3,425 reported proteins, 44 were recently found to be H_2O_2-responding proteins implicated in stress, glucose metabolism, and signal transduction. H_2O_2 resistance may include several proteins, such as intrinsic protein 1, superoxide dismutase (SOD), and fasciclin-like arabinogalactan protein (Ge et al., 2013).

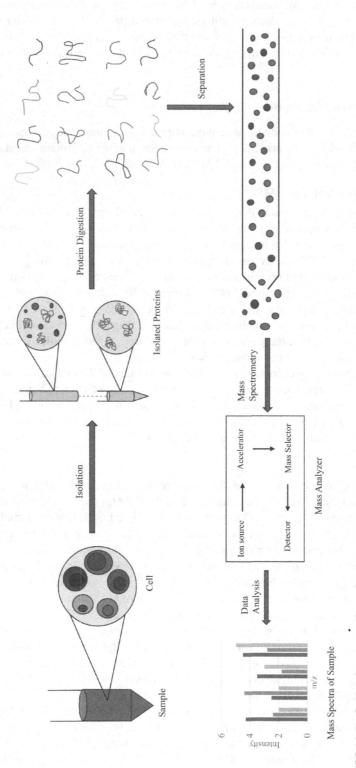

FIGURE 14.5 Study of proteome through mass spectrometry.

14.4.4.3 SILAC

SILAC is a quantitative proteomics method based on MS that relies on the metabolic labelling of the entire cellular proteome. Through the use of heavy or light amino acid labels, the proteomes of various cells raised in the culture of cells are distinguished via MS. The SILAC was created as a quick method to explore cell signalling, post-translation alterations, and the control of gene expression. SILAC is a crucial method for secretory pathways and proteins in a culture of cells as well (Ong et al., 2006).

Fundamentally and practically, SILAC is simple. It includes the growth of two or more cultures of cells, one growing in a media containing a "light" amino acid while the other is in a media containing a "heavy" amino acid. The heavier amino acid may substitute ^{15}N for ^{14}N, ^{13}C for ^{12}C, or ^{2}H for H. While there are no further chemical alterations when the heavier amino acid is incorporated into a peptide, there is a known mass transfer when compared to the peptide that includes the lighter amino acid (for instance, 6 Da in the instance of $^{13}C_6$-arginine). SILAC shares some similarities with other metabolic tagging methods that have been employed in biological research for several decades (Mann., 2006).

The quantitative proteome study of *A. thaliana* was performed using SILAC. Salicylic-acid-induced abiotic stress was studied concerning glutathione S-transferase expression and the resulting proteins were measured (Gruhler et al., 2005).

14.4.4.4 ICAT Labelling

ICAT is an isotope-labelling technique that quantifies proteins using chemically labelled reagents. The ICAT has often increased the quantity of proteins that may be studied and made it possible to precisely quantify and identify protein sequences from intricate combinations. The ICAT reagents include reactive groups, isotopically encoded linkers, and affinity tags for isolating labelled peptides (Shiio and Aebersold, 2006).

This strategy was first used by Gygi et al. (1999). Cysteine-containing proteins originating from two separate states are transformed using the ICAT reagent, and the transformed proteins in the analysed combination are quantified using MS and affinity chromatography after the two were mixed, proteolysed, and analysed. The novel aspect of this process is the utilisation of the ICAT reagent, which has three aspects: the first aspect is the reactive group that is specific to thiol that obtains the cysteine alkylation, and the second aspect is polyether linkers which enable the substitution of 8 H-atoms with their respective D-atoms, and the last aspect includes the biotin group that enables the isolation of the proteins that participate in the reaction with ICAT reagent (Hamdan and Righetti, 2002).

Protein substances that have undergone reversible oxidation may serve as signal transmitters and redox sensors for anti-stress signals. The function of proteins can be modified by oxidation because the thiol group present on the cysteine residue is susceptible to oxidative radicals. The approach used in conjunction with MS is beneficial to measure the redox proteins that contain the thiol group because ICAT reagents accurately react with the cysteine residues thiol group (Alvarez et al., 2012).

14.4.5 HIGH-THROUGHPUT METHODS

14.4.5.1 X-Ray Crystallography

The method of choice for determining the three-dimensional structure of the protein is X-ray crystallography. The size of the subunit that makes the crystal and the symmetry of the crystal packing are determined by processing the diffraction patterns that result from the exposure of the extremely purified crystallised samples to X-rays. Numerous topics like the viral system, protein-nucleic acid complexes, and immunological complexes may be studied using X-ray crystallography. Additionally, the precise information regarding the clarification of site-directed mutagenesis, drug design, enzyme mechanism, and the interaction between protein and ligand is provided by the three-dimensional structure of the protein (Smyth and Martin, 2000).

Membrane proteins are essential for physiological processes in every organism, including signal transduction, ion transport, and drug resistance. Understanding their three-dimensional structures is crucial for comprehending their roles, and this knowledge may also be used to help in the discovery of new drugs. In this regard, the most popular method for figuring out high-resolution three-dimensional membrane protein structures has been X-ray crystallography. However, effective protein extraction, crystal growth, stabilisation, and solubilisation are necessary for this method to be successful. For those working to crystallise membrane proteins, each of these processes might provide significant difficulties (Kermani, 2021).

Non-specific lipid-transferring proteins allow fatty acids, glycolipids, phospholipids, and steroids to flow across membranes. When maize non-specific lipid-transferring proteins were complexed with various ligands, the relative structure showed that the quantity of the hydrophobic regions varied according to the size of the attached ligands (Han et al., 2001).

The three-dimensional structure of the recombinant horseradish peroxidase (HRP) in association with benzohydroxamic acid (BHA) was disclosed by X-ray crystallography. The hydrophobic pocket next to the BHA aromatic ring and the charge density for the BHA were both found at the peroxidase catalytic site (Henriksen et al., 1998).

14.4.5.2 Nuclear Magnetic Resonance (NMR) Spectroscopy

NMR is a superior technique for examining molecular composition, protein behaviour, and folding. NMR spectroscopy often comprises many steps, each involving a distinct set of highly specialised procedures. To validate the structure, samples are produced, measurements are taken, and then interpretative procedures are used. The structure of a protein is crucial in several scientific fields, including homology modelling, functional genomics, and medication creation based on protein structure (Wiese et al., 2007).

To improve the sensitivity and resolution for high throughput profiling of proteins, NMR can be used with other techniques such as LC or ultra-high performance LC (UHPLC). Additionally, the comparison of structural data was collected with respect to the identification of metabolites in complicated mixtures (Wolfender et al., 2015).

For the purpose of identifying potential biomarkers for earlier treatment and prognosis, NMR in conjunction with UHPLC was developed to evaluate the metabolic abnormalities in patients with oesophageal cancer. When patients with oesophageal cancer were compared to healthy controls, the study found significant differences in the processes of glycolysis, ketogenesis, tricarboxylic acid cycle, and metabolism of lipids and amino acids (Zhang et al., 2013).

The carbon and nitrogen cycles depend on the breakdown of plant litter to supply the soil with the nutrients it needs and to produce CO_2 in the atmosphere. High-resolution-magic angle spinning (HR-MAS) NMR spectroscopy was employed to track the environmental damage of pine and wheat-grass residues, using ^{13}C- and ^{15}N-labelled plant components. The spectra showed that almost all plant tissues lost concentrated and hydrolysed tannin, whereas aromatic compounds, aliphatic compounds and cuticles and waxes, and minor amounts of carbohydrates survived (Kelleher et al., 2006).

14.4.6 Bioinformatics Based Method

Proteomics requires the use of bioinformatics; therefore, its applications have been growing along with the development of high-throughput techniques that rely on robust data processing. Novel techniques are being offered by this young and developing discipline to handle enormous and diverse proteome data and advance the discovery process (Vihinen, 2001). Over the past few years, the application of bioinformatics towards proteomics research has become much more popular. The creation of a unique algorithm that allows the analysis of larger amounts of data with more accuracy and specificity aids in the quantification and identification of proteins, making it possible to get detailed information about the expression of proteins. The key issue with these sorts of analyses is the administration of such a large amount of data. Finding connections among proteomic research and various other omics techniques, such as metabolomics and genomics, is still challenging. However, database technology and newly developed semantic statistical approaches are effective instruments that can be helpful to get beyond these restrictions (Aslam et al., 2017).

In order to obtain a specific collection of peptides for MS, the sample of proteins is isolated and degraded by utilising one or more proteases (Wiśniewski and Mann, 2012). To reduce the complexity of the samples or when the assessment of a particular protein subgroup is desirable, further processes, such as enrichment and separation, can be applied at the peptide or protein level (Altelaar and Heck, 2012; Lee et al., 2010; Schmidt et al., 2014). LC and MS are used to examine the resulting peptides. The extensive coverage of the proteome by shotgun MS or the quantitative evaluation of a specific group of proteins using targeted MS are both common methodologies (Picotti and Aebersold, 2012; Schmidt et al., 2014). The generated spectra include details about the sequence, which is crucial for identifying proteins. The resulting information can be shown as a three-dimensional image with the retention time (RT), mass-to-charge ratio, and peptide intensities together with fragmented spectra.

To obtain the chromatography peak, the frequency of the mass-to-charge ratio for a given peptide is displayed along with the RT. Peptides may be quantified using the area within this curve, whereas proteins can be recognised by their fragmented spectra. Proteomic data may be published in repositories that are also available for database searches (Riffle and Eng, 2009). The biggest proteome repositories, such as Peptide-Atlas, Proteome Commons, and PRIDE proteomics recognition database, allow immediate access to the majority of the data stored and are effective tools for data gathering (Desiere et al., 2005; Vizcaíno et al., 2012).

14.5 ROLE OF PROTEOMICS IN MEDICINAL PLANTS RESEARCH

Herbal remedy is a discipline that encompasses substances produced from plants to treat illness. Herbal medicine plays a significant part in the treatment of various ailments and has a very long history throughout the world. The science and technology of herbal medicines and pharmaceuticals using natural ingredients have advanced significantly over the past few decades. Herbal medications have a more comprehensive therapeutic effect than drugs with a single-constituent dosage and can raise the danger of cytotoxicity brought upon by chemotherapy (Al-Obaidi, 2021; Zaynab et al., 2018). To better understand the mechanism underlying the pharmacological nature of plants, proteomic approaches have been used to research the physiological changes of medicinal herbs and their effects on organisms as shown in Table 14.1. Only a small number of species of medicinal plants now possess comprehensive genomic data, although proteomic methods can still be used on

TABLE 14.1
Employed Proteomics Approaches in Medicinal Plant Research

S. No.	Medicinal Plants	Active Constituent/ Plant Extracts	Medicinal Properties	Employed Proteomics Approach	References
1.	*Elettaria cardamomum*	Cardamom plant extract/ y-bisabolene	Antiproliferative effects on tumour cells.	New anticancer drugs from y-bisabolene discovered through phospho-proteomics.	(Jou et al., 2015)
2.	*Curcuma longa*	Curcumin	Anti-inflammatory and anticancer	The proteomic analysis shows that curcumin can interact with hundreds of proteins to exert its anticancer effects on colon cancer.	(Wang et al., 2016)
3.	*Vanilla planifolia*	Vanillin and vanillic acid	Antibacterial and sedative agents	Both the vanillin precursor's coumaric acid and glucoside are formed at the early stages of plant shooting differentiation, according to proteomics research on the *Vanilla planifolia*.	(Palama et al., 2010)
4.	*Rosmarinus officinalis*	Rosemary plant extract	Anti-oxidant, antibacterial, and anticancer	The extract altered tens of proteins that trigger adaptive reactions to reduce stress in proteomic research on anticancer properties against colon cancer.	(De Oliveira et al., 2019)

(Continued)

TABLE 14.1 *(Continued)*

Employed Proteomics Approaches in Medicinal Plant Research

S. No.	Medicinal Plants	Active Constituent/ Plant Extracts	Medicinal Properties	Employed Proteomics Approach	References
5.	*Alpinia oxyphylla*	Oxyphylla A	Neuro-protective	A neuroprotective substance was found in the fruit of *Alpinia oxyphylla* utilising LC-MS/MS proteomic-based research.	(Li et al., 2016)
6.	*Artemisia annua*	Dihydroartemisinin	Antitumour	The subproteome-capturing technique revealed that energy metabolism imbalance in mitochondria causes dihydroartemisinin's anticancer action.	(Lu et al., 2012)
7.	*Allium sativum*	Allicin	Immunity booster, antimicrobial, and anticancer	The proteomic analysis indicated that allicin exhibits anticancer activity. It is created during garlic injury or crashing and has been shown to alter millions of proteins in human T lymphocyte cell (Jurkat T-cell) and mice fibroblast (L-929) cell lines by generating thioallylation.	(Iciek et al., 2009)
8.	Panax ginseng	Ginsenosides	Neurodegenerative and antidiabetic	Proteomic research has shown that ginseng roots start ginsenoside production when the plant enters a slow-growth stage.	(Ma et al., 2013; Xu et al., 2017)
9.	*Trigonella foenum-graecum*	Plant extract	Anti-cancerous effect on central nervous system (CNS) T-cell lymphomas	A proteomics investigation identified a set of proteins from the plant extract, having therapeutic anticancerous activity.	(Alsemari et al., 2014)
10.	*Hibiscus sabdariffa*	Roseltide	Neutrophil elastase inhibitors	Proteomic methods assisted in the identification and characterisation of a new peptide known as roseltide in combination with an NMR-based metabolomics approach to fight against neutrophil elastase related disorders.	(Loo et al., 2016)
11.	*Salvia miltiorrhiza*	Salvianolic acid	Antioxidant and improves the circulation of blood in the body	Evidences of their medicinal properties have been examined by applying the proteomics research on (A10) muscle cell line of rat by revealing it to plant extract.	(Hung et al., 2010)
12.	*Acanthopanax senticosus*	Plant extract	Anti-neuroinflammatory properties	LC-ESIMS/MS and 2D-DIGE proteomics technique was used to examine the cellular mechanisms behind the neuroinflammation effects of *Acanthopanax senticosus* plant extract.	(Jiang et al., 2015)

(Continued)

TABLE 14.1 *(Continued)*

Employed Proteomics Approaches in Medicinal Plant Research

S. No.	Medicinal Plants	Active Constituent/ Plant Extracts	Medicinal Properties	Employed Proteomics Approach	References
13.	*Nigella sativa*	Proteins	Immuno-modulatory	Four peaks were obtained in full fractionation when ion exchange chromatography (IEX)-based proteomics approach was used to separate proteins with immunomodulating activity.	(Haq et al., 1999)
14.	*Cynodon dactylon*	Leaf Extract	Anti-diabetic	MALDI-TOF mass spectrometry and 2-DE proteomics approaches were used to examine the impact of *Cynodon* leaf extracts on the proteome of rat liver, which is associated with diabetes mellitus.	(Karthik et al., 2012)
15.	*Glycine max*	Genistein	Anti-cancerous effect on breast cancer and mammary cancer of rats	Proteomic techniques were used to assess the action mechanism of genistein in preventing breast cancer and to control mammary cancer.	(Wang et al., 2011b)
16.	*Tripterygium wilfordii*	Triptolide	Anti-cancerous effect on colon cancer	Using MALDI-TOF and 2-DE proteomic technologies, the effects of triptolide treatment on the changes of proteome during the colon cancer were examined.	(Liu et al., 2012)
17.	*Scutellaria baicalensis* Georgi	Flavonoids	Anti-inflammatory properties	A proteomics technique was used to examine the anti-inflammatory effects of flavonoids extracted from the *Scutellaria baicalensis* Georgi in lipopolysaccharide-prompted L6 cells of skeletal muscles.	(Kim et al., 2014)
18.	*Huperzia serrata*	Huperzine A	Anti-inflammatory	Nano-LC-MS-based proteomics technique was applied to examine the impact of Huperzine A on various inflammatory effects.	(Tao et al., 2013)
19.	*Gastrodia elata* Blume	Dry rhizome of *Gastrodia elata* (tianma)	Depression	To examine the mechanism of action and the potential of Tianma for the treatment of neurodegenerative illnesses like Alzheimer's, proteomics techniques were used.	(Manavalan et al., 2012)
20.	*Isodon rubescens*	Oridonin	Anticancer	Proteomic characterisation of proteins responsible for oridonin's anticancer effects in hepatocarcinoma (HepG2) cells.	(Wang et al., 2011a)

species that lack this information. The development of technology to control, improve, and enhance the productivity of plants, decrease resistance to infection, and most specifically, provide proof of the therapeutic and pharmaceutical impact on human health, will all be aided by proteomic strategies that have been found to be helpful in elucidating the signalling pathways underlying vital medicinal plant traits (Al-Obaidi, 2021).

The benefit of proteomics is the ability to compile a wide range of physiological data from both therapeutic plants and the animals that ingest them. The discovery of important enzymes has made it easier to understand the intricate, multi-step biosynthetic processes that give rise to plant bioactive chemicals. The proteome studies of plants are used to characterise the external variables that impact the physiological changes brought upon by the stimuli. It was demonstrated that numerous biological mechanisms were involved in responses in animal cells supplied with natural compounds produced from plants (Hashiguchi et al., 2017).

Today, a wide range of disorders are treated using the proteomic technique. Since most protein functions are regulated by both protein localisation and post-translational changes of mRNA, proteomics is emerging as a new field that has the potential to revolutionise both biology and medicine (Zaynab et al., 2017; Zaynab et al., 2018). So, proteome studies are focused on elaborating protein expression and functions; to fully comprehend the operation of biological systems, proteomics will be integrated with other data, such as gene profiling (Manzoni et al., 2018). Integrated proteomic methods can precisely map a comprehensive understanding of Chinese traditional medicine (CTM). In the future, novel bioactive compounds and target molecules may be found using proteomic approaches. Both the pharmacological characteristics of CTM and its mode of action have been studied using proteomic approaches (Buriani et al., 2012).

Proteomics will be a crucial tool for CTM in the areas of toxicity studies, quality assurance, and standardisation of CTM formulations, which are the main areas where Chinese herbal medicine is used in the West. Similar to genomics and transcriptomics, proteomic studies have successfully characterised the mechanism of action of several CTM formulations, for example, the impact of SiWU decoctions on overall health (Guo et al., 2014). In understanding the implications of *Salvia miltiorrhiza* on atherosclerotic lesions (Hung et al., 2010), and qualities of *Ganoderma* in nerve damage, proteomic research has been undertaken to examine the effects of different treatments on cancer (Hung et al., 2010; Zhang et al., 2014).

Proteomics data has also been used to study the influence of the Shuanglong CTM formulation on the differentiation and usage of pluripotent cells in myocardial ischemia therapy (Fan et al., 2010). Despite the challenging interpretation of proteomic due to the intrinsic complexity of the ingredients in CTM decoctions, it is emerging as a crucial tool for investigating the various impacts of complicated herbal preparations, developing active functions, discovering bioactive compounds, molecular diagnosis, and providing a safe prescription for CTM treatments (Zaynab et al., 2018). *Pseudostellaria heterophylla*, *Moringa oleifera*, and *N. sativa* are three historically used medicinal herbs with published proteomics profiles (Alanazi et al., 2016).

14.6 CHALLENGES

The study of proteins is different from the study of nucleic acids in that it involves some challenges. Compared to DNA and RNA, working with proteins is more difficult. They have secondary and tertiary structure, which frequently needs to be maintained during analysis. Proteins can be broken down by enzymes, heat, light, or vigorous mixing. There are some proteins that are difficult to examine because they are poorly soluble (Cho, 2007). Unlike DNA, proteins cannot be amplified. About half of the total amount of protein in plasma is accounted for by albumin (55 mg/mL), while the remaining ten proteins account for 90%. Cytokines and other less prevalent proteins are often present at 1–5 pg/mL. Each proteomics technique can only study proteins that are 3–4 orders of magnitude or smaller, and typically at the higher concentration end of the spectrum. It is therefore necessary to remove extremely abundant proteins from plasma or serum in order to perform more in-depth proteomics experiments on low-abundant proteins. It is possible, however, that many consequentially significant biomarkers are lost in the process due to non-specific binding or co-removal of proteins/peptides associated with the highly prevalent carrier proteins. Removal of certain proteins, in particular albumin, has been demonstrated to cause a significant loss of cytokines (Granger et al., 2005).

Thulasiraman et al.(2005) employed ligand library beads to create the novel deep proteome strategy. It has been demonstrated that the use of equaliser beads in combination with a combinatorial library of ligands enables the access of several low-abundant proteins or polypeptides that are undetected by conventional analytical techniques. Since the population of beads is so diverse, most proteins in a sample should have a binding partner. Equivalent binding power exists in each bead. While trace proteins are focused on their unique affinity ligands, highly abundant proteins saturate their interaction partners and wash the excess proteins away. This variety of techniques may represent a significant advance in "mining below the tip of the iceberg" in order to find the "unseen proteome."

In order to address the sensitivity issue, Nettikadan et al.(2006) created ultra-microarrays, which combine the multiplexing capabilities, higher throughput, and cost-effectiveness of microarrays with the capacity to screen very small sample volumes. These microarrays have been found to have high specificity and sensitivity within the attomole range when used with pure proteins. This method permits the proteomics analysis of materials that are only occasionally made available, such as those obtained from laser capture microdissection, neonatal biopsy, micro specimens, and forensic samples. It appears that a lot of technical challenges need to be overcome before routine proteomics analysis is accomplished in the clinic. These obstacles are beginning to be resolved, though, with the standardisation of techniques and the sharing of proteomics data into open databases. The majority of proteomics technologies require sophisticated equipment, powerful computers, and expensive consumables. The integration of proteomics data with genomics and metabolomics data, as well as their functional interpretation in connection with clinical outcomes and epidemiology, will be another significant challenge (Kolch et al., 2005).

14.7   CONCLUSION AND FUTURE PROSPECTIVE

Proteomics has the potential to provide a breakthrough in the therapeutic monitoring and interpretation of protein-based diseases, which will contribute to future diagnostic tools. A proteome-based approach is imperative for improving current pharmacological knowledge by examining the mechanism of action of many herbal drugs used to treat specific diseases. Herbal medication treatment is complicated by issues such as accurate data processing, structural characterisation of relevant proteomes, and handling differently changed proteins throughout different disease stages. The development of technological and methodological advances in proteome research must be coordinated to have a meaningful impact in future. Adequate proteome characterisation will come from improvements in the creation of more efficient quantitative and high-throughput methodologies for proteome analysis. In this way, proteome changes can be reliably monitored throughout the proteomics process. Furthermore, proteome investigations will be aided by effective scientific collaboration, which will result in openly accessible and cost-free databases worldwide.

ACKNOWLEDGEMENT

We are grateful to the Director, Dayalbagh Educational Institute, Dayalbagh, Agra, for encouragement and kind support.

REFERENCES

Alanazi, M. A., Tully, M. P., & Lewis, P. J. (2016). A systematic review of the prevalence and incidence of prescribing errors with high-risk medicines in hospitals. Journal of Clinical Pharmacy and Therapeutics, 41(3), 239–245.

Al-Obaidi, J. R. (2021). Proteomics research in aromatic plants and its contribution to the nutraceuticals and pharmaceutical outcomes. In Medicinal and Aromatic Plants (pp. 223–239). Academic Press.

Alsemari, A., Alkhodairy, F., Aldakan, A., Al-Mohanna, M., Bahoush, E., Shinwari, Z., & Alaiya, A. (2014). The selective cytotoxic anti-cancer properties and proteomic analysis of *Trigonella foenum-graecum*. BMC Complementary and Alternative Medicine, 14(1), 114. https://doi.org/10.1186/1472-6882-14-114.

Altelaar, A. M., & Heck, A. J. (2012). Trends in ultrasensitive proteomics. Current Opinion in Chemical Biology, 16(1–2), 206–213.

Alvarez, S., Hicks, L. M., & Liu, Z. (2012). Redox protein characterization and quantification using ICAT-MS to investigate thiol-based regulatory mechanisms induced by oxidative stress in plants. Journal of Biomolecular Techniques: JBT, 23, S54.

Anderson, N. L., & Anderson, N. G. (1998). Proteome and proteomics: New technologies, new concepts, and new words. Electrophoresis, 19(11), 1853–1861.

Anderson, N. G., Matheson, A., & Anderson, N. L. (2001). Back to the future: The human protein index (HPI) and the agenda for post-proteomic biology. Proteomics, 1(1), 3–12.

Aparadh, V. T., Amol, V. P., & Karadge, B. A. (2012). Comparative analysis of seed and leaf proteins by SDS PAGE gel electrophoresis within *Cleome* species. International Journal of Advanced Life Sciences, 3, 50–58.

Aryal, U. K., Xiong, Y., McBride, Z., Kihara, D., Xie, J., Hall, M. C., & Szymanski, D. B. (2014). A proteomic strategy for global analysis of plant protein complexes. The Plant Cell, 26(10), 3867–3882.

Asandei, A., Di Muccio, G., Schiopu, I., Mereuta, L., Dragomir, I. S., Chinappi, M., & Luchian, T. (2020). Nanopore-based protein sequencing using biopores: Current achievements and open challenges. Small Methods, 4(11), 1900595.

Aslam, B., Basit, M., Nisar, M. A., Khurshid, M., & Rasool, M. H. (2017). Proteomics: Technologies and their applications. Journal of Chromatographic Science, 55(2), 182–196.

Auffray, C., Chen, Z., & Hood, L. (2009). Systems medicine: The future of medical genomics and healthcare. Genome Medicine, 1(1), 1–11.

Aviram, A., & Ratner, M. A. (1974). Molecular rectifiers. Chemical Physics Letters, 29(2), 277–283.

Azzoni, A. R., Takahashi, K., Woodard, S. L., Miranda, E. A., & Nikolov, Z. L. (2005). Purification of recombinant aprotinin produced in transgenic corn seed: Separation from CTI utilizing ion-exchange chromatography. Brazilian Journal of Chemical Engineering, 22, 323–330.

Banerjee, A., & Roychoudhury, A. (2017). Effect of salinity stress on growth and physiology of medicinal plants. In: Ghorbanpour, M., & Varma, A. (Eds.) Medicinal Plants and Environmental Challenges, Springer International Publishing, Cham, Switzerland, pp. 177–188.

Bauer, A., & Kuster, B. (2003). Affinity purification-mass spectrometry: Powerful tools for the characterization of protein complexes. European Journal of Biochemistry, 270(4), 570–578.

Bayley, H. (2015). Nanopore sequencing: From imagination to reality. Clinical Chemistry, 61(1), 25–31.

Bentley, D. R., Balasubramanian, S., Swerdlow, H. P., Smith, G. P., Milton, J., Brown, C. G., ... & Roe, P. M. (2008). Accurate whole human genome sequencing using reversible terminator chemistry. Nature, 456(7218), 53–59.

Biosciences, A. (2002). Ion exchange chromatohgraphy, principles and methods, Amercham Pharmacia. Biotech SE, 751.

Blackburn, M. R., Minkoff, B. B., & Sussman, M. R. (2022). Mass spectrometry-based technologies for probing the 3D world of plant proteins. Plant Physiology, 189(1), 12–22.

Blackstock, W. P., & Weir, M. P. (1999). Proteomics: Quantitative and physical mapping of cellular proteins. Trends in Biotechnology, 17(3), 121–127.

Bodzon-Kulakowska, A., Bierczynska-Krzysik, A., Dylag, T., Drabik, A., Suder, P., Noga, M., ... & Silberring, J. (2007). Methods for samples preparation in proteomic research. Journal of Chromatography B, 849(1-2), 1–31.

Böttcher, T., Pitscheider, M., & Sieber, S. A. (2010). Natural products and their biological targets: Proteomic and metabolomic labeling strategies. Angewandte Chemie International Edition, 49(15), 2680–2698.

Bradley, C. V., Williams, D. H., & Hanley, M. R. (1982). Peptide sequencing using the combination of Edman degradation, carboxypeptidase digestion and fast atom bombardment mass spectrometry. Biochemical and Biophysical Research Communications, 104(4), 1223–1230.

Braslavsky, I., Hebert, B., Kartalov, E., & Quake, S. R. (2003). Sequence information can be obtained from single DNA molecules. Proceedings of the National Academy of Sciences, 100(7), 3960–3964.

Brauer, E. K., Popescu, S. C., & Popescu, G. V. (2014). Experimental and analytical approaches to characterize plant kinases using protein microarrays. In Plant MAP Kinases (pp. 217–235). Humana Press, New York, NY.

Buriani, A., Garcia-Bermejo, M. L., Bosisio, E., Xu, Q., Li, H., Dong, X., ... & Hylands, P. J. (2012). Omic techniques in systems biology approaches to traditional Chinese medicine research: Present and future. Journal of Ethnopharmacology, 140(3), 535–544.

Calero, M., Rostagno, A., & Ghiso, J. (2012). Search for amyloid-binding proteins by affinity chromatography. In Amyloid Proteins (pp. 213–223). Humana Press.

Canales, R. D., Luo, Y., Willey, J. C., Austermiller, B., Barbacioru, C. C., Boysen, C., ... & Goodsaid, F. M. (2006). Evaluation of DNA microarray results with quantitative gene expression platforms. Nature Biotechnology, 24(9), 1115–1122.

Chandrasekhar, K., Dileep, A., Lebonah, D. E., & Pramoda Kumari, J. (2014). A short review on proteomics and its applications. International Letters of Natural Sciences, 12(1).

Cho, W. C. (2007). Proteomics technologies and challenges. Genomics, Proteomics & Bioinformatics, 5(2), 77–85.

Christendat, D., Yee, A., Dharamsi, A., Kluger, Y., Savchenko, A., Cort, J. R., ... & Arrowsmith, C. H. (2000). Structural proteomics of an archaeon. Nature Structural Biology, 7(10), 903–909.

Ciordia, S., Ríos, V. D. L., & Albar, J. P. (2006). Contributions of advanced proteomics technologies to cancer diagnosis. Clinical and Translational Oncology, 8(8), 566–580.

Coskun, O. (2016). Separation techniques: Chromatography. Northern Clinics of Istanbul, 3(2), 156.

Cressiot, B., Bacri, L., & Pelta, J. (2020). The promise of nanopore technology: Advances in the discrimination of protein sequences and chemical modifications. Small Methods, 4(11), 2000090.

Cuatrecasas, P., Wilchek, M., & Anfinsen, C. B. (1968). Selective enzyme purification by affinity chromatography. Proceedings of the National Academy of Sciences, 61(2), 636–643.

De Oliveira, J. R., Camargo, S. E. A., & De Oliveira, L. D. (2019). Rosmarinus officinalis L. (rosemary) as therapeutic and prophylactic agent. Journal of Biomedical Science, 26(1), 1–22.

Deamer, D., Akeson, M., & Branton, D. (2016). Three decades of nanopore sequencing. Nature Biotechnology, 34(5), 518–524.

Dekker, C., Tans, S. J., Oberndorff, B., Meyer, R., & Venema, L. C. (1997). STM imaging and spectroscopy of single copper phthalocyanine molecules. Synthetic Metals, 84(1–3), 853–854.

Desiere, F., Deutsch, E. W., Nesvizhskii, A. I., Mallick, P., King, N. L., Eng, J. K., ... & Aebersold, R. (2005). Integration with the human genome of peptide sequences obtained by high-throughput mass spectrometry. Genome Biology, 6(1), 1–12.

Determann, H. (2012). Gel Chromatography Gel Filtration Gel Permeation Molecular Sieves: A Laboratory Handbook. Springer Science & Business Media, New York.

Domon, B., & Aebersold, R. (2006). Mass spectrometry and protein analysis. Science, 312(5771), 212–217.

Dunn, M. J., & Burghes, A. H. (1986). High resolution two-dimensional polyacrylamide-gel electrophoresis. In Gel Electrophoresis of Proteins (pp. 203–261). Butterworth-Heinemann, Oxford, UK.

Dutta, S., & Bose, K. (2022). Protein Purification by Affinity Chromatography. In Textbook on Cloning, Expression and Purification of Recombinant Proteins (pp. 141–171). Springer, Singapore.

Ebhardt, H. A., Root, A., Sander, C., & Aebersold, R. (2015). Applications of targeted proteomics in systems biology and translational medicine. Proteomics, 15(18), 3193–3208.

Edman, P. (1950). Method for determination of the amino acid sequence in peptides. Acta Chemica Scandinavica, 4, 283–293.

Efferth, T. (2017, October). From ancient herb to modern drug: Artemisia annua and artemisinin for cancer therapy. In Seminars in Cancer Biology (Vol. 46, pp. 65–83). Academic Press, Cambridge, MA.

Eid, J., Fehr, A., Gray, J., Luong, K., Lyle, J., Otto, G., ... &Turner, S. (2009). Real-time DNA sequencing from single polymerase molecules. Science, 323(5910), 133–138.

Fan, X., Li, X., Lv, S., Wang, Y., Zhao, Y., & Luo, G. (2010). Comparative proteomics research on rat MSCs differentiation induced by Shuanglong Formula. Journal of Ethnopharmacology, 131(3), 575–580.

Feilner, T., Hultschig, C., Lee, J., Meyer, S., Immink, R. G., Koenig, A., ... & Kersten, B. (2005). High throughput identification of potential Arabidopsis mitogen-activated protein kinases Substrates S. Molecular & Cellular Proteomics, 4(10), 1558–1568.

Freeman, W. M., & Hemby, S. E. (2004). Proteomics for protein expression profiling in neuroscience. Neurochemical Research, 29(6), 1065–1081.

Gallagher, S. R. (2014). Overview of electrophoresis. Current Protocols Essential Laboratory Techniques, 8(1), 7–1.

Ge, P., Hao, P., Cao, M., Guo, G., Lv, D., Subburaj, S., ... & Yan, Y. (2013). iTRAQ-based quantitative proteomic analysis reveals new metabolic pathways of wheat seedling growth under hydrogen peroxide stress. Proteomics, 13(20), 3046–3058.

Godovac-Zimmermann, J., & Brown, L. R. (2001). Perspectives for mass spectrometry and functional proteomics. Mass Spectrometry Reviews, 20(1), 1–57.

Gowda, G. N., Zhang, S., Gu, H., Asiago, V., Shanaiah, N., & Raftery, D. (2008). Metabolomics-based methods for early disease diagnostics. Expert Review of Molecular Diagnostics, 8(5), 617–633.

Granger, J., Siddiqui, J., Copeland, S., & Remick, D. (2005). Albumin depletion of human plasma also removes low abundance proteins including the cytokines. Proteomics, 5(18), 4713–4718.

Gruhler, A., Schulze, W. X., Matthiesen, R., Mann, M., & Jensen, O. N. (2005). Stable isotope labeling of *Arabidopsis thaliana* cells and quantitative proteomics by mass spectrometry* S. Molecular & Cellular Proteomics, 4(11), 1697–1709.

Guo, Z., Man, Y., Wang, X., Jin, H., Sun, X., Su, X., ... & Mi, W. (2014). Levo-tetrahydropalmatine attenuates oxaliplatin-induced mechanical hyperalgesia in mice. Scientific Reports, 4(1), 1–4.

Gygi, S. P., Rist, B., Gerber, S. A., Turecek, F., Gelb, M. H., & Aebersold, R. (1999). Quantitative analysis of complex protein mixtures using isotope-coded affinity tags. Nature Biotechnology, 17(10), 994–999.

Hage, D. S., Anguizola, J. A., Bi, C., Li, R., Matsuda, R., Papastavros, E., ... & Zheng, X. (2012). Pharmaceutical and biomedical applications of affinity chromatography: Recent trends and developments. Journal of Pharmaceutical and Biomedical Analysis, 69, 93–105.

Hall, D. A., Ptacek, J., & Snyder, M. (2007). Protein microarray technology. Mechanisms of Ageing and Development, 128(1), 161–167.

Hamdan, M., & Righetti, P. G. (2002). Modern strategies for protein quantification in proteome analysis: Advantages and limitations. Mass Spectrometry Reviews, 21(4), 287–302.

Han, G. W., Lee, J. Y., Song, H. K., Chang, C., Min, K., Moon, J., ... & Suh, S. W. (2001). Structural basis of non-specific lipid binding in maize lipid-transfer protein complexes revealed by high-resolution X-ray crystallography. Journal of Molecular Biology, 308(2), 263–278.

Haq, A., Lobo, P. I., Al-Tufail, M., Rama, N. R., & Al-Sedairy, S. T. (1999). Immunomodulatory effect of *Nigella sativa* proteins fractionated by ion exchange chromatography. International Journal of Immunopharmacology, 21(4), 283–295.

Harris, D. C. (2012). Exploring Chemical Analysis, Macmillan, London, UK.

Harry, J. L., Wilkins, M. R., Herbert, B. R., Packer, N. H., Gooley, A. A., & Williams, K. L. (2000). Proteomics: Capacity versus utility. Electrophoresis: An International Journal, 21(6), 1071–1081.

Hashiguchi, A., Tian, J., & Komatsu, S. (2017). Proteomic contributions to medicinal plant research: From plant metabolism to pharmacological action. Proteomes, 5(4), 35.

Heinrich, M., Barnes, J., Prieto-Garcia, J., Gibbons, S., & Williamson, E. M. (2017). Fundamentals of Pharmacognosy and Phytotherapy. E-Book. Elsevier Health Sciences.

Henriksen, A., Schuller, D. J., Meno, K., Welinder, K. G., Smith, A. T., & Gajhede, M. (1998). Structural interactions between horseradish peroxidase C and the substrate benzhydroxamic acid determined by X-ray crystallography. Biochemistry, 37(22), 8054–8060.

Hernandez, E. T., Swaminathan, J., Marcotte, E. M., & Anslyn, E. V. (2017). Solution-phase and solid-phase sequential, selective modification of side chains in KDYWEC and KDYWE as models for usage in single-molecule protein sequencing. New Journal of Chemistry, 41(2), 462–469.

Hirano, H., & Shirakawa, J. (2022). Recent developments in Phos-tag electrophoresis for the analysis of Phosphoproteins in proteomics. Expert Review of Proteomics, 19(2), 103–114.

Hisanaga, S. I., Krishnankutty, A., & Kimura, T. (2022). In vivo analysis of the Phosphorylation of tau and the tau protein kinases Cdk5-p35 and GSK3β by using Phos-tag SDS–PAGE. Journal of Proteomics, 262, 104591.

Howorka, S., & Siwy, Z. (2009). Nanopore analytics: Sensing of single molecules. Chemical Society Reviews, 38(8), 2360–2384.

Hu, Z. L., Huo, M. Z., Ying, Y. L., & Long, Y. T. (2021). Biological nanopore approach for single-molecule protein sequencing. Angewandte Chemie, 133(27), 14862–14873.

Hung, Y. C., Wang, P. W., & Pan, T. L. (2010). Functional proteomics reveal the effect of *Salvia miltiorrhiza* aqueous extract against vascular atherosclerotic lesions. Biochimica et Biophysica Acta (BBA)-Proteins and Proteomics, 1804(6), 1310–1321.

Hurwitz, S. N., Rider, M. A., Bundy, J. L., Liu, X., Singh, R. K., & Meckes, D. G. Jr (2016). Proteomic profiling of NCI-60 extracellular vesicles uncovers common protein cargo and cancer type-specific biomarkers. Oncotarget, 7(52), 86999.

Iciek, M., Kwiecień, I., & Włodek, L. (2009). Biological properties of garlic and garlic-derived organosulfur compounds. Environmental and Molecular Mutagenesis, 50(3), 247–265.

Islam, N., Lonsdale, M., Upadhyaya, N. M., Higgins, T. J., Hirano, H., & Akhurst, R. (2004). Protein extraction from mature rice leaves for two-dimensional gel electrophoresis and its application in proteome analysis. Proteomics, 4(7), 1903–1908.

Jain, M., Fiddes, I. T., Miga, K. H., Olsen, H. E., Paten, B., & Akeson, M. (2015). Improved data analysis for the MinION nanopore sequencer. Nature Methods, 12(4), 351–356.

Jain, M., Olsen, H. E., Paten, B., & Akeson, M. (2016). The Oxford Nanopore MinION: Delivery of nanopore sequencing to the genomics community. Genome Biology, 17(1), 1–11.

Jiang, T., Wang, Z., Xia, T., Zhao, X., Jiang, L., & Teng, L. (2015). Quantitative proteomics analysis for effect of *Acanthopanax senticosus* extract on neuroinflammation. Pakistan Journal of Pharmaceutical Sciences, 28, 308–313.

Jou, Y. J., Chen, C. J., Liu, Y. C., Way, T. D., Lai, C. H., Hua, C. H., ... & Lin, C. W. (2015). Quantitative phosphoproteomic analysis reveals γ-bisabolene inducing p53-mediated apoptosis of human oral squamous cell carcinoma via HDAC2 inhibition and ERK1/2 activation. Proteomics, 15(19), 3296–3309.

Jovanovic, S., Barac, M., Macej, O., Vucic, T., & Lacnjevac, C. (2007). SDS-PAGE analysis of soluble proteins in reconstituted milk exposed to different heat treatments. Sensors, 7(3), 371–383.

Jungbauer, A., & Hahn, R. (2009). Ion-exchange chromatography. Methods in Enzymology, 463, 349–371.

Karthik, D., Ilavenil, S., Kaleeswaran, B., & Ravikumar, S. (2012). Analysis of modification of liver proteome in diabetic rats by 2d electrophoresis and MALDI-TOF-MS. Indian Journal of Clinical Biochemistry, 27(3), 221–230.

Kasianowicz, J. J., Brandin, E., Branton, D., & Deamer, D. W. (1996). Characterization of individual poly-nucleotide molecules using a membrane channel. Proceedings of the National Academy of Sciences of the United States of America, 93, 13770–13773.

Kelleher, B. P., Simpson, M. J., & Simpson, A. J. (2006). Assessing the fate and transformation of plant residues in the terrestrial environment using HR-MAS NMR spectroscopy. Geochimica et Cosmochimica Acta, 70(16), 4080–4094.

Kermani, A. A. (2021). A guide to membrane protein X-ray crystallography. The FEBS Journal, 288(20), 5788–5804.

Kim, S. W., Gupta, R., Lee, S. H., Min, C. W., Agrawal, G. K., Rakwal, R., ... & Kim, S. T. (2016). An integrated biochemical, proteomics, and metabolomics approach for supporting medicinal value of *Panax ginseng* fruits. Frontiers in Plant Science, 7, 994.

Kim, J. A., Nagappan, A., Park, H. S., Venkatarame Gowda Saralamma, V., Hong, G. E., Yumnam, S., ... & Kim, G. S. (2014). Proteome profiling of lipopolysaccharide induced L6 rat skeletal muscle cells response to flavonoids from *Scutellaria baicalensis* Georgi. BMC Complementary and Alternative Medicine, 14(1), 1–10.

Kolch, W., Mischak, H., & Pitt, A. R. (2005). The molecular make-up of a tumour: Proteomics in cancer research. Clinical Science, 108(5), 369–383.

Kolch, W., & Pitt, A. (2010). Functional proteomics to dissect tyrosine kinase signalling pathways in cancer. Nature Reviews Cancer, 10(9), 618–629.

Kollerová, E., Glasa, M., & Šubr, Z. W. (2008). Western Blotting analysis of the Plum pox virus capsid protein. Journal of Plant Pathology, S19–S22.

Kosak, S. T., & Groudine, M. (2004). Gene order and dynamic domains. Science, 306(5696), 644–647.

Krishna, R. G., & Wold, F. (1993). Post-translational modification of proteins. Advances in Enzymology and Related Areas of Molecular Biology, 67, 265–298.

Kroksveen, A. C., Jaffe, J. D., Aasebø, E., Barsnes, H., Bjørlykke, Y., Franciotta, D., ... & Berven, F. S. (2015). Quantitative proteomics suggests decrease in the secretogranin-1 cerebrospinal fluid levels during the disease course of multiple sclerosis. Proteomics, 15(19), 3361–3369.

Kurien, B. T., & Scofield, R. H. (2006). Western Blotting. Methods, 38(4), 283–293.

Lee, Y., & Kwon, J. S. (2022). Microfluidic free-flow electrophoresis: A promising tool for protein purification and analysis in proteomics. Journal of Industrial and Engineering Chemistry, 109, 79–99.

Lee, Y. H., Tan, H. T., & Chung, M. C. (2010). Subcellular fractionation methods and strategies for proteomics. Proteomics, 10(22), 3935–3956.

Lequin, R. M. (2005). Enzyme immunoassay (EIA)/Enzyme-linked immune sorbent assay (ELISA). Clinical Chemistry, 51(12), 2415–2418.

Liao, T. T., Xiang, Z., Zhu, W. B., & Fan, L. Q. (2009). Proteome analysis of round-headed and normal spermatozoa by 2-D fluorescence difference gel electrophoresis and mass spectrometry. Asian Journal of Andrology, 11(6).

Li, X., Bai, H., Wang, X., Li, L., Cao, Y., Wei, J., & Liu, G. (2011). Identification and validation of rice reference proteins for western blotting. Journal of Experimental Botany, 62(14), 4763–4772.

Li, K. W., & Geraerts, W. P. M. (1997).Neuropeptide Protocols (eds Irvine, G. B. & Williams, C. H.; pp.17–26). Humana Press, New York.

Li, G., Zhang, Z., Quan, Q., Jiang, R., Szeto, S. S., Yuan, S., ... & Chu, I. K. (2016). Discovery, synthesis, and functional characterization of a novel neuroprotective natural product from the fruit of *Alpinia oxyphylla* for use in Parkinson's disease through LC/MS-based multivariate data analysis-guided fractionation. Journal of Proteome Research, 15(8), 2595–2606.

Liu, Y., Song, F., Wu, W. K., He, M., Zhao, L., Sun, X., ... & Peng, K. (2012). Triptolide inhibits colon cancer cell proliferation and induces cleavage and translocation of 14-3-3 epsilon. Cell Biochemistry and Function, 30(4), 271–278.

Longeon, A., Peduzzi, J., Barthelemy, M., Corre, S., Nicolas, J. L., & Guyot, M. (2004). Purification and partial identification of novel antimicrobial protein from marine bacterium *Pseudoalteromonas* species strain X153. Marine Biotechnology, 6(6), 633–641.

Long, Y., & Zhang, M. (2009). Self-assembling bacterial pores as components of nanobiosensors for the detection of single peptide molecules. Science in China Series B: Chemistry, 52(6), 731–733.

Loo, S., Kam, A., Xiao, T., Nguyen, G. K., Liu, C. F., & Tam, J. P. (2016). Identification and characterization of roseltide, a knottin-type neutrophil elastase inhibitor derived from *Hibiscus sabdariffa*. Scientific Reports, 6(1), 1–16.

Lu, H., Giordano, F., & Ning, Z. (2016). Oxford Nanopore MinION sequencing and genome assembly. Genomics, Proteomics & Bioinformatics, 14(5), 265–279.279.

Lu, J. J., Yang, Z., Lu, D. Z., Wo, X. D., Shi, J. J., Lin, T. Q., ... &Tang, L. H. (2012). Dihydroartemisinin-induced inhibition of proliferation in BEL-7402 cells: An analysis of the mitochondrial proteome. Molecular Medicine Reports, 6(2), 429–433.

Lv, K., Zhang, M., Zhang, W., Hao, J. Q., & Zheng, C. X. (2015). Separation of ovule proteins during female gametophyte cellularization of 'Pinus tabuliformis' using 2D-DIGE. Plant Omics, 8(2), 106–111.

Ma, R., Sun, L., Chen, X., Jiang, R., Sun, H., & Zhao, D. (2013). Proteomic changes in different growth periods of ginseng roots. Plant Physiology and Biochemistry, 67, 20–32.

Manavalan, A., Ramachandran, U., Sundaramurthi, H., Mishra, M., Sze, S. K., & Hu, J. M., ... & Heese, K. (2012). *Gastrodia elata* Blume (tianma) mobilizes neuro-protective capacities. International Journal of Biochemistry and Molecular Biology, 3(2), 219.

Mann, M. (2006). Functional and quantitative proteomics using SILAC. Nature reviews Molecular Cell Biology, 7(12), 952–958.

Manzoni, C., Kia, D. A., Vandrovcova, J., Hardy, J., Wood, N. W., Lewis, P. A., & Ferrari, R. (2018). Genome, transcriptome and proteome: The rise of omics data and their integration in biomedical sciences. Briefings in Bioinformatics, 19(2), 286–302.

Marouga, R., David, S., & Hawkins, E. (2005). The development of the DIGE system: 2D fluorescence difference gel analysis technology. Analytical and Bioanalytical Chemistry, 382(3), 669–678.

Marsoni, M., Vannini, C., Campa, M., Cucchi, U., Espen, L., & Bracale, M. (2005). Protein extraction from grape tissues by two-dimensional electrophoresis. Vitis, 44(4), 181–186.

McConkey, D. J., Lee, S., Choi, W., Tran, M., Majewski, T., Lee, S., ... & Czerniak, B. (2010). Molecular genetics of bladder cancer: Emerging mechanisms of tumor initiation and progression. Urologic Oncology: Seminars and Original Investigations, 28(4), 429–440.

Meller, A., Nivon, L., & Branton, D. (2001). Voltage-driven DNA translocations through a nanopore. Physical Review Letters, 86(15), 3435.

Milani, M., & Sparavigna, A. (2017). The 24-hour skin hydration and barrier function effects of a hyaluronic 1%, glycerin 5%, and *Centella asiatica* stem cells extract moisturizing fluid: An intra-subject, randomized, assessor-blinded study. Clinical, Cosmetic and Investigational Dermatology, 10, 311.

Miyashita, M., Presley, J. M., Buchholz, B. A., Lam, K. S., Lee, Y. M., Vogel, J. S., & Hammock, B. D. (2001). Attomole level protein sequencing by Edman degradation coupled with accelerator mass spectrometry. Proceedings of the National Academy of Sciences, 98(8), 4403–4408.

Monti, M., Cozzolino, M., Cozzolino, F., Vitiello, G., Tedesco, R., Flagiello, A., & Pucci, P. (2009). Puzzle of protein complexes in vivo: A present and future challenge for functional proteomics. Expert Review of Proteomics, 6(2), 159–169.

Monti, M., Orru, S., Pagnozzi, D., & Pucci, P. (2005). Functional proteomics. Clinica Chimica Acta, 357(2), 140–150.

Moult, J., Fidelis, K., Kryshtafovych, A., Schwede, T., & Tramontano, A. (2014). Critical assessment of methods of protein structure prediction (CASP)—Round x. Proteins: Structure, Function, 82, 1–6.

Mumtaz, M. W., Hamid, A. A., Akhtar, M. T., Anwar, F., Rashid, U., & AL-Zuaidy, M. H. (2017). An overview of recent developments in metabolomics and proteomics–phytotherapic research perspectives. Frontiers in Life Science, 10(1), 1–37.

Nettikadan, S., Radke, K., Johnson, J., Xu, J., Lynch, M., Mosher, C., & Henderson, E. (2006). Detection and quantification of protein biomarkers from fewer than 10 cells. Molecular & Cellular Proteomics, 5(5), 895–901.

Novakova, K., Sedo, O., & Zdrahal, Z. (2011). Mass spectrometry characterization of plant phosphoproteins. Current Protein and Peptide Science, 12(2), 112–125.

Ong, S. E., Blagoev, B., Kratchmarova, I., Foster, L. J., Andersen, J. S., & Mann, M. (2006). Stable isotope labeling by amino acids in cell culture for quantitative proteomics. In Cell Biology (pp. 427–436). Academic Press, Cambridge, MA.

Osaki, T., Suzuki, H., Le Pioufle, B., & Takeuchi, S. (2009). Multichannel simultaneous measurements of single-molecule translocation in α-hemolysin nanopore array. Analytical Chemistry, 81(24), 9866–9870.

Palama, T. L., Menard, P., Fock, I., Choi, Y. H., Bourdon, E., Govinden-Soulange, J., ... & Kodja, H. (2010). Shoot differentiation from protocorm callus cultures of *Vanilla planifolia* (Orchidaceae): Proteomic and metabolic responses at early stage. BMC Plant Biology, 10(1), 1–18.

Pandey, A., & Mann, M. (2000). Proteomics to study genes and genomes. Nature, 405(6788), 837–846.

Pandey, M. M., Rastogi, S., & Rawat, A. K. S. (2013). Indian Traditional ayurvedic system of medicine and nutritional supplementation. Evidence-Based Complementary and Alternative Medicine, 2013, 1–12.

Pandey, R., Tiwari, R. K., & Shukla, S. S. (2016). Omics: A newer technique in herbal drug standardization & quantification. Journal of Young Pharmacists, 8(2), 76–81.

Patel, V. J., Thalassinos, K., Slade, S. E., Connolly, J. B., Crombie, A., Murrell, J. C., & Scrivens, J. H. (2009). A comparison of labeling and label-free mass spectrometry-based proteomics approaches. Journal of Proteome Research, 8(7), 3752–3759.

Petrovska, B. B. (2012). Historical review of medicinal plants' usage. Pharmacognosy Reviews, 6(11), 1–5.

Peukert, W., Kaspereit, M., Hofe, T., & Gromotka, L. (2022). Size exclusion chromatography (SEC). In Particle Separation Techniques (pp. 409–447). Elsevier, Amsterdam, The Netherlands.

Picotti, P., & Aebersold, R. (2012). Selected reaction monitoring–based proteomics: Workflows, potential, pitfalls and future directions. Nature Methods, 9(6), 555–566.

Popescu, S. C., Snyder, M., & Dinesh-Kumar, S. (2007). *Arabidopsis* protein microarrays for the high-throughput identification of protein-protein interactions. Plant Signalling & Behavior, 2(5), 416–420.

Porath, J. (1997). From gel filtration to adsorptive size exclusion. Journal of Protein Chemistry, 16(5), 463–468.

Qi, Y., Chen, X., Chan, C. Y., Li, D., Yuan, C., Yu, F., ... & Lai, L. (2008). Two-dimensional differential gel electrophoresis/analysis of diethylnitrosamine induced rat hepatocellular carcinoma. International Journal of Cancer, 122(12), 2682–2688.

Rabilloud, T. (2002). Two-dimensional gel electrophoresis in proteomics: Old, old fashioned, but it still climbs up the mountains. PROTEOMICS: International Edition, 2(1), 3–10.

Raharjo, T. J., Widjaja, I., & Roytrakul, S., & Verpoorte, R. (2004). Comparative proteomics of *Cannabis sativa* plant tissues. Journal of Biomolecular Techniques: JBT, 15(2), 97–106.

Ratner, M. (2013). A brief history of molecular electronics. Nature Nanotechnology, 8(6), 378–381.

Reed, M. A., Zhou, C., Muller, C. J., Burgin, T. P., & Tour, J. M. (1997). Conductance of a molecular junction. Science, 278(5336), 252–254.

Regnier, F. E. (1983). High-performance liquid chromatography of biopolymers. Science, 222(4621), 245–252.

Reisinger, V., & Eichacker, L. A. (2007). How to analyze protein complexes by 2D blue native SDS-PAGE. Proteomics, 7(S1), 6–16.

Restrepo-Pérez, L., Joo, C., & Dekker, C. (2018). Paving the way to single-molecule protein sequencing. Nature Nanotechnology, 13(9), 786–796.

Riffle, M., & Eng, J. K. (2009). Proteomics data repositories. Proteomics, 9(20), 4653–4663.

Rosenberg, J. M., & Utz, P. J. (2015). Protein microarrays: A new tool for the study of autoantibodies in immunodeficiency. Frontiers in Immunology, 6, 138.

Roychoudhury, A., & Bhowmik, R. (2021a). Understanding the mechanistic functioning of bioactive compounds in medicinal plants. In: Aftab, T., & Hakeem, K. (Eds.) Medicinal and Aromatic Plants, Springer Nature, Switzerland, Pp. 159–184.

Roychoudhury, A., & Bhowmik, R. (2021b). State-of-the-art technologies for improving the quality of medicinal and aromatic plants. In: Aftab, T., & Hakeem, K. (Eds.) Medicinal and Aromatic Plants, Springer Nature, Switzerland, pp. 593–627.

Roychoudhury, A., Datta, K., & Datta, S. K. (2011). Abiotic stress in plants: From genomics to metabolomics. In: Tuteja, N., Gill, S. S., & Tuteja, R. (Eds.) Omics and Plant Abiotic Stress Tolerance, Bentham Science Publishers, Sharjah, United Arab Emirates, pp. 91–120.

Šali, A. (1998). 100,000 protein structures for the biologist. Nature Structural Biology, 5(12), 1029–1032.

Sanchez, R. O. B. E. R. T. O., & Sali, A. (1998). Large-scale protein structure modeling of the *Saccharomyces cerevisiae* genome. Proceedings of the National Academy of Sciences, 95(23), 13597–13602.

Schmidt, A., Forne, I., & Imhof, A. (2014). Bioinformatic analysis of proteomics data. BMC Systems Biology, 8(2), 1–7.

Shao, L. I., & Zhang, B. (2013). Traditional Chinese medicine network pharmacology: Theory, methodology and application. Chinese Journal of Natural Medicines, 11(2), 110–120.

Shiio, Y., & Aebersold, R. (2006). Quantitative proteome analysis using isotope-coded affinity tags and mass spectrometry. Nature Protocols, 1(1), 139–145.

Shimonishi, Y., Hong, Y. M., Kitagishi, T., Matsuo, T., Matsuda, H., & Katakuse, I. (1980). Sequencing of peptide mixtures by Edman degradation and field-desorption mass spectrometry. European Journal of Biochemistry, 112(2), 251–264.

Smith, J. B. (2001). Peptide Sequencing by Edman Degradation. eLS, New York Plaza, NY.

Smyth, M. S., & Martin, J. H. J. (2000). X -ray crystallography. Molecular Pathology, 53(1), 8–14.

Cragg, G. M., Boyd, M. R., Cardellina, J. H., Newman, D. J., Snader, K. M., & McCloud, T. G. (1994). Ethnobotany and the search for new drugs. In Ciba Foundation Symposium (Vol. 185, pp. 178–196). Wiley & Sons, Chichester, UK.

Stojanoski, N. (1999). Development of health culture in Veles and its region from the past to the end of the 20th century. Veles: Society of science and Art, 13, 34.

Sutandy, F. R., Qian, J., Chen, C. S., & Zhu, H. (2013). Overview of protein microarrays. Current Protocols in Protein Science, 72(1), 2711–2716.

Tannu, N. S., & Hemby, S. E. (2006). Two-dimensional fluorescence difference gel electrophoresis for comparative proteomics profiling. Nature Protocols, 1(4), 1732–1742.

Tao, Y., Fang, L., Yang, Y., Jiang, H., Yang, H., Zhang, H., & Zhou, H. (2013). Quantitative proteomic analysis reveals the neuroprotective effects of huperzine A for amyloid beta treated neuroblastoma N 2a cells. Proteomics, 13(8), 1314–1324.

Thomas Pannakal, S., Eilstein, J., Prasad, A., Ekhar, P., Shetty, S., Peng, Z., ... & Roy, N. (2022). Comprehensive characterization of naturally occurring antioxidants from the twigs of mulberry (*Morus alba*) using on-line high-performance liquid chromatography coupled with chemical detection and high-resolution mass spectrometry. Phytochemical Analysis, 33(1), 105–114.

Thulasiraman, V., Lin, S., Gheorghiu, L., Lathrop, J., Lomas, L., Hammond, D., & Boschetti, E. (2005). Reduction of the concentration difference of proteins in biological liquids using a library of combinatorial ligands. Electrophoresis, 26(18), 3561–3571.

Tirumalai, R. S., Chan, K. C., Prieto, D. A., Issaq, H. J., Conrads, T. P., & Veenstra, T. D. (2003). Characterization of the low molecular weight human serum proteome* S. Molecular & Cellular Proteomics, 2(10), 1096–1103.

Turgeon, R. T., & Bowser, M. T. (2009). Micro free-flow electrophoresis: Theory and applications. Analytical and Bioanalytical Chemistry, 394(1), 187–198.

Van Emon, J. M. (2016). The omics revolution in agricultural research. Journal of Agricultural and Food Chemistry, 64(1), 36–44.

Varongchayakul, N., Song, J., Meller, A., & Grinstaff, M. W. (2018). Single-molecule protein sensing in a nanopore: A tutorial. Chemical Society Reviews, 47(23), 8512–8524.

Vélez-Bermúdez, I. C., Wen, T. N., Lan, P., & Schmidt, W. (2016). Isobaric tag for relative and absolute quantitation (iTRAQ)-based protein profiling in plants. In Plant Proteostasis (pp. 213–221). Humana Press, New York, NY.

Vihinen, M. (2001). Bioinformatics in proteomics. Biomolecular Engineering, 18(5), 241–248.

Vizcaíno, J. A., Côté, R. G., Csordas, A., Dianes, J. A., Fabregat, A., Foster, J. M., ... & Hermjakob, H. (2012). The proteomics ID entifications (PRIDE) database and associated tools: Status in 2013. Nucleic Acids Research, 41(D1), D1063–D1069.

Voedisch, B., & Thie, H. (2010). Size exclusion chromatography. In Antibody Engineering (pp. 607–612). Springer, Berlin, Heidelberg.

Wang, J., Betancourt, A. M., Mobley, J. A., & Lamartiniere, C. A. (2011b). Proteomic discovery of genistein action in the rat mammary gland. Journal of Proteome Research, 10(4), 1621–1631.

Wang, Y., Gu, L. Q., & Tian, K. (2018). The aerolysin nanopore: From peptidomic to genomic applications. Nanoscale, 10(29), 13857–13866.

Wang, L., McLeod, H. L., & Weinshilboum, R. M. (2011c). Genomics and drug response. New England Journal of Medicine, 364(12), 1144–1153.

Wang, Z. Q., Xu, X. Y., Gong, Q. Q., Xie, C., Fan, W., Yang, J. L., ... & Zheng, S. J. (2014). Root proteome of rice studied by iTRAQ provides integrated insight into aluminum stress tolerance mechanisms in plants. Journal of Proteomics, 98, 189–205.

Wang, H., Ye, Y., Pan, S. Y., Zhu, G. Y., Li, Y. W., Fong, D. W., & Yu, Z. L. (2011a). Proteomic identification of proteins involved in the anticancer activities of oridonin in HepG2 cells. Phytomedicine, 18(2–3), 163–169.

Wang, J., Zhang, J., Zhang, C. J., Wong, Y. K., Lim, T. K., Hua, Z. C., ... & Lin, Q. (2016). In situ proteomic profiling of curcumin targets in HCT116 colon cancer cell line. Scientific Reports, 6(1), 1–8.

Wasinger, V. C., Cordwell, S. J., Cerpa-Poljak, A., Yan, J. X., Gooley, A. A., Wilkins, M. R., ... & Humphery-Smith, I. (1995). Progress with gene-product mapping of the Mollicutes: *Mycoplasma genitalium*. Electrophoresis, 16(1), 1090–1094.

Weaver, B. A. (2014). How Taxol/paclitaxel kills cancer cells. Molecular Biology of the Cell, 25(18), 2677–2681.

Wiese, S., Reidegeld, K. A., Meyer, H. E., & Warscheid, B. (2007). Protein labeling by iTRAQ: A new tool for quantitative mass spectrometry in proteome research. Proteomics, 7(3), 340–350.

Wilkins, J. (2014). Galen's Simple Medicines: Problems in Ancient Herbal Medicine. Critical Approaches to the History of Western Herbal Medicine: from Classical Antiquity to the Early Modern Period. A&C Black, New York City USA, pp. 173–190.

Wilkins, M. R., Sanchez, J. C., Gooley, A. A., Appel, R. D., Humphery-Smith, I., Hochstrasser, D. F., & Williams, K. L. (1996). Progress with proteome projects: Why all proteins expressed by a genome should be identified and how to do it. Biotechnology and Genetic Engineering Reviews, 13(1), 19–50.

Wiśniewski, J. R., & Mann, M. (2012). Consecutive proteolytic digestion in an enzyme reactor increases depth of proteomic and phosphoproteomic analysis. Analytical Chemistry, 84(6), 2631–2637.

Wolfender, J. L., Marti, G., Thomas, A., & Bertrand, S. (2015). Current approaches and challenges for the metabolite profiling of complex natural extracts. Journal of Chromatography A, 1382, 136–164.

Xu, W., Choi, H. K., & Huang, L. (2017). State of *Panax ginseng* research: A global analysis. Molecules, 22(9), 1518.

Yan, S. K., Liu, R. H., Jin, H. Z., Liu, X. R., Ye, J., Shan, L., & Zhang, W. D. (2015). "Omics" in pharmaceutical research: Overview, applications, challenges, and future perspectives. Chinese Journal of Natural Medicines, 13(1), 3–21.

Yates III, J. R. (2011). A century of mass spectrometry: From atoms to proteomes. Nature Methods, 8(8), 633–637.

Ying, Y. L., & Long, Y. T. (2019). Nanopore-based single-biomolecule interfaces: From information to knowledge. Journal of the American Chemical Society, 141(40), 15720–15729.

Yokoyama, S., Hirota, H., Kigawa, T., Yabuki, T., Shirouzu, M., Terada, T., ... & Kuramitsu, S. (2000). Structural genomics projects in Japan. Nature Structural Biology, 7(11), 943–945.

Zabrouskov, V., Giacomelli, L., van Wijk, K. J., & McLafferty, F. W. (2003). A new approach for plant proteomics: Characterization of chloroplast proteins of *Arabidopsis thaliana* by top-down mass spectrometry. Molecular & Cellular Proteomics, 2(12), 1253–1260.

Zaynab, M., Fatima, M., Abbas, S., Sharif, Y., Jamil, K., Ashraf, A., ... & Batool, W. (2018). Proteomics approach reveals importance of herbal plants in curing diseases. Molecular Microbiology, 1(1), 23–28.

Zaynab, M., Kanwal, S., Abbas, S., Fida, F., Islam, W., Qasim, M., Rehman, N., Fida, F., Furqan, M., Rizwan, M. and Anwar, M., (2017). Bioinformatics tools in agriculture: An update. PSM Biological Research, 2(3), 111–116.

Zhang, X., Xu, L., Shen, J., Cao, B., Cheng, T., Zhao, T., ... & Zhang, H. (2013).Metabolic signatures of esophageal cancer: NMR-based metabolomics and UHPLC-based focused metabolomics of blood serum; Biochimica et Biophysica Acta, 1832(8): 1207–1216.

Zhang, W., Zhang, Q., Deng, W., Li, Y., Xing, G., Shi, X., & Du, Y. (2014). Neuroprotective effect of pretreatment with *Ganoderma lucidum* in cerebral ischemia/reperfusion injury in rat hippocampus. Neural Regeneration Research, 9(15), 1446–1452.

15 Proteomics-based Strategies to Develop Food Crops with Enhanced Nutritional Quality

Namisha Sharma, Seema Pradhan, and Shuvobrata Majumder

15.1 INTRODUCTION

The human population is continuously increasing, and soon we will have 10 billion mouths to feed on Earth. Initiatives and resources are being built up worldwide for the supply of enough food, but satiating hunger alone will not be sufficient. Ample food that is also nutritious is the only way we can fight both hunger and hidden hunger (malnutrition), specifically where diversified foods are not available due to demographic and economic challenges (Majumder et al., 2019). Countries like Africa, South Asia and Latin America are most affected by hidden hunger. Currently, one out of three people in the world is suffering from malnutrition and related health issues like iron deficiency anaemia (IDA), vitamin A deficiency (VAD) and zinc deficiency (FAO, 2015). Continuous biotechnological interventions for nutritionally improved food production have been held to address food and nutrition security (Majumder et al., 2019; Datta and Datta, 2020; Roychoudhury, 2020a; Datta et al., 2021).

Plant yield and its nutritional quality are complex biological mechanisms that could be better understood by using modern multi-omics approaches (Figure 15.1). Next-generation sequencing and single-molecule, real-time sequencing technologies for RNA sequencing have created unrivalled prospects for transcriptomics-based biomarker identification. However, mRNA levels do not give a complete picture of how cells function (Roychoudhury, 2020b). Numerous interactions between proteins and their metabolites are necessary for the majority of cellular processes, including plant stress management and nutritional quality. In addition, proteins are subjected to post-translational modifications (PTMs), frequently necessary to enable function which cannot be predicted from the genetic sequences or from expression patterns of mRNA transcripts. Biological functions of an individual protein depend on its PTM, cellular location, interaction partners and environment (epigenetics). As a result, direct measurements of protein abundance are important along with the transcriptomics data, and such combined analysis came under the omics analysis called proteomics. The term proteomics, coined by Marc Wilkins, 1994, means the identification of the protein species synthesised within an organism with emphasis on their function and their role in cellular metabolic activities (Wilkins et al., 1996; Jorrín-Novo et al., 2015).

Proteomics is a cutting-edge, high-throughput method that offers comprehensive details about complex molecular systems in plants. To adapt to the shifting climatic conditions, crops have developed a variety of strategies. Complex protein combinations and their interactions inside and outside of cells may now be completely understood, thanks to highly sophisticated equipment and separation procedures. Furthermore, finding proteins that demonstrate quantitative or qualitative differences between control and test samples is made possible by modern proteomic analysis (Roychoudhury et al., 2011). These proteins might be essential for plant nutrition and biotic and abiotic stress management. High-throughput proteome analysis can be used to gain information on protein relative abundance, but functional studies using knockout mutants, enzyme assays, methods for studying protein–protein interactions, etc., provide vital information on the biological roles of proteins. Modern genetic/genomic techniques, improved gel-free/label-free proteomics and a wide

DOI: 10.1201/b23255-15

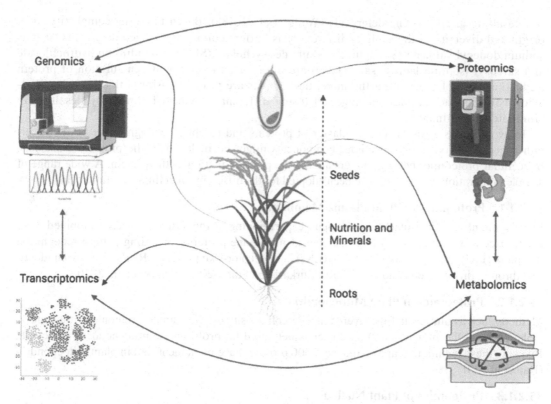

FIGURE 15.1   Multi-omics approaches involved in biofortification research projects. Four major omics–genomics, transcriptomics, proteomics and metabolomics – are interconnected with each other, and the combined information is important in the development of biofortified crops. This image was created using BioRender.com.

range of extremely sophisticated bioinformatic tools have all made it possible to fully comprehend complicated protein/gene operational cascades in many crops. In this chapter, we have systematically reviewed the advancement of proteomic techniques and their application in plant science with a special focus on biofortification.

15.2   PROTEOMICS TECHNIQUES

There are three components of proteomics for the identification and quantification of proteins. It starts with protein extraction and sample preparation, then selective quantification of individual peptides using mass spectrometry (MS) and finally bioinformatic tools to analyse the data and correlate it with genomic or metabolite databases. In many cases, a tissue-specific proteomics study is essential in identifying transporters and elongation factors. For example, protein from root tissue is a good source for metal homeostasis-related protein identifications. Similarly, effects of biotic and abiotic stresses on plants and their consequences on protein expression at the subcellular level can also be studied through different tissue- or organ-specific proteomics analyses.

15.2.1   PROTEIN EXTRACTION AND PREPARATION OF THE SAMPLES

In any proteomics technique, protein extraction and sample preparation are the most important processes (Komatsu and Jorrin-Novo, 2021). Sample preparation involves (i) collecting plant samples, (ii) protein extraction, (iii) digestion of peptides using different enzymes and

(iv) desalting and removing detergents (Wang et al., 2018a). Based upon the complexity, size, origin and diversity of the sample, different lysis buffers comprising varied detergents such as sodium dodecyl sulphate (SDS), urea, sodium deoxycholate (SDC), guanidinium hydrochloride and sarcosyl (sodium lauroyl sarcosinate) are used for extraction and denaturation of protein (reduction of thiol groups). Furthermore, the samples are reduced (using dithiothreitol), alkylated (using iodoacetamide) and digested (using different enzymes) followed by desalting to eliminate contaminants.

To avoid biases towards specific classes of proteins and to obtain enough target proteins, certain tissues, cell types or subcellular components might be included in the proteomic analysis. Subcellular proteomics helps in understanding the proper function of the protein. A wide range of sample preparation techniques have been adopted for the different subcellular fractions (Table 15.1).

15.2.1.1 Proteomics of Plant Plasma Membrane

Plasma membrane is isolated by two-phase partitioning using ultracentrifuge combined with carrier-free electrophoresis. A large number of membrane proteins, including pattern recognition receptors (PRR), receptor-like kinases (RLKs) and receptor-like proteins (RLPs), have been identified through this method (Cao et al., 2016; Burkart and Stahl, 2017; Iwasaki et al., 2020).

15.2.1.2 Proteomics of Plant Mitochondria

Methods such as ultracentrifuge layered in a Percoll density gradient and QProteome Mitochondria Isolation Kit (Qiagen, Germany) have been widely used for profiling the mitochondrial proteome. By using these techniques, approximately 2000 proteins have been identified in plant mitochondria (Rao et al., 2017).

15.2.1.3 Proteomics of Plant Nucleus

The study on nuclear proteome has led to the identification of approximately 5000 nuclear proteins, most of them being transcription factors, kinases and phosphatases, harbouring nucleus localisation signals in their peptide sequences (Yin et al., 2016). It has been reported that during exposure to abiotic and biotic stress the nuclear proteome is majorly involved in signal transduction, reactive oxygen species (ROS)-scavenging proteins, chromatin modifications and regulatory proteins involved in ubiquitinylation-mediated protein degradation proteins (Rajamäki et al., 2020; Barua et al., 2019; Lamelas et al., 2020).

15.2.1.4 Proteomics of Plant Cell Wall

The cell wall is the physical barrier that is involved in stress perceiving and signalling. During environmental stress, biochemical modifications occur in the cell wall proteins which constitute approximately 10% of the cell wall mass and are responsible for remodelling the cell wall and communication with other organelles. There are different types of cell wall proteomes: secretome, in which secreted proteins are studied; apoplastic proteome, in which proteins in the cell wall matrix and apoplastic fluid are studied by vacuum infiltration-centrifugation (VIC); and, lastly, the proteins localised in plasmodesmata are studied (tubular structures in the plant cell wall). Approximately 800 cell wall proteins have been identified in Arabidopsis through proteome analysis, among which 61 are plasmodesmata-associated proteins (Duruflé et al., 2017; Kraner et al., 2017).

15.2.1.5 Proteomics of Plant Chloroplast

The chloroplast apparatus is the metabolic factory that conducts photosynthesis, generates pigments and hormones (salicylic acid and jasmonic acid) and senses metabolic and stress stimuli. Chloroplast proteins act as coordinators for the retrograde signalling between chloroplast and nucleus and regulate the stress response (Tamburino et al., 2017). Isolation of chloroplasts through Percoll density gradient centrifugation is well established, and 2190 proteins have been identified in tomato plants using this method (Bhattacharya et al., 2020).

TABLE 15.1

Methods for Isolation of Sub-Cellular Organelles for Proteomics Analysis

Subcellular Organelles	Purification Technique	Plant	Organ	Number of Proteins Identified	Reference
Cell wall	Cell wall purification by using sodium acetate and sucrose.	*Marchantia polymorpha*	Thallus	409	Kolkas et al. (2022)
	Cell wall proteins extracted using $CaCl_2$, EGTA and lithium chloride-complemented buffers	*Medicago sativa*	Stem	458	Printz et al. (2015)
	Purification of N-glycoproteins using concanavalin A affinity chromatography	*Arabidopsis thaliana*	Stem	992	Minic et al. (2007)
	Vacuum infiltration-centrifugation technique	*Zea mays*	Leaf and root	863	Niu et al. (2020)
Nucleus	Plant nuclei extraction kit	*Glycine max*	Root	268	Murashita et al. (2021)
	Formaldehyde-mediated cross-linking followed by flow cytometric sorting	*Hordeum vulgare*	Root	800	Petrovská et al. (2014)
	Extraction using Honda buffer with high sucrose concentration	*A. thaliana*	Cell suspension	1539	Goto et al. (2019)
Plasma membrane	Two-polymer-based system followed by free-flow electrophoresis	*A. thaliana*	Seedling	1000	de Michele et al. (2016)
	Mem-PER Plus Extraction Kit	*G. max*	Root	268	Murashita et al. (2021)
Apoplastic fluid	Vacuum infiltration-centrifugation	*Gossypium barbadense*	Root	68	Han et al. (2019); Li et al. (2016)
		Solanum tuberosum	Leaf	1257	Abreha et al. (2021)
Mitochondria	Ultracentrifugation in Percoll density gradient	*S. tuberosum*	Tuber	1060	Salvato et al. (2014)
	Centrifugation followed by 1D-blue native PAGE	*A. thaliana*	Leaf	1359	Senkler et al. (2017)
	Affinity purification of plant mitochondria (Mito-AP)	*A. thaliana*	Seedling	619	Niehaus et al. (2020)
Chloroplast	Sorbitol-based medium	*Malus domestica*	Shoots	–	Morkunaite-Haimi et al. (2018)
	NETN buffer, anti-acetyl lysine antibody beads	*G. max*	Leaf	1538	Li et al. (2018)
	Percoll density gradient	*Solanum lycopersicum*	Leaf	2186	Bhattacharya et al. (2020)
Ribosome	Asymmetric flow field-flow fractionation	*Nicotiana benthamiana*	Seedling	–	Pitkänen et al. (2014)
	Sucrose density gradient centrifugation	*A. thaliana*	Leaf, root and seed tissue	414	Firmino et al. (2020)
	Ribosome extraction kit	*H. vulgare*	Roots	–	Martinez-Seidel et al. (2021)

15.2.2.6 Post-translational Modifications Effects on Protein Isolation

Apart from cellular localisation, another aspect that needs to be taken into consideration during protein isolation is the PTMs within the protein. The PTMs such as protein phosphorylation, ubiquitination, deubiquitination, SUMOylation and glycosylation are reversible modifications and play a critical role in determining the function, structure and subcellular localisation of a protein (Ghatak et al., 2017). The traditional methods such as ultracentrifugation, solid phase extraction and solvent precipitation lack selectively and therefore hinder the enrichment of PTMs.

Protein glycosylation is one of the most important modifications involved in determining the physiological functions and biological pathways of the proteins in higher eukaryotes. For instance, proteome analysis identified that protein glycosylation regulates fungus pathogenesis, reduced fertility and seed germination (Chen et al., 2022). Techniques such as lectin affinity chromatography, boronic acid affinity chromatography (BAAC) and hydrophilic interaction liquid chromatography (HILIC) have been extensively used for the enrichment of glycoproteins. HILIC-based techniques have been reported to be highly sensitive, easy to operate and time-saving. It is based on the hydrophilic interaction between N-glycans and modified HILIC materials such as zwitterionic groups, carbohydrate groups, magnetic metal-organic framework (MOFs) and covalent organic framework (COF). The use of these materials for glycoprotein enrichment has been extensively reviewed recently (Xie et al., 2022).

Apart from glycoproteome, phosphoproteome is the most extensively explored PTMs. Phosphorylation mainly occurs on serine, threonine, histidine or tyrosine residues of the protein and plays an essential regulatory role in determining several biological phenomena such as signal transduction, proliferation cell growth and development (Huber, 2007; Li et al., 2022). Recently, approximately 1000 phosphopeptides were identified through phosphoproteomic analysis (Chen et al., 2021). Several methods such as molecular imprinting (MIP), immobilised metal affinity chromatography (IMAC), functional ligand-binding identification by Tat-based recognition of associating proteins, polymer-supported metal-ion-affinity capture and metal oxide affinity chromatography (MOAC) have been frequently used for this purpose. Among these, the concept of IMAC is based on used immobilisation of metal ions (Sn^{4+}, Cu^{2+}, Al^{3+}, Fe^{3+}) on IMAC material using chelating agents such as nitrilotriacetic acid (NTA), adenosine triphosphate (ATP) and polydopamine (PDA). These metal ions help in the enrichment of phosphoproteins due to their affinity towards the phosphate group (Wang et al., 2019). Apart from these two PTMs, several other PTMs have also been studied for their role in plant development and growth and in response to different environmental stresses. The techniques for the sample preparation and enrichment have been summarised in Table 15.2.

15.2.2.7 Quality Control and Analysis of Isolated Protein

After the extraction of plant protein, there can be three strategies for the analysis of proteins: bottom-up proteomics (BUP), middle-down proteomics (MDP) and top-down proteomics (TUP). The BUP involves the digestion of protein into peptides using different proteases, whereas in TUP, proteins are not digested and it involves direct analysis of intact protein. In MDP, the long peptides are generated using proteases having a low frequency and sensitivity (Cassidy et al., 2021). Among these strategies, the BUP approach is the most popular, as it is a feasible experimental strategy. It is because peptides are easily separated by liquid chromatography as compared to proteins which aid in the high-throughput analysis of numerous proteins from complex biological samples (Dupree et al., 2020).

There are two main approaches for proteolytic digestion in BUP. First, the in-gel digestion method requires separating the protein samples through SDS-polyacrylamide gel electrophoresis and then digesting the proteins embedded in the gel pieces. Second, the in-solution method involves the isolation of plant protein using chaotropic reagents (like urea), followed by protein precipitation and

TABLE 15.2

Techniques Used for Identification of PTMs in Plant Proteomic Studies

PTMs	Identification Technique	Plant	Organ	Number of Proteins Identified	Reference
Phosphorylation	Nickel-nitriloacetic acid beads	Maize	Leaf	692	Fan et al. (2021)
	PolyMAC-Ti (polymer-based metal ion affinity capture using titanium)	Tomato	leaf	550	Hsu et al. (2018)
Glycosylation	Lectin affinity chromatography	Common beans	Leaf	35	Zadraznik et al. (2017)
	Glycopeptide hydrophilic interaction chromatography-MS (HILIC) enrichment	Wheat	Leaf	173	Chang et al. (2021)
Methylation	Affinity enrichment of lysine-methylated peptides	Tomato	Fruit	241	Xiao et al. (2022)
		Arabidopsis	Seedling	–	Serre et al. (2020)
Ubiquitination	PTM Scan ubiquitin remnant motif K-ε-GG kit	Maize	Kernel	881	Fan et al. (2021)
		Rice	Seedling	1376	Chen et al. (2018)
	Anti-ubiquitin antibody	Tomato	Leaf	652	Zhang et al. (2021)
		Nicotiana tabacum	Leaf	1963	Zhan et al. (2020)
Acetylation	Anti-acetylation antibody	Nicotiana benthamiana	Leaf	1964	Yuan et al. (2021)
	Chromatography-based acetyl peptide separation	Arabidopsis thaliana	Seedlings	2638	Liu et al. (2018)
SUMOylation	Affinity purification of His-tagged SUMO variants	A. thaliana	Leaf	261	Ingole et al. (2021)

digestion in the denaturing solution. This gel-free method comprises multidimensional protein identification technology (MudPIT) strategy for the two-dimensional separation of peptides (Zhang et al., 2010). The main protease used for this digestion is trypsin which has cleavage specificity towards Arg and Lys (C-terminal) residues. Trypsin is mostly used in BUP; however, combining trypsin with other proteases can intensely enhance the sequence coverage by generating overlapping peptides (Sun et al., 2022). Table 15.3 lists the proteases used for parallel or sequential digestion of protein samples. Software tools such as ProteaseGuru (Miller et al., 2021), DeepDigest (Yang et al., 2021) and MetaMorpheus (Miller et al., 2019) assist in the selection of protease based on the protein sample.

At last, to remove the small molecules and salts added in the sample during extraction, enrichment and digestion, and desalting of the samples is performed. This can be done through dialysis, ion-exchange, protein precipitation and gel electrophoresis. Among all these methods, precipitation of the sample by using organic solvents (acetone, trichloroacetic acid [TCA] or methanol/chloroform) is the most simple and less expensive method (Alvarez and Naldrett, 2016). Desalted samples are analysed using liquid chromatography (LC) coupled with tandem MS (LC-MS/MS).

TABLE 15.3

List of Proteases for Proteome Profiling

Protease	Type	Cleavage Site	Use
Trypsin	Serine protease	C-terminal of K and R	Proteome sequence profiling
Chymotrypsin	Serine protease	C-terminal of F, Y, L, W and M	To digest transmembrane regions of membrane proteins
ArgC	Serine protease	C-terminal of R	To investigate PTMs
LysC	Serine protease	C-terminal of K	Proteome sequence profiling
AspN	Metalloprotease	N-terminal D	Proteome sequence profiling
LysN	Metalloprotease	N-terminal of K	Proteome sequence profiling
GluC	Serine protease	C-terminal of D and E	Produce large peptides and used for middle-down proteomics
Pepsin	Aspartic protease	C-terminal of Y, F and W	Determining disulfide bonds by MS
WaLP	Serine protease	C-terminal of T, V, A, S and M	Study of membrane protein sequences
MaLP	Serine protease	C-terminal of M, L, F, Y, T and V	Study of membrane protein sequences
LysargiNase	Metalloprotease	N-terminal of R and K	For methylation, phosphorylation, ubiquitinylation and acetylation profiling
Legumain	Cysteine protease	C-terminal of K and R	For glycosylation profiling

15.2.3 Instrumentation Involved in Proteomics Study

Proteomics depends mainly on mass spectrometric techniques like LC-MS, MALDI-TOF MS or MS/MS to obtain data required for the identification, quantification and characterisation of individual peptides from complex plant lysates. MS is a powerful tool that consists of three components. First, it uses electrospray ionisation (ESI) or matrix-assisted laser desorption/ionisation (MALDI) or surface-enhanced laser desorption/ionisation (SELDI) to ionise the peptides and proteins. These ions are then operated by electromagnetic fields. In the second step, these ionised molecules are separated on the basis of their mass-to-charge ratio by using mass analysers such as time-of-flight (TOF) or ion trap or Quadrupole Fourier-transform ion cyclotron (FTIC). Finally, these ions are detected by an ion detection system, and the mass-to-charge ratio of ionised molecules is used to calculate the molecular weight of proteins.

The advantage of using MS-based proteomics is its ability to highlight variations in the accumulation of protein among different biological samples by performing quantitative analysis. The strategies for quantitative proteomics can be clustered into two groups –unlabelled and labelled methods. In unlabelled or labelled free quantification (LFQ), the digested and desalted peptides are not treated for any chemical modifications before MALDI-TOF/MS. In LFQ, the abundance of peptides between biological samples is compared using the data-dependent acquisition (DDA) mode which selects the most abundant peptide for detection in MS/MS (Hart-Smith et al., 2017). However, DDA can exclude low abundant peptides and thus limit the number of peptides identified and might lead to irreproducible results (Stead et al., 2008). An alternative data acquisition strategy that has improved the depth of proteome coverage, detection of peptides from low-abundance protein and reproducibility is data-independent acquisition (DIA) (Zhang et al., 2020). In this strategy, multiple precursor ions are continuously collected for detection in MS/MS.

Compared to LFQ, labelled methods have higher accuracy and can be used for processing small amounts of multiple samples. In the labelling method, the protein or derived peptide is labelled using isotope-coded isobaric reagents. Several robust protein labelling techniques have been developed such as tandem mass tag (TMT), stable isotopic labelling with amino acids in cell culture (SILAC) and isobaric tag for relative and absolute quantitation (iTRAQ). In plant proteomics research, the iTRAQ-based multiplex labelling system has been extensively used. It depends on the labelling of protein with isobaric tags (8-plex and 4-plex) in such a way that different labels add the same mass to peptides across different biological samples. It can be used for the identification and quantitation (relative and absolute) of up to 18 samples without affecting the PTMs (Li et al., 2021). These tags label all the primary amines of N-termini and side chains of peptides, primarily lysines. It has been used for quantitative analysis of plant proteomics (Velez-Bermudez et al., 2016), organelle proteomics such as membrane (Adav et al., 2011) and chloroplast (Singh et al., 2022) and plant pathogen analysis (Wang et al., 2018b).

15.2.4 Bioinformatic Tools in Proteomics

High-throughput method-based proteomic studies generate a huge amount of data, and bioinformatic tools are essential for managing, processing and analysing heterogeneous proteomics data. Table 15.4 lists the various bioinformatic tools required for proteomic analysis. MS-based proteomics data is obtained in the form of a peak list which has the mass-to-charge ratio and intensity

TABLE 15.4
Various Bioinformatic Tools Required for Proteomic Analysis

Software	Function	Link
Proteome Discoverer	Database search algorithms	https://www.thermofsher.com/order/catalog/product/OPTON-30810
ProteinPilot	Database search algorithms	https://sciex.com/products/software/proteinpilot-software
MassHunter	Database search algorithms	https://www.agilent.com/en/product/softw are-informatics/mass-spectrometry-software/ data-analysis
MASCOT	Database search algorithms	https://www.matrixscience.com/server.html (Perkins et al. 1999)
MaxQuant	Database search algorithms	https://www.maxquant.org/ (Cox and Mann 2008)
Scaffold	Database search algorithms	https://www.proteomesoftware.com/products/scafold-5 (Searle 2010)
Protein Prospector	Database search algorithms	https://prospector.ucsf.edu/prospector/ mshome.htm (Chalkley et al. 2005)
Andromeda	Database search algorithms	https://bioinformaticshome.com/tools/proteomics/descriptions/Andromeda.html#gsc.tab (Tyanova et al., 2016)
SEQUEST	Database search algorithms	https://proteomicsresource.washington.edu/protocols06/sequest.php (Brodbelt and Russell et al., 2015)
PEAKS	De novo peptide sequencing	https://www.bioinfor.com/peaks-studio/ (Ma et al., 2003)
SWPepNovo	De novo peptide sequencing	https://bio.tools/SWPepNovo (Li et al., 2019)
UniNovo	De novo peptide sequencing	http://proteomics.ucsd.edu/Software/UniNovo.html (Jeong et al., 2013)
ProteomeGenerator	Hybrid identification approach	https://github.com/jtpoirier/proteomegenerator (Cifani et al., 2018)
DirecTag	Hybrid identification approach	http://fenchurch.mc.vanderbilt.edu (Tabb et al., 2008)
MetaMorpheus	PTM identification	https://smith-chem-wisc.github.io/MetaMorpheus/ (Solntsev et al., 2018)
PTMselect	PTM identification	https://sites.google.com/site/fredsoftwares/products/pepalign (Perchey et al., 2019)

of fragmented ions. To analyse and interpret the MS-based proteomics data, several search engines are available that generate the list of putative peptides based on homology with the protein sequence available in plant genomic databases like National Centre for Biotechnology Information (NCBI), European Bioinformatics Institute (EBI), Ensembl, Gramene Phytozome and UniProt. One of the most extensively used search engines for label-free proteomics is Mascot by Matrix Science Inc. It identifies protein by comparing the experimental spectra obtained by MS with the database of known proteins. In this software, the search can be modified to identify certain PTMs based on the mass shift. It uses a scoring system that is based on the MOlecular Weight SEarch (MOWSE) algorithm where the threshold value is set as 30 (Perkins et al., 1999). The protein sequence with a value greater than or equal to 30 is a significant match. Other examples of identification software that support both label-free and labelled quantification are MaxQuant, PEAKS, Comet, Sequest and PANDA. Among these, PEAKS allows cross-species search for the identification of mutations within the peptide (Ma et al., 2003).

If the peptide/protein is non-annotated and the corresponding protein is not available in the database, proteogenomics tools are used. In this approach, the sequence of mRNA obtained from RNA-seq data is translated (using six-frame translation, expressed sequence tags and ab initio gene prediction), and a customised protein database is generated which has all the theoretically possible peptide sequences (Slavoff et al., 2013). Some of the software used for producing personalised protein sequence databases are customProDB, Galaxy Integrated Omics, QUILTS, Mutation DB and MSProGene. A huge amount of data is generated by MS, and several proteome repositories are available where these datasets are accessible, for example, PRIDE, ExPASy, omicX, OBRC, Proteome Commons and PeptideAtlas.

15.3   STRATEGISING BIOFORTIFICATION IN CROPS THROUGH PROTEOMICS

Biofortification of crop plants has never been without its challenges. However, the most prominent problem has been the development of a standardised work plan to bring about successful improvement of the nutritive value of a crop plant. To illustrate, we would like to first mention the overall pathway of mineral uptake and localisation as explained in their review by Carvalho and Vasconcelos (2013). The uptake of minerals from soil is usually performed by the roots which then relay these to the shoot through the xylem. The process is facilitated by a transpiration pull, leading to the deposition of minerals in the leaves. The leaves become sinks for the nutrients and must transit to a source, a phenomenon that happens when the carbon accumulated by photosynthesis exceeds the essential requirement threshold. However, the molecular mechanisms that regulate this process are still unclear and murk the understanding of how these nutrients translocate to other edible parts like fruits, seeds and grains. Therefore, the challenge of biofortification in crop plants requires a concerted effort on all the "omics" fronts. Genomics and transcriptomics have been instrumental in the identification of genes that are differentially regulated in response to nutrient availability and uptake (Wu et al., 2005; del Pozo et al., 2010). However, it is the field of proteomics that has helped predict which transcripts would actually be utilised for protein synthesis.

The basic idea of using proteomics to bring about successful biofortification in plants involves close monitoring of the protein turnover that would be instrumental in identifying the molecules that actually affect the end results. Protein dynamics could be the best way to gain some important insights into the process of nutrient assimilation in plants. This hypothesis has been a premise for many studies. Targeted proteomics has been used to identify protein QTLs (pQTLs) which can predict the outcome of genetic crosses (Rödiger and Baginsky, 2018). Similar studies have also identified pQTLs in potato (Acharjee et al., 2018). Another strategy using proteomics has been demonstrated by dos Santos-Donado et al. (2021), where they took a reverse approach and identified the proteins responsible for high-quality biofortified QPM maize in a bid to add to the knowledge of protein dynamics involved in the nutritive quality of crops. The conceptualisation and execution of these methods for developing biofortified crops can be better explained through some examples.

15.3.1   CAROTENOID BIOFORTIFICATION IN RICE (GOLDEN RICE)

The biofortification of rice with carotenoids has perhaps been one of the most popular success stories in the field of crop science. The concept of transforming rice grains, which naturally do not produce carotenoid compounds, was ground-breaking and would go on to become a yardstick to measure other biofortified crops in terms of their reproducibility and utility. The entire tale of carotenoid-enriched golden rice development involved almost all the major aspects of plant biology, including standardisation of rice transformation, selection of the genes encoding enzymes of beta-carotene biosynthesis pathway from daffodil, selection of specific promoters to target accumulation of beta-carotene in seeds and many more subtle re-calibrations of experiments, which ultimately led to worldwide recognition of golden rice (Ye et al., 2000; Datta et al., 2007; Datta et al., 2021; Majumder et al., 2022).

Since these methods focus on increasing the carotenoid content and its sequestration in the desired edible part of the plant, there is also a need to address the proteomics of carotenoid accumulation, i.e., how the plant makes space for the extra carotenoids. In one of their publications, Torres-Montilla and Rodriguez-Concepcion (2021) have emphasised the need to take protein profiling into consideration while designing experiments for biofortification to ensure that researchers account for the mechanisms that regulate carotenoid biosynthesis, storage as well as degradation and also ensure that the change is reflected at the level of protein. Prolonged storage of golden rice can deteriorate its carotenoid level (Datta et al., 2021). This deterioration is mainly lipoxygenase (LOX) driven that causes lipid peroxidation. Hydroperoxy fatty acids are produced by LOX which in turn oxidise carotenoids (Gayen et al., 2015). LOX also causes the decolourisation of rice seeds during storage. Seed-specific downregulation of the *lox* gene could be helpful in the prolonged storage of golden rice, bringing stability in seed quality and better carotenoid preservation. This hypothesis has been proved by RNAi-mediated *lox* gene silencing in golden rice (Gayen et al., 2014).

Commercialisation of "Golden Rice" was highly approved by the world-renowned and distinguished agencies for international food safety regulation –the Food Standards Australia, New Zealand, Health Canada and the United States Food and Drug Administration (Datta and Datta, 2020). The first country to initiate golden rice cultivation was the Philippines (Wu et al., 2021). The perception is gradually building positively towards commercialising golden rice with the continuous availability of favourable biosafety data (Owens et al., 2018).

15.3.2   IRON AND ZINC FORTIFICATION IN PLANTS

Many studies have emphasised the need for linking transcriptional and translational regulatory mechanisms to ensure successful biofortification in crop plants (Zaghum et al., 2022). Iron and zinc form the most widely chosen elements for biofortification in crops, given the widespread iron deficiency in the world population (Kassebaum et al., 2014). The attempts in iron and zinc fortification using various molecular biology methods have been successful in rice (Vasconcelos et al., 2003; Wirth et al., 2009; Paul et al., 2014; Majumder et al., 2019). However, similar approaches have not been quite as effective in enhancing Fe and Zn content in wheat where the root-shoot barrier and the lack of clarity about the accumulation of these elements in the grain are major bottle-necks (Borrill et al., 2014). Researchers have reported that heavy metal stress is responsible for adverse effects on major metabolic pathways like photosynthesis, respiration, along with nitrogen and sulphur metabolism and that proteomics has a crucial role in deciphering the changes in the protein profile of a plant under such conditions (Hossain and Komatsu, 2013). These changes also affect any attempt at biofortification since any duress due to heavy metal toxicity could lead to significant shifts in protein dynamics which ultimately affect the efficacy of the transgenes (most often Fe/Zn transporters) introduced for biofortification. For example, abiotic stresses, like drought and some heavy metal stresses like Cd toxicity, are alleviated by gibberellic acid (GA). In Arabidopsis, it was observed that GA could effectively reduce the effects of Cd toxicity by reducing NO levels, which, in turn, downregulated the expression of an iron transporter, *IRT1* (Zhu et al., 2012). Therefore, it is prudent to study targeted or global proteomic changes in plants for any successful biofortification attempts.

15.3.3 FORTIFYING CEREALS WITH VITAMIN B

Vitamin B deficiency in humans manifests in the form of various neurological and physiological disorders, including epilepsy, inflammation and abnormalities in immune response (di Salvo et al., 2012; Ghosh et al., 2019). Therefore, there have been efforts to improve the content of vitamin B_6 in crops that are high on the biofortification priority index (http://www.ifpri.org/tools/bpimapping-tool). These crops include cassava, rice and wheat which account for the majority of the world's dietary consumption (Fudge et al., 2017). Transgenic rice was successfully developed for enhanced vitamin B_1 after constitutively expressing the genes for vitamin B_1 biosynthesis. However, most of this was lost after grain polishing, which implied that the accumulation of the vitamin is a complex process (Dong et al., 2016; Ghosh et al., 2019). However, in a previous report, biofortification of rice with vitamin B_9 had found more success in this regard since the researchers improved the stability of folate with a folate-binding protein (Blancquaert et al., 2015). These observations emphasise the need for the involvement of proteomics, along with metabolomics to bring about successful biofortification events.

15.4 CURRENT LIMITATION OF PROTEOMICS-BASED ANALYSIS AND RECOMMENDATIONS

For both basic and applied research, gas chromatography (GC)-MS and other proteomic techniques have made significant advancements in the field of protein analysis. MS-based proteomics has made tremendous development in the identification and quantification of proteins. Recent research on subcellular proteomics and PTMs has tremendously aided the understanding of the intricate relationship between improving agricultural output and combating environmental stresses. Presently, the proteins and peptides with molecular weight up to 30kDa can be identified using MS. For sample preparation, there are standardised protocols and a wide range of database search engines. Even if a proteomics experiment is conducted correctly, the biggest obstacle still to be overcome is an exhaustive database search within the proteomics dataset. Improvements in proteome bioinformatics analysis, such as developing new software that enables recognition of more peptides that match previously unassigned MS/MS spectra, can address false positive findings, false negative results and unassigned spectra.

Another emerging area which needs attention is the integration of metaproteomics and metagenomics. Data from genomics, transcriptomics, proteomics and metabolomics when combined can help us gain a more complete knowledge of the various components influencing the changing environment. The ability to query the collected data with multi-omics datasets in system biology investigations presents perhaps the biggest hurdle for future proteomics and metabolomics research. The large-scale disciplines, including genomics, transcriptomics, proteomics, metabolomics and lipidomics, are now all covered by high-throughput sequencing and MS. With this much data, a multi-omics software platform with customised pipelines is a prerequisite for the integration of the massive datasets from all of those techniques.

Furthermore, proteomics data should be validated and combined with other traditional and contemporary (omics) methodologies in order to overcome technological limitations and analytical biases. Integrating proteomics with other fields such as structural biology (NMR microscopy, X-ray crystallography and cryo-electron microscopy (cryo-EM) service) is important. We have just touched on a small portion of the fundamental principle of biology, including metabolites and mRNA. The most appropriate approach is probably to concentrate on the massive amounts of data that today's powerful technology has produced.

15.5 CONCLUSION

Plant nutrition is a complex trait that is regulated by many biological and environmental factors. Plant biology research is far from completely utilising the full potential of proteomics, and apart from Arabidopsis, not much is known about the proteomes of other plant species. This field of plant science research is rapidly improving. In the coming decade, proteomics study should be focused on analysis of PTMs and interactomics, performing targeted and untargeted proteomics, data validation and integration of proteomics data with other techniques. This combined knowledge will help us understand complex traits, including nutritional improvement of crops. Such understanding is the necessity of the hour to develop biofortified crops to fight against the global hidden hunger and also to secure quality foods for all.

ACKNOWLEDGEMENTS

Authors acknowledge help and support provided by Late Dr. Ajay Kumar Parida, Ex-director, DBT-Institute of Life Sciences, Bhubaneswar, Odisha, India for smooth running of the research facility. Authors also acknowledge "M. K. Bhan Young Researcher Fellowship Award" (BT/HRD/MK-YRFP/50/17/2021) by the Department of Biotechnology, Government of India to Dr. Shuvobrata Majumder.

REFERENCES

Abreha, K. B., Alexandersson, E., Resjö, S., Lankinen, Å, Sueldo, D., Kaschani, F., Kaiser, M., van der Hoorn, R. A. L., Levander, F., & Andreasson, E. (2021). Leaf apoplast of field-grown potato analyzed by quantitative proteomics and activity-based protein profiling. International Journal of Molecular Sciences, 22(21), 12033.

Acharjee, A., Chibon, P. V., Kloosterman, B., America, T., Renaut, J., Maliepaard., C., & Richard, G. F. V. (2018). Genetical genomics of quality related traits in potato tubers using proteomics. BMC Plant Biology, 18, 20.

Adav, S. S., Ng, C. S., & Sze, S. K. (2011). iTRAQ-based quantitative proteomic analysis of *Thermobifida fusca* reveals metabolic pathways of cellulose utilization. Journal of Proteomics, 74(10), 2112–2122.

Alvarez, S., & Naldrett, M. J. (2016). Plant structure and specificity challenges and preparation considerations for proteomics. In: Mirzaei, H., Carrasco, M. (Eds.). Modern Proteomics—Sample Preparation, Analysis and Practical Applications (pp. 63–82). Springer, Switzerland

Barua, P., Lande, N. V., Subba, P., Gayen, D., Pinto, S., Keshava Prasad, T. S., Chakraborty, S., & Chakraborty, N. (2019). Dehydration-responsive nuclear proteome landscape of chickpea (*Cicer arietinum* L.) reveals phosphorylation-mediated regulation of stress response. Plant, Cell & Environment, 42(1), 230–244.

Bhattacharya, O., Ortiz, I., & Walling, L. L. (2020). Methodology: An optimized, high-yield tomato leaf chloroplast isolation and stroma extraction protocol for proteomics analyses and identification of chloroplast co-localizing proteins. Plant Methods, 16, 131.

Blancquaert, D., Van Daele, J., Strobbe, S., Kiekens, F., Storozhenko, S., De Steur, H., Gellynck, X., Lambert, W., Stove, C., & Van Der Straeten, D. (2015). Improving folate (vitamin B9) stability in biofortified rice through metabolic engineering. Nature Biotechnology, 33(10), 1076–1078.

Borrill, P., Connorton, J. M., Balk, J., Miller, A. J., Sanders, D., & Uauy, C. (2014). Biofortification of wheat grain with iron and zinc: Integrating novel genomic resources and knowledge from model crops. Frontiers in Plant Science, 5, 53.

Brodbelt, J. S., & Russell, D. H. (2015). Focus on the 20-year anniversary of SEQUEST. Journal of the American Society for Mass Spectrometry, 26(11), 1797–1798.

Burkart, R. C., & Stahl, Y. (2017). Dynamic complexity: Plant receptor complexes at the plasma membrane. Current Opinion in Plant Biology, 40, 15–21.

Cao, J., Yang, C., Li, L., Jiang, L., Wu, Y., Wu, C., Bu, Q., Xia, G., Liu, X., Luo, Y., & Liu, J. (2016). Rice plasma membrane proteomics reveals *Magnaporthe oryzae* promotes susceptibility by sequential activation of host hormone signaling pathways. Molecular Plant-Microbe Interactions, 29(11), 902–913.

Carvalho, S. M. P., & Vasconcelos, M. W. (2013). Producing more with less: Strategies and novel technologies for plant-based food biofortification. Food Research International, 54(1), 2013, 961–971.

Cassidy, L., Kaulich, P. T., Maaß, S., Bartel, J., Becher, D., & Tholey, A. (2021). Bottom-up and top-down proteomic approaches for the identification, characterization, and quantification of the low molecular weight proteome with focus on short open reading frame-encoded peptides. Proteomics, 21(23–24), e2100008.

Chalkley, R. J., Baker, P. R., Hansen, K. C., Medzihradszky, K. F., Allen, N. P., Rexach, M., & Burlingame, A. L. (2005). Comprehensive analysis of a multidimensional liquid chromatography mass spectrometry dataset acquired on a quadrupole selecting, quadrupole collision cell, time-of-flight mass spectrometer: I. How much of the data is theoretically interpretable by search engines? Molecular & Cellular Proteomics: MCP, 4(8), 1189–1193.

Chang, Y., Zhu, D., Duan, W., Deng, X., Zhang, J., Ye, X., & Yan, Y. (2021). Plasma membrane N-glycoproteome analysis of wheat seedling leaves under drought stress. International Journal of Biological Macromolecules, 193, 1541–1550.

Chen, X. L., Liu, C., Tang, B., Ren, Z., Wang, G. L., & Liu, W. (2020). Quantitative proteomics analysis reveals important roles of N-glycosylation on ER quality control system for development and pathogenesis in *Magnaporthe oryzae*. PLoS Pathogens, 16(2), e1008355.

Chen, X. L., Xie, X., Wu, L., Liu, C., Zeng, L., Zhou, X., Luo, F., Wang, G. L., & Liu, W. (2018). Proteomic analysis of ubiquitinated proteins in rice (*oryza sativa*) after treatment with pathogen-associated molecular pattern (PAMP) elicitors. Frontiers in Plant Science, *9*, 1064.

Chen, Y. H., Shen, H. L., Chou, S. J., Sato, Y., & Cheng, W. H. (2022). Interference of arabidopsisN-acetylglucosamine-1-Puridylyl transferase expression impairs protein N-glycosylation and induces ABA-mediated salt sensitivity during seed germination and early seedling development. Frontiers in Plant Science, 13, 903272.

Chen, Y., Wang, Y., Yang, J., Zhou, W., & Dai, S. (2021). Exploring the diversity of plant proteome. Journal of Integrative Plant Biology, 63(7), 1197–1210.

Cifani, P., Dhabaria, A., Chen, Z., Yoshimi, A., Kawaler, E., Abdel-Wahab, O., Poirier, J. T., & Kentsis, A. (2018). Proteome Generator: A framework for comprehensive proteomics based on de novo transcriptome assembly and high-accuracy peptide mass spectral matching. Journal of Proteome Research, 17(11), 3681–3692.

Cox, J., & Mann, M. (2008). MaxQuant enables high peptide identification rates, individualized p.p.b.-range mass accuracies and proteome-wide protein quantification. Nature Biotechnology, 26(12), 1367–1372.

Datta, S. K., & Datta, K. (2020). Golden rice. In: Oliveira, A. C-de., Pegoraro, C., & Viana, V. E. (Eds.) The Future of Rice Demand: Quality Beyond Productivity. Springer, Cham. https://doi.org/10.1007/978-3-030-37510-2_6

Datta, S. K., Datta, K., Parkhi, V., Rai, M., Baisakh, N., Sahoo, G., Rehana, S., Bandyopadhyay, A., Alamgir, M., Ali, M. S., Abrigo, E., Oliva, N., & Torrizo, L. (2007). Golden rice: Introgression, breeding, and field evaluation. Euphytica, 154, 271–278.

Datta, S. K., Majumder, S., & Datta, K. (2021). Molecular breeding for improved b-carotene synthesis in golden rice: Recent progress and future perspectives. In: Hossain, M. A., Hassan, L., Iftekharuddaula, K. Md., Kumar, A., & Henry, R. (Eds.) Molecular Breeding for Rice Abiotic Stress Tolerance and Nutritional Quality, First Edition. John Wiley & Sons, pp. 287–303.

de Michele, R., McFarlane, H. E., Parsons, H. T., Meents, M. J., Lao, J., González Fernández-Niño, S. M., Petzold, C. J., Frommer, W. B., Samuels, A. L., & Heazlewood, J. L. (2016). Free-flow electrophoresis of plasma membrane vesicles enriched by two-phase partitioning enhances the quality of the proteome from Arabidopsis seedlings. Journal of Proteome Research, 15(3), 900–913.

del Pozo, T., Cambiazo, V., & González, M. (2010). Gene Expression profiling analysis of copper homeostasis in *Arabidopsis thaliana*. Biochemical and Biophysical Research Communications, 393(2), 248–252.

di Salvo, M. L., Safo, M. K., & Contestabile, R. (2012). Biomedical aspects of pyridoxal 5'-phosphate availability. Frontiers in Bioscience (Elite Edition), 4(3), 897–913.

Dong, W., Thomas, N., Ronald, P. C., & Goyer, A. (2016). Overexpression of Thiamin biosynthesis genes in rice increases leaf and unpolished grain Thiamin content but not resistance to *Xanthomonas oryzae* pv. oryzae. Frontiers in Plant Science, 7, 616.

dos Santos-Donado, P. R., Donado-Pestana, C. M., Kawahara, R., Rosa-Fernandes, L., Palmisano, G., & Finardi-Filho, F. (2021). Comparative analysis of the protein profile from biofortified cultivars of quality protein maize and conventional maize by gel-based and gel-free proteomic approaches. LWT, 138, 110683.

Dupree, E. J., Jayathirtha, M., Yorkey, H., Mihasan, M., Petre, B. A., & Darie, C. C. (2020). A critical review of Bottom-up proteomics: The good, the bad, and the future of this field. Proteomes, 8(3), 14.

Duruflé, H., Clemente, H. S., Balliau, T., Zivy, M., Dunand, C., & Jamet, E. (2017). Cell wall proteome analysis of *Arabidopsis thaliana* mature stems. Proteomics, 17(8). https://doi.org/10.1002/pmic.201600449.

Fan, W., Zheng, H., & Wang, G. (2021). Proteomic analysis of ubiquitinated proteins in maize immature kernels. Journal of Proteomics, 243, 104261.

Firmino, A., Gorka, P., Graf, M., Skirycz, A., Martinez-Seidel, A., Zander, F., Kopka, K., & Beine-Golovchuk, J. (2020). Separation and paired proteome profiling of plant chloroplast and cytoplasmic ribosomes. Plants (Basel, Switzerland), 9(7), 892.

Food and Agriculture Organization FAO; International Fund for Agricultural Development IFAD; World Food Programme WFP. The State of Food Insecurity in the World 2015. Food and Agriculture Organization of the United Nations: Rome, Italy, 2015.

Fudge, J., Mangel, N., Gruissem, W., Vanderschuren, H., & Fitzpatrick, T. B. (2017). Rationalising vitamin B6 biofortification in crop plants. Current Opinion in Biotechnology, 44, 130–137.

Gayen, D., Ali, N., Ganguly, M., Paul, S., Datta, K., & Datta, S. K. (2014). RNAi mediated silencing of *lipoxygenase* gene to maintain rice grain quality and viability during storage. Plant Cell, Tissue and Organ Culture, 118, 229–243.

Gayen, D., Ali, N., Sarkar, S. N., Datta, S. K., & Datta, K. (2015). Down regulation of lipoxygenase gene reduces degradation of carotenoids of golden rice during storage. Planta, 252, 353–363.

Ghatak, A., Chaturvedi, P., & Weckwerth, W. (2017). Cereal crop proteomics: Systemic analysis of crop-droughtstressresponsestowardsmarker-assistedselectionbreeding. Frontiers in Plant Science, 8, 757.

Ghosh, S., Datta, K., & Datta, S. K. (2019). Rice vitamins. In: Bao, J. (Ed.) Rice, Chemistry and Technology, 4th edition. Elsevier, Rice, pp 195–220. https://doi.org/10.1016/B978-0-12-811508-4.00007-1

Goto, C., Hashizume, S., Fukao, Y., Hara-Nishimura, I., & Tamura, K. (2019). Comprehensive nuclear proteome of Arabidopsis obtained by sequential extraction. Nucleus, 10(1), 81–92.

Han, L. B., Li, Y. B., Wang, F. X., Wang, W. Y., Liu, J., Wu, J. H., Zhong, N. Q., Wu, S. J., Jiao, G. L., Wang, H. Y., & Xia, G. X. (2019). The cotton apoplastic protein CRR1 stabilizes chitinase 28 to facilitate defense against the fungal pathogen *Verticillium ahlia*. The Plant Cell, 31(2), 520–536.

Hart-Smith, G., Reis, R. S., Waterhouse, P. M., & Wilkins, M. R. (2017). Improved quantitative plant proteomics via the combination of targeted and untargeted data acquisition. Frontiers in Plant Science, 8, 1669.

Hossain, Z., & Komatsu, S. (2013). Contribution of proteomic studies towards understanding plant heavy metal stress response. Frontiers in Plant Science, 3, 310.

Hsu, C. C., Zhu, Y., Arrington, J. V., Paez, J. S., Wang, P., Zhu, P., Chen, I. H., Zhu, J. K., & Tao, W. A. (2018). Universal plant phosphoproteomics workflow and its application to tomato signaling in response to cold stress. Molecular & Cellular Proteomics, 17(10), 2068–2080.

Huber, S. C. (2007). Exploring the role of protein phosphorylation in plants: From signalling to metabolism. Biochemical Society Transactions, 35(1), 28–32.

Ingole, K. D., Dahale, S. K., & Bhattacharjee, S. (2021). Proteomic analysis of SUMO1-SUMOylome changes during defense elicitation in Arabidopsis. Journal of Proteomics, 232, 104054.

Iwasaki, Y., Itoh, T., Hagi, Y., Matsuta, S., Nishiyama, A., Chaya, G., Kobayashi, Y., Miura, K., & Komatsu, S. (2020). Proteomics analysis of plasma membrane fractions of the root, leaf, and flower of rice. International Journal of Molecular Sciences, 21(19), 6988.

Jeong, K., Kim, S., & Pevzner, P. A. (2013). UniNovo: A universal tool for de novo peptide sequencing. Bioinformatics (Oxford, England), 29(16), 1953–1962.

Jorrín-Novo, J. V., Pascual, J., Sánchez-Lucas, R., Romero-Rodríguez, M. C., Rodríguez-Ortega, M. J., Lenz, C., & Valledor, L. (2015). Fourteen years of plant proteomics reflected in proteomics: Moving from model species and 2DE-based approaches to orphan species and gel-free platforms. Proteomics, 15(5–6), 1089–1112.

Kassebaum, N. J., Jasrasaria, R., Naghavi, M., Wulf, S. K., Johns, N., Lozano, R., Regan, M., Weatherall, D., Chou, D. P., Eisele, T. P., Flaxman, S. R., Pullan, R. L., Brooker, S. J., & Murray, C. J. (2014). A systematic analysis of global anemia burden from 1990 to 2010. Blood, 123(5), 615–624.

Kolkas, H., Balliau, T., Chourré, J., Zivy, M., Canut, H., & Jamet, E. (2022). The cell wall proteome of *Marchantia polymorpha* reveals specificities compared to those of flowering plants. Frontiers in Plant Science, 12, 765846.

Komatsu, S., & Jorrin-Novo, J. V. (2021). Plant proteomic research 3.0: Challenges and perspectives. International Journal of Molecular Sciences, 22(2), 766.

Kraner, M. E., Müller, C., & Sonnewald, U. (2017). Comparative proteomic profiling of the choline transporter-like1 (CHER1) mutant provides insights into plasmodesmata composition of fully developed *Arabidopsis thaliana* leaves. The Plant Journal, 92(4), 696–709.

Lamelas, L., Valledor, L., Escandón, M., Pinto, G., Cañal, M. J., & Meijón, M. (2020). Integrative analysis of the nuclear proteome in *Pinus radiata* reveals thermopriming coupled to epigenetic regulation. Journal of Experimental Botany, 71(6), 2040–2057.

Li, C., Li, K., Li, K., Xie, X., & Lin, F. (2019). SWPepNovo: An efficient de novo peptide sequencing tool for large-scale MS/MS spectra analysis. International Journal of Biological Sciences, 15(9), 1787–1801.

Li, H., Liao, Y., Zheng, X., Zhuang, X., Gao, C., & Zhou, J. (2022). Shedding light on the role of phosphorylation in plant autophagy. FEBS Letters, 596(17), 2172–2185.

Li, J., Cai, Z., Bomgarden, R. D., Pike, I., Kuhn, K., Rogers, J. C., Roberts, T. M., Gygi, S. P., & Paulo, J. A. (2021). TMTpro-18plex: The expanded and complete set of TMTpro reagents for sample multiplexing. Journal of Proteome Research, 20(5), 2964–2972.

Li, X., Rehman, S. U., Yamaguchi, H., Hitachi, K., Tsuchida, K., Yamaguchi, T., Sunohara, Y., Matsumoto, H., & Komatsu, S. (2018). Proteomic analysis of the effect of plant-derived smoke on soybean during recovery from flooding stress. Journal of Proteomics, 181, 238–248.

Li, Y. B., Han, L. B., Wang, H. Y., Zhang, J., Sun, S. T., Feng, D. Q., Yang, C. L., Sun, Y. D., Zhong, N. Q., & Xia, G. X. (2016). The thioredoxin GbNRX1 plays a crucial role in homeostasis of apoplastic reactive oxygen species in response to verticillium dahlia infection in cotton. Plant Physiology, 170(4), 2392–2406.

Liu, S., Yu, F., Yang, Z., Wang, T., Xiong, H., Chang, C., Yu, W., & Li, N. (2018). Establishment of dimethyl labeling-based quantitative acetylproteomics in Arabidopsis. Molecular & Cellular Proteomics, 17(5), 1010–1027.

Ma, B., Zhang, K., Hendrie, C., Liang, C., Li, M., Doherty-Kirby, A., & Lajoie, G. (2003). PEAKS: Powerful software for peptide de novo sequencing by tandem mass spectrometry. Rapid Communications in Mass Spectrometry, 17(20), 2337–2342.

Majumder, S., Datta, K., & Datta, S. K. (2019). Rice biofortification: High iron, zinc, and vitamin—A to fight against "hidden hunger. Agronomy, 9, 803.

Majumder, S., Datta, K., & Datta, S. K. (2022) Transgenics for biofortification with special reference to rice. In: Kumar, S., Dikshit, H. K., Mishra, G. P., & Singh, A. (Eds.) Biofortification of Staple Crops. Springer, Singapore. pp. 439–460. https://doi.org/10.1007/978-981-16-3280-8_17

Martinez-Seidel, F., Suwanchaikasem, P., Nie, S., Leeming, M. G., Pereira Firmino, A. A., Williamson, N. A., Kopka, J., Roessner, U., & Boughton, B. A. (2021). Membrane-enriched proteomics link ribosome accumulation and proteome reprogramming with cold acclimation in barley root meristems. Frontiers in Plant Science, 12, 656683.

Miller, R. M., Ibrahim, K., & Smith, L. M. (2021). ProteaseGuru: A tool for protease selection in bottom-up proteomics. Journal of Proteome Research, 20(4), 1936–1942.

Miller, R. M., Millikin, R. J., Hoffmann, C. V., Solntsev, S. K., Sheynkman, G. M., Shortreed, M. R., & Smith, L. M. (2019). Improved protein inference from multiple protease bottom-up mass spectrometry data. Journal of Proteome Research, 18(9), 3429–3438.

Minic, Z., Jamet, E., Négroni, L., Arsene der Garabedian, P., Zivy, M., & Jouanin, L. (2007). A sub-proteome of Arabidopsis thaliana mature stems trapped on Concanavalin A is enriched in cell wall glycoside hydrolases. Journal of Experimental Botany, 58(10), 2503–2512.

Morkunaite-Haimi, S., Vinskiene, J., Staniene, G., & Haimi, P. (2018). Efficient isolation of chloroplasts from in vitro shoots of Malus and Prunus. Zemdirbyste, 105(2), 171–176.

Murashita, Y., Nishiuchi, T., Rehman, S. U., & Komatsu, S. (2021). Subcellular proteomics to understand promotive effect of plant-derived smoke solution on soybean root. Proteomes, 9(4), 39.

Niehaus, M., Straube, H., Künzler, P., Rugen, N., Hegermann, J., Giavalisco, P., Eubel, H., Witte, C. P., & Herde, M. (2020). Rapid affinity purification of tagged plant mitochondria (Mito-AP) for metabolome and proteome analyses. Plant Physiology, 182(3), 1194–1210.

Niu, L., Liu, L., & Wang, W. (2020). Digging for stress-responsive cell wall proteins for developing stress-resistant maize. Frontiers in Plant Science, 11, 576385.

Owens, B. (2018). Golden rice is safe to eat, says FDA. Nature Biotechnol, 36, 559–560.

Paul, S., Ali, N., Datta, S. K., & Datta, K. (2014). Development of an iron-enriched high-yielding indica rice cultivar by introgression of a high-iron trait from transgenic iron-biofortified rice. Plant Foods for Human Nutrition, 69, 203–208.

Perchey, R. T., Tonini, L., Tosolini, M., Fournié, J. J., Lopez, F., Besson, A., & Pont, F. (2019). PTMselect: Optimization of protein modifications discovery by mass spectrometry. Scientific Reports, 9(1), 4181.

Perkins, D. N., Pappin, D. J., Creasy, D. M., & Cottrell, J. S. (1999). Probability-based protein identification by searching sequence databases using mass spectrometry data. Electrophoresis, 20(18), 3551–3567.

Petrovská, B., Jeřábková, H., Chamrád, I., Vrána, J., Lenobel, R., Uřinovská, J., Sebela, M., & Doležel, J. (2014). Proteomic analysis of barley cell nuclei purified by flow sorting. Cytogenetic and Genome Research, 143(3), 78–86.

Pitkänen, L., Tuomainen, P., & Eskelin, K. (2014). Analysis of plant ribosomes with asymmetric flow field-flow fractionation. Analytical and Bioanalytical Chemistry, 406(6), 1629–1637.

Printz, B., Dos Santos Morais, R., Wienkoop, S., Sergeant, K., Lutts, S., Hausman, J. F., & Renaut, J. (2015). An improved protocol to study the plant cell wall proteome. Frontiers in Plant Science, 6, 237.

Rajamäki, M. L., Sikorskaite-Gudziuniene, S., Sarmah, N., Varjosalo, M., & Valkonen, J. P. T. (2020). Nuclear proteome of virus-infected and healthy potato leaves. BMC Plant Biology, 20(1), 355.

Rao, R. S., Salvato, F., Thal, B., Eubel, H., Thelen, J. J., & Møller, I. M. (2017). The proteome of higher plant mitochondria. Mitochondrion, 33, 22–37.

Rödiger, A., & Baginsky, S. (2018). Tailored use of targeted proteomics in plant-specific applications. Frontiers in Plant Science, 9, 1204.

Roychoudhury, A. (2020a). Agronomic and genetic biofortification of rice grains with microelements to assure human nutritional security. SF Journal of Agricultural and Crop Management, 1, 1005.

Roychoudhury, A. (2020b). Next generation sequencing: Prospects in plant breeding and crop improvement. SF Journal of Agricultural and Crop Management, 1, 1004.

Roychoudhury, A., Datta, K., & Datta, S. K. (2011). Abiotic stress in plants: From genomics to metabolomics. In: Tuteja, N., Gill, S. S., & Tuteja, R. (Eds.) Omics and Plant Abiotic Stress Tolerance, Bentham Science Publishers, Sharjah, UAE, pp. 91–120.

Salvato, F., Havelund, J. F., Chen, M., Rao, R. S., Rogowska-Wrzesinska, A., Jensen, O. N., Gang, D. R., Thelen, J. J., & Møller, I. M. (2014). The potato tuber mitochondrial proteome. Plant Physiology, 164(2), 637–653.

Searle, B. C. (2010). Scaffold: A bioinformatic tool for validating MS/MS-based proteomic studies. Proteomics, 10(6), 1265–1269.

Senkler, J., Senkler, M., Eubel, H., Hildebrandt, T., Lengwenus, C., Schertl, P., Schwarzländer, M., Wagner, S., Wittig, I., & Braun, H. P. (2017). The mitochondrial complexome of *Arabidopsis thaliana*. The Plant Journal, 89(6), 1079–1092.

Serre, N. B. C., Sarthou, M., Gigarel, O., Figuet, S., Corso, M., Choulet, J., Rofidal, V., Alban, C., Santoni, V., Bourguignon, J., Verbruggen, N., & Ravanel, S. (2020). Protein lysine methylation contributes to modulating the response of sensitive and tolerant Arabidopsis species to cadmium stress. Plant, Cell & Environment, 43(3), 760–774.

Singh, R. K., Muthamilarasan, M., & Prasad, M. (2022). SiHSFA2e regulated expression of SisHSP21.9 maintains chloroplast proteome integrity under high temperature stress. Cellular and Molecular Life Sciences: CMLS, 79(11), 580.

Slavoff, S. A., Mitchell, A. J., Schwaid, A. G., Cabili, M. N., Ma, J., Levin, J. Z., Karger, A. D., Budnik, B. A., Rinn, J. L., & Saghatelian, A. (2013). Peptidomic discovery of short open reading frame-encoded peptides in human cells. Nature Chemical Biology, 9(1), 59–64.

Solntsev, S. K., Shortreed, M. R., Frey, B. L., & Smith, L. M. (2018). Enhanced global post-translational modification discovery with MetaMorpheus. Journal of Proteome Research, 17(5), 1844–1851.

Stead, D. A., Paton, N. W., Missier, P., Embury, S. M., Hedeler, C., Jin, B., Brown, A. J., & Preece, A. (2008). Information quality in proteomics. Briefings in Bioinformatics, 9(2), 174–188.

Sun, B., Liu, Z., Liu, J., Zhao, S., Wang, L., & Wang, F. (2022). The utility of proteases in proteomics, from sequence profiling to structure and function analysis. Proteomics, 23, e2200132.

Tabb, D. L., Ma, Z. Q., Martin, D. B., Ham, A. J., & Chambers, M. C. (2008). DirecTag: Accurate sequence tags from peptide MS/MS through statistical scoring. Journal of Proteome Research, 7(9), 3838–3846.

Tamburino, R., Vitale, M., Ruggiero, A., Sassi, M., Sannino, L., Arena, S., Costa, A., Batelli, G., Zambrano, N., Scaloni, A., Grillo, S., & Scotti, N. (2017). Chloroplast proteome response to drought stress and recovery in tomato (*Solanum lycopersicum* L.). BMC Plant Biology, 17(1), 40.

Torres-Montilla, S., & Rodriguez-Concepcion, M. (2021). Making extra room for carotenoids in plant cells: New opportunities for biofortification. Progress in Lipid Research, 84, 101128.

Tyanova, S., Temu, T., & Cox, J. (2016). The MaxQuant computational platform for mass spectrometry-based shotgun proteomics. Nature Protocols, 11(12), 2301–2319.

Vasconcelos, M., Datta, K., Oliva, N., Khalekuzzaman, M., Torrizo, L., Krishnan, S., Oliveira, M., Goto, F., & Datta, S. K. (2003). Enhanced iron and zinc accumulation in transgenic rice with the *ferritin* gene. Plant Science, 164, 371–378.

Vélez-Bermúdez, I. C., Wen, T. N., Lan, P., & Schmidt, W. (2016). Isobaric tag for relative and absolute quantitation (iTRAQ)-based protein profiling in plants. Methods in Molecular Biology (Clifton, N.J.), 1450, 213–221.

Wang, B., Liu, B., Yan, Y., Tang, K., & Ding, C. F. (2019). Binary magnetic metal-organic frameworks composites: A promising affinity probe for highly selective and rapid enrichment of mono- and multi-phosphopeptides. Microchimica Acta, 186(12), 832.

Wang, F. X., Luo, Y. M., Ye, Z. Q., Cao, X., Liang, J. N., Wang, Q., Wu, Y., Wu, J. H., Wang, H. Y., Zhang, M., Cheng, H. Q., & Xia, G. X. (2018b). iTRAQ-based proteomics analysis of autophagy-mediated immune responses against the vascular fungal pathogen verticillium dahliae in Arabidopsis. Autophagy, 14(4), 598–618.

Wang, W. Q., Jensen, O. N., Møller, I. M., Hebelstrup, K. H., & Rogowska-Wrzesinska, A. (2018a). Evaluation of sample preparation methods for mass spectrometry-based proteomic analysis of barley leaves. Plant Methods, 14, 72.

Wang, X., Hu, H., Li, F., Yang, B., Komatsu, S., & Zhou, S. (2021). Quantitative proteomics reveals dual effects of calcium on radicle protrusion in soybean. Journal of Proteomics, 230, 103999.

Wilkins, M. R., Sanchez, J. C., Gooley, A. A., Appel, R. D., Humphery-Smith, I., Hochstrasser, D. F., & Williams, K. L. (1996). Progress with proteome projects: Why all proteins expressed by a genome should be identified and how to do it. Biotechnology & Genetic Engineering Reviews, 13, 19–50.

Wirth, J., Poletti, S., Aeschlimann, B., Yakandawala, N., Drosse, B., Osorio, S., Tohge, T., Fernie, A. R., Gunther, D., & Gruissem, W., et al. (2009). Rice endosperm iron biofortification by targeted and synergistic action of nicotianamine synthase and ferritin. Plant Biotechnology Journal. 7, 631–644.

Wiśniewski, J. R., Zougman, A., Nagaraj, N., & Mann, M. (2009). Universal sample preparation method for proteome analysis. Nature Methods, 6(5), 359–362.

Wu, F., Wesseler, J., Zilberman, D., Russell, R. M., Chen, C., & Dubock, A. D. (2021). Allow golden rice to save lives. Proceedings of the National Academy of Sciences of the United States of America, 118(51), e2120901118.

Wu, H., Li, L., Du, J., Yuan, Y., Cheng, X., & Ling, H. Q. (2005). Molecular and biochemical characterization of the Fe(III) chelate reductase gene family in *Arabidopsis thaliana*. Plant & Cell Physiology, 46(9), 1505–1514.

Xiao, L., Liang, H., Jiang, G., Ding, X., Liu, X., Sun, J., Jiang, Y., Song, L., & Duan, X. (2022). Proteome-wide identification of non-histone lysine methylation in tomato during fruit ripening. Journal of Advanced Research, *42*, 177–188.

Xie, Z., Feng, Q., Zhang, S., Yan, Y., Deng, C., & Ding, C. F. (2022). Advances in proteomics sample preparation and enrichment for phosphorylation and glycosylation analysis. Proteomics, 22, e2200070.

Yadeta, K. A., Elmore, J. M., Creer, A. Y., Feng, B., Franco, J. Y., Rufian, J. S., He, P., Phinney, B., & Coaker, G. (2017). A cysteine-rich protein kinase associates with a membrane immune complex and the cysteine residues are required for cell death. Plant Physiology, 173(1), 771–787.

Yang, J., Gao, Z., Ren, X., Sheng, J., Xu, P., Chang, C., & Fu, Y. (2021). DeepDigest: Prediction of protein proteolytic digestion with deep learning. Analytical Chemistry, 93(15), 6094–6103.

Ye, X., Al-Babili, S., Kloti, A., Zhang, J., Lucca, P., Beyer, P., & Potrykus, I. (2000). Engineering the provitamin a (β-carotene) biosynthetic pathway into (carotenoid-free) rice endosperm. Science, 287, 303–305.

Yin, X., & Komatsu, S. (2016). Plant nuclear proteomics for unraveling physiological function. New Biotechnology, 33(5), 644–654.

Yuan, B., Liu, T., Cheng, Y., Gao, S., Li, L., Cai, L., Yang, J., Chen, J., & Zhong, K. (2021). Comprehensive proteomic analysis of lysine acetylation in *Nicotiana benthamiana* after sensing CWMV infection. Frontiers in Microbiology, 12, 672559.

Zadraznik, T., Moen, A., Egge-Jacobsen, W., Meglič, V., & Šuštar-Vozlič, J. (2017). Towards a better understanding of protein changes in common bean under drought: A case study of N-glycoproteins. Plant Physiology and Biochemistry, 118, 400–412.

Zaghum, M. J., Ali, K., & Teng, S. (2022). Integrated genetic and omics approaches for the regulation of nutritional activities in rice (*Oryza sativa* L.). Agriculture, 12(11), 1757.

Zhan, H., Song, L., Kamran, A., Han, F., Li, B., Zhou, Z., Liu, T., Shen, L., Li, Y., Wang, F., & Yang, J. (2020). Comprehensive proteomic analysis of lysine ubiquitination in seedling leaves of *Nicotiana tabacum*. ACS Omega, 5(32), 20122–20133.

Zhang, F., Ge, W., Ruan, G., Cai, X., & Guo, T. (2020). Data-independent acquisition mass spectrometry-based proteomics and software tools: A glimpse in 2020. Proteomics, 20(17–18), e1900276.

Zhang, X., Fang, A., Riley, C. P., Wang, M., Regnier, F. E., & Buck, C. (2010). Multi-dimensional liquid chromatography in proteomics—A review. Analytica Chimica Acta, 664(2), 101–113.

Zhang, Y., Lai, X., Yang, S., Ren, H., Yuan, J., Jin, H., Shi, C., Lai, Z., & Xia, G. (2021). Functional analysis of tomato CHIP ubiquitin E3 ligase in heat tolerance. Scientific Reports, 11(1), 1713.

Zhu, X. F., Jiang, T., Wang, Z. W., Lei, G. J., Shi, Y. Z., Li, G. X., & Zheng, S. J. (2012). Gibberellic acid alleviates cadmium toxicity by reducing nitric oxide accumulation and expression of IRT1 in *Arabidopsis thaliana*. Journal of Hazardous Materials, 239–240, 302–307.

16 Proteomic Approaches to Understand Post-translational Modifications of Proteins in Plants

Asmita Pal and Rita Kundu

16.1 INTRODUCTION TO POST-TRANSLATIONAL MODIFICATION OF PROTEINS IN PLANTS

Proteins, being the ultimate functional units of a cell, undergo various modifications after their translation as crucial regulatory processes. These 'post-translational modifications' (PTMs) are implemented as a regulatory step to control the biological activities, localization, as well as interactions of proteins with other cellular proteins and/or other constituents. Therefore, identifying the PTMs of proteins has gained an important spot in both animal and plant proteomics research. Since plant proteomics are relatively less explored than animal proteomics, identifying these post-translational marks in plant science can reveal many unknown facts. Moreover, these protein PTMs are counted among the quickest and earliest plant responses to the environmental/extrinsic fluctuations or intrinsic triggers, thus making these PTMs occupy an important position in plant science research (Roychoudhury et al. 2011).

In chemical terminology, PTMs can be defined as covalent modifications of one or more amino acids either by proteolytic cleavage or by the addition of different groups like acetyl, phosphoryl, glycosyl, methyl, etc., which leads to vast deviations in the structural and functional properties of the respective protein. These deviations can be reversible or irreversible depending on the type of modification; e.g., covalent modifications usually tend to be reversible in nature, while proteolytic processes are irreversible. The significance of PTMs lies in the fact that these structural modifications affect protein functionality, thus regulating crucial cellular processes (Singh and Roychoudhury 2021). The life span of a protein, its solubility, folding dynamics, localization, and interactive ability—all are dependent on the structure of the protein regulated by PTMs. In turn, protein behaviours like enzyme assembly/function, cytoskeleton stability, cell–cell or cell–matrix interactions, molecular trafficking, and receptor activation are also critically affected (Ramazi and Zahiri 2021). In consequence, these modifications impact cellular processes like signalling pathways, gene expression regulation, membrane trafficking, DNA repair, cell cycle control, and others. The subcellular site of these protein modifications includes organelles, viz., nucleus, cytoplasm, endoplasmic reticulum (ER), and Golgi apparatus (Blom et al. 2004).

In terms of the tools and technologies devised to study these protein PTMs, affinity enrichment followed by mass spectrometry (MS)-based identification is commonly followed. Combination of immunoprecipitation along with MS is being considered a more effective strategy (Larsen et al. 2006). Nonetheless, such wide-scaled identification of PTMs is expensive as well as strenuous. Therefore, in the past few years, comparatively inexpensive and less vigorous computational technologies for predicting PTMs have attracted a lot of scientific attention. Hence, this chapter focuses on the different types of PTMs discovered and studied till date, their implication in plant biology, computational methods to predict PTMs, different databases available, and the

DOI: 10.1201/b23255-16

243

experimental tools devised to identify the PTMs along with the future scope of PTMs in plant science research.

16.2   VARIOUS POST-TRANSLATIONAL MODIFICATIONS IN PLANTS AND THEIR REGULATORY FUNCTION

More than 400 different types of PTMs have been noted till date, including both animal and plant systems. PTMs are a common cellular event in all eukaryotic cells but have been majorly studied in model organisms like human, mouse, *Arabidopsis*, rat, yeast, and *Escherichia coli*. According to dbPTM 2022 information, around 2,777,000 PTM substrate sites have been identified from existing databases by text mining and manual literature curation. Among them, around 2,235,000 entries have been confirmed through experiments, corresponding to 72 PTM types (Li et al. 2022). This was accomplished due to the large-scale rapid development of MS-based proteomics. The broadly categorized PTMs till date comprise phosphorylation, acetylation, glycosylation, methylation, hydroxylation, ubiquitination, SUMOylation, carbonylation, deamidation, nitrosylation, etc. (Banerjee and Roychoudhury 2018) as shown in Figure 16.1.

Among these, it has been evidently noted that some of the PTMs occur more frequently than the other types and have been studied in greater detail. According to the dbPTM database and as stated by Ramazi and Zahiri (2021), the three main PTMs include phosphorylation, acetylation, and ubiquitination, which encompass more than 90% of all the reported PTMs. Therefore, it can be concluded that every amino acid experiences a minimum of three variable PTMs. Lys undergoes the largest (15) number of PTMs with more than 100,000 sites of Lys modifications mapped in over 10,000 proteins (Wang and Cole 2020). Even Cys and Ser residues have been found to be modified by at least ten PTM types. Ultimately, phosphorylation of Ser residue has been reported to be the most common type. The different types of PTMs of all the amino acids reported in plants are listed in Table 16.1.

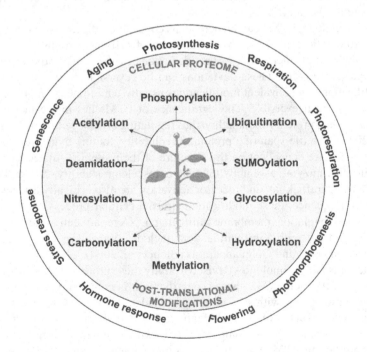

FIGURE 16.1   PTMs in plants. The different types of PTMs reported in plants till date have been shown in the inner circle, comprising the total cellular proteome of a plant cell. The outer cell depicts the various metabolic pathways and cellular processes regulated by these PTMs in general.

TABLE 16.1

Details on Different Types of PTMs of Proteins in Plants

PTM Type	Amino Acids Involved	Reversibility	Cellular Processes Effected
Phosphorylation	Ser (most common), Thr, Tyr, His, Arg, Lys, Asp, Glu, Cys	Reversible	Photosynthesis, stress response (biotic and abiotic), light and temperature sensing, senescence, seed germination, flowering
Glycosylation	Asn (most common), Ser, Thr	Reversible	Stress tolerance (through flavonoid biosynthesis), fruit ripening, leaf development, flowering, seed germination and development, root and root hair growth, development of trichomes
Acetylation	Lys (most common), Ser, Ala, Met, Gly, Thr, Asp	Reversible	Plant–pathogen interaction, abiotic stress tolerance, photosynthesis, light and temperature signalling, cellular patterning of the root epidermis, ABA signalling
Ubiquitination	Lys	Reversible	Almost all aspects of plant growth and development, flowering, hormone response, stress response, photomorphogenesis, circadian rhythm, senescence
Methylation	Lys (most common), Arg, His, Glu, Asp, Gln, Asn, Cys	Reversible	Temperature signalling, root development, auxin and ABA signalling, photosynthesis, and environmental stress response
Hydroxylation	Pro, Lys, Asn, Asp, His	Irreversible	Root hair expansion, hypoxia sensing
SUMOylation	Lys	Reversible	Environmental stress response, plant development, light regulation, flowering through hormone pathways
Carbonylation	Lys, Pro, Arg, Thr, Cys, His	Irreversible	Seed biology, fruit ripening, stress response
Deamidation	Asn-to-Asp and isoAsp Gln-to-Glu	Irreversible	Protein stability, ageing and denaturation Seed germination vigour, aquaporin activity during abiotic stress response
Nitrosylation	Cys, Tyr	Irreversible	Guard cell signalling

16.2.1 PHOSPHORYLATION

Phosphorylation is the most frequently reversible PTM in a plant cell, acting as a vital regulatory step associated with multiple biochemical and developmental processes as well as stress response mechanisms (Roychoudhury et al. 2008). Amino acids most commonly phosphorylated include the hydroxylated amino acids like serine (S), tyrosine (Y), and threonine (T).

Examples of growth and developmental processes include photosynthetic activities, which are tightly regulated by light- and redox-mediated protein phosphorylation mechanisms. The photosynthetic cascade proteins like D1, D2, and CP43 belonging to photosystem II (PSII), PsbH, and the light-harvesting polypeptides in the light-harvesting complex II (LHCII) were found to be phosphorylated, along with their phosphorylated sites mapped. Even a previously uncharacterized protein, TPS9 (12 kDa), was detected to be phosphorylated from the spinach thylakoid membranes (Carlberg et al. 2003). All the mitogen-activated protein kinases (MAPK) acting through phosphorylation-dephosphorylation mechanism have also shown immense role in plant growth and development (Xu and Zhang 2015). Another report by Chan et al. (2017) also showed that SnRK1-induced FUS3 phosphorylation is a necessary requirement for embryogenesis and integration of environmental signals. Even phosphorylation of brassinosteroid-insensitive 1 (BRI1), the receptor of brassinosteroids (BRs) at specific sites, was noted to have a differential effect on growth and development (Wang et al. 2016). Moreover, BR-dependent phosphorylation was reported to affect the PIF4 transcriptional activity with direct impact on diurnal hypocotyl growth (Bernardo-García et al. 2014). A very interesting fact is that plasma membrane H⁺-ATPase is also regulated by phosphorylation (Haruta et al. 2015). Several other reports have also been made regarding the diverse role of phosphorylation on growth and development in plants.

Similarly, protein phosphorylation has also been found to be employed as a regulatory module in the defence mechanism against any types of environmental stressors—abiotic and biotic both (Roychoudhury et al. 2013). For example, AtPhos43 was found to be phosphorylated in response to flagellin or chitin fragments' treatment (Peck et al. 2001). Even in the case of salt-stressed wheat, maize, and sugar beet, along with drought-stressed wheat and beans, S/T and Y phosphorylated proteins have been studied, which show direct impact on crop productivity (Feki et al. 2011; Hu et al. 2013; Yu et al. 2016; Luo et al. 2019). Many other environmental stress situations have also been reported and reviewed to include phosphorylation mechanism as a critical regulatory event (Damaris and Yang 2021). Despite such critical role of phosphorylation, no protein tyrosine kinase genes were observed in plant cells for a long time. Nemoto et al. (2015) reported the presence of calcium-dependent protein kinase (CPK)-related protein kinases (CRK2 and CRK3) as Tyr kinases in *Arabidopsis*, which phosphorylates the Tyr residues of β-tubulin and other transcription factors like ethylene response factor 13 (ERF13), WRKY DNA-binding protein 14 (WRKY14), ERF/AP2 transcription factor 2.6 (RAP2.6), and cryptochrome-interacting basic-helix-loop-helix 5 (CIB5). Another study by Miyamoto et al. (2019) also found a major role of *Arabidopsis* CRK2 and CRK3 in defence responses against the herbivore *Spodoptera litura*. However, protein tyrosine phosphatases were discovered long back in crop plants like tomatoes (Luan et al. 2001). Less abundant Y-phosphorylated proteins (4.2% in *Arabidopsis*, 2.9% in rice, and 1.3% in *Medicago*) have been observed to be vital in plant stress responses (Sugiyama et al. 2008; Grimsrud et al. 2010; Nakagami et al. 2010). A study showed that ectopic induction of plant and fungi atypical dual-specificity phosphatase (PFA-DSP1) can fortify tobacco plants against drought stress, referring to the role of phosphorylation-dephosphorylation regulatory mechanism in stress response in plants (Liu et al. 2012). Another report depicted that in cold-acclimatized chestnut trees, DSP4 gets up-regulated, again suggesting the involvement of phosphorylation-mediated cold stress defence (Berrocal-Lobo et al. 2011). Even CPK2 kinase overexpression leads to drought sensitivity in barley plants (Cieśla et al. 2016). Critical dissection of phosphorylation targets of CPK2-like kinases along with the associated pathways may further enlighten the significance of phosphorylation mechanisms.

16.2.2 GLYCOSYLATION

Glycosylation is one of the most abundant and necessary PTM forms after phosphorylation, mostly targeting surface proteins and secretory pathway proteins. Glycosylation can exist as N-glycosylation, O-glycosylation, or glycosylphosphatidylinositol (GPI) anchors. The N- and O-glycosylations can both take place at the Asp (D) residues. However, N- and O-glycosylations have been exclusively observed at S/T and Y residues, respectively, as well.

16.2.2.1 N-Glycosylation

Secretory pathway proteins are tagged with N-glycan molecules, i.e. they are N-glycosylated in ER, and using these N-glycans as the delivery tags, they are correctly delivered to their subcellular sites (Aebi et al. 2010). In fact, many studies have reported that stress tolerance in rice is dependent on these delivery systems (Harmoko et al. 2016). These glycan moiety additions lead to the formation of heterogeneous sugar chains with variable branching patterns. The pattern recognition receptors (PRRs), i.e. the plant receptor kinases have been reported to essentially require N-glycosylation to be able to induce plant immunity (Häweker et al. 2010). Single-step editing of sugar chains has been found to exert a significant role in the seed development of *Raphanus sativus* (Mega 2005). The procedure for N-glycosylation of proteins as mediated by a conserved oligosaccharyl transferase (OT) complex begins with the *en bloc* transfer of pre-assembled core oligosaccharide structures ($Glc_3Man_9GlcNAc_2$) to nascent polypeptide chains (to the NH_2-group of asparagine residues) emerging from the translocation channel of ER. Following this, glucosidases I and II remove the terminal glucose residues in the ER lumen, which are then folded via proper mechanism to form mature complex N-glycans. In plants, the multifaceted N-glycans are composed of the core containing α1, 3-Fuc, and β1, 2-Xyl residues, which are structurally discrete from their animal equivalents due to the glycosyl transferases present exclusively in plants. Plants produce comparatively less complex and less diverse population of oligosaccharides. Nevertheless, some characteristic complex N-glycan modifications are well preserved in higher plants and also in not-so-closely related bryophytes like *Physcomitrella patens* (Wilson et al. 2001; Viëtor et al. 2003). This hints at probable evolutionary restraints preventing the loss of N-glycan tags.

16.2.2.2 O-Glycosylation

O-glycosylation modification is primarily different in plant systems. In plants, a single Gal molecule is attached to Ser residues on certain proteins. Moreover, arabinose chains and complex arabinogalactans are observed on hydroxyproline amino acids of the proteins found in cell walls (Taylor et al. 2012; Tryfona et al. 2012). A bioinformatics study by Showalter et al. (2010) identified around 166 hydroxyproline-rich glycoproteins, including 85 putative arabinogalactan proteins. A report by Niemann et al. (2015) identified ROCK1 as an ER-localized transporter of UDP-GlcNAc and UDP-GalNAc in plants like *Arabidopsis thaliana*, which led to the idea of an anonymous protein glycosylation mechanism in plants as a possible regulator of protein quality. This transport activity is seemingly involved in cytokinin response regulation in *A. thaliana*, although experimental verifications and details regarding this are yet to be revealed. This mode of PTM in plants has not been studied in greater detail and hence represents a huge scope to be probed in future.

16.2.2.3 Glycosyl Phosphatidylinositol (GPI) Modification

The anchoring of GPI chains at the C-terminal end of protein chains is another commonly observed amendment taking place in ER. GPI-anchors help in the asymmetric distribution of membrane proteins to establish and maintain cell polarity. A study in the *A. thaliana* genome, as early as in 2003 by Borner et al. had combined proteomic analysis with bioinformatics to identify around 248 putative GPI-anchored proteins. This GPI-anchor: Man α (1–2) Man α (1–6) Man α (1–4) GlcN-inositol is synthesized stepwise and transferred *en bloc* by the GPI-transamidase complex (Kinoshita 2014). The variation in the glycan moiety attached depending on the species, cells, and proteins is still not clear. Long back, putative plant orthologs of varied GPI biosynthesis-related proteins were marked in *A. thaliana* and rice plants (Eisenhaber et al. 2003; Ellis et al. 2010) but are still pending functional characterization. A study by Lalanne et al. (2004) found male-specific defects in heterozygous mutants lacking *A. thaliana* GPI-GlcNAc transferases (SETH1 and SETH2). Similarly, plants lacking mammalian α-1,4-mannosyltransferase PIG-M (termed PEANUT1) depicted reduction in GPI-anchored proteins and also turned out to be seedling lethal (Gillmor et al. 2005). Even the mutated *A. thaliana* homologue of α-1,2-mannosyltransferase PIG-B (APTG1) depicted identical phenotypes with male fertility severely affected along with the embryos turning lethal (Dai et al. 2014). Even in a female gametophyte, LORELEI (LRE), a GPI-anchored protein was identified with ovule functions having supportive role in female fertility (Liu et al. 2016). The maize *roothairless3* gene encoding a putative GPI-anchored, monocot-specific COBRA-like protein significantly affected the crop yield (Hochholdinger et al. 2008). Another study depicted *Arabidopsis* COBRA-LIKE 10, a GPI-anchored protein, as being in charge of the directional growth of pollen tubes (Li et al. 2013). The study by Simpson et al. (2009) identified a family of small glycosylphosphatidylinositol (GPI)-linked proteins localized to plasmodesmata with callose binding activity *in vitro*. These studies establish the fact that accurate assembly of GPI-core oligosaccharide is a vital element to ensure somatic and reproductive development.

16.2.3 Ubiquitination

Ubiquitination is a cellular process designed for the purpose of protein degradation. In this process, covalent conjugation of ubiquitin (Ub) to the lysine residues acts as a mark above the target proteins for proteasomal degradation. As expected, this ubiquitination-mediated protein degradation surely plays a very essential role in almost all cellular processes. However, the significance of the Ub-proteasome pathway in plants has gained the limelight in past few years, majorly with respect to cellular homeostasis, growth and development, hormone response, and stress response in plants (Moon et al. 2004). The complexity of this ubiquitination process can be gauged from the fact that the *Arabidopsis* genome encodes more than 1400 (more than 5% of total proteome) pathway components (Smalle and Vierstra 2004).

In plants, the ubiquitin proteasomal system of protein degradation has been found to be associated with every phase of plant life cycle, including embryogenesis, photomorphogenesis, organogenesis, and also in response to stress. Studies noted increase in *Arabidopsis* UBQ14 transcript during heat stress (Sun and Callis 1997), multiple polyubiquitin genes in response to high temperatures in tobacco, potato, and maize (Christensen et al. 1992; Garbarino et al. 1992; Genschik et al. 1992). A mono-ubiquitin gene, when overexpressed, strengthened stress tolerance with absolutely no adverse effect on growth and development (Guo et al. 2008). A wheat polyubiquitin-overexpressed transgenic tobacco showed significantly better survivability in cold, high salinity, and drought conditions (Khan et al. 2019). In fact, deficiencies in 26S proteasome (i.e. its RP subunits) affect stress tolerance ability at various conditions. Mutated RP subunits lessen the rate of ubiquitin-dependent proteolysis, and also affected the abiotic stress tolerance (Smalle and Vierstra 2004; Ueda et al. 2004; Kurepa et al. 2008). For example, *Arabidopsis rpn10-1*, *rpn1a-4*, and *rpn1a-5* mutants failed to exhibit salt stress tolerance (Smalle et al. 2003; Wang et al. 2009), *rpn10-1* mutants became much more sensitive to UV rays and DNA damaging agents (Smalle et al. 2003); *rpn1a-4*, *rpn1a-5*, *rpn10-1*, *rpn12a-1*, and *rpt2a-2* mutants failed to tolerate heat shock (Kurepa et al. 2008; Wang et al. 2009). A recent study showed that tissue-specific ubiquitination by IPA1 (INTERACTING PROTEIN1) regulates plant architecture in rice (Wang et al. 2017). With regard to immunity response, studies have depicted a crosstalk mechanism between ubiquitination and other modifications, like phosphorylation (Zhang and Zeng 2020).

16.2.4 Acetylation

Acetylation is another very important PTM of cellular proteins, catalysed by different N-terminal and lysine acetyltransferases. During acetylation, an acetyl group is contributed by acetyl-coenzyme A to be co- or post-translationally joined to the side chains of amino acids. Acetylation is of basically three types: Nα-acetylation, Nε-acetylation, and O-acetylation, the first being the only irreversible one. These three forms of acetylation occur on many amino acid residues, with Lys acetylation being the most popular one. Among these, while Nε-acetylation is biologically more significant than others, Nα-acetylation is more abundant, at around 72% in *A. thaliana* (Bienvenut et al. 2012). The cellular processes regulated by acetylation-deacetylation interplay include chromatin stability, protein-protein interaction, cell cycle control, metabolic pathways, nuclear transport, and actin nucleation, all of which showed a massive impact on many aspects of plant growth and developmental physiology.

Acetylation of histone molecules is one of the most crucial events regulating the fate of a cell since it correlates with activated transcription of a particular genomic segment. Upon acetylation, lysine residue (in the histone molecule) drops its positive charge and thus loses its salt-bridge-forming ability with the negatively charged phosphate backbone of DNA, leading to disruption of the DNA-histone bond and the formation of an open, accessible chromatin structure. This open DNA region gets more easily accessed by transcription factors and/or chromatin remodellers, leading to increased gene transcription. The *Arabidopsis* genome comprises 12 histone acetyltransferase

(HAT) genes, among which GCN5 is mostly reported for its role in plant development and stress response (Vlachonasios et al. 2021). Even 16 different HDACs have been noted in *A. thaliana* (Shen et al. 2019). PTMs of histones have been found to regulate many important stages in plant's life cycle, including thermomorphogenesis. For example, HDA15 and HDA9 are reported to act in thermomorphogenesis. They can interact with transcription factors HFR1 or PWR, causing a change in the acetylation level of downstream genes (Shen et al. 2019; Van Der Woude et al. 2019). N-acetylation has been reported to impact the sorting and location of a few particular proteins, their interactive abilities, and their stability (Starheim et al. 2012).

The other form of acetylation, i.e. ε-amino lysine acetylation, extends beyond transcriptional regulation by histone modification. Along with lysine acetylases and lysine deacetylases, proteins recognizing and binding acetyl-lysines are also critically important. Majority of these PTMs can be retreated by sirtuins, like SIRT1, SIRT3, and SIRT5 (Zheng 2020). Studies show that GCN5 silencing dampens thermotolerance in *Arabidopsis* since heat stress-responsive genes get suppressed (Hu et al. 2015). Hu et al. observed GCN5 to hoard around the promoter sites of HSFA3 and UVH6 genes, which regulate the H3K9 and H3K14 acetylation amounts. In fact, GCN5-mediated thermotolerance was concluded to be a conserved feature in plants since wheat TaGCN5 could restore the thermotolerance defect in the *gcn5 Arabidopsis* mutants (Hu et al. 2015). Furthermore, HD2C in *Arabidopsis* has also been reported for its direct role in heat stress response. HD2C can directly interact with SWI3B (a SWI/SNF complex subunit) to regulate the heat-responsive genes and the H4K16Ac levels of HSFA3 and HSP101 (Buszewicz et al. 2016). Many vital proteins, like Calvin cycle enzymes (RuBisCO and others), chloroplast membrane proteins, tricarboxylic acid cycle proteins, and matrix-exposed proteins of the mitochondrial electron transport chain, were found to be acetylated (Finkemeier et al. 2011; König et al. 2014; Papanicolaou et al. 2014). As a part of the crosstalk mechanism, protein acetylation can suppress the phosphorylation output by acetylating GRF/14-3-3 proteins in plants, causing vital cellular consequences (Guo et al. 2022).

Besides histones, many other proteins, like α-tubulin, etc., are also imperative lysine acetylation sites, particularly on Lys40 by α-tubulin acetyltransferase 1 (ATAT1), NAA10, GCN5, or elongator protein 3 (ELP3). Deacetylation takes place by histone deacetylase 5/6 (HDAC5/6) or sirtuin 2 (SIRT2) (Gardiner 2019). This specifically occurs in the microtubule lumen, and acetylated kinesin motor proteins get preferentially targeted. The significance of acetylated tubulin lies in the fact that it has a role in response to environmental toxins, including bisphenol A (BPA) and hexavalent chromium (Adamakis et al. 2019). Depending on the degree of tubulin acetylation, legume cortical MTs also show depolymerization or stabilization upon treatment with chromium (Eleftheriou et al. 2016). In the roots of *Lens culinaris*, chromium also interrupts mitosis by acetylation-mediated stabilization of MTs (Eleftheriou et al. 2016). The abundance of acetylated tubulin in the cotyledons of young *Brassica rapa* and in the leaf margins suggests a possible role in growth and development (Nakagawa et al. 2013). Tubulin acetyltransferases such as ELP3 (part of the ELONGATOR complex) are found in *Arabidopsis*, and acetylated microtubules have been reported in angiosperms (Pereira et al. 2018; Gardiner 2019). Similar to animals and fungi, acetylated tubulin can promote autophagy under stress (Olenieva et al. 2019).

16.2.5 Hydroxylation

Although not much is known about the hydroxylation of amino acid residues in plants, a major PTM reported comprises proline hydroxylation by prolyl 4-hydroxylase (P4H). Hydroxyproline residues have been reported in proteins like TobHypSys, TomHypSys, PSY1, TDIF, CEP1, CLV3, CLE2, and RGF1 (Pearce et al. 2001; Pearce and Ryan 2003; Ito et al. 2006; Kondo et al. 2006; Amano et al. 2007; Ohyama et al. 2008; Ohyama et al. 2009; Matsuzaki et al. 2010). In TobHypSys, TomHypSys, PSY1, CLV3, and CLE2, these hydroxyproline residues get additionally altered with a pentose sugar, viz. L-arabinose. TDIF was initially known to inhibit trans-differentiation of dispersed *Zinnia* mesophyll cells into tracheary elements (Ito et al. 2006). Additional *in vivo* studies further observed that TDIF/CLE41/CLE44 peptides in the phloem and neighbouring cells can be identified by TDR/PXY in the plasma membrane of procambial cells. This promotes proliferation of procambial cells and inhibits differentiation into xylem cells (Hirakawa et al. 2008). Another hydroxyproline-containing peptide, CEP1, found in the lateral root primordia, can arrest root growth when overexpressed or externally applied. However, hydroxylation as a PTM has not been explored much till date.

16.2.6 METHYLATION

The methylome world in cells has its own diversity and complexity and is one of the most regular mechanisms to fine-tune growth and stress response in plants (Serre et al. 2018). In fact, methylation can crosstalk with other PTMs like phosphorylation and/or S-nitrosylation (Higashimoto et al. 2007). Histones are mostly reported for the effect of methylation since it can regulate gene expression, i.e. either activate or suppress them. For example, histone H3 lysine 9 (H3K9), H3K27, and H4K20 methylations result in transcriptional repression, but H3K4 and H3K36 methylations, on the other hand, cause activation (Yu et al. 2009). However, cellular proteins other than histone are also methylated, majorly at lysine and arginine residues, which are mostly identified in mammals till date. Protein lysine (K) methyltransferases (PKMTs) catalyse methyl group addition from S-adenosylmethionine (AdoMet) as the donor to the target lysine residue. Mono-, di-, or tri-methylation can create distinctive signatures acting as the docking sites for the effector proteins, resulting in activation or attenuation of the particular genes. Similarly, PRMT (protein arginine methyltransferases) catalyse methylation on arginine (R) residues (Alban et al. 2014). Alban et al. (2014) also identified Lys and Arg methylation sites in stromal and membrane chloroplast proteins, thus proving the crucial regulatory involvement of methylation in photosynthesis (Calvin cycle and light reaction). These PRMTs have been found to be evolutionarily conserved between monocot and dicot lineages, with functions reported in growth, flowering time, and pre-mRNA splicing (Pei et al. 2007; Deng et al. 2010). Ribosomal proteins appear to be methylated; hence, methylation has been deemed a necessary requirement for ribosomal RNA maturation and consequent function. This was proved by a *prmt3* mutant of *Arabidopsis* (Hang et al. 2014).

16.2.7 SUMOYLATION

SUMO (small ubiquitin-like modifier) proteins are a very recently identified group of proteins, named so because of their similarity to ubiquitin, viz., in size and attachment process to other proteins. Nevertheless, SUMOylation differs functionally and is a reversible form of PTM (unlike ubiquitination) noted for its supportive role in defence response (Raorane et al. 2013; Augustine and Vierstra 2018). In *Arabidopsis*, around 400 SUMOylated proteins have been identified till date, mostly transcription factors involved in growth, development, and environmental cue signalling (Mazur et al. 2017). Similar to ubiquitination, in SUMOylation also, E3 ligase proteins are the most important proteins for tagging the target TFs, such as the SUMO E3 ligase SIZ1, which SUMOylates and inactivates a bZIP-type TF ABI5, to negatively regulate ABA signalling in *Arabidopsis* (Miura and Hasegawa 2009). However, the same SIZ1 when attached to the TF MYB30 helps stabilize the TF and leads to successful germination (Zheng et al. 2012). The same SIZ1 homologue in apple (MdSIZ1) SUMOylates MdPHR1, a Pi-starvation-responsive MYB TF (Zhang et al. 2019). SUMO proteins interact with SUMO-targeted Ub ligases (STUbLs) to reduce the amount of cycling Dof factor 2 (CDF2) and thus amplify CONSTANS to promote flowering (Elrouby et al. 2013). Under water-limiting stress, auxin response factor 7 (ARF7) upon SUMOylation loses its DNA-attaching ability. This activates the repressor Aux/IAA3 to cause changes in hydro-patterning and lateral root development (Orosa-Puente et al. 2018). OsSIZ2, another SUMO E3 ligase in rice, is involved in N-assimilation routes, possibly by targeting nitrate reductases and also anther dehiscence and seed size (Pei et al. 2019).

In *Arabidopsis*, heat stress resulted in SUMOylation of AtBAG7 [B-cell lymphoma2 (Bcl2)-associated athanogene], thus allowing it to move from ER to the nucleus. AtBAG7 was also found to interact with WRKY29 to stimulate the expression of heat stress genes (Li et al. 2017). In biotic stress response mechanism also, SUMOylation has been reported to activate proteins during infection and limit them otherwise. Upon infection in *Arabidopsis*, the same SIZ1 SUMOylates TPR1, preventing it from interacting with HDAC19, thereby allowing DND1 and DND2, the hypersensitive immune regulators, to express (Niu et al. 2019). SUMOylation also facilitates the dissociation of NPR1 from the WRKY70 repressor to enhance immunity (Saleh et al. 2015). Like ubiquitination, SUMOylation also shows crosstalk ability with other PTMs. For example, SUMOylation stabilizes NPR1, while phosphorylation degrades it (Saleh et al. 2015). Similarly, SUMOylation and phosphorylation compete for the same amino acid residue in CESTA, a homologue of brassinosteroid-enhanced expression (BEE) (Khan et al. 2014). Another interesting fact is that SUMOylation can aim proteasomes in a feedback loop, which helps to maintain the protein turnover in a cell. AtMMS21, a SUMO E3 ligase, has been found to be a vital requirement for the stability of RPT2a, a 26S proteasome subunit (Yu et al. 2019).

16.2.8 OTHER PTM TYPES

Many other PTM types also exist which have not been discussed in detail, like carbonylation, nitrosylation, and deamidation, among others. Briefly, carbonylation can be defined as an ROS-induced PTM involving metal-catalysed oxidation of the side chains of certain amino acid residues. Carbonylation generally has been linked to marking the proteins for proteolytic degradation, but when proteins become severely carbonylated, they tend to aggregate owing to increased hydrophobicity. Many metabolic enzymes and cellular pathways have been linked to carbonylation, such as fruit ripening, stress response, and seed germination (Friso and van Wijk 2015). Deamidation of amino acid residues is also another very slow irreversible PTM form where Asn is converted to Asp and isoAsp, and Gln is converted to Glu. Deamidation can lead to loss of stability and degradation of proteins, with consequences for longevity and stress repair in plants. For nitrosylation, NO reacting with superoxide and GSH generates ONOO and S-nitrosylated GSH, respectively (also termed reactive nitrogen species, RNS), which react with Cys thiols to form S-nitrosothiols or with Tyr residues to form 3-nitrotyrosine. This form of PTM also shows significant implications in plant metabolism. Although other PTM types could not be discussed in detail, future studies shall hopefully enlighten greater understanding.

16.3 PROTEOMIC TECHNIQUES AND TOOLS

Proteomic studies have made a geometric progression, since the success of numerous genome sequencing projects. It has taken up an important spot in functional genomics. Initially, proteomics was defined as the 'large-scale identification of the wide array of protein content or proteome of a cell or tissue', but at present, proteomics refers to the large-scale study of any protein-related information, including the PTMs of proteins that regulate the functional diversity of proteins (Roychoudhury et al. 2011). Thus, PTM-based proteomics comprises the identification of the modified proteins, precise mapping of the modified amino acid residues, and their quantification and functional characterization. Proteomic techniques can be categorized under top–down and bottom–up proteomics. Although bottom–up proteomics is comparatively more widely followed, top–down or middle–down proteomics can result in a better understanding of the different combinations of PTMs on the same protein. Bottom–up proteomics refers to the proteolytic or chemical digestion of proteins into small fragments prior to MS analysis, whereas in top–down proteomics, undigested proteins are directly injected and analysed by a mass spectrometer. Less complex protein mixtures can be analysed by top–down proteomics with the help of a high-resolution (HR) mass spectrometer and are usually not regarded as high

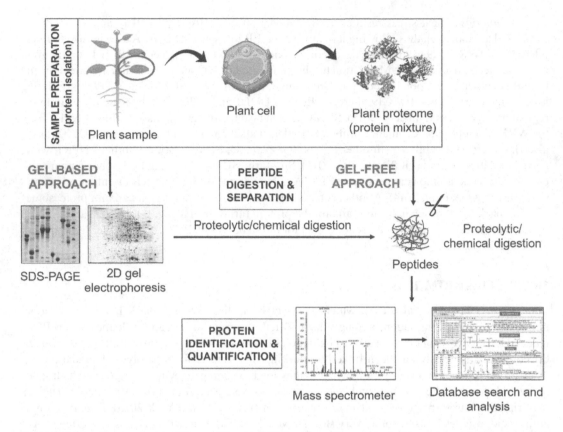

FIGURE 16.2 Proteomics workflow to identify PTMs. A step-by-step depiction of the workflow is usually followed for proteomic analysis. The first step undergoes sample preparation, which comprises the isolation of the total protein content of the plant cells or from different organelles as and when required. In the next step, the protein content of the cells is separated into peptides either by a gel-based approach or by a gel-free method. In the gel-free technique, the total protein content of the cells is digested (chemically/enzymatically) followed by identification using a mass spectrometer. In the gel-based technique, the protein content is first separated by electrophoresis (1D or 2D gels) and then subjected to digestion into peptides and mass spectrometry-based identification. Lastly, the post-translational modifications are identified through different databases available.

throughput. Figure 16.2 depicts proteomics workflow followed for the detection of different PTMs in plants.

PTM identification is a highly sensitive task and requires large amounts of proteins. Till date, around 400 variations of PTMs have been marked, and newer ones regularly come up through scientific studies (Ramazi and Zahiri 2021). However, the complexity of proteomic techniques has been now-a-days reduced by the help of affinity enrichment to increase the protein concentration and chromatographic fractionation for clear separation of individual proteins. The use of two-dimensional gel electrophoresis (2-DE) is an extensively applied proteomic technology. The fractionated protein contents are resolved in the gel for their different modification states and subjected to further evaluation. In the case of phosphorylated proteins, scientists have undertaken autoradiographic technique (32^P or 33^P) and immunoblot analysis using anti-phosphoamino acid antibodies (Luo and Wirth 1993; Astoul et al. 2003). Similarly, for glycoproteins, methylated protein forms and radioactive and non-radioactive staining methods have been devised. Following this, MS analysis helps to chemically characterize the PTMs. Examples of MS analysis include matrix-assisted laser desorption/ionization time-of-flight (MALDI-TOF) MS, electrospray ionization (ESI) tandem MS (MS/MS), liquid chromatography (LC) MS/MS, etc. However, mapping of PTMs is quite challenging, majorly owing to ion suppression, purity and instability of the modified peptides, sequence coverage, and quantification (Kwon et al. 2006). In today's date, LC-MS/MS-based shotgun approaches along with label-free approaches, are taking over the age-old gel-based technologies. Moreover, label-free approaches are currently preferred for quantification. Lately devised HR and fast-scanning instruments like Orbitraps Mass Spectrometer have enabled such fast techniques.

16.3.1 Phosphoproteomics

Phosphoproteomics analyses or identification of phosphorylation proteins and their associated information are being increasingly attempted in plant system, both *in vitro* and *in vivo*. Phosphorylation sites are being mapped, and web resources are being developed for plant-specific phosphoproteomic data storage and analysis. Consequently, computational prediction methods are being developed to increase sensitivity and specificity for methods trained on less plant-specific data. However, these studies are challenged by the low stoichiometry and low ionization efficiency of phosphopeptides. Prior to MS analysis, the modified proteins and/or peptides need to be enriched. The regularly used enrichment practices are summarized in Table 16.2.

Immobilized metal affinity chromatography (IMAC) jointly with strong cation exchange (SCX) is the most extensively used enrichment method, wherein affinity binding between negatively charged phosphate moiety and positively charged metal ions is employed (Fíla and Honys 2012). IMAC-SCX resulted in the discovery of three times more phosphopeptides as compared to SCX or IMAC alone (Trinidad et al. 2006). The initial SCX-IMAC application by Nühse et al. (2004) in plants identified 283 phosphopeptides successfully. Polymer-based metal-ion affinity capture (PolyMAC) is another highly selective variant of IMAC. PolyMAC-titanium (Ti) and PolyMAC-zirconium (Zr) ion affinity chromatographies were successful in the identification of 5386 unique phosphopeptides (Wang et al. 2013). Metal dioxides like titanium dioxide (TiO_2) and zirconium dioxide (ZrO_2) are the most popular choices; TiO_2 enriches multiple-phosphorylated peptides, while ZrO_2 enriches singly-phosphorylated peptides. The metal dioxide-based enrichment technique can be combined with other peptide fractionation methods as well. For instance, Yang et al. (2013)

TABLE 16.2

Phosphoprotein Enrichment Techniques for Plant Systems

S. No.	Enrichment Technique	Chemical Nature
1	Immunoprecipitation	Employs antibodies raised against phosphorylated amino acids for both phosphoproteins/phosphopeptides
2	Immobilized metal affinity chromatography (IMAC)	Employs a resin matrix (made up of iminodiacetic acid and nitrilotriacetic acid) with linked positively charged metal ions (to attract negatively charged phosphate group)
3	Phos-Tag (considered as an IMAC variation)	Employs Phos-Tag matrix secured in agarose layer. The matrix contains two Zn atoms able to accept two electrons each from the phosphate moiety
4	Metal oxide affinity chromatography (MOAC)	Employs a matrix of metal oxides or hydroxides, like titanium dioxide for phosphopeptides and aluminium hydroxide for phosphoprotein enrichment
5	Sequential elution from IMAC (SIMAC)	This method is particularly used for enrichment of phosphopeptides, which utilizes MOAC and IMAC together. It helps to separate both multiple and single phosphorylated peptides
6	Hydroxyapatite chromatography	This uses hydroxyapatite as a matrix
7	Chemical modification	Types include carbodiimide condensation along with bead fishing, β-elimination along with Michael addition, oxidation-reduction condensation, α-diazo substituted resin, and carbodiimide condensation using dendrimer
8	Phosphopeptide precipitation	Using calcium phosphate precipitation from a solution

combined TiO_2 enrichment and hydrophilic interaction liquid chromatography (HILIC) together to reveal 2305 phosphopeptides belonging to 964 proteins in wheat. A HILIC variant called electrostatic repulsion hydrophilic interaction chromatography (ERLIC) has also been successful (Gan et al. 2008; Loroch et al. 2015).

Comparative phosphoproteomic approaches can be either gel-based or gel-free, stable isotope-labelled or label-free. Firstly, for gel-based techniques, two-dimensional gel electrophoresis (2D GE) is a common method to resolve thousands of proteins on the basis of the changes in isoelectric point and molecular weight. Then, for detection, Pro-Q Diamond fluorescent stain that can selectively stain phosphoproteins is widely applied. As an example, Shin et al. (2007) applied 2-DE and Pro-Q to identify the differentially phosphorylated proteins in wild-type and *snk2.8* mutant *Arabidopsis* plants, wherein the putative substrates of SnRK2.8 were revealed (Shin et al. 2007). Next, gel-free methods encompass stable isotope labelling, for example, stable isotope labelling of amino acids in cell culture (SILAC). The first SILAC in plants used ^{15}N in *Arabidopsis* suspension cells (Benschop et al. 2007). Another such method employs multiplex isobaric tags to isolated proteins or digested peptides *in vitro*, viz. isobaric tags for relative and absolute quantification (iTRAQ) and tandem mass tags (TMT). Due to its ability to multiplex up to ten samples at a time and the enrichment effect for low-abundance proteins, iTRAQ/TMT labelling is an extremely popular technique in present day (Jones et al. 2006). Unlike labelled methods, label-free approaches help analyse unlimited samples together. One such method is based on precursor ion peak intensity/area, and the other is based on the number of MS/MS spectra attained for a peptide (also called 'spectral counting'). Both of these methods have been applied in plant phosphoproteomics. For instance, Reiland et al. (2011) characterized a thylakoid-associated kinase STN8 in cyclic electron flow. For targeted phosphopeptides, multiple reaction monitoring (MRM) has been used (Glinski and Weckwerth 2006).

Applications of phosphoproteomics in plant biology research are manifold, like in the dissection of signal transduction through the identification of many key proteins like kinases, transcription factors, and ubiquitin ligases. For example, four isoforms of Pto-interacting-like kinase 1 (PTI1) were identified from maize root plasma membrane, which depicted upregulation in response to variable iron conditions (Hopff et al. 2013). Hoehenwarter et al. (2013), while working on complex *A. thaliana* protein samples, combined $Al(OH)_3$-based metal oxide affinity chromatography with TiO_2-based metal oxide affinity chromatography for the purpose of phosphopeptide enrichment. The effort turned out to be successful; many novel phosphorylation sites were unravelled along with 141 MAPK substrate candidates, including time for coffee (TIC) and non-phototropic hypocotyl 3 (NPH3) proteins. Phosphoproteomic studies by Umezawa et al. (2013) identified phosphorylation networks in ABA signalling in ABA and dehydration stress-treated *Arabidopsis*. Molecules like SnRK2s, MAPKs, and CPKs were dissected out. A meta-analysis from 27 phosphoproteomic-based publications in *Arabidopsis* comprised of 60,366 phosphopeptides matched to 8141 non-redundant proteins.

Some studies have highlighted the significance of subcellular compartments for PTMs, and herein, a few examples have been cited. Jones et al. (2009) identified 416 phosphopeptides from 345 proteins from nuclei-enriched fractions of suspension cell cultures and seedlings of *A. thaliana*. In fact, subcellular proteomics have been linked to plant stress defence mechanisms. An increase in similar phosphoproteins (viz. diphosphorylated peptide of PsbH) in *Arabidopsis* and rice in response to salt stress revealed an important connection between the species (Chang et al. 2012). In *Arabidopsis*, the thylakoid protein kinase 7 (STN7) and STN8 kinases were also identified to participate in light-regulated phosphorylation of thylakoid membrane and stroma proteins (Ingelsson and Vener 2012). Another study by Han et al. (2014) used PolyMAC phosphopeptide enrichment and gel-free proteomics to identify 933 phosphorylated peptides belonging to 413 proteins in rice embryos during early germination stages, out of which 149 phosphorylated proteins were significantly altered during germination. A TiO_2 enrichment along with an LC-MS/MS study of *Brachypodium distachyon* identified 1470 phosphorylation sites in 950 proteins, containing 58 transcription factors, 84 protein kinases, 8 protein phosphatases, and 6 cellulose synthases (Lv et al. 2014). This helped to establish a phosphatase-centred network related to rapid vegetative growth. The currently in-use tandem MS upholds large-scale discoveries of new phosphorylation sites and phosphoproteins. Such studies analysing phosphorylation sites in many eukaryotic proteins helped to unravel molecular details, like that phosphorylation probability is higher towards the end of protein sequences (greater in C-terminus); and a large proportion (51.3%) of the occupied sites show phosphorylation within a close distance of ten amino acid residues. Thus, with time, the world of phosphoproteins and their regulatory roles will slowly unravel.

16.3.2 GLYCOSYLATION-BASED PROTEOMIC TECHNIQUES

Several strategies have come up for the structural characterization of the glycan bone of a glycoprotein. While an immunoblot strategy (post-1D or 2D electrophoresis gels) uses probes (lectins or antibodies) specific for glycoprotein oligosaccharide moieties, the other strategy is to chemically characterize the monosaccharide composition to identify the linked glycan. Among the probes, very few are available for the detection of plant glycans, available commercially in a biotinylated form. After the isolation of the glycoprotein, the glycan can be cleaved by chemical or enzymatic treatment, followed by electrophoresis or MS. On one hand, the cleavage type will notify about the nature of the linked glycans; on the other hand, the protein mass difference in MS will enlighten the glycan structure. Upon identification, the glycoproteins need to be checked if they are possibly secreted, i.e., whether they bear any signal peptide, if any potential N-glycosylation site? The deglycosylation step required for glycoprotein purification, i.e. to remove high-mannose and complex N-glycans, is catalysed by PNGase F. However, its effect is restricted on complex N-glycans in plants, owing to α-1-3 fucose linked to the N-glycan core. Certain softwares like Predotar, Target P, or iPSORT help in *in silico* identification of the likely subcellular site of the glycoproteins. Another server called PROSITE detects N-glycosylation sites within a protein.

A different approach nurses cells with a synthetic monosaccharide analogue tagged with a chemical reporter capable of reacting with a probe (e.g., a fluorophore). However, this technology is less applied in the plant kingdom owing to the presence of the impermeable cell wall and also due to the very less availability of diverse sugar analogues to replace the endogenous sugars. One of the latest approaches devised by Wu et al. (2020) follows the isolation of single glycan molecules by mass-selective, soft-landing electrospray ion beam deposition and imaging by a low-temperature scanning tunnelling microscope. Although glycoproteomic research has been progressing steadily, the precise N-glycan composition is still not discovered. Starting from 2004 till date, only a few studies could successfully purify glycoproteins to be able to reveal their glycan structural variations (Tan et al. 2004; Tryfona et al. 2010, 2012, 2014; Pfeifer et al. 2020). A few years down the lane, a broader understanding of the connection between N- and O-glycosylation pathways along with their application in plant cell biology and systems biology will be established.

16.3.3 PROTEOMICS FOR OTHER PTMs

For acetylated and methylated peptides too, affinity enrichment techniques followed by MS identification are ensued. Affinity enrichment is also followed for ubiquitinated protein identification. Moreover, an *in vivo* ubiquitination assay is also carried out to detect E3-mediated protein ubiquitination. This assay detects interaction between the substrate and the E3 ligase and the effect of E3 ligase on substrate ubiquitination and degradation by MG132 (proteasome inhibitor). MS-based identification can detect hydroxylated proteins since hydroxylation-mediated increase in molecular weight due to the addition of 16 Da oxygen atoms is detectable.

Enrichment of SUMOylated proteins is a comparatively new research domain and has been seeing rapid progress. Either monoclonal SUMO antibodies are being used for immunoprecipitation or cells expressing epitope-tagged SUMO isoforms are being studied for their proteomic diversity by using epitope-specific peptides for elution or affinity purification with peptides containing SUMO interaction motifs, specifically to capture polySUMOylated proteins (Budhiraja et al. 2009; Elrouby and Coupland 2010; Miller et al. 2010). Yeast two-hybrid assays are followed to identify SUMO-associating proteins like GTE3 from bromodomain and extra terminal (BET) domain protein family (Garcia-Dominguez et al. 2008). Budhiraja et al. (2009) used affinity purification and MS-based identification using tagged AtSUM1/3/5 in plants and revealed 15 probable SUMO targets functioning in RNA-dependent and chromatin-related processes. Another study undertook systemic yeast two-hybrid screening using the SUMO E2 activating enzyme SCE and the SUMO-specific protease ESD4, wherein 238 potential targets and 124 SUMOylated proteins were detected. Miller et al. (2010) also performed affinity purification, enrichment with Ni-NTA affinity chromatography, and anti-SUMO antibody affinity chromatography, followed by LC-MS/MS with protein extracts from His-H89R SUMO1 sum1-1sum2-1-expressing transient plants to ultimately result in the identification of 357 potential SUMO targets.

16.3.4 COMPUTATIONAL TOOLS IN PROTEOMICS

Computational approach in the field of proteomics has been applied from two discrete angles: one is for bioinformatics prediction, and the second is for the analysis of MS-based proteomics datasets. Both of these techniques are quite demanding. Hence, with the days, this approach has seen tremendous growth owing to a surplus increase in tools, resources, and databases. Well-known online resources include OMICtools, the popular server ms-utils.org, software repositories such as Comprehensive R Archive Network (CRAN), Bioconductor, GitHub, and also Pacific Northwest National Laboratory (PNNL)'s Integrative Omics open source software collections. Post-data acquisition typically uses commercially available tools, including Proteome Discoverer, Galaxy-P, Scaffold, and freely available softwares such as MaxQuant and Skyline. Moreover, many other computer softwares exist for multiple functions, as listed in Table 16.3.

Lately devised QC (quality check) tools for MS-based proteomics include QuaMeter, OpenMS, proteoQC, PTXQC, SProCoP, SimpatiQCo, and iMonDB. The statistical analysis tools include ProteoSign, HiQuant, DAPAR, ProStaR, SWATH2stats, Percolator, BioInfra.

TABLE 16.3

Computational Tools in Proteomics

S. No.	Computational Tools	Function
1	Panorama AutoQC	To assess the performed LC-MS/MS
2	DeepRT	Deep learning-based software to predict peptide retention time ahead of runs
3	Ursgal	As a Python interface, integrates multiple database search engines and performs statistical post-processing algorithms
4	Perseus	For protein quantification, interaction, and PTM data
5	MaxQuant	Helps in quantitative label-free proteomics and isobaric labelling methods
6	OpenMS	For flexible and reproducible analysis of high-throughput MS data
7	PatternLab	Integrates published computational environments for the analysis of shotgun proteomic data
8	TRIC	For automatic generation of quantitative proteomics data matrices
9	ACTG (Amino aCids To Genome)	Maps peptides to a genome
10	PTXQC (Proteomics Quality Control)	R-based QC tool for bottom-up LC-MS data generated by the MaxQuant.
11	RIPPER	For MS-based label-free relative quantification for proteomics data
12	PGCA (Protein Group Code Algorithm)	To connect groups from multiple experimental runs by forming global protein groups from related local groups

Prot, fCI, and proBAMsuite. Databases usually followed for peptide identification from MS/MS data include Ensemble, RefSeq, and UniProt (either non-redundant or organism-specific). A data search would mean a search against the 'organism-specific' Swiss-Prot and Swiss-Prot+TrEMBL databases along with a database of common contaminants. The most popular sequence databases include the National Center for Biotechnology Information (NCBI), the European Molecular Biology Laboratory (EMBL), and the DNA Data Bank of Japan (DDBJ)—all integrated together via the International Nucleotide Sequence Database Collaboration (INSDC), the UniProt Knowledge database (UniProtKB), and the Joint Genome Institute's genome portal (JGI). Frequently used shotgun proteomics database search engines include SEQUEST, MASCOT, X! Tandem, Andromeda, MS-GF+, MS Amanda, OMSSA, COMET, and MyriMatch.

For plant science, particularly few platforms exist, like Plant PTM viewer (Willems et al. 2019), wherein approximately 370,000 PTM sites have been mapped for 19 types of protein modifications in plant proteins from five different species.

16.4 CONCLUSION

Undeniably, there still remains a lot to be unravelled in the field of plant proteomics. Although research did progress, challenges still persist at almost all levels, starting from sample preparation to protein quantification, to MS to data analysis, and most importantly, biological interpretation. More specifically, contamination-based errors during sample preparation, generating bigger fragments for analysis, inability to capture low abundance proteins, difficulty in detection of membrane proteins, false discovery rate, among other problems, are particularly worrying. Nevertheless, the proteomic research is a constantly evolving research domain with multiple databases, analytical and bioinformatics softwares being developed daily along with the incessant effort to improvise the technologies and turn them more specific for plant science. Irrespective of the challenges, scientists across the world envisage the development of a broad proteomic atlas across all plant tissues and cell types.

REFERENCES

Adamakis, I. D. S., Panteris, E., & Eleftheriou, E. P. (2019). Tubulin acetylation mediates bisphenol A effects on the microtubule arrays of *Allium cepa* and *Triticum turgidum*. *Biomolecules*, 9(5), 185.

Aebi, M., Bernasconi, R., Clerc, S., & Molinari, M. (2010). N-glycan structures: Recognition and processing in the ER. *Trends in Biochemical Sciences*, 35(2), 74–82.

Alban, C., Tardif, M., Mininno, M., Brugière, S., Gilgen, A., Ma, S., … & Ravanel, S. (2014). Uncovering the protein lysine and arginine methylation network in *Arabidopsis* chloroplasts. *PLoS One*, 9(4), e95512.

Amano, Y., Tsubouchi, H., Shinohara, H., Ogawa, M., & Matsubayashi, Y. (2007). Tyrosine-sulfated glycopeptide involved in cellular proliferation and expansion in *Arabidopsis*. *Proceedings of the National Academy of Sciences*, 104(46), 18333–18338.

Astoul, E., Laurence, A. D., Totty, N., Beer, S., Alexander, D. R., & Cantrell, D. A. (2003). Approaches to define antigen receptor-induced serine kinase signal transduction pathways. *Journal of Biological Chemistry*, 278(11), 9267–9275.

Augustine, R. C., & Vierstra, R. D. (2018). SUMOylation: Re-wiring the plant nucleus during stress and development. *Current Opinion in Plant Biology*, 45, 143–154.

Banerjee, A., & Roychoudhury, A. (2018). The gymnastics of epigenomics in rice. *Plant Cell Reports*, 37, 25–49.

Benschop, J. J., Mohammed, S., O'Flaherty, M., Heck, A. J., Slijper, M., & Menke, F. L. (2007). Quantitative phosphoproteomics of early elicitor signaling in *Arabidopsis*. *Molecular & Cellular Proteomics*, 6(7), 1198–1214.

Bernardo-García, S., de Lucas, M., Martínez, C., Espinosa-Ruiz, A., Daviere, J. M., & Prat, S. (2014). BR-dependent phosphorylation modulates PIF4 transcriptional activity and shapes diurnal hypocotyl growth. *Genes & Development*, 28(15), 1681–1694.

Berrocal-Lobo, M., Ibanez, C., Acebo, P., Ramos, A., Perez-Solis, E., Collada, C., … & Allona, I. (2011). Identification of a homolog of *Arabidopsis* DSP4 (SEX4) in chestnut: Its induction and accumulation in stem amyloplasts during winter or in response to the cold. *Plant, Cell & Environment*, *34*(10), 1693–1704.

Bienvenut, W. V., Sumpton, D., Martinez, A., Lilla, S., Espagne, C., Meinnel, T., & Giglione, C. (2012). Comparative large scale characterization of plant versus mammal proteins reveals similar and idiosyncratic N-α-acetylation features. *Molecular & Cellular Proteomics*, *11*(6). https://doi.org/10.1074/mcp. M111.015131

Blom, N., Sicheritz-Pontén, T., Gupta, R., Gammeltoft, S., & Brunak, S. (2004). Prediction of post-translational glycosylation and phosphorylation of proteins from the amino acid sequence. *Proteomics*, *4*(6), 1633–1649.

Budhiraja, R., Hermkes, R., Muller, S., Schmidt, J., Colby, T., Panigrahi, K., … & Bachmair, A. (2009). Substrates related to chromatin and to RNA-dependent processes are modified by *Arabidopsis* SUMO isoforms that differ in a conserved residue with influence on desumoylation. *Plant Physiology*, *149*(3), 1529–1540.

Buszewicz, D., Archacki, R., Palusiński, A., Kotliński, M., Fogtman, A., Iwanicka-Nowicka, R., … & Koblowska, M. K. (2016). HD2C histone deacetylase and a SWI/SNF chromatin remodelling complex interact and both are involved in mediating the heat stress response in *Arabidopsis*. *Plant, Cell & Environment*, *39*(10), 2108–2122.

Carlberg, I., Hansson, M., Kieselbach, T., Schröder, W. P., Andersson, B., & Vener, A. V. (2003). A novel plant protein undergoing light-induced phosphorylation and release from the photosynthetic thylakoid membranes. *Proceedings of the National Academy of Sciences*, *100*(2), 757–762.

Chan, A., Carianopol, C., Tsai, A. Y. L., Varatharajah, K., Chiu, R. S., & Gazzarrini, S. (2017). SnRK1 phosphorylation of FUSCA3 positively regulates embryogenesis, seed yield, and plant growth at high temperature in *Arabidopsis*. *Journal of Experimental Botany*, *68*(15), 4219–4231.

Chang, F., Hsu, J. L., Hsu, P. H., Sheng, W. A., Lai, S. J., Lee, C., … & Chen, C. C. (2012). Comparative phosphoproteomic analysis of microsomal fractions of *Arabidopsis thaliana* and *Oryza sativa* subjected to high salinity. *Plant Science*, *185*, 131–142.

Christensen, A. H., Sharrock, R. A., & Quail, P. H. (1992). Maize polyubiquitin genes: Structure, thermal perturbation of expression and transcript splicing, and promoter activity following transfer to protoplasts by electroporation. *Plant Molecular Biology*, *18*(4), 675–689.

Cieśla, A., Mituła, F., Misztal, L., Fedorowicz-Strońska, O., Janicka, S., Tajdel-Zielińska, M., … & Sadowski, J. (2016). A role for barley calcium-dependent protein kinase CPK2a in the response to drought. *Frontiers in Plant Science*, *7*, 1550.

Dai, X. R., Gao, X. Q., Chen, G. H., Tang, L. L., Wang, H., & Zhang, X. S. (2014). Abnormal pollen tube guidance1, an endoplasmic reticulum-localized mannosyltransferase homolog of glycosylphosphatidylinositol10 in yeast and phosphatidylinositol glycan anchor biosynthesis b in human, is required for *Arabidopsis* pollen tube micropylar guidance and embryo development. *Plant Physiology*, *165*(4), 1544–1556.

Damaris, R. N., & Yang, P. (2021). Protein phosphorylation response to abiotic stress in plants. *Plant Phosphoproteomics*, *2358*, 17–43.

Deng, X., Gu, L., Liu, C., Lu, T., Lu, F., Lu, Z., … & Cao, X. (2010). Arginine methylation mediated by the *Arabidopsis* homolog of PRMT5 is essential for proper pre-mRNA splicing. *Proceedings of the National Academy of Sciences*, *107*(44), 19114–19119.

Eisenhaber, B., Wildpaner, M., Schultz, C. J., Borner, G. H., Dupree, P., & Eisenhaber, F. (2003). Glycosylphosphatidylinositol lipid anchoring of plant proteins. Sensitive prediction from sequence-and genome-wide studies for *Arabidopsis* and rice. *Plant Physiology*, *133*(4), 1691–1701.

Eleftheriou, E. P., Adamakis, I. D. S., & Michalopoulou, V. A. (2016). Hexavalent chromium-induced differential disruption of cortical microtubules in some *Fabaceae* species is correlated with acetylation of α-tubulin. *Protoplasma*, *253*(2), 531–542.

Ellis, M., Egelund, J., Schultz, C. J., & Bacic, A. (2010). Arabinogalactan-proteins: Key regulators at the cell surface? *Plant Physiology*, *153*(2), 403–419.

Elrouby, N., Bonequi, M. V., Porri, A., & Coupland, G. (2013). Identification of *Arabidopsis* SUMO-interacting proteins that regulate chromatin activity and developmental transitions. *Proceedings of the National Academy of Sciences*, *110*(49), 19956–19961.

Elrouby, N., & Coupland, G. (2010). Proteome-wide screens for small ubiquitin-like modifier (SUMO) substrates identify *Arabidopsis* proteins implicated in diverse biological processes. *Proceedings of the National Academy of Sciences*, *107*(40), 17415–17420.

Feki, K., Quintero, F. J., Pardo, J. M., & Masmoudi, K. (2011). Regulation of durum wheat Na+/H+ exchanger TdSOS1 by phosphorylation. *Plant Molecular Biology*, *76*(6), 545–556.

Fíla, J., & Honys, D. (2012). Enrichment techniques employed in phosphoproteomics. *Amino Acids*, *43*(3), 1025–1047.

Finkemeier, I., Laxa, M., Miguet, L., Howden, A. J., & Sweetlove, L. J. (2011). Proteins of diverse function and subcellular location are lysine acetylated in *Arabidopsis*. *Plant Physiology*, *155*(4), 1779–1790.

Friso, G., & van Wijk, K. J. (2015). Posttranslational protein modifications in plant metabolism. *Plant Physiology*, *169*(3), 1469–1487.

Gan, C. S., Guo, T., Zhang, H., Lim, S. K., & Sze, S. K. (2008). A comparative study of electrostatic repulsion-hydrophilic interaction chromatography (ERLIC) versus SCX-IMAC-based methods for phosphopeptide isolation/enrichment. *Journal of Proteome Research*, *7*(11), 4869–4877.

Garbarino, J. E., Rockhold, D. R., & Belknap, W. R. (1992). Expression of stress-responsive ubiquitin genes in potato tubers. *Plant Molecular Biology*, *20*(2), 235–244.

Garcia-Dominguez, M., March-Diaz, R., & Reyes, J. C. (2008). The PHD domain of plant PIAS proteins mediates sumoylation of bromodomain GTE proteins. *Journal of Biological Chemistry*, *283*(31), 21469–21477.

Gardiner, J. (2019). Posttranslational modification of plant microtubules. *Plant Signaling & Behavior*, *14*(10), e1654818.

Genschik, P., Parmentier, Y., Durr, A., Marbach, J., Criqui, M. C., Jamet, E., & Fleck, J. (1992). Ubiquitin genes are differentially regulated in protoplast-derived cultures of *Nicotiana sylvestris* and in response to various stresses. *Plant Molecular Biology*, *20*(5), 897–910.

Gillmor, C. S., Lukowitz, W., Brininstool, G., Sedbrook, J. C., Hamann, T., Poindexter, P., & Somerville, C. (2005). Glycosylphosphatidylinositol-anchored proteins are required for cell wall synthesis and morphogenesis in *Arabidopsis*. *The Plant Cell*, *17*(4), 1128–1140.

Glinski, M., & Weckwerth, W. (2006). The role of mass spectrometry in plant systems biology. *Mass Spectrometry Reviews*, *25*(2), 173–214.

Grimsrud, P. A., den Os, D., Wenger, C. D., Swaney, D. L., Schwartz, D., Sussman, M. R., ... & Coon, J. J. (2010). Large-scale phosphoprotein analysis in *Medicago truncatula* roots provides insight into in vivo kinase activity in legumes. *Plant Physiology*, *152*(1), 19–28.

Guo, J., Chai, X., Mei, Y., Du, J., Du, H., Shi, H., ... & Zhang, H. (2022). Acetylproteomics analyses reveal critical features of lysine-ε-acetylation in *Arabidopsis* and a role of 14-3-3 protein acetylation in alkaline response. *Stress Biology*, *2*(1), 1–19.

Guo, Q., Zhang, J., Gao, Q., Xing, S., Li, F., & Wang, W. (2008). Drought tolerance through overexpression of monoubiquitin in transgenic tobacco. *Journal of Plant Physiology*, *165*(16), 1745–1755.

Han, C., He, D., Li, M., & Yang, P. (2014). In-depth proteomic analysis of rice embryo reveals its important roles in seed germination. *Plant and Cell Physiology*, *55*(10), 1826–1847.

Hang, R., Liu, C., Ahmad, A., Zhang, Y., Lu, F., & Cao, X. (2014). *Arabidopsis* protein arginine methyltransferase 3 is required for ribosome biogenesis by affecting precursor ribosomal RNA processing. *Proceedings of the National Academy of Sciences*, *111*(45), 16190–16195.

Harmoko, R., Yoo, J. Y., Ko, K. S., Ramasamy, N. K., Hwang, B. Y., Lee, E. J., ... & Lee, K. O. (2016). N-glycan containing a core α1,3-fucose residue is required for basipetal auxin transport and gravitropic response in rice (*Oryza sativa*). *New Phytologist*, *212*(1), 108–122.

Haruta, M., Gray, W. M., & Sussman, M. R. (2015). Regulation of the plasma membrane proton pump (H+-ATPase) by phosphorylation. *Current Opinion in Plant Biology*, *28*, 68–75.

Häweker, H., Rips, S., Koiwa, H., Salomon, S., Saijo, Y., Chinchilla, D., ... & von Schaewen, A. (2010). Pattern recognition receptors require N-glycosylation to mediate plant immunity 2. *Journal of Biological Chemistry*, *285*(7), 4629–4636.

Higashimoto, K., Kuhn, P., Desai, D., Cheng, X., & Xu, W. (2007). Phosphorylation-mediated inactivation of coactivator-associated arginine methyltransferase 1. *Proceedings of the National Academy of Sciences*, *104*(30), 12318–12323.

Hirakawa, Y., Kondo, Y., & Fukuda, H. (2010). TDIF peptide signaling regulates vascular stem cell proliferation via the WOX4 homeobox gene in *Arabidopsis*. *The Plant Cell*, *22*(8), 2618–2629.

Hochholdinger, F., Wen, T. J., Zimmermann, R., Chimot-Marolle, P., Da Costa e Silva, O., Bruce, W., ... & Schnable, P. S. (2008). The maize (*Zea mays* L.) roothairless3 gene encodes a putative GPI-anchored, monocot-specific, COBRA-like protein that significantly affects grain yield. *The Plant Journal*, *54*(5), 888–898.

Hoehenwarter, W., Thomas, M., Nukarinen, E., Egelhofer, V., Röhrig, H., Weckwerth, W., ... & Beckers, G. J. (2013). Identification of novel in vivo MAP kinase substrates in *Arabidopsis thaliana* through use of tandem metal oxide affinity chromatography. *Molecular & Cellular Proteomics*, *12*(2), 369–380.

Hopff, D., Wienkoop, S., & Lüthje, S. (2013). The plasma membrane proteome of maize roots grown under low and high iron conditions. *Journal of Proteomics*, *91*, 605–618.

Hu, Y., Guo, S., Li, X., & Ren, X. (2013). Comparative analysis of salt-responsive phosphoproteins in maize leaves using T i4+-IMAC enrichment and ESI-Q-TOF MS. *Electrophoresis*, *34*(4), 485–492.

Hu, Z., Song, N., Zheng, M., Liu, X., Liu, Z., Xing, J., ... & Sun, Q. (2015). Histone acetyltransferase GCN 5 is essential for heat stress-responsive gene activation and thermotolerance in *Arabidopsis*. *The Plant Journal*, *84*(6), 1178–1191.

Ingelsson, B., & Vener, A. V. (2012). Phosphoproteomics of *Arabidopsis* chloroplasts reveals involvement of the STN7 kinase in phosphorylation of nucleoid protein pTAC16. *FEBS Letters*, *586*(9), 1265–1271.

Ito, Y., Nakanomyo, I., Motose, H., Iwamoto, K., Sawa, S., Dohmae, N., & Fukuda, H. (2006). Dodeca-CLE peptides as suppressors of plant stem cell differentiation. *Science*, *313*(5788), 842–845.

Jones, A. M., Bennett, M. H., Mansfield, J. W., & Grant, M. (2006). Analysis of the defence phosphoproteome of *Arabidopsis thaliana* using differential mass tagging. *Proteomics*, *6*(14), 4155–4165.

Jones, A. M., MacLean, D., Studholme, D. J., Serna-Sanz, A., Andreasson, E., Rathjen, J. P., & Peck, S. C. (2009). Phosphoproteomic analysis of nuclei-enriched fractions from *Arabidopsis thaliana*. *Journal of Proteomics*, *72*(3), 439–451.

Khan, S., Anwar, S., Yu, S., Sun, M., Yang, Z., & Gao, Z. Q. (2019). Development of drought-tolerant transgenic wheat: Achievements and limitations. *International Journal of Molecular Sciences*, *20*(13), 3350.

Khan, M., Rozhon, W., Unterholzner, S. J., Chen, T., Eremina, M., Wurzinger, B., ... & Poppenberger, B. (2014). Interplay between phosphorylation and SUMOylation events determines CESTA protein fate in brassinosteroid signalling. *Nature Communications*, *5*(1), 1–10.

Kinoshita, T. (2014). Biosynthesis and deficiencies of glycosylphosphatidylinositol. *Proceedings of the Japan Academy, Series B*, *90*(4), 130–143.

Kondo, T., Sawa, S., Kinoshita, A., Mizuno, S., Kakimoto, T., Fukuda, H., & Sakagami, Y. (2006). A plant peptide encoded by CLV3 identified by in situ MALDI-TOF MS analysis. *Science*, *313*(5788), 845–848.

König, A. C., Hartl, M., Boersema, P. J., Mann, M., & Finkemeier, I. (2014). The mitochondrial lysine acetylome of *Arabidopsis*. *Mitochondrion*, *19*, 252–260.

Kurepa, J., Toh-e, A., & Smalle, J. A. (2008). 26S proteasome regulatory particle mutants have increased oxidative stress tolerance. *The Plant Journal*, *53*(1), 102–114.

Kwon, S. J., Choi, E. Y., Choi, Y. J., Ahn, J. H., & Park, O. K. (2006). Proteomics studies of post-translational modifications in plants. *Journal of Experimental Botany*, *57*(7), 1547–1551.

Lalanne, E., Michaelidis, C., Moore, J. M., Gagliano, W., Johnson, A., Patel, R., ... & Twell, D. (2004). Analysis of transposon insertion mutants highlights the diversity of mechanisms underlying male progamic development in *Arabidopsis*. *Genetics*, *167*(4), 1975–1986.

Larsen, M. R., Trelle, M. B., Thingholm, T. E., & Jensen, O. N. (2006). Analysis of posttranslational modifications of proteins by tandem mass spectrometry: Mass spectrometry for proteomics analysis. *Biotechniques*, *40*(6), 790–798.

Li, Y., Williams, B., & Dickman, M. (2017). *Arabidopsis* B-cell lymphoma2 (Bcl-2)-associated athanogene 7 (BAG 7)-mediated heat tolerance requires translocation, sumoylation and binding to WRKY 29. *New Phytologist*, *214*(2), 695–705.

Li, S., Ge, F. R., Xu, M., Zhao, X. Y., Huang, G. Q., Zhou, L. Z., ... & Zhang, Y. (2013). *Arabidopsis* COBRA-LIKE 10, a GPI-anchored protein, mediates directional growth of pollen tubes. *The Plant Journal*, *74*(3), 486–497.

Li, Z., Li, S., Luo, M., Jhong, J. H., Li, W., Yao, L., ... & Lee, T. Y. (2022). dbPTM in 2022: An updated database for exploring regulatory networks and functional associations of protein post-translational modifications. *Nucleic Acids Research*, *50*(D1), D471–D479.

Liu, X., Castro, C., Wang, Y., Noble, J., Ponvert, N., Bundy, M., ... & Palanivelu, R. (2016). The role of LORELEI in pollen tube reception at the interface of the synergid cell and pollen tube requires the modified eight-cysteine motif and the receptor-like kinase FERONIA. *The Plant Cell*, *28*(5), 1035–1052.

Liu, B., Fan, J., Zhang, Y., Mu, P., Wang, P., Su, J., ... & Wang, H. (2012). OsPFA-DSP1, a rice protein tyrosine phosphatase, negatively regulates drought stress responses in transgenic tobacco and rice plants. *Plant Cell Reports*, *31*(6), 1021–1032.

Loroch, S., Schommartz, T., Brune, W., Zahedi, R. P., & Sickmann, A. (2015). Multidimensional electrostatic repulsion–hydrophilic interaction chromatography (ERLIC) for quantitative analysis of the proteome and phosphoproteome in clinical and biomedical research. *Biochimica et Biophysica Acta (BBA)- Proteins and Proteomics*, *1854*(5), 460–468.

Luan, S., Ting, J., & Gupta, R. (2001). Protein tyrosine phosphatases in higher plants. *New Phytologist*, *151*(1), 155–164.

Luo, X., Han, C., Deng, X., Zhu, D., Liu, Y., & Yan, Y. (2019). Identification of phosphorylated proteins in response to salt stress in wheat embryo and endosperm during seed germination. *Cereal Research Communications, 47*(1), 53–66.

Luo, L. D., & Wirth, P. J. (1993). Consecutive silver staining and autoradiography of 35S and 32P-labeled cellular proteins: Application for the analysis of signal transducing pathways. *Electrophoresis, 14*(1), 127–136.

Lv, D. W., Subburaj, S., Cao, M., Yan, X., Li, X., Appels, R., ... & Yan, Y. M. (2014). Proteome and phospho-proteome characterization reveals new response and defense mechanisms of *Brachypodium distachyon* leaves under salt stress. *Molecular & Cellular Proteomics, 13*(2), 632–652.

Matsuzaki, Y., Ogawa-Ohnishi, M., Mori, A., & Matsubayashi, Y. (2010). Secreted peptide signals required for maintenance of root stem cell niche in *Arabidopsis. Science, 329*(5995), 1065–1067.

Mazur, M. J., Spears, B. J., Djajasaputra, A., Van Der Gragt, M., Vlachakis, G., Beerens, B., ... & Van den Burg, H. A. (2017). *Arabidopsis* TCP transcription factors interact with the SUMO conjugating machinery in nuclear foci. *Frontiers in Plant Science, 8*, 2043.

Mega, T. (2005). Glucose trimming of N-glycan in endoplasmic reticulum is indispensable for the growth of *Raphanus sativus* seedling (kaiware radish). *Bioscience, Biotechnology, and Biochemistry, 69*(7), 1353–1364.

Miller, M. J., Barrett-Wilt, G. A., Hua, Z., & Vierstra, R. D. (2010). Proteomic analyses identify a diverse array of nuclear processes affected by small ubiquitin-like modifier conjugation in *Arabidopsis. Proceedings of the National Academy of Sciences, 107*(38), 16512–16517.

Miura, K., & Hasegawa, P. M. (2009). Sumoylation and abscisic acid signaling. *Plant Signaling & Behavior, 4*(12), 1176–1178.

Miyamoto, T., Uemura, T., Nemoto, K., Daito, M., Nozawa, A., Sawasaki, T., & Arimura, G. I. (2019). Tyrosine kinase-dependent defense responses against herbivory in *Arabidopsis. Frontiers in Plant Science, 10*, 776.

Moon, J., Parry, G., & Estelle, M. (2004). The ubiquitin-proteasome pathway and plant development. *The Plant Cell, 16*(12), 3181–3195.

Nakagami, H., Sugiyama, N., Mochida, K., Daudi, A., Yoshida, Y., Toyoda, T., ... & Shirasu, K. (2010). Large-scale comparative phosphoproteomics identifies conserved phosphorylation sites in plants. *Plant Physiology, 153*(3), 1161–1174.

Nakagawa, U., Kamemura, K., & Imamura, A. (2013). Regulated changes in the acetylation of α-tubulin on Lys40 during growth and organ development in fast plants, *Brassica rapa* L. *Bioscience, Biotechnology, and Biochemistry, 77*(11), 2228–2233.

Nemoto, K., Takemori, N., Seki, M., Shinozaki, K., & Sawasaki, T. (2015). Members of the plant CRK superfamily are capable of trans- and autophosphorylation of tyrosine residues. *Journal of Biological Chemistry, 290*(27), 16665–16677.

Niemann, M. C., Bartrina, I., Ashikov, A., Weber, H., Novák, O., Spíchal, L., ... & Werner, T. (2015). *Arabidopsis* ROCK1 transports UDP-GlcNAc/UDP-GalNAc and regulates ER protein quality control and cytokinin activity. *Proceedings of the National Academy of Sciences, 112*(1), 291–296.

Niu, D., Lin, X. L., Kong, X., Qu, G. P., Cai, B., Lee, J., & Jin, J. B. (2019). SIZ1-mediated SUMOylation of TPR1 suppresses plant immunity in *Arabidopsis. Molecular Plant, 12*(2), 215–228.

Nühse, T. S., Stensballe, A., Jensen, O. N., & Peck, S. C. (2004). Phosphoproteomics of the *Arabidopsis* plasma membrane and a new phosphorylation site database. *The Plant Cell, 16*(9), 2394–2405.

Ohyama, K., Ogawa, M., & Matsubayashi, Y. (2008). Identification of a biologically active, small, secreted peptide in *Arabidopsis* by in silico gene screening, followed by LC-MS-based structure analysis. *The Plant Journal, 55*(1), 152–160.

Ohyama, K., Shinohara, H., Ogawa-Ohnishi, M., & Matsubayashi, Y. (2009). A glycopeptide regulating stem cell fate in *Arabidopsis thaliana. Nature Chemical Biology, 5*(8), 578–580.

Olenieva, V., Lytvyn, D., Yemets, A., Bergounioux, C., & Blume, Y. (2019). Tubulin acetylation accompanies autophagy development induced by different abiotic stimuli in *Arabidopsis thaliana. Cell Biology International, 43*(9), 1056–1064.

Orosa-Puente, B., Leftley, N., Von Wangenheim, D., Banda, J., Srivastava, A. K., Hill, K., ... & Bennett, M. J. (2018). Root branching toward water involves posttranslational modification of transcription factor ARF7. *Science, 362*(6421), 1407–1410.

Papanicolaou, K. N., O'Rourke, B., & Foster, D. B. (2014). Metabolism leaves its mark on the powerhouse: Recent progress in post-translational modifications of lysine in mitochondria. *Frontiers in Physiology, 5*, 301.

Pearce, G., Moura, D. S., Stratmann, J., & Ryan, C. A. (2001). Production of multiple plant hormones from a single polyprotein precursor. *Nature*, *411*(6839), 817–820.

Pearce, G., & Ryan, C. A. (2003). Systemic signaling in tomato plants for defense against herbivores: Isolation and characterization of three novel defense-signaling glycopeptide hormones coded in a single precursor gene. *Journal of Biological Chemistry*, *278*(32), 30044–30050.

Peck, S. C., Nuhse, T. S., Hess, D., Iglesias, A., Meins, F., & Boller, T. (2001). Directed proteomics identifies a plant-specific protein rapidly phosphorylated in response to bacterial and fungal elicitors. *The Plant Cell*, *13*(6), 1467–1475.

Pei, W., Jain, A., Ai, H., Liu, X., Feng, B., Wang, X., ... & Sun, S. (2019). OsSIZ2 regulates nitrogen homeostasis and some of the reproductive traits in rice. *Journal of Plant Physiology*, *232*, 51–60.

Pei, Y., Niu, L., Lu, F., Liu, C., Zhai, J., Kong, X., & Cao, X. (2007). Mutations in the Type II protein arginine methyltransferase AtPRMT5 result in pleiotropic developmental defects in *Arabidopsis*. *Plant Physiology*, *144*(4), 1913–1923.

Pereira, J. A., Yu, F., Zhang, Y., Jones, J. B., & Mou, Z. (2018). The *Arabidopsis* elongator subunit ELP3 and ELP4 confer resistance to bacterial speck in tomato. *Frontiers in Plant Science*, *9*, 1066.

Pfeifer, L., Shafee, T., Johnson, K. L., Bacic, A., & Classen, B. (2020). Arabinogalactan-proteins of *Zostera marina* L. contain unique glycan structures and provide insight into adaption processes to saline environments. *Scientific Reports*, *10*(1), 8232.

Ramazi, S., & Zahiri, J. (2021). Post-translational modifications in proteins: Resources, tools and prediction methods. *Database*, *2021*. https://doi.org/10.1093/database/baab012

Raorane, M. L., Mutte, S. K., Varadarajan, A. R., Pabuayon, I. M., & Kohli, A. (2013). Protein SUMOylation and plant abiotic stress signaling: In silico case study of rice RLKs, heat-shock and Ca^{2+}-binding proteins. *Plant Cell Reports*, *32*(7), 1053–1065.

Reiland, S., Finazzi, G., Endler, A., Willig, A., Baerenfaller, K., Grossmann, J., ... & Baginsky, S. (2011). Comparative phosphoproteome profiling reveals a function of the STN8 kinase in fine-tuning of cyclic electron flow (CEF). *Proceedings of the National Academy of Sciences*, *108*(31), 12955–12960.

Roychoudhury, A., Datta, K., & Datta, S. K. (2011). Abiotic stress in plants: From genomics to metabolomics. In: Tuteja, N., Gill, S. S., & Tuteja, R. (Eds.). Omics and Plant Abiotic Stress Tolerance, Bentham Science Publishers, UAE, pp. 91–120.

Roychoudhury, A., Gupta, B., & Sengupta, D. N. (2008). Trans-acting factor designated OSBZ8 interacts with both typical abscisic acid responsive elements as well as abscisic acid responsive element-like sequences in the vegetative tissues of indica rice cultivars. *Plant Cell Reports*, *27*(4), 779–794.

Roychoudhury, A., Paul, S., & Basu, S. (2013). Cross-talk between abscisic acid-dependent and abscisic acid-independent pathways during abiotic stress. *Plant Cell Reports*, *32*(7), 985–1006.

Saleh, A., Withers, J., Mohan, R., Marqués, J., Gu, Y., Yan, S., ... & Dong, X. (2015). Posttranslational modifications of the master transcriptional regulator NPR1 enable dynamic but tight control of plant immune responses. *Cell Host & Microbe*, *18*(2), 169–182.

Serre, N. B., Alban, C., Bourguignon, J., & Ravanel, S. (2018). An outlook on lysine methylation of non-histone proteins in plants. *Journal of Experimental Botany*, *69*(19), 4569–4581.

Shen, Y., Lei, T., Cui, X., Liu, X., Zhou, S., Zheng, Y., ... & Zhou, D. X. (2019). *Arabidopsis* histone deacetylase HDA 15 directly represses plant response to elevated ambient temperature. *The Plant Journal*, *100*(5), 991–1006.

Shin, R., Alvarez, S., Burch, A. Y., Jez, J. M., & Schachtman, D. P. (2007). Phosphoproteomic identification of targets of the *Arabidopsis* sucrose nonfermenting-like kinase SnRK2.8 reveals a connection to metabolic processes. *Proceedings of the National Academy of Sciences*, *104*(15), 6460–6465.

Showalter, A. M., Keppler, B., Lichtenberg, J., Gu, D., & Welch, L. R. (2010). A bioinformatics approach to the identification, classification, and analysis of hydroxyproline-rich glycoproteins. *Plant Physiology*, *153*(2), 485–513.

Simpson, C., Thomas, C., Findlay, K., Bayer, E., & Maule, A. J. (2009). An *Arabidopsis* GPI-anchor plasmodesmal neck protein with callose binding activity and potential to regulate cell-to-cell trafficking. *The Plant Cell*, *21*(2), 581–594.

Singh, A., & Roychoudhury, A. (2021). Gene regulation at transcriptional and post transcriptional levels to combat salt stress in plants. *Physiologia Plantarum*, *173*, 1556–1572.

Smalle, J., Kurepa, J., Yang, P., Emborg, T. J., Babiychuk, E., Kushnir, S., & Vierstra, R. D. (2003). The pleiotropic role of the 26S proteasome subunit RPN10 in *Arabidopsis* growth and development supports a substrate-specific function in abscisic acid signaling. *The Plant Cell*, *15*(4), 965–980.

Smalle, J., & Vierstra, R. D. (2004). The ubiquitin 26S proteasome proteolytic pathway. *Annual Review of Plant Biology*, *55*(1), 555–590.

Starheim, K. K., Gevaert, K., & Arnesen, T. (2012). Protein N-terminal acetyltransferases: When the start matters. *Trends in Biochemical Sciences, 37*(4), 152–161.

Sugiyama, N., Nakagami, H., Mochida, K., Daudi, A., Tomita, M., Shirasu, K., & Ishihama, Y. (2008). Large-scale phosphorylation mapping reveals the extent of tyrosine phosphorylation in *Arabidopsis*. *Molecular Systems Biology, 4*(1), 193.

Sun, C. W., & Callis, J. (1997). Independent modulation of *Arabidopsis thaliana* polyubiquitin mRNAs in different organs and in response to environmental changes. *The Plant Journal, 11*(5), 1017–1027.

Tan, P. V., Taranenko, N. I., Laiko, V. V., Yakshin, M. A., Prasad, C. R., & Doroshenko, V. M. (2004). Mass spectrometry of N-linked oligosaccharides using atmospheric pressure infrared laser ionization from solution. *Journal of Mass Spectrometry, 39*(8), 913–921.

Taylor, C. B., Talib, M. F., McCabe, C., Bu, L., Adney, W. S., Himmel, M. E., ... & Beckham, G. T. (2012). Computational investigation of glycosylation effects on a family 1 carbohydrate-binding module. *Journal of Biological Chemistry, 287*(5), 3147–3155.

Trinidad, J. C., Specht, C. G., Thalhammer, A., Schoepfer, R., & Burlingame, A. L. (2006). Comprehensive identification of phosphorylation sites in postsynaptic density preparations. *Molecular & Cellular Proteomics, 5*(5), 914–922.

Tryfona, T., Liang, H. C., Kotake, T., Kaneko, S., Marsh, J., Ichinose, H., ... & Dupree, P. (2010). Carbohydrate structural analysis of wheat flour arabinogalactan protein. *Carbohydrate Research, 345*(18), 2648–2656.

Tryfona, T., Liang, H. C., Kotake, T., Tsumuraya, Y., Stephens, E., & Dupree, P. (2012). Structural characterization of *Arabidopsis* leaf arabinogalactan polysaccharides. *Plant Physiology, 160*(2), 653–666.

Tryfona, T., Theys, T. E., Wagner, T., Stott, K., Keegstra, K., & Dupree, P. (2014). Characterisation of FUT4 and FUT6 α-(1 → 2)-fucosyltransferases reveals that absence of root arabinogalactan fucosylation increases *Arabidopsis* root growth salt sensitivity. *PLoS One, 9*(3), e93291.

Ueda, M., Matsui, K., Ishiguro, S., Sano, R., Wada, T., Paponov, I., ... & Okada, K. (2004). The HALTED ROOT gene encoding the 26S proteasome subunit RPT2a is essential for the maintenance of *Arabidopsis* meristems. *Development, 131*(9), 2101–11.

Umezawa, T., Sugiyama, N., Takahashi, F., Anderson, J. C., Ishihama, Y., Peck, S. C., & Shinozaki, K. (2013). Genetics and phosphoproteomics reveal a protein phosphorylation network in the abscisic acid signaling pathway in *Arabidopsis thaliana*. *Science Signaling, 6*(270), rs8–rs8.

Van Der Woude, L. C., Perrella, G., Snoek, B. L., Van Hoogdalem, M., Novák, O., Van Verk, M. C., ... & Van Zanten, M. (2019). HISTONE DEACETYLASE 9 stimulates auxin-dependent thermomorphogenesis in *Arabidopsis thaliana* by mediating H2A. Z depletion. *Proceedings of the National Academy of Sciences, 116*(50), 25343–25354.

Viëtor, R., Loutelier-Bourhis, C., Fitchette, A. C., Margerie, P., Gonneau, M., Faye, L., & Lerouge, P. (2003). Protein N-glycosylation is similar in the moss *Physcomitrella patens* and in higher plants. *Planta, 218*(2), 269–275.

Vlachonasios, K., Poulios, S., & Mougiou, N. (2021). The histone acetyltransferase GCN5 and the associated coactivators ADA2: From evolution of the SAGA complex to the biological roles in plants. *Plants, 10*(2), 308.

Wang, Z. A., & Cole, P. A. (2020). The chemical biology of reversible lysine post-translational modifications. *Cell Chemical Biology, 27*(8), 953–969.

Wang, S., Kurepa, J., & Smalle, J. A. (2009). The *Arabidopsis* 26S proteasome subunit RPN1a is required for optimal plant growth and stress responses. *Plant and Cell Physiology, 50*(9), 1721–1725.

Wang, Q., Wang, S., Gan, S., Wang, X., Liu, J., & Wang, X. (2016). Role of specific phosphorylation sites of *Arabidopsis* brassinosteroid-insensitive 1 receptor kinase in plant growth and development. *Journal of Plant Growth Regulation, 35*(3), 755–769.

Wang, P., Xue, L., Batelli, G., Lee, S., Hou, Y. J., Van Oosten, M. J., ... & Zhu, J. K. (2013). Quantitative phosphoproteomics identifies SnRK2 protein kinase substrates and reveals the effectors of abscisic acid action. *Proceedings of the National Academy of Sciences, 110*(27), 11205–11210.

Wang, J., Yu, H., Xiong, G., Lu, Z., Jiao, Y., Meng, X., ... & Li, J. (2017). Tissue-specific ubiquitination by IPA1 INTERACTING PROTEIN1 modulates IPA1 protein levels to regulate plant architecture in rice. *The Plant Cell, 29*(4), 697–707.

Willems, P., Horne, A., Van Parys, T., Goormachtig, S., De Smet, I., Botzki, A., ... & Gevaert, K. (2019). The plant PTM viewer, a central resource for exploring plant protein modifications. *The Plant Journal, 99*(4), 752–762.

Wilson, I. B., Zeleny, R., Kolarich, D., Staudacher, E., Stroop, C. J., Kamerling, J. P., & Altmann, F. (2001). Analysis of Asn-linked glycans from vegetable foodstuffs: Widespread occurrence of Lewis a, core α1, 3-linked fucose and xylose substitutions. *Glycobiology, 11*(4), 261–274.

Wu, X., Delbianco, M., Anggara, K., Michnowicz, T., Pardo-Vargas, A., Bharate, P., … & Kern, K. (2020). Imaging single glycans. *Nature*, *582*(7812), 375–378.

Xu, J., & Zhang, S. (2015). Mitogen-activated protein kinase cascades in signaling plant growth and development. *Trends in Plant Science*, *20*(1), 56–64.

Yang, F., Melo-Braga, M. N., Larsen, M. R., Jørgensen, H. J., & Palmisano, G. (2013). Battle through signaling between wheat and the fungal pathogen *Septoria tritici* revealed by proteomics and phosphoproteomics. *Molecular & Cellular Proteomics*, *12*(9), 2497–2508.

Yu, Y., Bu, Z., Shen, W. H., & Dong, A. (2009). An update on histone lysine methylation in plants. *Progress in Natural Science*, *19*(4), 407–413.

Yu, B., Li, J., Koh, J., Dufresne, C., Yang, N., Qi, S., … & Li, H. (2016). Quantitative proteomics and phosphoproteomics of sugar beet monosomic addition line M14 in response to salt stress. *Journal of Proteomics*, *143*, 286–297.

Yu, M., Meng, B., Wang, F., He, Z., Hu, R., Du, J., … & Yang, C. (2019). A SUMO ligase AtMMS21 regulates activity of the 26S proteasome in root development. *Plant Science*, *280*, 314–320.

Zhang, Y., & Zeng, L. (2020). Crosstalk between ubiquitination and other post-translational protein modifications in plant immunity. *Plant Communications*, *1*(4), 100041.

Zhang, R. F., Zhou, L. J., Li, Y. Y., You, C. X., Sha, G. L., & Hao, Y. J. (2019). Apple SUMO E3 ligase MdSIZ1 is involved in the response to phosphate deficiency. *Journal of Plant Physiology*, *232*, 216–225.

Zheng, W. (2020). The plant sirtuins. *Plant Science*, *293*, 110434.

Zheng, Y., Schumaker, K. S., & Guo, Y. (2012). Sumoylation of transcription factor MYB30 by the small ubiquitin-like modifier E3 ligase SIZ1 mediates abscisic acid response in *Arabidopsis thaliana*. *Proceedings of the National Academy of Sciences*, *109*(31), 12822–12827.

17 Understanding Post-translational Protein Modification through Proteomics Tools

Sampat Nehra, Raj Kumar Gothwal, Aruna Shekhar N. C.,
Parul Sinha, Erica Zinnia Nehra, Alok Kumar Varshney,
Pooran Singh Solanki, and Purnendu Ghosh

17.1 INTRODUCTION

Proteomics is the study of protein on a vast scale. In general, it refers to just about any large-scale examination of protein mixtures, which is frequently performed without previous knowledge of the identification of the proteins in the samples. In recent decades, mass-spectrometry (MS)-based proteomics has evolved as a powerful method for high-throughput protein detection and analysis. With the completion of genome sequencing efforts for a wide range of creatures from all kingdoms of life, protein identification in tandem mass spectrometric investigations relying upon peptide fragmentation patterns has evolved into a nearly automated operation. The introduction of soft protein ionisation technologies, such as electrospray ionisation (ESI) or matrix-assisted laser desorption ionisation (MALDI), has contributed to the success of protein MS (MALDI). This accomplishment was recognised in 2002 when John Fenn and Koichi Tanaka were awarded the Nobel Prize in Chemistry. However, without the knowledge acquired from multiple full genome sequencing initiatives, as well as effective methods for peptide sequence identification from fragmentation spectra (Eng et al., 1994; Pevzner et al., 2001), proteomic research would be much more challenging today. Recent and ongoing advances in mass analyser and fragmentation technologies (Hu et al., 2005; McAlister et al., 2007; Zhang et al., 2009) have contributed to MS-based proteomics becoming a commonly utilised technique. Proteomics was mostly a qualitative subject. Typical proteomic investigations produced lists of proteins identified in a specific tissue or protein complex but provided no more information regarding abundance, distribution or stoichiometry. In contrast, quantitative techniques like microarrays or quantitative PCR have become extensively employed for analysing gene expression, and the capacity of large-scale genomics has yielded fresh insights into many areas of development and physiology, particularly in plants.

However, enzymatic processes and signalling cascades are ultimately dependent on protein function. Protein synthesis and degradation influence protein quantity, which may be independent of transcriptional regulation (Piques et al., 2009). Furthermore, transcript abundance analysis does not capture post-translational changes, isoforms or splice variants. The technological basis for analysing proteome complexity is provided by modern proteomic techniques. Protein mixtures may now be regularly defined in terms of the proteins present in the sample, but quantitative studies are required for biological interpretation.

The area of quantitative MS-based proteomics is still in its early stages, with new and improved instruments being created virtually on an annual basis.

The majority of the early advances in quantitative MS-based proteomics applications were driven by studies on yeast and mammalian cell lines (Blagoev et al., 2003, 2004; Schulze et al., 2005).

Quantitative proteomic techniques have also aided in the characterisation of protein complexes (Andersen et al., 2002, 2003; Rinner et al., 2007) and in the identification of actual interaction partners for a specific bait protein over background proteins (Blagoev et al., 2003; Schulze and Mann, 2004). However, in plant physiology, MS-based proteomics is no longer just a descriptive tool. Alternatively, quantitative proteomics has been used to study multiple facets of organelle biology, growth control and signalling.

Pioneering investigations of distinct plant subproteomes have discovered potential proteins that are phosphorylated selectively under different stress circumstances (Benschop et al., 2007; Niittyla et al., 2007) or during a light-dark cycle (Reiland et al., 2009). Protein abundance variations in reaction to heat shock (Palmblad et al., 2008) or leaf senescence (Hebeler et al., 2008) were tracked, and protein turnover of photosynthetic proteins was tracked employing pulse-chase labelling in conjunction with protein MS (Nowaczyk et al., 2006). Organelle proteomes were studied using either fractions of separated proteins in a sucrose density gradient (Dunkley et al., 2004, 2006; Sadowski et al., 2006) or by focusing on specific pure subproteomes such as chloroplasts (Kleffmann et al., 2007; Majeran et al., 2005; Peltier et al., 2000, 2006; Reiland et al., 2009) or plasma membranes and their microdomains (Kierszniowska et al., 2009a,b; Nelson et al., 2006).

Finally, quantitative proteomics approaches have added to the toolbox of methodologies for studying regulatory systems in plants in connection to whole plant factors such as growth or development (Piques et al., 2009; Raffaele et al., 2009; Stanislas et al., 2009). When compared to other species, namely, yeast and humans, plant biology research is still far from fully using the promise of proteomics, and a number of difficulties, mostly technological, need to be overcome.

The proteome of many plant species researched so far, with a significant portion on *Arabidopsis thaliana* (Lunn, 2007) and rice (Liu et al., 2006). The study of plant development (Jung et al., 2006), the impact of hormones and signalling molecules (Tuskan et al., 2006), post-translational modifications (PTMs) (Tuskan et al., 2006) and protein interactions have been the subject of a few contributions (Foster et al., 2006). The practically exclusive platform used in plant proteome study is 2-DE (particularly IEF-SDS-PAGE) linked to MS. The use of "second generation" proteomic techniques like multi-dimensional protein identification technology (MudPIT), gel-free protein separation techniques and quantitative proteomics methods like DIGE, isotope-coded affinity tags (ICAT), iTRAQ and stable isotope labelling by amino acids in cell culture (SILAC) are still anecdotal.

The examination of the phosphoproteomes (Olsen et al., 2006) in *Arabidopsis*, Medicago *truncatula*, barley, tobacco and, to a lesser degree, the redox proteome (Schöneich and Sharov, 2006) are mainly included under post-translational studies in plants (Cvetkovska et al., 2006; Ito et al., 2007; Laugesen et al., 2006; Rossignol, 2006; Ströher and Dietz, 2006). Global proteome techniques have begun to address other PTMs, such as ubiquitination and ubiquitination-like modifications, which have been widely explored in mammals and yeast (Brennan et al., 2006; del Riego et al., 2006; Ghesquiere et al., 2006; Gerber et al., 2006; Maor et al., 2007; Pedrioli et al., 2006; Roth et al., 2006).

Studying the quantitative and qualitative alterations in the phosphoproteome in tobacco cells in response to lipopolysaccharides has revealed new details about the signal perception and transduction processes behind induced innate immunity. Early on in the reaction, the phosphoproteome was altered, affecting proteins such as the G-protein, Ca21/calmodulin-dependent, W-ATPase, thioredoxin and 14-3-3, among others (Gerber et al., 2006).

A detailed examination of the phosphorylation pattern of the plastid ATP-synthase beta subunit isoforms in a distinct research carried out in barley (del Riego et al., 2006) found various degrees of phosphorylation, with Ser and Thr phosphorylation sites identified. With the use of iTRAQ (Jones et al., 2007), quantitative alterations have been documented in the *Arabidopsis* phosphoproteome in response to *Pseudomonas syringae*. They discovered five proteins that may be phosphorylated as a result of the plant's basic defence response, including the major subunit of RuBisCo, a suspected p23 cochaperone, heat shock protein 81 and a plastid-associated protein.

17.2 PLANTS AND POST-TRANSLATIONAL MODIFICATIONS

Proteomics has turned out to be a significant area of functional genomics with the completion of genome sequencing programmes and the advancement of analytical tools for protein characterisation. The comprehensive identification of every protein species in a cell or tissue was the original goal of proteomics. Analysis of the numerous functional characteristics of proteins that experience PTMs is now required, which complicates the applications. PTMs are covalent processes that add and remove functional groups in fundamental structures of proteins in a way that is sequence-specific. Examples of these processes include phosphorylation, acylation, glycosylation, nitration and ubiquitination (Mann and Jensen, 2003; Seo and Lee, 2004). Proteins undergo structural changes as a result of these alterations, which also affect protein functions, subcellular localisation, stability and interactions with other chemicals and proteins. Thus, PTMs greatly enhance the complexity and dynamics of proteins, resulting in the complicated control of biological activities.

Even though PTMs play crucial roles in biological processes, research on PTMs has not always been practical. A very sensitive approach and enormous quantities of proteins are needed for the identification of PTMs (Jensen, 2004). When isolated, a single protein typically has a relatively limited number of proteins in a particular modification state because a single protein frequently exhibits a heterogeneous population of proteins with various PTMs at several locations that are transitory and dynamic in nature. Utilising techniques like affinity enrichment and chromatographic fractionation will help to lessen the complexity. For the proteomic study of PTMs, two-dimensional gel electrophoresis (2DE) has been frequently used. It typically offers significant quantities once the proteins have been pre-fractionated depending on the PTMs to be discovered, allowing protein species in various modification states to be resolved in the gel and subjected to an examination of the modifications. In the gel, the changed proteins may be seen using certain techniques that are appropriate for various PTMs (Patton, 2002). Autoradiography of incorporated 32P or 33P and western analysis with antiphosphoamino acid antibodies are often utilised for the identification of phosphorylated proteins (Astoul et al., 2003; Gronborg et al., 2002; Luo and Wirth, 1993).

Through the treatment with phosphatase or alkaline hydrolysis of phosphate esters, phosphoproteins may also be uniquely identified in gels (Debruyne, 1983; Yamagata et al., 2002). Similar to this, the techniques for detecting glycoproteins, proteolytic changes, nitrosylation and methylation have been developed using radioactive or non-radioactive staining (Patton, 2002).

Following the identification of modified proteins, PTMs of the proteins are characterised by MS analysis using liquid chromatography (LC) MS/MS, ESI tandem MS (MS/MS) and matrix-assisted laser desorption/ionisation time-of-flight (MALDI-TOF) MS (Jensen, 2004). Peptides are released when isolated proteins are digested chemically or by enzymes, and the changed peptides can then be recovered using chromatographic techniques or affinity purification. The detection of the changed peptides is aided by isotope labelling.

Since alterations either make the molecular mass of the changed amino acid residue larger or less, localising the modifications is theoretically straightforward. The mapping of PTMs is difficult due to a number of technical issues, including ion suppression, the stability and purity of the changed peptides, sequence coverage and quantification (Mann and Jensen, 2003). The sensitivity and accuracy for the identification of PTMs would increase with the introduction of chemical tagging and enrichment technologies along with an improvement in MS apparatus.

Studies on PTMs in plants have been few, and those that have been done have solely focused on phosphorylation, GPI modification and ubiquitination. This contrasts with the advancements achieved in animal proteomics on PTMs. The PTM investigations in plants are included here. Additionally, the sample preparation procedures of each study are discussed.

17.3 PHOSPHORYLATION

One of the most significant and well-studied PTMs controlling cellular signalling systems is protein phosphorylation. Phosphoproteins in the chloroplast thylakoid and plasma membranes have been the main targets in research on the plant PTMs. Immobilised metal affinity chromatography (IMAC) is frequently employed to enrich phosphoproteins after shaving the surface-exposed portions of thylakoid and plasma membrane proteins for examination (Posewitz and Tempst, 1999). The negatively charged phosphate groups are strongly bound by the positively charged metal ions (Fe^{3+}, Ga^{3+}) of IMAC.

For plants to manage their photosynthetic processes, protein phosphorylation under the control of light and redox is essential. Tryptic peptides were released from the surface of *Arabidopsis* thylakoids in order to analyse phosphoproteins in the chloroplast thylakoids. Phosphopeptides were then enriched by IMAC and identified by MALDI-TOF MS and ESI MS/MS (Vener et al., 2001). A variety of phosphorylated proteins, including D1, D2 and CP43 of PSII as well as the peripheral protein PsbH and the light-harvesting polypeptide LCHII, were identified and their phosphorylation locations effectively mapped. These investigations were expanded to characterise a 12-kDa phosphoprotein (TPS9) isolated from spinach thylakoid membranes (Carlberg et al., 2003). Three threonine residues in TSP9 were shown to be phosphorylated, according to extensive MS investigations.

Later experiments used trypsin to break the surface-exposed peptides from the *Arabidopsis* thylakoid membranes and then methylation of the acidic residues to enhance the ability of the phosphopeptides to specifically attach to IMAC. ESI MS/MS was used to sequence the enriched phosphopeptides (Hansson and Vener, 2003). Three previously unknown phosphorylation sites were discovered and later attributed to three proteins, PsaD, CP29 and a unique protein called TMP14 in PSI, in addition to the five known phosphorylation sites in PSII proteins. The same method was used to identify the phosphorylation and acetylation sites on the unprocessed transit peptide in thylakoid membranes isolated from the green alga *Chlamydomonas reinhardtii* (Turkina et al., 2004).

In pathogen response, such as plant-pathogen interactions, gene expression and defence signalling, protein phosphorylation is crucial (Xing et al., 2002). A suspension culture of *Arabidopsis* was pulse-labelled with ^{32}P, and the phosphorylated proteins were seen by 2DE to help comprehend the early processes brought about by pathogen treatment. By using nano-ESI MS/MS, it was possible to identify one of these proteins, AtPhos43, that was phosphorylated in a matter of minutes after being exposed to flagellin or chitin fragments (Peck et al., 2001). AtPhos43 is a new protein with ankyrin repeats, and that FLS2, a receptor-like kinase implicated in flagellin perception, is necessary for phosphorylation of AtPhos43 following flagellin treatment.

Plasma membrane proteins are involved in the start and modulation of several signalling pathways, as well as the regulation of cell-cell contacts throughout developmental processes and reactions to the environment. Identification of signalling mechanisms and phosphoproteins at the plasma membrane is of significant interest in this area. The "shaving" technique was extended for membrane proteomics in a concept for large-scale phosphoproteomics of the plasma membrane (Nühse et al., 2003). Strong anion exchange chromatography was added before IMAC, which reduced the complexity of IMAC-purified phosphopeptides and produced a much better coverage of monophosphorylated peptides. Trypsin digestion of cytoplasmic face-out vesicles was paired with IMAC and LC-MS/MS.

Through the use of these techniques, 200 plasma-membrane proteins from *Arabidopsis* were shown to include more than 300 phosphorylation sites, and considerably more than 50 of those sites were mapped onto receptor-like kinases (Nühse et al., 2004). To identify common themes around the sites, the properties of phosphorylation sites and their conservation in sequence were examined. These analyses offer the guidelines for determining the substrate specificity and phosphorylation sites for kinases in plants. The vast volume of data is a useful resource that has made it possible to create phosphorylation site prediction techniques.

One of the PTMs, GPI anchoring, is used to attach several cell surface proteins to the extracellular membrane. A mechanism for the asymmetric distribution of membrane proteins that develops and maintains cell polarity is provided by GPI membrane anchors (Chatterjee and Mayor, 2001).

Genomic and proteomic investigations have revealed GPI-anchored proteins (GAPs), which have been demonstrated to be involved in a variety of cellular processes including cell signalling, adhesion, matrix remodelling and pathogen response in plants (Borner et al., 2002). Proteoglycans of the extracellular matrix called arabinogalactan proteins (AGPs) have a role in the growth and development of plants (Schultz et al., 2000). According to predictions, AGPs will be temporarily or permanently anchored to the outer surface of the plasma membrane by a GPI molecule. AGPs were pre-deglycosylated, and the smaller, more traditional, AGPs were divided up into AG-peptides (Schultz et al., 2004). The deglycosylated AG-peptides range in length from 10 to 17 residues and are thus analysed directly without tryptic digestion.

Eight AG-peptides have their exact cleavage sites for the C-terminal GPI anchor signal and the N-terminal endoplasmic reticulum secretion signal identified using tandem MS/MS and MALDI-TOF MS, respectively.

A total of 167 putative GAPs were discovered through a database study utilising a computer-based technique created for the discovery of GAPs (Eisenhaber et al., 1999), in addition to the 43 candidates that were previously disclosed (Borner et al., 2002). B-1,3-glucanases, metallo- and aspartyl proteases, glycerophosphodiesterases, phytocyanins, multicopper oxidases, extensins, plasma membrane receptors and lipid-transfer proteins were among the homologs found in the expected GAP. It was hypothesised that many proteins with unknown functions, including several fasciclin-like proteins and many AG peptides, would be GPI anchored. Following extensive proteome research, plant GAPs were discovered (Borner et al., 2003).

To create GAP-rich fractions from *Arabidopsis* callus cells, Triton X-114 phase partitioning and treatment with phosphatidylinositol-specific phospholipase C (Pi-PLC) were utilised (Hooper et al., 1987). When the GPI anchor was cut with Pi-PLC, a distinctive transition from the hydrophobic detergent-rich phase to the hydrophilic aqueous phase was observed. It is widely acknowledged that the separated proteins are GAPs as this method of fractionating GPI anchoring of proteins is well-established. B-1,3 glucanases, phytocyanins, fasciclin-like AGPs, receptor-like proteins, hedgehog-interacting-like proteins, suspected glycerophosphodiesterases, lipid transfer-like protein, COBRA-like protein, SKU5 and SKS1 are a few of the 30 GAPs that were discovered using LC-MS/MS. These findings confirmed their earlier bioinformatics research for identifying GAPs in the *Arabidopsis* protein database.

Using the validated GAPs from the proteomic study, the search methodology and genomic annotation were improved. As a result, an updated in silico screen identified 64 more candidates, bringing the total number of predicted GAPs in *Arabidopsis* to 248. Using the same experimental technique, 44 GAPs were discovered in an *Arabidopsis* membrane preparation in additional proteomic investigations (Elortza et al., 2003).

17.4 UBIQUITINATION

The covalent attachment of ubiquitin (Ub) to Lys residues is the functional aspect of ubiquitination that targets proteins for destruction by the proteasome (Glickman and Ciechanover, 2002). More and more evidence suggests that ubiquitination plays a significant role in the control of a variety of biological activities (Welchman et al., 2005). In recent years, the regulatory functions of the Ub-proteasome pathway in plants have received more attention (Smalle and Vierstra, 2004).

One important regulatory mechanism for numerous cellular processes, such as homeostasis, growth, development, hormone response and stress response in plants, is the regulation of protein breakdown (Devoto et al., 2003; Moon et al., 2004).

Proteomic investigation has focused on the Ub-proteasome pathway to find Ub- and Ub-like proteins (Ubls)-system components in yeast and mammals (Denison et al., 2005). The majority of researches so far have coupled MS/MS approaches with affinity purification methods. Ub and Ubl substrates have been found by purification using epitope tags fused to Ub and Ubls (Peng et al., 2003; Wohlschlegel et al., 2004). Additionally, using this technique, additional interacting proteins in the Ub-proteasomal complexes are isolated (Verma et al., 2000).

Bioinformatic study has established the magnitude and complexity of the Ub-26S proteasome system in plants (Smalle and Vierstra, 2004; Vierstra, 2003). The Ub-26S proteasome pathway is thought to be involved in the majority of cellular functions in plants, since the *Arabidopsis* genome is expected to encode more than 1400 components (>5% of the entire proteome). The discovery of substrates and other interacting molecules, as well as our comprehension of their precise activities in the system, would be improved by future studies in conjunction with thorough proteomic analyses.

17.5   DIFFERENTIAL PROTEOMICS STRATEGIES

The majority of quantitative proteomics experimental designs compare a stressed or disturbed condition to a non-stressed or undisturbed reference sample. Several technologies and procedures have been introduced during the previous decade, leaving the biologist with an often befuddling array of techniques, each with unique benefits and limitations in their respective situations. However, the choice of quantitation technique is less significant than the practical experience of the method, which results in strong technical repeatability (Turck et al., 2007). Furthermore, biological variation must be addressed in a suitably repeated experimental design, and this, along with economic concerns, may impact strategy selection. Because there is no amplification stage like the polymerase chain reaction in microarray research, sample size is generally the limiting issue in proteomic investigations. Thus, enrichment and purification of subproteomes are frequently important procedures before comparing protein abundances. Furthermore, the difficulty in getting appropriate quantities of samples may impact the selection of certain quantitative procedures.

17.6   QUANTITATION USING GELS

Originally, two-dimensional (2D) gels were thought to be the best approach for visualising variations among protein samples taken from various environments or tissues. Complex protein combinations might be efficiently resolved, and detecting variations in band or spot intensity was natural. On single 2D gels, it is now feasible to see over 10,000 dots corresponding to over 1000 proteins. Traditional 2D gel experiments, in contrast, do not immediately link protein separation and differential analysis to the determination of the protein underlying a specific spot. Individual spots may indeed contain more than one protein in many situations; however, this may only be identified if differential spots are removed and evaluated by MS (Gygi et al., 2000). With the introduction of MS-based peptide sequencing technology in the mid-1990s (Wilm et al., 1996), 2D gel separation and quantitative analysis could be combined by protein identification. In this context, 2D-gel-based quantification remains appealing and has previously been used in a number of plant biology topics spanning from stress response analysis (Riccardi et al., 1999) to cell type or organelle characterisation (Kleffmann et al., 2007; Majeran et al., 2008; Peltier et al., 2006).

Accurate reproducibility of 2D gels is frequently a barrier in quantitative experiments. As a result, important breakthroughs in gel-based quantification may occur when fluorescent dyes are used to mark various protein samples that were subsequently differentiated on the same gel (Ünlü et al., 1997). This so-called DIGE method has significantly increased the quantitative accuracy of 2D gels. The DIGE method has been used effectively to investigate phosphorylation reactions in plant plasma membrane proteins in response to brassinosteroid treatment (Deng et al., 2007), as well as to compare light- and dark-adapted proteomes of chloroplast thylakoid lumen (Granlund et al., 2009). Recent advances in the use of DIGE technology in conjunction with blue native gel electrophoresis (rather than isoelectric focusing) as a first-dimension separation provide new perspectives and applications, particularly in comparative as well as structural survey of protein complexes or the allocation of complexes to subcellular fractions (Heinemeyer et al., 2009).

17.7   QUANTIFICATION USING MASS SPECTROMETRY

Because of the varied physical and chemical characteristics of distinct tryptic peptides, protein MS employing LC-MS/MS is not quantifiable. Even though peptides belong to the same protein, changes in charge state, peptide length, amino acid makeup or PTMs cause significant variances in ion intensities. Thus, for reliable quantification utilising ion intensities, comparisons between different samples can only be made using the same peptide mass-to-charge ratios (m/z) collected during LC-MS/MS studies under the same general circumstances. As a result, all MS-based quantitative methods are inevitably relative comparisons of one or more samples, and comparative quantitation is now only possible with careful experimental design and appropriate data analysis methodologies (Hu et al., 2005). A variety of such comparison tactics have grown in popularity, and they may be classified either as stable-isotope-labelling strategies or label-free approaches.

Relative standard deviations (RSD) for stable-isotope labelling techniques based on survey scan quantification are often less than 10% (Gygi et al., 1999; Ong et al., 2003). Quantification without stable isotopes, instead of methods based on peak intensities, may often offer quantitative accuracies of less than 30%. (Andersen et al., 2003). The quantitative precision of label-free techniques utilising spectral counting or derived indices, on the other hand, can be as low as 50% RSD (Old et al., 2005).

17.8   TECHNIQUES WITHOUT LABELS

Differential proteomic studies that use label-free quantification compare more samples based on ion intensities of identical proteins or the number of recorded spectra for each protein. Samples with label-free comparisons should ideally be run sequentially on the same LC-MS/MS setup to eliminate changes in ion intensities caused by variances in system configuration (column characteristics, temperature), allowing for perfect reproduction of retention periods. Label-free techniques are less costly, can be used on any biological material and have a high proteome coverage of quantified proteins since almost any protein recognised by one or more peptide spectra may be quantified. Aside from these benefits, blending various proteomes does not enhance the overall complexity of such sample. Label-free techniques frequently possess a high analytical depth and dynamic range, making them advantageous when substantial, global protein alterations among treatments were predicted. However, especially when it comes to spectral count, the benefit of extensive proteome coverage can come at the expense of rather low accuracy (Old et al., 2005).

17.9   COUNTING THE SPECTRUM AND DERIVING INDICES

Protein-based techniques allow for relative quantification of protein quantities within and across samples, as well as being quick and simple to use. Furthermore, an almost infinite quantity of samples may be compared. The quantity of peptide-identifying spectra attributed to each protein is used as a quantitative measure in the spectrum count techniques (Liu et al., 2004). This quantification approach is justified by the fact that even more plentiful peptides as well as proteins were sampled more frequently in fragments ion scans than low-abundance peptides and proteins. Apparently, the output of spectrum counting is determined by the data-dependent acquisition parameters of the mass spectrometer. Different dynamic exclusion settings (Wang and Li, 2008), in particular, alter the linear range of quantitation and the amount of proteins to be measured; the optimal parameters would rely on sample complexity. The most notable drawback of spectrum counts is that they perform extremely poorly with low-quantity proteins and few spectra. The method of spectrum count method accuracy degrades, particularly when using low-abundance proteins, since every spectrum is marked with the number "1" regardless of its ion intensities. To address this issue, a method that employs the overall mean of total ion quantity from all fragment spectra which indicate a protein as a quantifiable metric has been proposed. As a result, its linear dynamic spectrum for quantitation may be greatly expanded (Asara et al., 2008).

The empirical link between the number of detected spectra or peptides for a specific protein and overall protein abundance in the sample was utilised to quantify the absolute concentration of each protein within the sample. Its exponentially modified protein abundance index (PAI) (emPAI) is determined by dividing the number of observed spectra for each protein by the number of potentially observable peptides, a percentage known as a PAI (Rappsilber et al., 2002). PAI is then multiplied by 10 to get an exponentially modified value, known as the emPAI index (Ishihama et al., 2005). The absolute protein expression (APEX) index was calculated using a very similar method (Lu et al., 2007). The protein concentrations assessed by emPAI indices corresponded extremely closely to the protein concentrations predicted by enzymatic activities (Piques et al., 2009). The indices emPAI and APEX are thereby calculated measurements of absolute protein content in a particular sample based on mass spectrometric analytical characteristics. Their predictive value is almost definitely at least as excellent as traditional protein staining quantification.

The APEX index was used to generate an *Arabidopsis* proteome PAI (Baerenfaller et al., 2008) and a chloroplast protein abundance map (Zybailov et al., 2008). The emPAI index has been used to quantify protein abundances of enzymes involved and ribosomal proteins throughout a 24-hour period (Piques et al., 2009). Spectrum counting has been employed to examine drought stress response in *Medicago* root nodules (Larrainzar et al., 2007), as well as to compare protein amazing array in mesophyll and stack sheath chloroplasts (Majeran et al., 2005, 2008). Spectrum count, in conjunction with high mass precision precursor alignment, has aided in identifying variant-specific proteome alterations among potato types (Hoehenwarter et al., 2008).

17.10   ION INTENSITIES AND PROTEIN CORRELATION PROFILING USING PEPTIDES

Peptide-based techniques quantify each detected peptide ion species using averaged, normalised ion intensities. The elevation or volume of a peak with a particular m/z represents the number of ions of that specific mass identified within a given time interval. This procedure of finding the peak volume is known as ion extraction, and it produces an extracted ion chromatogram of the specific ion species. Such extracted ion chromatograms may be generated for each m/z throughout all LC-MS/MS runs inside an experiment, as well as the resulting peak volumes are then able to be quantitatively compared. Due to variances in the ionisation efficiency of distinct peptide species, only a single ion species may be compared between multiple samples. This aspect greatly complicates the computing work required to utilise total ion abundance as just a comparison metric. One major issue in conventional MS/MS investigations is the fact that parent ion survey scans are interrupted by fragment ion scan events (MS/MS), resulting in discontinuous coverage of such peptide ion peaks. Depending on the fragmentation duty cycle duration of an instrument, this results in more or less data points gathered over the elution time of each ion peak. As a result, the optimal balance between the acquisition of the questionnaire and fragment spectra must be found experimentally for every instrument and for varied sample complexity. While numerous fragment ion scans (MS2) are required for comprehensive peptide sequencing and detection of as many peptides and proteins as feasible in complicated mixtures, survey scans (MS) need a robust quantitative readout of ion intensities. Better quantitative precision will invariably come at the expense of decreased proteome coverage, as well as vice versa. Because ion concentration must be especially used in comparison among two samples if the same ion species is used, and because two different LC-MS/MS experiments of the same complex protein extract sample generally get a crossover of roughly 60% just on peptide tier, the above method could outcome in really poor coverage of prevalent ions among two samples. This issue is exacerbated when comparing more than two samples. To circumvent these limitations, a technique known as protein correlation profiling was developed, which aligns the total ion chromatograms of distinct samples. Usually, ion species for which fragment spectra (and consequently peptide sequence determination) have been acquired are associated based on their chromatographic retention durations. Through using relative retention correlation information, peaks which been fragmented in just one of the samples may be recognised inside the survey scan spectra of the remaining tests, even if no fragment spectrum is provided. Extracted ion chromatograms from both samples may thus be utilised in statistical means to understand retention time and precise mass. This approach improves the amount of proteins available for measurement by up to 40%. The main disadvantage of the protein correlation profiling method consists of the computational process of ion chromatogram extraction, and the arrangement of different chromatographic profiles. Insertion of respective "missing" ion chromatograms is significantly more difficult and usually necessitates a significant level of quality check and confirmation when analysed with counting simple spectrum; label-free quantification has emerged as a particularly enticing alternative with the advent of new high-precision mass spectrometers, since improved mass accuracy enhances the accuracy of tracing peptides among samples due to narrower mass-to-charge ratio that characterises each peptide peak (Hoehenwarter et al., 2008). Nevertheless, the strong repeatability of retention time values across multiple LC-MS/MS runs remains critical for accuracy in label-free quantification utilising peptide ion intensities.

A detailed examination of the various label-free peptide-based approaches found that when at least four spectra per protein were utilised for quantification, test variance was less than twice and protein ratios were within 95% confidence intervals (Old et al., 2005). The elimination of highly abundant proteins from either the sample improved repeatability and linearity (Wang et al., 2006). On average, evaluations of spectral counting approaches to quantification methods that are based on abundant supply show that both are well suited to distinguishing protein abundance variations. The spectral counting helped in the detection of additional proteins that changed in abundance (better coverage), whereas quantification relying on peptide ion concentration resulted in improved accuracy of predicted protein ratio (Old et al., 2005). Several recent investigations have demonstrated that both approaches of label-free quantitation are complementary (Wienkoop et al., 2006); therefore, the decision here between the two methods may be dependent on the expertise of the experimental group (Wong et al., 2008). Profiling approaches based on ion concentration were originally used to identify the proteome of a human centrosome (Andersen et al., 2003), then expanded to perform all those protein profiles on mouse liver (Foster et al., 2006), then subsequently utilised to characterise mouse liver peroxisomes (Wiese et al., 2007). Label-free peptide quantification combined with statistical profiling approaches allowed for the definition of protein-protein interactions during pull-down studies above background proteins (Rinner et al., 2007). In plant biology, protein correlation profiling has been used to investigate phosphorylation time patterns in *Arabidopsis* seedlings in response to sucrose administration (Niittyla et al., 2007).

17.11   LABELLING FOR STABLE ISOTOPES

Stable isotope labelling techniques are rooted in the fact that perhaps a peptide labelled with a stable isotope varies solely in mass from an unlabelled peptide but displays the same chemical characteristics during chromatography. The label can be added at different stages of sample preparation. The label is delivered to the entire organism or cell via the growth media in metabolic labelling, whereas in chemical labelling, it is really attached to proteins or tryptic peptides by a chemical reaction. After tryptic digestion, synthetic tagged standardised peptides are introduced to the extract. The fundamental limitation of stable-isotope-based approaches for protein measurement in complex samples is signal interference produced by co-eluting the components of equal mass. As a result, the most effective strategy to improve statistical analysis is to reduce sample intricacy prior to LC-MS/MS analysis by increasing chromatographic gradient durations or by biochemical fractionation.

[^{15}N]-labelling was originally employed in proteomics to analyse protein phosphorylation in bacteria (Oda et al., 1999). The elevated quantitative protein approach that relies on metabolic labelling was created in mammalian cell cultures (SILAC) employing stable-isotope-labelled essential amino acids (Ong et al., 2002). This technique has since been applied to other areas of mammalian signalling biology, including protein-protein interactions (Blagoev et al., 2003; Schulze and Mann, 2004; Selbach and Mann, 2006), protein dynamics (Blagoev et al., 2004; Olsen et al., 2006), the influence of micro-RNA production on global protein levels (Selbach et al., 2008) and small molecule interactions involving proteins (Ong et al., 2009). SILAC also works effectively in bacteria or particular yeast strains that have developed auxotrophy for the tagged amino acid (de Godoy et al., 2008; Gruhler et al., 2008). In plants, SILAC only achieved label incorporation of around 70% (Gruhler et al., 2005), which is insufficient for many global proteomics applications. *Chlamydomonas* is the only organism in the plant kingdom that has been successfully SILAC tagged using auxotrophic mutations (Naumann et al., 2007).

However, because plants are autotrophic organisms, they may be readily metabolically tagged by feeding labelled inorganic chemicals in the form of [^{15}N]-containing salts, as initially established in NMR investigations (Ippel et al., 2004). Full [^{15}N]-labelling in multicellular organisms was initially accomplished in *Drosophila* and *Caenorhabditis elegans* for proteomics applications by feeding the animals [^{15}N]-labelled yeast or bacteria (Krijgsveld et al., 2003). The tagging of plant cell cultures with [^{15}N] for large-scale proteome analysis was independently shown (Engelsberger et al., 2006), and a comprehensive data analysis procedure was devised (Palmblad et al., 2007).

In stable-isotope approaches, quantification is based on extracted ion chromatograms of survey scans including the pair of labelled (heavy) and unlabelled (light) peptide isoforms. Because the isotope label has little effect on the physicochemical characteristics of the peptides, the heavy and light versions frequently co-elute from the chromatographic gradient. A typical peak has a width of 10–30 seconds (250 nL min^{-1}, 75-m ID, 10-cm C18 column). Depending on the operating parameters of the mass spectrometer, one survey scan may be acquired every 1–5 seconds, allowing the sampling of many MS full scan spectra during the time course of the eluting peak. Each of these MS spectra represents a single observation of the peptide pair, and the ratio and standard deviation for each peptide are calculated by averaging the ratios derived from numerous such MS spectra. Based on the collected ion chromatograms, there are two methods for calculating an abundance ratio between heavy and light forms. The ratio between heavy and light forms is computed at each survey scan event over the ion peak in one technique, and the individual ratios are then averaged to get a peptide ratio. Each measured ion will have an average ratio and a standard deviation in this scenario. The alternate approach involves calculating the ratio of the peak volumes over the complete extracted ion chromatogram for the heavy and light forms of the peptide. This approach is less impacted by minor changes in chromatographic elution between heavy and light peaks. Because of the huge number of tagged atoms integrated into the peptide sequence, more isobaric amino acid variants are formed, resulting in greater ambiguity in the sequence-matching process (Nelson et al., 2007). Furthermore, the isotope clusters of [^{15}N]-labelled peptides are wider, particularly for longer peptides. Because most quantitation methods employ the first isotope, the accuracy with which peptides are measured is also affected by peptide length and amino acid composition (Gouw et al., 2008). However, methods for accounting for discrepancies in the isotopic envelope of both heavy and light peptide isoforms have indeed been described (Whitelegge et al., 2004). In general, depending on the instrumentation, the dynamic range across which reliable quantitation is attainable employing [^{15}N]-labelling spans one to two orders of magnitude (Venable et al., 2007). This implies that increases less than tenfold are frequently outside the linear range and will therefore be over or underestimated. A comprehensive evaluation of the use of complete vs. partial labelling indicated that, in general, both the procedures are equivalent in terms of dynamic range and accuracy. While partial labelling makes the automatic identification of labelled and unlabelled peptide pairings more difficult, it allows the measurement of more peptides throughout the whole dynamic range.

17.12 LABELLING OF CHEMICALS

Chemical labelling is similar to metabolic [^{15}N]-labelling in that the label is added to the isolated proteins or peptides by a chemical reaction, such as with sulfhydryl groups or amine groups, or through acetylation or esterification of amino acid residues (Ong and Mann, 2005). The isotope label can also be inserted into the peptide chain during the tryptic digestion enzymatic activity by adding H_2 [^{18}O] to the peptide cleavage sites (Yao et al., 2001). In contrast to [^{15}N] tagging, the mass change between heavy and light peptides is continuous during quantification. It should be emphasised, however, that the mass difference between the heavy and light forms of the peptide should be at least 4 Da in order to clearly differentiate the isotopomer clusters of the heavy and light forms of the peptide. Because the isotopomer cluster grows in size with increasing peptide mass, tiny labels, such as [^{18}O] generated by a tryptic digest in H_2 [^{18}O], become limiting for bigger peptides. The isotope-coded affinity tag (ICAT), which attaches to the sulfhydryl groups of cysteine residues, is a popular chemical isotope marker (Gygi et al., 1999). It is a valuable method for studying the oxidation or reduction state of proteins, but because cysteine is a rare amino acid, the number of peptides that can be tagged and measured by sulfhydryl labels is limited.

It is a valuable method for studying the oxidation or reduction state of proteins, but because cysteine is a rare amino acid, the number of peptides that can be tagged and measured by sulfhydryl labels is limited.

17.13 TAGS FOR ISOBARIC MASSES

Isobaric mass tagging (Thompson et al., 2003) differs from the previous technique in that the mass tags are initially added to generate labelled peptides having the same total mass that co-elute in LC. The distinct mass tags can only be utilised after peptide fragmentation. Due to the fact that each tag adds the same total mass to a particular peptide, each peptide species produces just one peak during LC, even when two or more samples are combined. As a result, there will be only one peak in the survey MS scan, and only one m/z will be separated for fragmentation. When the mass tags are fragmented, they separate. The fragments are in the low-mass region, which is not often covered by peptide fragment ions. As a quantitative readout, the intensity ratio of the various reporter ions is employed.

The fragment spectra, rather than the survey scans, are used for quantification in isobaric mass tagging. As a result, quantitative accuracy is dependent on the isolation width of precursor ions for fragmentation, because all ions separated within that window contribute to fragments in the reporter ion mass ranges. It is also worth noting that in fragment scans, only a single fragment spectrum per peptide is often accessible, but in quantitation based on survey scans, many data points spanning the eluting peptide peak are typically sampled. The commercially available isobaric mass TAGs iTRAQ (Ross et al., 2004) and TMT (Sadowski et al., 2006) are introduced to a protein of relevance at the level of tryptic peptIDES. iTRAQ has been frequently utilised in plant proteomics to compare different time points after elicitor administration to examine phosphoproteomic responses (Jones et al., 2006). Protein breakdown in chloroplasts (Rudella et al., 2006) and alterations in chloroplast proteomes (Kleffmann et al., 2007) were investigated. Maize (Majeran et al., 2008) and *Brassica* (Majeran et al., 2008) chloroplast proteomes were compared in different cell types (Zhu et al., 2009). Organelle proteomes (Dunkley et al., 2006) and endomembranous proteomes (Sadowski et al., 2006) were defined in an elegant research by differential mass tagging of consecutive fractions over continuous sucrose gradients and assigning displays to profiles of known marker proteins.

17.14 STANDARD PEPTIDES

In 1983, the utilisation of stable-isotope-labelled standard peptides was first disclosed (Desiderio and Kai, 1983). However, only lately has increased analytical throughput on current tandem mass spectra enabled the widespread use of synthetic isotope-labelled peptides as an absolute quantification (AQUA) standard (Gerber et al., 2003). Such focused studies of individual proteins over a large variety of samples are highly efficient when combined using multiple selected reaction monitoring (SRM) on a triplequad mass spectrometer. The collection of data on residence time, peptide mass and fragment ion mass provides great specificity to the specific target peptide, and the linear range for quantification is expanded up to five orders of magnitude due to exceptionally low noise levels in the SRM spectra (Kirpatrick et al., 2005).

The concentration of the native peptide in the sample may be estimated by adding known amounts of the labelled standard peptide to the sample. However, because sample preparation stages may result in losses or enrichments that are not addressed by the AQUA approach, the quantity of protein identified by AQUA may not reflect the genuine expression levels of this protein in the tissue.

Standard peptides have been utilised in plant proteomics to track the abundance variations of distinct isoforms of sucrose phosphate synthase in *Arabidopsis* (Lehmann et al., 2008) and *Medicago* root nodules during drought stress (Wienkoop et al., 2008a,b). Target locations for phosphorylation in trehalose phosphate synthase isoforms were investigated using conventional peptides in *in vitro* kinase experiments (Glinski et al., 2003).

17.15   DIFFICULTIES IN DATA PROCESSING

Regardless of the quantitative approach used, quantitative proteomic data are often complicated and of varying quality. The fundamental problem is that even today's most capable mass spectrometers cannot sample and fragment every peptide ion found in complicated materials. As a result, only a fraction of the peptides and proteins found in a sample may be identified (Aebersold and Mann, 2003). Furthermore, due to a number of quality concerns (discussed below), the proportion of detected peptides and proteins that can be quantified is much less. As a result, for the successful interpretation of proteomic datasets, meticulous experimental design comprising phases of protein separation, enrichment and purification is required.

17.16   FROM SPECTRA TO QUANTITATIVE DATA

The process in quantitative MS techniques necessitates the extraction of quantitative information from either a survey scan or fragment spectra, as well as qualitative information for peptide identification from fragment spectra. Most importantly, manual identification and quantification of peptides is necessary. Manual validation allows you to analyse sequence assignment to spectra and assess the quality of the quantitative data in terms of signal-to-noise ratio, the existence of interference peaks and isotope label incorporation. In many situations, the identification and quantification processes are performed independently and are then connected at the level of individual spectra. While advanced protein identification algorithms have been available for some time [Sequest (21), Mascot (91), X!Tandem (14), Omssa (25), InsPect (109)], the development of robust workflows and algorithms to extract quantitative information from multidimensional proteomics experiments using MS has only recently begun. Generic formats, such as mzData, mzXML or pepXML, enable the independent employment of software algorithms to change data without regard for compatibility. A full overview of several publicly accessible softwares for MS data analysis may be obtained elsewhere (Mueller et al., 2008). Intensity ratios of high-quality spectra are often averaged to produce peptide abundance ratios, and peptide abundance ratios are averaged to produce protein abundance ratios. Because many peptide sequences might match more than one protein, assigning peptides to particular proteins takes careful caution. It is consequently critical for proper quantification to include just those peptides that are unique to a certain protein, known as proteotypic peptides. This is significant because various protein isoforms can be variably controlled, causing peptide ratios of conserved peptides to diverge from peptide ratios of proteotypic peptides (Ong and Mann, 2005).

The following factors must be examined in regard to the nature of the biological question posed and the quantitation technique utilised throughout the experimental planning stage, as well as during the extraction of primary data.

a. Several factors influence the quantitative accuracy of peptide ratios in general. Technical factors like instrument resolution, sensitivity and scan speed, as well as LC (peak width), all have a substantial impact on quantitation quality. The adoption of high-resolution and high-mass-accuracy equipment will undoubtedly boost confidence in protein identification, but it will also boost quantitative confidence by allowing for narrower isolation widths and fewer peak interferences on complete scans (Olsen et al., 2004, 2005; Zubarev and Mann, 2007).

b. The quantity of data points available throughout the eluting peak affects quantitative accuracy at the peptide level. More MS spectra boost confidence in the quantitative results. This is true for all quantification techniques that rely on peptide ion intensities.

c. Ion intensities have a significant impact on the quantitative accuracy of single peptides. Peptides with high ion intensities are more precisely measured, whereas peptides with low ion intensities have a significantly wider range of ratio variations. Although the majority of peptide ratios in a sample 1:1 mixture are within the predicted range, outliers in a real biological experiment are typically deemed to be the intriguing candidates and must thus be thoroughly confirmed.

d. Overlap of unrelated signals with isotope clusters either of members of stable isotope peptide pairs or peaks to be examined in label-free protein correlations profiling causes further issues. As a result, the unrelated peak may contribute to the peptide ratio. This issue becomes less obvious when high-resolution sensors are employed to quantify complicated samples.

e. Depending on the size of the peptide, the isotopic envelopes of the unlabelled and tagged versions may overlap. This issue is determined by the mass differential of the included stable-isotope label. Larger peptides are likely to have more overlap, but smaller mass variations between heavy and light labels tend to cause more overlap. Quantitation without labels is unaffected.

f. In stable-isotope labelling procedures, chromatographic separation of heavy and light peptide pairs might result in unique ionisation circumstances for each member of the peptide pair, implying that ion intensities cannot be compared from the same survey scan spectrum. Rather, peak areas must be computed individually for each partner, and the ratio of these summed peak areas of extracted ion chromatograms must be calculated. Chromatographic separation has little effect on isobaric mass tags, which are quantified using a fragment ion scan.

g. Using spectrum count data can result in zero counts of a specific protein in one sample, but it may be found in another. This makes calculating a fold change difficult and results in datasets with missing values. Similarly, if only one partner of a peptide pair is identified (i.e., just the labelled form or only the unlabelled form), no ratio can be established and the true abundance level of the single partner cannot be reliably assessed.

h. The accuracy of the protein ratio is mostly determined by accurate peptide quantification, but it is also affected by the number of quantified peptides per protein and the number of proteotypic peptides for each protein. Protein ratios derived from more than one contributing peptide are more accurate than protein ratios derived from only one peptide. Although the vast majority of proteins are unaffected by this issue, proteins measured based on a single peptide must be extensively validated, as the possibilities of inaccurate quantification are quite high.

i. To address some of these issues, particularly in the case of stable-isotope labelling, reciprocal experimental designs that distinguish treatment effects from both labelling effects and biological variation have been proposed. Prior to sequence assignment, reciprocal studies employing metabolic [^{15}N] labelling were utilised to discover differentially regulated candidate proteins by database search (Wang et al., 2002). A thorough statistical approach for identifying treatment-responsive proteins vs. biological variation in two proteomes has also been created (Kierszniowska et al., 2009a,b). The reciprocal design of [^{15}N] labelling experiments has been used in phosphoproteomic investigations of elicitor administration (Benschop et al., 2007), the discovery of plant sterol-rich regions (Kierszniowska et al., 2009a,b) and proteome alterations in leaf senescence studies (Hebeler et al., 2008).

17.17 STATISTICAL EVALUATION

Experiments employing quantitative proteomics often aim to either analyse changes in protein abundance between a set of treated samples and a control condition or to define protein complexes or particular subproteomes in comparison to background proteins. In most situations, the goal of the data analysis techniques is to define "deviating" proteins that are eventually designated "responsive candidates" in the biological context under examination. When protein complexes or organelle fractions are defined, several proteins with comparable quantitative behaviour are regarded to be members of the same complex or subproteome as opposed to other proteins with variable quantitative behaviour (Andersen et al., 2003; Dunkley et al., 2004, 2006; Foster et al., 2006). Normalisation, which is a fundamental step in quantitative investigations, is generally the first step in primary data processing. It impacts the outcome by correcting for technical issues such as sample mixing problems, inadequate isotope incorporation or changes in ionisation across different LC-MS/MS studies. Total ion counts, total number of spectra or average ratios of the most abundant proteins are frequently employed as the foundation for normalisation, assuming that the majority of proteins will remain stable and hence may be utilised for normalisation. A control combination of labelled and unlabelled untreated samples is ideally employed in stable-isotope-labelling procedures to investigate the biological fluctuation of ratios and mixing mistakes (Kierszniowska et al., 2009a,b). Log-transformation of ratios is a frequent step used to align variances. Following that, the log-converted data are utilised in exploratory plots that display the average log abundance on the x-axis and the log fold change across conditions on the y-axis. Using such charts, linear and nonlinear biases in data may be recognised, allowing normalisation processes to be benchmarked.

17.18 TREATING CONTINUOUS DATA

Scaling the individual log intensity values using the global median value or the average derived from a selection of proteins (Krijgsveld et al., 2003), for example, might occasionally offer sufficient normalisation. Other potentially more powerful approaches, such as quantile normalisation (Callister et al., 2006; Higgs et al., 2005), variance stabilisation (Kreil et al., 2004) or a "spectral index" incorporating several properties of each data point, have been benchmarked for label-free proteomics (Kultima et al., 2009; Griffin et al., 2010).

Alternatively, data can be standardised by using a locally weighted scatter plot smoothing process (Xia et al., 2007), which efficiently removes biases that are dependent on measurement values such as ion intensities. However, systematic bias caused by analysis order (Kultima et al., 2009) or data (Karpievitch et al., 2009) should be reduced by the experimental design (Hu et al., 2005).

After the datasets have been normalised, the next step is to detect differences between the different situations. Statistical tests such as the student's t-test and the Wilcoxon rank sum test can be used to examine the datasets (Benschop et al., 2007). Tests that account for experimental confidence in quantification should give higher statistical power for small sample numbers (Kierszniowska et al., 2009a,b). When comparing more than two groups, it is typically better to employ ANOVA-type analyses, many of which may be conducted using the free R programme.

17.19 HANDLING DISCRETE DATA

Because spectral count data is similar to data from serial analysis of gene expression (SAGE), statistical methods and concepts established for SAGE can be applied. However, proteomics-specific issues must be considered, such as the fact that the quantity of visible peptides is not the same for all proteins or that not all peptides can be identified with the same level of confidence. Such effects can be compensated for in continuous data by weighting the influence of peptide ions by their confidence (Cox and Mann, 2008; Li et al., 2003).

Methods for assessing count data that have been employed in the proteomics area include the goodness of fit (G-test), Fisher's exact test and the AC test, all of which performed quite similarly (Zhang et al., 2006). The G-test has been expanded to include more than two criteria (Zhang et al., 2006). Because these tests are designed to take into consideration the entire number of spectra obtained every run, this parameter does not require normalisation. Extending beyond these count-based statistical techniques usually necessitates normalisation that particularly attempts to reduce the technical implications of spectral count data. Methods that consider detection confidence, protein length or the number of observable peptides (for example, the emPAI and APEX values given above) are appropriate. Using these derivations provides values that, when compared across tests, behave quite similarly to the data. The standard deviation is affected by protein abundance, and this tendency may be explicitly accounted for in statistical analysis when using R (Pavelka et al., 2008). More advanced models, which take into account potentially biassing parameters such as protein length and general count abundance, can also be used (Choi et al., 2008).

17.20 GENERAL PROBLEMS

The major difficulty in comparative proteomics so far has been accounting for missing data due to insufficient proteome coverage in peptide fragmentation. The random sequencing of peptides through a mass spectrometer does not allow the fragmentation of every peptide in a sample. Low-abundance peptides and peptides with low ionisation efficiency, in particular, are likely to be fragmented in only a few of the numerous samples examined, resulting in missing quantitative results. Statistical procedures, on the other hand, frequently need entire datasets. Estimating missing data is a typical workaround; however, this has an impact on statistical assessment. The number of proteins that may be statistically analysed is greatly reduced when incomplete datasets are excluded from the study. Improving peptide separation and fractionation, as well as improving chromatography, has an impact on quantitative coverage (Eriksson and Fenyö, 2007). Finally, regardless of how the data are created, contemporary proteomics methods generate vast volumes of data from a single experiment, necessitating the need for many statistical analyses. Correcting p-values for multiple testing requires employing family-wise error rate methods or false discovery rate methodologies. When working with proteomics data, the latter sort of correction approach is frequently preferred.

17.21 ANALYSIS OF THE OUTCOMES

When several measurements from biological samples are obtained, insights can be gleaned by subjecting the data to more specialised analysis and visualisation processes. Data clustering by treatment frequently exposes global patterns or may aid in identifying variables. There are other clustering techniques available; however, either hierarchical (Baerenfaller et al., 2008; Hu, 2005) or k-means clustering (Rinner et al., 2007) is often utilised, possibly because these algorithms are supported by many software applications. Proteins can also be classified by clustering based on their behaviour in several tests or over time.

Often, either the groups of proteins in a cluster or all differentially expressed proteins are classified and statistically analysed for the enrichment of biological categories (Baerenfaller et al., 2008; Larrainzar et al., 2007; Nelson et al., 2006). Many tools are accessible to the plant community for these analyses, including TAIR's GO thin classification; the Classification SuperViewer (Provart and Zhu, 2003) for *Arabidopsis*; and PageMan (Usadel et al., 2006), which are available for many plant species. These strategies aid in the evaluation and categorisation of novel protein activities.

Principal component analysis, which aims to retain the maximum variance in a low-dimensional space, and independent component analysis (Larrainzar et al., 2007), which attempts to discover separate components in the dataset, are often the used procedures for projecting data into a 2D environment. The projected data typically gives a visible distinction between conditions such as stressed and unstressed plants, several genotypes or various treatments (Wienkoop et al., 2008a,b). When several replicates from various plant samples are available, proteins specific for a certain condition/tissue can be discovered by data inspection or machine learning (Baerenfaller et al., 2008).

17.22   PERSPECTIVES

Although proteomics has been able to provide a full perspective of quantitative changes in subproteomes such as organelles for some time, proteomics datasets are often substantially smaller than microarray datasets and are biassed against lower abundance proteins. Global abundance studies in *Saccharomyces cerevisiae* indicated a bell-shaped distribution of proteins covering about six orders of magnitude in abundance (Ghaemmaghami et al., 2003), but contemporary LC-MS/MS techniques can only cover around three to four orders of magnitude in complex samples (de Godoy et al., 2006). Most notably, the robust and repeated detection of low-abundance proteins across replicates is still far from common. The fundamental drawback in quantitative proteomics is thus the poor proteome coverage between different samples, resulting in datasets with missing results in many studies.

These restrictions provide important challenges for the researcher, as only usable proteome fractionation and intelligent strategies for protein target enrichment can maximise coverage of the relevant proteome. Improvements in scan, speed and sensitivity of mass spectrometer, as well as new breakthroughs in chromatographic separation, will almost certainly contribute to wider proteome coverage in the future. Notwithstanding these constraints, differential proteomics approaches are becoming more common in plant biology. Most researchers now use simple pair-wise comparisons of tissue types or treatments, but more complicated experimental designs will develop in the future. The application of data-mining technologies to plant biology will be the difficulty in large-scale quantitative proteomics investigations. Well-designed experiments and hypotheses, along with high-quality MS, are anticipated to make a considerable contribution to our understanding of protein function in plant growth and development, both at the global and in-depth levels.

Some of the points that are noteworthy include:

1. Comparative proteomic techniques based on stable-isotope-labelled quantification provide the best accuracy and lowest relative standard deviations. However, increasing sample complexity may compromise proteome coverage.
2. Label-free quantitative techniques enable low-cost comparison of a large number of samples or conditions. Because quantitative precision is lower than in stable-isotope techniques, this strategy is particularly appealing for projected significant differences.
3. Full scan quantification provides exact statistics over many scans of the chromatographic peak of each peptide ion.
4. Noise levels in fragment ion quantification are often low. However, only one scan per peptide is frequently accessible.
5. The amount of peptides contributing to protein quantification increases its accuracy. Protein quantification, based on a single peptide, must be thoroughly validated.
6. Statistical treatment of the data is required, and despite the intricacy of the data, various tools are available for data normalisation and differential expression identification.

17.23 CONCLUSION AND FUTURE PERSPECTIVES

The extensive identification of proteins expressed in a cell or tissue was the original goal of proteomics. Studies of functional characteristics of proteins, which are frequently controlled by PTMs of proteins, have been included in the scope of the objective. To quickly identify PTMs worldwide and their modification locations, a variety of approaches have been developed and put to use. Profiling PTM dynamic changes is still a long-term endeavour, but it will eventually provide the groundwork for a thorough knowledge of the signalling networks controlling intricate biological processes. Advanced proteomics technologies, in particular advancements in detection and selective isolation procedures and MS equipment, have considerable potential. However, there is a need for improved proteome coverage as a result of speedier instrumentation and the widespread use of peptide and protein separation methods during sample preparation. Within the proteomic community, we will need to define quality requirements for quantitative analysis. Good experimental design based on particular biological hypotheses will be critical to our functional knowledge of protein function in plant growth and development regulatory mechanisms. For improved biological interpretation of large-scale proteomics datasets, further development of statistical methodologies and data-mining procedures in conjunction with modelling approaches is required.

REFERENCES

Aebersold, R., & Mann, M. (2003). Mass spectrometry-based proteomics. *Nature, 422*(6928), 198–207.

Andersen, J. S., Lyon, C. E., Fox, A. H., Leung, A. K., Lam, Y. W., Steen, H., ... & Lamond, A. I. (2002). Directed proteomic analysis of the human nucleolus. *Current Biology, 12*(1), 1–11.

Andersen, J. S., Wilkinson, C. J., Mayor, T., Mortensen, P., Nigg, E. A., & Mann, M. (2003). Proteomic characterization of the human centrosome by protein correlation profiling. *Nature, 426*(6966), 570–574.

Asara, J. M., Christofk, H. R., Freimark, L. M., & Cantley, L. C. (2008). A label-free quantification method by MS/MS TIC compared to SILAC and spectral counting in a proteomics screen. *Proteomics, 8*(5), 994–999.

Astoul, E., Laurence, A. D., Totty, N., Beer, S., Alexander, D. R., & Cantrell, D. A. (2003). Approaches to define antigen receptor-induced serine kinase signal transduction pathways. *Journal of Biological Chemistry, 278*, 9267–9275.

Baerenfaller, K., Grossmann, J., Grobei, M. A., Hull, R., Hirsch-Hoffmann, M., Yalovsky, S., ... & Baginsky, S. (2008). Genome-scale proteomics reveals Arabidopsis thaliana gene models and proteome dynamics. *Science, 320*(5878), 938–941.

Benschop, J. J., Mohammed, S., O'Flaherty, M., Heck, A. J., Slijper, M., & Menke, F. L. (2007). Quantitative phosphoproteomics of early elicitor signaling in Arabidopsis. *Molecular & Cellular Proteomics, 6*(7), 1198–1214.

Blagoev, B., Kratchmarova, I., Ong, S. E., Nielsen, M., Foster, L. J., & Mann, M. (2003). A proteomics strategy to elucidate functional protein-protein interactions applied to EGF signaling. *Nature Biotechnology, 21*(3), 315–318.

Blagoev, B., Ong, S. E., Kratchmarova, I., & Mann, M. (2004). Temporal analysis of phosphotyrosine-dependent signaling networks by quantitative proteomics. *Nature Biotechnology, 22*(9), 1139–1145.

Borner, G. H., Lilley, K. S., Stevens, T. J., & Dupree, P. (2003). Identification of glycosylphosphatidylinositol-anchored proteins in Arabidopsis. A proteomic and genomic analysis. *Plant Physiology, 132*, 568–577.

Borner, G. H., Sherrier, D. J., Stevens, T. J., Arkin, I. T., & Dupree, P. (2002). Prediction of glycosylphosphatidylinositol-anchored proteins in Arabidopsis. A genomic analysis. *Plant Physiology, 129*, 486–499.

Brennan, J. P., Miller, J. I. A., Fuller, W., Wait, R., Begum, S., Dunn, M. J., & Eaton, P. (2006). The utility of N,N-biotinyl glutathione disulfide in the study of protein S-glutathiolation. Molecular and Cellular Proteomics, 5(2), 215–225. https://doi. org/10.1074/mcp.M500212-MCP200

Callister, S. J., Barry, R. C., Adkins, J. N., Johnson, E. T., Qian, W. J., Webb-Robertson, B. J. M., ... & Lipton, M. S. (2006). Normalization approaches for removing systematic biases associated with mass spectrometry and label-free proteomics. *Journal of Proteome Research, 5*(2), 277–286.

Carlberg, I., Hansson, M., Kieselbach, T., Schroder, W. P., Andersson, B., & Vener, A. V. (2003). A novel plant protein undergoing light-induced phosphorylation and release from the photosynthetic thylakoid membranes. Proceedings of the National Academy of Sciences, USA 100, 757–762.

Chatterjee, S., & Mayor, S. (2001). The GPI-anchor and protein sorting. *Cellular and Molecular Life Sciences*, *58*, 1969–1987.

Choi, H., Fermin, D., & Nesvizhskii, A. I. (2008). Significance analysis of spectral count data in label-free shotgun proteomics. *Molecular & Cellular Proteomics*, *7*(12), 2373–2385.

Cox, J., & Mann, M. (2008). MaxQuant enables high peptide identification rates, individualized ppb-range mass accuracies and proteome-wide protein quantification. *Nature Biotechnology*, *26*(12), 1367–1372.

Cvetkovska, M., Rampitsch, C., Bykova, N., & Xing, T. (2006). GeNOmic analysis of MAP kinase cascades in Arabidopsis defense responses. *Plant Molecular Biology of Reproduction*, *23*, 331–343.

de Godoy, L. M. F., Olsen, J. V., Cox, J., Nielsen, M. L., Hubner, N. C., Fröhlich, F., ... & Mann, M. (2008). Comprehensive mass spectrometry-based proteome quantification of haploid versus diploid yeast. *Nature* 455:1251–55

de Godoy, L. M. F., Olsen, J. V., de Souza, G. A., Li, G., Mortensen, P., & Mann, M. (2006). Status of complete proteome analysis by mass spectrometry: SILAC labeled yeast as a model system. *Genome Biology*, *7*, R50.

Debruyne, I. (1983). Staining of alkali-labile phosphoproteins and alkaline phosphatases on polyacrylamide gels. *Analytical Bio-Chemistry*, *133*, 110–115.

del Riego, G., Casano, L. M., Martín, M., & Sabater, B. (2006). Multiple phosphorylation sites in the beta subunit of thylakoid ATP synthase. *Photosynthesis Research*, *89*(1), 11–18. https://doi.org/10.1007/s11120-006-9078-4

Deng, Z., Zhang, X., Tang, W., Oses-Prieto, J. A., Suzuki, N., Gendron, J.M., ...& Wang, Z.-Y. (2007). A proteomic study of brassinosteroid response in Arabidopsis. *Molecular & Cellular Proteomics* 6:2058–71

Denison, C., Kirkpatrick, D. S., & Gygi, S. P. (2005). Proteomic insights into ubiquitin and ubiquitin-like proteins. *Current Opinion in Chemical Biology*, *9*, 69–75.

Desiderio, D. M., & Kai, M. (1983). Preparation of stable isotope-incorporated peptide internal standards for field desorption mass spectrometric C-terminal sequencing technique. *Biomedical Mass Spectrometry*, *10*, 471–479.

Devoto, A., Muskett, P. R., & Shirasu, K. (2003). Role of ubiquitination in the regulation of plant defence against pathogens. *Current Opinion in Plant Biology*, *6*, 307–311.

Dunkley, T. P., Hester, S., Shadforth, I. P., Runions, J., Weimar, T., Hanton, S. L., ... & Lilley, K. S. (2006). Mapping the Arabidopsis organelle proteome. *Proceedings of the National Academy of Sciences of the United States of America* 103:6518–6523

Dunkley, T. P., Watson, R., Griffin, J. L., Dupree, P., & Lilley, K. S. (2004). Localization of organelle proteins by isotope tagging (LOPIT). *Molecular & Cellular Proteomics*, *3*, 1128–34.

Elortza, F., Nühse, T. S., Foster, L. J., Stensballe, A., Peck, S. C., & Jensen, O. N. (2003). Proteomic analysis of glycosylphosphatidylinositol-anchored membrane proteins. *Molecular and Cellular Proteomics*, *2*, 1261–1270.

Engelsberger, W. R., Erban, A., Kopka, J., & Schulze, W. X. (2006). Metabolic labeling of plant cell cultures with K15NO3 as a tool for quantitative analysis of proteins and metabolites. *Plant Methods*, *2*(1), 1–11.

Eng, J. K., McCormack, A. L., & Yates, J. R. (1994). An approach to correlate tandem mass spectral data of peptides with amino acid sequences in a protein database. *Journal of the American Society for Mass Spectrometry*, *5*(11), 976–989.

Eriksson, J., & Fenyö, D. (2007). Improving the success rate of proteome analysis by modeling protein-abundance distributions and experimental designs. *Nature Biotechnology*, *25*, 651–55.

Foster, L. J., de Hoog, C. L., Zhang, Y., Zhang, Y., Xie, X., Mootha, V. K., & Mann, M. (2006). A mammalian organelle map by protein correlation profiling. *Cell*, *125*(1), 187–199. https://doi.org/10.1016/j.cell.2006.03.022

Geer, L. Y., Markey, S. P., Kowalak, J. A., Wagner, L., Xu, M., Maynard, D. M., ... & Bryant, S. H. (2004). Open mass spectrometry search algorithm. *Journal of Proteome Research* 3:958–64

Gerber, I. B., Laukens, K., Witters, E., & Dubery, I. A. (2006). Lipopo-lysaccharide-responsive phosphoproteins in Nicotiana tabacumcells. *Plant Physiology and Biochemistry*, *44*(5–6), 369–379. https://doi.org/10.1016/j.plaphy.2006.06.015

Gerber, S. A., Rush, J., Stemman, O., Kirschner, M. W., & Gygi, S. P. (2003). Absolute quantification of proteins and phosphoproteins from cell lysates by tandem MS. *Proceedings of the National Academy of Sciences*, *100*(12), 6940–6945.

Ghaemmaghami, S., Huh, W. K., Bower, K., Howson, R. W., Belle, A., Dephoure, N., ... & Weismann, J. S. (2003). Global analysis of protein expression in yeast. *Nature* 425:737–41

Ghesquiere, B., Van Damme, J., Martens, L., Vandekerckhove, J., & Gevaert, K. (2006). Proteome-wide characterization of N-glycosylation events by diagonal chromatography. *Journal of Proteome Research*, *5*, 2438–2447.

Glickman, M. H., & Ciechanover, A. (2002). The ubiquitin-proteasome proteolytic pathway: Destruction for the sake of construction. *Physiological Reviews, 82*, 373–428.

Glinski, M., Romeis, T., Witte, C. P., Wienkoop, S., & Weckwerth, W. (2003). Stable isotope labeling of phosphopeptides for multiparallel kinase target analysis and identification of phosphorylation sites. *Rapid Communications in Mass Spectrometry, 17*, 1579–84.

Gouw, J. W., Tops, B. B. J., Mortensen, P., Heck, A. J., & Krijgsveld, J. (2008). Optimizing identification and quantitation of 15 N-labeled proteins in comparative proteomics. *Analytical Chemistry, 80*, 7796–80331.

Granlund, I., Hall, M., Kieselbach, T., & Schröder, W. P. (2009). Light induced changes in protein expression and uniform regulation of transcription in the thylakoid lumen of Arabidopsis thaliana. *PLoS One, 4*(5), e5649.

Griffin, N. M., Jingui, Y., Long, F., Oh, P., Shore, S., Li, Y., … & Schnitzer, J. E. (2010). Label-free, normalized quantification of complex mass spectrometry data for proteomic analyses. *Nature Biotechnology.* 28:83–91

Gronborg, M., Kristiansen, T. Z., Stensballe, A., Andersen, J. S., Ohara, O., Mann, M., Jensen, O. N., & Pandey, A. (2002). A mass spectrometry-based proteomic approach for identification of serine/threonine-phosphorylated proteins by enrichment with phospho-specific antibodies: Identification of a novel protein, Frigg, as a protein kinase a substrate. *Molecular and Cellular Proteomics 1*, 517–527.

Gruhler, A., Schulze, W. X., Matthiesen, R., Mann, M., & Jensen, O. N. (2005). Stable isotope labeling of Arabidopsis thaliana cells and quantitative proteomics by mass spectrometry* S. *Molecular & Cellular Proteomics, 4*(11), 1697–1709.

Gygi, S. P., Corthals, G. L., Zhang, Y., Rochon, Y., & Aebersold, R. (2000). Evaluation of two-dimensional gel electrophoresis-based proteome analysis technology. *Proceedings of the National Academy of Sciences, 97*(17), 9390–9395.

Gygi, S. P., Rist, B., Gerber, S. A., Turecek, F., Gelb, M. H., & Aebersold, R. (1999). Quantitative analysis of complex protein mixtures using isotope-coded affinity tags. *Nature Biotechnology, 17*(10), 994–999.

Hansson, M., & Vener, A. V. (2003). Identification of three previously unknown in vivo protein phosphorylation sites in thylakoid membranes of Arabidopsis thaliana. *Molecular and Cellular Proteomics, 2*, 550–559.

Hebeler, R., Oeljeklaus, S., Reidegeld, K. A., Eisenacher, M., Stephan, C., Sitek, B., … & Bettina, W. (2008). Study of early leaf senescence in *Arabidopsis thaliana* by quantitative proteomics using reciprocal 14 N/15 N labeling and difference gel electrophoresis. *Molecular & Cellular Proteomics* 7:108–120

Heinemeyer, J., Scheibe, B., Schmitz, U. K., & Braun, H. P. (2009). Blue native DIGE as a tool for comparative analyses of protein complexes. *Journal of Proteomics, 72*, 539–44.

Higgs, R. E., Knierman, M. D., Gelfanova, V., Butler, J. P., & Hale, J. E. (2005). Comprehensive label-free method for the relative quantification of proteins from biological samples. *Journal of Proteome Research, 4*, 1442–50.

Hoehenwarter, W., van Dongen, J. T., Wienkoop, S., Steinfath, M., Hummel, J., Erban, A., …& Weckwerth, W. (2008). A rapid approach for phenotype-screening and database independent detection of cSNP/protein polymorphism using mass accuracy precursor alignment. *Proteomics, 8*(20), 4214–4225.

Hooper, N. M., Low, M. G., & Turner, A. J. (1987). Renal dipeptidase is one of the membrane proteins released by phosphatidylinositol-specific phospholipase C. *Biochemical Journal, 244*, 465–469.

Hu, J., Coombes, K. R., Morris, J. S., & Baggerly, K. A. (2005). The importance of experimental design in proteomic mass spectrometry experiments: Some cautionary tales. *Briefings in Functional Genomics, 3*(4), 322–331.

Ippel, J. H., Pouvreau, L., Kroef, T., Gruppen, H., Versteeg, G., van den Putten, P., … & van Mierlo, C. P. (2004). In vivo uniform 15N-isotope labelling of plants: Using the greenhouse for structural proteomics. *Proteomics, 4*(1), 226–234.

Ishihama, Y., Oda, Y., Tabata, T., Sato, T., Nagasu, T., Rappsilber, J., & Mann, M. (2005). Exponentially modified protein abundance index (emPAI) for estimation of absolute protein amount in proteomics by the number of sequenced peptides per proteins. *Molecular & Cellular Proteomics, 4*(9), 1265–1272.

Ito, J., Heazlewood, J. L., & Millar, A. H. (2007). The plant mitochondrial proteome and the challenge of defining the post-translational modifications responsible for signalling and stress effects on respiratory functions. *Physiologia Plantarum, 129*(1), 207–224. https://doi.org/10.1111/j.1399-3054.2006.00795.x

Jensen, O. N. (2004). Modification-specific proteomics: Characterization of post-translational modifications by mass spectrometry. *Current Opinion in Chemical Biology, 8*, 33–41.

Jones, A. M., Bennett, M. H., Mansfield, J. W., & Grant, M. (2006). Analysis of the defence phosphoproteome of *Arabidopsis thaliana* using differential mass tagging. *Proteomics, 6*, 4155–6549.

Jones, A. M. E., Thomas, V., Bennett, M. H., Mansfield, J., & Grant, M. (2007). Modifications to the Arabidopsis defense proteome occur prior to significant transcriptional change in response to inoculation with *Pseudomonas syringae*. *Plant Physiology, 142*, 1603–1620.

Jung, Y. H., Lee, J. H., Agrawal, G. K., Rakwal, R., Kim, J. A., Shim, J. K., ... & Jwa, N. S. (2006). Differential expression of defense/stress-related marker proteins in leaves of a unique rice blast lesion mimic mutant (blm). *Journal of Proteome Research, 5*, 2547–2553

Karpievitch, Y. V., Taverner, T., Adkins, J. N., Callister, S. J., Anderson, G. A.,, Smith, R.D., Dabney, A R. (2009). Normalization of peak intensities in bottom-up MS-based proteomics using singular value decomposition. *Bioinformatics* 25:2573–8050.

Kierszniowska, S., Seiwert, B., & Schulze, W. X. (2009a). Definition of Arabidopsis sterol-rich membrane microdomains by differential treatment with methyl-ß-cyclodextrin and quantitative proteomics. *Molecular & Cellular Proteomics, 8*, 612–2351.

Kierszniowska, S., Walther, D., & Schulze, W. X. (2009b). Ratio-dependent significance thresholds in reciprocal 15N-labeling experiments as a robust tool in detection candidate proteins responding to biological treatment. *Proteomics, 9*, 19162452.

Kirpatrick, D. S., Gerber, S. A., & Gygi, S. P. (2005). The absolute quantification strategy: A general procedure for the quantification of proteins and post-translational modifications. *Methods, 35*, 26573.

Kleffmann, T., von Zychlinski, A., Russenberger, D., Hirsch-Hoffmann, M., Gehrig, P., Gruissem, W., & Baginsky, S. (2007). Proteome dynamics during plastid differentiation in rice. *Plant Physiology, 143*(2), 912–923.

Kreil, D. P., Karp, N. A., & Lilley, K. S. (2004). DNA microarray normalization methods can remove bias from differential protein expression analysis of 2D difference gel electrophoresis results. *Bioinformatics, 20*(13), 2026–2034.

Krijgsveld, J., Ketting, R. F., Mahmoudi, T., Johansen, J., Artal-Sanz, M., Verrijzer, C. P., ... & Heck, A. J. R. (2003). Metabolic labeling of *C. elegans* and *D. melanogaster* for quantitative proteomics. *Nature Biotechnology*. 8:927–931

Kultima, K., Nilsson, A., Scholz, B., Rossbach, U. L., Fa¨lth, M., & Andre´n, P. E. (2009). Development and evaluation of normalization methods for label-free relative quantification of endogenous peptides. *Molecular & Cellular Proteomics, 8*, 2285–2295.

Larrainzar, E., Wienkoop, S., Weckwerth, W., Ladrera, R., Arrese-Igor, C., & Gonzalez, E. M. (2007). *Medicago truncatula* root nodule proteome analysis reveals differential plant and bacteroid responses to drought stress. *Plant Physiology, 144*, 1495–507.

Laugesen, S., Messinese, E., Hem, S., Pichereaux, C., Grat, S., Ranjeva, R., Rossignol, M., & Bono, J. J. (2006). Phosphoproteins analysis in plants: A proteomic approach. *Phytochemistry, 67*(20), 2208–2214. https://doi.org/10.1016/j.phytochem.2006.07.010

Lehmann, U., Wienkoop, S., & Weckwerth, W. (2008). If the antibody fails—A mass Western approach. *The Plant Journal, 55*, 1039–1046.

Li, B., Takahashi, D., Kawamura, Y., & Uemura, M. (2018). Plasma membrane proteomics of Arabidopsis suspension-cultured cells associated with growth phase using nano-LC-MS/MS. In *Plant Membrane Proteomics* (pp. 185–194). Humana Press, New York, NY.

Liu, Y., Lamkemeyer, T., Jakob, A., & Mi, G. H., Zhang, F., Nordheim, A., & Hochholdinger, F.(2006). Comparative proteome analyses of maize (*Zea mays* L.) primary roots prior to lateral root initiation reveal differential protein expression in the lateral root initiation mutant rum1. *Proteomics, 6*, 4300–4308.

Liu, H., Sadygov, R. G., & Yates, J. R. I. (2004). A model for random sampling and estimation of relative protein abundance in shotgun proteomics. *Analytical Chemistry, 76*, 4193–4201.

Li, X. J., Zhang, H., Ranish, J. A., & Aebersold, R. (2003). Automated statistical analysis of protein abundance ratios from data generated by stable-isotope dilution and tandem mass spectrometry. *Analytical Chemistry, 75*(23), 6648–6657.

Lunn, J. E. (2007). Compartmentation in plant metabolism. *Journal of Experimental Botany, 1*, 35–47.

Luo, L., & Wirth, P. J. (1993). Consecutive silver staining and autoradiography of 35S and 32P-labeled cellular proteins: Application for the analysis of signal transducing pathways. *Electrophoresis, 14*, 127–136.

Lu, P., Vogel, C., Wang, R., Yao, X., & Marcotte, E. M. (2007). Absolute protein expression profiling estimates the relative contributions of transcriptional and translational regulation. *Nature Biotechnology, 25*(1), 117–124.

Majeran, W., Cai, Y., Sun, Q., & van Wijk, K. J. (2005). Functional differentiation of bundle sheath and mesophyll maize chloroplasts determined by comparative proteomics. *The Plant Cell, 17*(11), 3111–3140.

Majeran, W., Zybailov, B., Ytterberg, A. J., Dunsmore, J., Sun, Q., & Van Wijk, K. J. (2008). Consequences of C4 differentiation for chloroplast membrane proteomes in maize mesophyll and bundle sheath cells. *Molecular & Cellular Proteomics, 7*, 1609–1638.

Mann, M., & Jensen, O. N. (2003). Proteomic analysis of post-translational modifications. *Nature Biotechnology, 21*, 255–261.

Maor, R., Jones, A., Nühse, T. S., Studholme, D. J., Peck, S. C., & Shirasu, K. (2007). Multidimensional protein identification technology (Mud-PIT) analysis of ubiquitinated proteins in plants. *Molecular and Cellular Proteomics, 6*(4), 601–610. https://doi.org/10.1074/mcp.M600408-MCP200

McAlister, G. C., Phanstiel, D., Good, D. M., Berggren, W. T., & Coon, J. J. (2007). Implementation of electron-transfer dissociation on a hybrid linear ion trap–orbitrap mass spectrometer. *Analytical Chemistry, 79*(10), 3525–3534.

Moon, J., Parry, G., & Estelle, M. (2004). The ubiquitin-proteasome pathway and plant development. *The Plant Cell, 16*, 3181–3195.

Mueller, L. N., Brusniak, M.-Y., Mani, D. R., & Aebersold, R. (2008). An assessment of software solutions for the analysis of mass spectrometry based quantitative proteomics data. *Journal of Proteome Research, 7*, 51–61.

Naumann, B., Busch, A., Allmer, J., Ostendorf, E., Zeller, M., Kirchhoff, H., & Hippler, M. (2007). Comparative quantitative proteomics to investigate the remodeling of bioenergetic pathways under iron deficiency in *Chlamydomonas reinhardtii. Proteomics, 7*(21), 3964–3979.

Nelson, C. J., Hegeman, A. D., Harms, A. C., & Sussman, M. R. (2006). A quantitative analysis of Arabidopsis plasma membrane using trypsin-catalyzed 18O Labeling* S. *Molecular & Cellular Proteomics, 5*(8), 1382–1395.

Nelson, C. J., Huttlin, E. L., Hegeman, A. D., Harms, A. C., & Sussman, M. R. (2007). Implications of 15N-metabolic labeling for automated peptide identification in Arabidopsis thaliana. *Proteomics, 7*(8), 1279–1292.

Nelson, T., Tausta, S. L., Gandotra, N., & Liu, T. (2006). Laser micro-dissection of plant tissue: What you see is what you get. *Annual Review of Plant Biology, 57*, 181–201. https://doi.org/10.1146/annurev.arplant.56.032604.144138

Niittyla, T., Fuglsang, A. T., Palmgren, M. G., Frommer, W. B., & Schulze, W. X. (2007). Temporal analysis of sucrose-induced phosphorylation changes in plasma membrane proteins of Arabidopsis. *Molecular & Cellular Proteomics, 6*(10), 1711–1726.

Nowaczyk, M. M., Hebeler, R., Schlodder, E., Meyer, H. E., Warscheid, B., & Rögner, M. (2006). Psb27, a cyanobacterial lipoprotein, is involved in the repair cycle of photosystem II. *The Plant Cell, 18*(11), 3121–3131.

Nühse, T. S., Stensballe, A., Jensen, O. N., & Peck, S. C. (2003). Large-scale analysis of in vivo phosphorylated membrane proteins by immobilized metal ion affinity chromatography and mass spectrometry. *Molecular and Cellular Proteomics, 2*, 1234–1243.

Nühse, T. S., Stensballe, A., Jensen, O. N., & Peck, S. C. (2004). Phosphoproteomics of the Arabidopsis plasma membrane and a new phosphorylation site database. *The Plant Cell, 16*, 2394–2405.

Oda, Y., Huang, K., Cross, F. R., Cowburn, D., & Chait, B. T. (1999). Accurate quantitation of protein expression and site-specific phosphorylation. *Proceedings of the National Academy of Sciences of the United States of America, 96*, 6591–96.

Old, W. M., Meyer-Arendt, K., Aveline-Wolf, L., Pierce, K. G., Mendoza, A., Sevinsky, J. R., ... & Ahn, N. G. (2005). Comparison of label-free methods for quantifying human proteins by shotgun proteomics. *Molecular & Cellular Proteomics* 4:1487–502

Olsen, J. V., Blagoev, B., Gnad, F., Macek, B., Kumar, C., Mortensen, P., & Mann, M. (2006). Global, in vivo, and site-specific phosphorylation dynamics in signaling networks. *Cell, 127*(3), 635–648. https://doi.org/10.1016/j.cell.2006.09.026

Olsen, J. V., de Godoy, L. M. F., Li, G., Macek, B., Mortensen, P., Reinhold, P., ... & Mann, M. (2005). Parts per million mass accuracy on an orbitrap mass spectrometer via lock mass injection into a C-trap. *Molecular & Cellular Proteomics* 4:2010–2021

Olsen, J. V., Ong, S. E., & Mann, M. (2004). Trypsin cleaves exclusively C-terminal to arginine and lysine residues. *Molecular & Cellular Proteomics, 3*, 608–614.

Ong, S.-E., Blagoev, B., Kratchmarova, I., Kristensen, D. B., Steen, H., Pandey, A., & Mann, M. (2002). Stable isotope labeling by amino acids in cell culture, SILAC, as a simple and accurate approach to expression proteomics. *Molecular & Cellular Proteomics* 1:376–386

Ong, S.-E., Kratchmarova, I., & Mann, M. (2003). Properties of 13 C-substituted arginine in stable isotope labeling by amino acids in cell culture (SILAC). *Journal of Proteome Research, 2*, 173–181.

Ong, S.-E., & Mann, M. (2005). Mass spectrometry-based proteomics turns quantitative. *Nature Chemical Biology, 1*, 252–262.

Ong, S.-E., Schenone, M., Margolin, A. A., Li, X., & Do, K., Doud, M.K., ... & Carr, S.A. (2009). Identifying the proteins to which small-molecule probes and drugs bind in cells. *Proceedings of the National Academy of Sciences of the United States of America* 106:4617–4622.

Palmblad, M., Bindschedler, L. V., & Cramer, R. (2007). Quantitative proteomics using uniform 15N-labeling, MASCOT, and the trans-proteomic pipeline. *Proteomics, 7*, 3462–3469.

Palmblad, M., Mills, D. J., & Bindschedler, L. V. (2008). Heat-shock response in *Arabidopsis thaliana* explored by multiplexed quantitative proteomics using differential metabolic labeling. *Journal of Proteome Research, 7*, 780–785.

Patton, W. F. (2002). Detection technologies in proteome analysis. *Journal of Chromatography B: Analytical Technologies in the Biomedical and Life Sciences, 771*, 3–31.

Pavelka, N., Fournier, M. L., Swanson, S. K., Pelizzola, M., Ricciardi-Castagnoli, P., Florens, L., & Washburn, M. P. (2008). Statistical similarities between transcriptomics and quantitative shotgun proteomics data. *Molecular & Cellular Proteomics, 7*(4), 631–644.

Peck, S. C., Nühse, T. S., Hess, D., Iglesias, A., Meins, F., & Boller, T. (2001). Directed proteomics identifies a plant-specific protein rapidly phosphorylated in response to bacterial and fungal elicitors. *The Plant Cell, 13*, 1467–1475.

Pedrioli, P. G., Raught, B., Zhang, X. D., Rogers, R., Aitchison, J., Matunis, M., & Aebersold, R. (2006). Automated identification of SUMOylation sites using mass spectrometry and SUMmOn pattern recognition software. *Nature Methods, 3*(7), 533–539. https://doi.org/10.1038/nmeth891

Peltier, J. B., Cai, Y., Sun, Q., Zabrouskov, V., Giacomelli, L., Rudella, A., ... & van Wijk, K. J. (2006). The oligomeric stromal proteome of Arabidopsis thaliana Chloroplasts* S. *Molecular & Cellular Proteomics, 5*(1), 114–133.

Peltier, J. B., Friso, G., Kalume, D. E., Roepstorff, P., Nilsson, F., Adamska, I., & van Wijka, K. J. (2000). Proteomics of the chloroplast: Systematic identification and targeting analysis of lumenal and peripheral thylakoid proteins. *The Plant Cell, 12*(3), 319–341.

Peng, J., Schwartz, D., Elias, J. E., Thoreen, C. C., Cheng, D., Marsischky, G., Roelofs, J., Finley, D., & Gygi, S. P. (2003). A proteomics approach to understanding protein ubiquitination. *Nature Biotechnology, 21*, 921–926.

Pevzner, P. A., Mulyukov, Z., Dancik, V., & Tang, C. L. (2001). Efficiency of database search for identification of mutated and modified proteins via mass spectrometry. *Genome Research, 11*(2), 290–299.

Piques, M., Schulze, W. X., Höhne, M., Usadel, B., Gibon, Y., Rohwer, J., & Stitt, M. (2009). Ribosome and transcript copy numbers, polysome occupancy and enzyme dynamics in Arabidopsis. *Molecular Systems Biology, 5*(1), 314.

Posewitz, M. C., & Tempst, P. (1999). Immobilized gallium(III) affinity chromatography of phosphopeptides. *Analytical Chemistry, 71*, 2883–2892.

Provart, N., & Zhu, T. (2003). A browser-based functional classification SuperViewer for Arabidopsis genomics. *Current Protocols in Molecular Biology, 2003*, 271–273.

Raffaele, S., Bayer, E., Lafarge, D., Cluzet, S., & German Retana, S., Boubekeur, T., ... & Mongrand, S. (2009). Remorin, a Solanaceae protein resident in membrane rafts and plasmodesmata, impairs potato virus X movement. *Plant Cell* 21:1541–1555

Rappsilber, J., Ryder, U., Lamond, A. I., & Mann, M. (2002). Large-scale proteomic analysis of the human spliceosome. *Genome Research, 12*(8), 1231–1245.

Reiland, S., Messerli, G., Baerenfaller, K., Gerrits, B., Endler, A., Grossmann, J., ... & Baginsky, S. (2009). Large-scale Arabidopsis phosphoproteome profiling reveals novel chloroplast kinase substrates and phosphorylation networks. *Plant Physiology, 150*, 889–903

Riccardi, F., Gazeau, P., de Vienne, D., & Zivy, M. (1999). Protein changes in response to progressive water deficit in maize. Quantitative variation and polypeptide identification. *Plant Physiology, 117*, 1253–1263.

Rinner, O., Mueller, L. N., Hubálek, M., Müller, M., Gstaiger, M., & Aebersold, R. (2007). An integrated mass spectrometric and computational framework for the analysis of protein interaction networks. *Nature Biotechnology, 25*(3), 345–352.

Ross, P. L., Huang, Y. N., Marchese, J. N., Williamson, B., Parker, K., Hattan, S., ... & Pappin, D. J. (2004). Multiplexed protein quantitation in *Saccharomyces cerevisiae* using amine-reactive isobaric tagging reagents. *Molecular & Cellular Proteomics, 3*(12), 1154–1169.

Rossignol, M. (2006). Proteomic analysis of phosphorylated proteins. *CurrentOpinionin Plant Biology, 9*(5), 538–543. https://doi.org/10.1016/j.pbi.2006.07.004

Roth, A. F., Wan, J., Green, W. N., Yates, J. R., & Davis, N. G. (2006). Proteomic identification of palmi-toylated proteins. *Methods*, *40*(2), 135–142. https://doi.org/10.1016/j.ymeth.2006.05.026

Rudella, A., Friso, G., Alonso, J. M., Ecker, J. R., & Van Wijk, K. J. (2006). Downregulation of ClpR2 leads to reduced accumulation of the ClpPRS protease complex and defects in chloroplast biogenesis in Arabidopsis. *The Plant Cell*, *18*(7), 1704–1721.

Sadowski, P. G., Dunkley, T. P., Shadforth, I. P., Dupree, P., Bessant, C., Griffin, J. L., & Lilley, K. S. (2006). Quantitative proteomic approach to study subcellular localization of membrane proteins. *Nature Protocols*, *1*(4), 1778–1789.

Saito, A., Nagasaki, M., Oyama, M., Kozuka-Hata, H., Semba, K., Sugano, S., ... & Miyano, S. (2007). AYUMS: An algorithm for completely automatic quantitation based on LC-MS/MS proteome data and its application to the analysis of signal transduction. *BMC Bioinformatics*, *8*, 15

Schöneich, C., & Sharov, V. S. (2006). Mass spectrometry of protein modifications by reactive oxygen and nitrogen species. *Free Radical Biology and Medicine*, *41*(10), 1507–1520. https://doi.org/10.1016/j.freeradbiomed.2006.08.013

Schultz, C. J., Ferguson, K. L., Lahnstein, J., & Bacic, A. (2004). Post-translational modifications of arabino-galactan-peptides of *Arabidopsis thaliana*. *Journal of Biological Chemistry*, *279*, 45503–45511.

Schultz, C. J., Johnson, K. L., Currie, G., & Bacic, A. (2000). The classical arabinogalactan protein gene family of *Arabidopsis*. *The Plant Cell*, *12*, 1751–1768.

Schulze, W. X., Deng, L., & Mann, M. (2005). The phosphotyrosine interactome of the ErbB-receptor kinase family. *Molecular Systems Biology*, *1*, E1–13.

Schulze, W. X., & Mann, M. (2004). A novel proteomic screen for peptide-protein interactions. *Journal of Biological Chemistry*, *279*, 10756–64.

Selbach, M., & Mann, M. (2006). Protein interaction screening by quantitative immunoprecipitation combined with knockdown (QUICK). *Nature Methods*, *3*(12), 981–983.

Selbach, M., Schwanhäusser, B., Thierfelder, N., Fang, Z., Khanin, R., & Rajewsky, N. (2008). Widespread changes in protein synthesis induced by microRNAs. *Nature*, *455*(7209), 58–63.

Seo, J., & Lee, K. J. (2004). Post-translational modifications and their biological functions: Proteomic analysis and systematic approaches. *Journal of Biochemistry and Molecular Biology*, *37*, 35–44.

Smalle, J., & Vierstra, R. D. (2004). The ubiquitin 26S proteasome proteolytic pathway. *Annual Review of Plant Biology*, *55*, 555–590.

Stanislas, T., Bouyssie, D., Rossignol, M., Vesa, S., Fromentin, J., Morel, J., ... & Simon-Plas, F. (2009). Quantitative proteomics reveals a dynamic association of proteins to detergent resistant membranes upon elicitor signaling in tobacco. *Molecular & Cellular Proteomics*, *8*, 2186–98

Ströher, E., & Dietz, K. J. (2006). Concepts and approaches towards understanding the cellular redox proteome. *Plant Biology*, *8*(4), 407–418. https://doi.org/10.1055/s-2006-923961

Thompson, A., Schäfer, J., Kuhn, K., Kienle, S., Schwarz, J., Schmidt, G., ... & Hamon, C. (2003). Tandem mass tags: A novel quantification strategy for comparative analysis of complex protein mixtures by MS/MS. *Analytical Chemistry*, *75*(8), 1895–1904.

Turck, C. W., Falick, A. M., Kowalak, J. A., Lane, W. S., Lilley, K. S., Phinney, B. S., ... & Yates, N. A. (2007). The association of biomolecular resource facilities proteomics research group 2006 study: Relative protein quantitation. *Molecular & Cellular Proteomics*, *6*(8), 1291–1298.

Turkina, M. V., Villarejo, A., & Vener, A. V. (2004). The transit peptide of CP29 thylakoid protein in *Chlamydomonas reinhardtii* is not removed but undergoes acetylation and phosphorylation. *FEBS Letters*, *564*, 104–108.

Tuskan, G. A., DiFazio, S., Jansson, S., Bohlmann, J., Grigoriev, I., ... & Rokhsar, D. (2006). The genome of black cottonwood, *Populus trichocarpa* (Torr and Gray). *Science*, *313*(5793), 1596–1604. https://doi.org/10.1126/science.1128691

Tuskan, G. A., DiFazio, S., Jansson, S., Bohlmann, J., Grigoriev, I., Hellsten, U., ... & Rokhsar, D. (2006). The genome of black cottonwood, *Populus trichocarpa* (Torr and Gray). *Science*, *313*(5793), 1596–1604. https://doi.org/10.1126/science.1128691

Ünlü, M., Morgan, M. E., & Minden, J. S. (1997). Difference gel electrophoresis. A single gel method for detecting changes in protein extracts. *Electrophoresis*, *18*(11), 2071–2077.

Usadel, B., Nagel, A., Steinhauser, D., Gibon, Y., Bläsing, O. E., Redestig, H., ... & Stitt, M. (2006). PageMan: An interactive ontology tool to generate, display, and annotate overview graphs for profiling experiments. *BMC Bioinformatics*, *7*, 535

Venable, J. D., Wohlschlegel, J., McClatchy, D. B., Park, S. K., & Yates, J. R. I. (2007). Relative quantification of stable isotope labeled peptides using a linear ion trap-orbitrap hybrid mass spectrometer. *Analytical Chemistry*, *79*, 3056–3064.

Vener, A. V., Harms, A., Sussman, M. R., & Vierstra, R. D. (2001). Mass spectrometric resolution of reversible protein phosphorylation in photosynthetic membranes of *Arabidopsis thaliana*. *Journal of Biological Chemistry*, 276, 6959–6966.

Verma, R., Chen, S., Feldman, R., Schieltz, D., Yates, J., Dohmen, J., & Deshaies, R. J. (2000). Proteasomal proteomics: Identification of nucleotide-sensitive proteasome-interacting proteins by mass spectrometric analysis of affinity-purified proteasomes. *Molecular Biology of the Cell*, 11, 3425–3439.

Vierstra, R. D. (2003). The ubiquitin/26S proteasome pathway, the complex last chapter in the life of many plant proteins. *Trends in Plant Science*, 8, 135–142.

Wang, N., & Li, L. (2008). Exploring the precursor ion exclusion feature of liquid chromatography electrospray ionization quadrupole time-of-flight mass spectrometry for improving protein identification in shotgun proteome analysis. *Analytical Chemistry*, 80, 4696–4710.

Wang, Y. K., Ma, Z., Quinn, D. F., & Fu, E. W. (2002). Inverse 15N-metabolic labeling/mass spectrometry for comparative proteomics and rapid identification of protein markers/targets. *Rapid Communications in Mass Spectrometry*, 16(14), 1389–1397.

Wang, G., Wu, W. W., Zeng, W., Chou, C. L., & Shen, R. F. (2006). Label-free protein quantification using LC-coupled ion trap or FT mass spectrometry: Reproducibility, linearity, and application with complex proteomes. *Journal of Proteome Research*, 5, 1214–1223.

Welchman, R. L., Gordon, C., & Mayer, R. J. (2005). Ubiquitin and ubiquitin-like proteins as multifunctional signals. *Nature Reviews Molecular Cell Biology*, 6, 599–609.

Whitelegge, J. P., Katz, J. E., Pikhakari, K. A., Hale, R., Aguilera, R., Gómez, S.M., ... & Vermaas, W. (2004). Subtle modification of isotope ratio proteomics; An integrated strategy for expression proteomics. *Phytochemistry* 65:1507–1515

Wienkoop, S., Larrainzar, E., Glinski, M., Gonzalez, E. M., Arrese-Igor, C., & Weckwerth, W. (2008a). Absolute quantification of *Medicago truncatula* sucrose synthase isoforms and N-metabolism enzymes in symbiotic root nodules and the detection of novel nodule phosphoproteins by mass spectrometry. *Journal of Experimental Botany*, 59, 3307–3315.

Wienkoop, S., Larrainzar, E., Niemann, M., Gonzalez, E. M., Lehmann, U., & Weckwerth, W. (2006). Stable isotope-free quantitative shotgun proteomics combined with sample pattern recognition for rapid diagnostics. *Journal of Separation Science*, 29, 2793–2801.

Wienkoop, S., Morgenthal, K., Wolschin, F., Scholz, M., Selbig, J., & Weckwerth, W. (2008b). Integration of metabolomic and proteomic phenotypes—Analysis of data-covariance dissects starch and RFO metabolism from low and high temperature compensation response in *Arabidopsis thaliana*. *Molecular & Cellular Proteomics*, 7, 1725–1736.

Wiese, S., Gronemeyer, T., Ofman, R., Kunze, M., Grou, C. P., Almeida, J. A., ... & Warscheid, B. (2007). Proteomics characterization of mouse kidney peroxisomes by tandem mass spectrometry and protein correlation profiling. *Molecular & Cellular Proteomics* 6:2045–57

Wilm, M., Shevchenko, A., Houthaeve, T., Breit, S., Schweigerer, L., Fotsis, T., & Mann, M. (1996). Femtomole sequencing of proteins from polyacrylamide gels by nano-electrospray mass spectrometry. *Nature* 379:466–469.

Wohlschlegel, J. A., Johnson, E. S., Reed, S. I., & Yates, III J.R. (2004). Global analysis of protein sumoylation in *Saccharomyces cerevisiae*. *Journal of Biological Chemistry*, 279, 45662–45668.

Wong, J. W., Sullivan, M. J., & Cagney, G. (2008). Computational methods for the comparative quantification of proteins in label-free LCn-MS experiments. *Brief Bioinformatics*, 9, 15665.

Xia, Q., Wang, T., Park, Y., Lamont, R. J., & Hackett, M. (2007). Differential quantitative proteomics of *Porphyromonas gingivalis* by linear ion trap mass spectrometry: non-label methods comparison, q-values and LOWESS curve fitting. *International Journal of Mass Spectrometry*, 259, 105–116.

Xing, T., Ouellet, T., & Miki, B. L. (2002). Towards genomic and proteomic studies of protein phosphorylation in plant-pathogen interactions. *Trends in Plant Science*, 7, 224–230.

Yamagata, A., Kristensen, D. B., Takeda, Y., Miyamoto, Y., Okada, K., Inamatsu, M., & Yoshizato, K. (2002). Mapping of phosphorylated proteins on two-dimensional polyacrylamide gels using protein phosphatase. *Proteomics*, 2(9), 1267–1276.

Yao, X., Freas, A., Ramirez, J., Demirev, P. A., & Fenselau, C. (2001). Proteolytic 18O labeling for comparative proteomics: Model studies with two serotypes of adenovirus. *Analytical Chemistry*, 73, 2836–2842.

Zhang, Y., Ficarro, S. B., Li, S., & Marto, J. A. (2009). Optimized Orbitrap HCD for quantitative analysis of phosphopeptides. *Journal of the American Society for Mass Spectrometry*, 20, 1425–1434.

Zhang, B., VerBerkmoes, N. C., Langston, M. A., Uberbacher, E., Hettich, R. L., & Samatova, N. F. (2006). Detecting differential and correlated protein expression in label-free shotgun proteomics. *Journal of Proteome Research, 5*, 2909–2918.

Zhu, M., Dai, S., McClung, S., Yan, X., & Chen, S. (2009). Functional differentiation of *Brassica napus* guard cells and mesophyll cells revealed by comparative proteomics. *Molecular & Cellular Proteomics, 8*, 752–766.

Zubarev, R., & Mann, M. (2007). On the proper use of mass accuracy in proteomics. *Molecular & Cellular Proteomics, 6*, 377–381.

Zybailov, B., Rutschow, H., Friso, G., Rudella, A., & Emanuelsson, O., Sun, Qi, & Wijk, K. J. V. (2008). Sorting signals, N-terminal modifications and abundance of the chloroplast proteome. *PLos ONE* 3:e1994

Index

Printed in the United States
by Baker & Taylor Publisher Services